Advances in
CANCER RESEARCH

Volume 96

Genomics in Cancer Drug Discovery and Development

Advances in
CANCER
RESEARCH

Volume 96

Series Editors

George F. Vande Woude
Van Andel Research Institute
Grand Rapids
Michigan

George Klein
Microbiology and Tumor Biology Center
Karolinska Institute
Stockholm
Sweden

Edited by

Garret M. Hampton
Sr. Dir. Biochemistry and Biomarker Development
Celgene Signal Research, San Diego
California, USA

Karol Sikora
Professor of Cancer
Medicine, Imperial College
Hammersmith Hospital, London, and
Medical Director Cancer Partners UK

ELSEVIER

AMSTERDAM • BOSTON • HEIDELBERG • LONDON
NEW YORK • OXFORD • PARIS • SAN DIEGO
SAN FRANCISCO • SINGAPORE • SYDNEY • TOKYO
Academic Press is an imprint of Elsevier

Academic Press is an imprint of Elsevier
525 B Street, Suite 1900, San Diego, California 92101-4495, USA
84 Theobald's Road, London WC1X 8RR, UK

This book is printed on acid-free paper. ∞

For information on all Elsevier Academic Press publications
visit our Web site at www.books.elsevier.com

ISBN-13: 978-0-12-006696-4
ISBN-10: 0-12-006696-3

PRINTED IN THE UNITED STATES OF AMERICA
07 08 09 10 9 8 7 6 5 4 3 2 1

Contents

Contributors to Volume 96 xi
Introduction xv

Biomarker Discovery in Epithelial Ovarian Cancer by Genomic Approaches
Samuel C. Mok, Kevin M. Elias, Kwong-Kwok Wong, Kae Ho, Tomas Bonome, and Michael J. Birrer

I. Introduction 2
II. Genomic Approaches in Biomarker Discovery 5
References 18

Mass Spectrometry: Uncovering the Cancer Proteome for Diagnostics
Da-Elene van der Merwe, Katerina Oikonomopoulou, John Marshall, and Eleftherios P. Diamandis

I. Current Cancer Biomarkers 24
II. Early Detection 25
III. The Need for New Diagnostic Strategies 27
IV. Mass Spectrometry 29
V. Mass Spectrometry-Based Diagnostics 32
VI. Current Limitations of Diagnostic Mass Spectrometry 37
VII. Suggestions for Future Progress 42
VIII. Future Direction 43
References 43

Microarrays to Identify New Therapeutic Strategies for Cancer
Christopher Sears and Scott A. Armstrong

I. Introduction 52
II. Microarray Technologies 52

III. Microarrays in Drug Development 59
IV. Microarrays to Direct the Use of Cancer Therapeutics 60
V. Identification of Therapeutic Targets in Distinct Disease Subtypes 65
VI. Conclusions 70
References 71

The Application of siRNA Technology to Cancer Biology Discovery

Uta Fuchs and Arndt Borkhardt

I. Introduction 75
II. The Mechanism of RNAi 76
III. Transcriptional Gene Silencing by siRNAs 80
IV. siRNAs Delivery: Strategies and Difficulties 81
V. RNAi as Discovery Tool in Cancer Biology 84
VI. RNAi Screens 87
VII. Limitations of siRNAs as Cancer Therapeutics: Not Related to Delivery Problems 95
VIII. Summary 96
References 96

Ribozyme Technology for Cancer Gene Target Identification and Validation

Qi-Xiang Li, Philip Tan, Ning Ke, and Flossie Wong-Staal

I. Brief Biology of Ribozymes 103
II. Ribozymes as Tools for Gene Inactivation 105
III. Ribozymes as Tools in Gene Target Discovery and Validation 109
IV. Ribozyme-Based Genomic Technology in Cancer Gene Target Discovery and Validation 113
V. Summary 137
References 139

Cancer Cell-Based Genomic and Small Molecule Screens

Jeremy S. Caldwell

I. Introduction 146
II. Drugs, Druggability, and Target Validation 147
III. Post-genomic Discovery of Novel Targets 148
IV. Small Molecule Screens in Oncology 155
V. Conclusions 167
References 168

Tumor Antigens as Surrogate Markers and Targets for Therapy and Vaccines

Angus Dalgleish and Hardev Pandha

I. Introduction: The Immune System and the Concept of Tumor Antigens 175
II. How Tumor Antigens Are Recognized 176
III. Cancer Antigens 179
IV. Novel TAA Identification Techniques 180
V. Effective Targeting of TAA in the Clinic 182
VI. TAAs as Therapeutic Targets 184
VII. Cancer Vaccines 184
VIII. Tumor Antigens and Cancer Vaccines 185
IX. Tumor Antigens as Surrogate Markers and Targets for Therapy and Vaccines 186
X. Summary and Future Directions 187
References 188

Practices and Pitfalls of Mouse Cancer Models in Drug Discovery

Andrew L. Kung

I. Why Are Animal Models Needed? 191
II. Model Types: Tumor Locations 193
III. Tumor Models: Cell Types 198
IV. Study Endpoints 199
V. Animal Modeling in the Post-genomics Age 203
VI. Conclusions 207
References 207

Pharmacodynamic Biomarkers for Molecular Cancer Therapeutics

Debashis Sarker and Paul Workman

I. Introduction 214
II. Types of Biomarkers 216
III. The Pharmacological Audit Trail 218
IV. Methodological Issues 220
V. Rationale for Use of PD Markers to Facilitate Drug Development 222
VI. The Need for Multidisciplinary and Broad-Based Collaborative Research and Development 229

VII. Biomarker Methodology 230
VIII. Examples of PD Biomarkers for Specific New Drug Classes 237
IX. Combining Chemotherapy with Molecularly Targeted Agents 252
X. Conclusions and Future Perspective 253
References 255

Biomarker Assay Translation from Discovery to Clinical Studies in Cancer Drug Development: Quantification of Emerging Protein Biomarkers
Jean W. Lee, Daniel Figeys, and Julian Vasilescu

I. Introduction 270
II. Biomarkers and Cancer Drug Development 272
III. Biomarker Research Challenges: Technology Translation 275
IV. Biomarker Research Challenges: Cultural and Process Translation 281
V. Clinical Qualification 291
VI. Conclusions and Perspectives 294
References 296

Molecular Optical Imaging of Therapeutic Targets of Cancer
Konstantin Sokolov, Dawn Nida, Michael Descour, Alicia Lacy, Matthew Levy, Brad Hall, Su Dharmawardhane, Andrew Ellington, Brian Korgel, and Rebecca Richards-Kortum

I. Introduction 300
II. Role of Molecular-Specific Optical Imaging of Carcinogenesis 302
III. Cancer Biomarkers and Therapeutic Targets 304
IV. Optical Technologies 306
V. Optically Active Contrast Agents 311
VI. Delivery of Contrast Agents *In Vivo* 322
VII. Molecular-Specific Optical Imaging 324
VIII. Future Directions: "Smart" Contrast Agents 331
IX. Conclusions 335
References 336

Personalized Medicine for Cancer: From Molecular Signature to Therapeutic Choice
Karol Sikora

I. Summary 345
II. Introduction 346
III. Prevention and Screening 351
IV. Detecting Cancer 353

V. New Treatment Approaches 354
VI. The Development of Personalized Medicine 357
VII. Barriers to Innovation 361
VIII. Patient's Experience 363
IX. Conclusions 365
References 369

Cancer Drug Approval in the United States, Europe, and Japan

R. A. V. Milsted

I. USA 371
II. Europe 381
III. Japan 384
IV. Discussion 387

Index 393

Contributors

Numbers in parentheses indicate the pages on which the authors' contributions begin.

Scott A. Armstrong, Department of Pediatric Oncology, Dana-Farber Cancer Institute, Harvard Medical School, Boston, Massachusetts 02115 (51)

Michael J. Birrer, Cell and Cancer Biology Branch, National Cancer Institute, NIH, Bethesda, Maryland 20892 (1)

Tomas Bonome, Cell and Cancer Biology Branch, National Cancer Institute, NIH, Bethesda, Maryland 20892 (1)

Arndt Borkhardt, Dr. von Haunersches Kinderspital, Ludwig Maximilians Universität München, München, Germany (75)

Jeremy S. Caldwell, Genomics Institute of the Novartis Research Foundation, San Diego, California 92121 (145)

Angus Dalgleish, St. George's University of London, Cranmer, London SW17 0RE, United Kingdom (175)

Michael Descour, Optical Sciences Center, The University of Arizona, Tucson, Arizona 85724 (299)

Su Dharmawardhane, Department of Anatomy and Cell Biology, Universidad Central del Caribe, School of Medicine, Bayamon, Puerto Rico 00960 (299)

Eleftherios P. Diamandis, Department of Pathology and Laboratory Medicine, Mount Sinai Hospital, Toronto, Ontario M5G1X5, Canada; Department of Laboratory Medicine and Pathobiology, University of Toronto, Toronto, Ontario M5G1L5, Canada (23)

Kevin M. Elias, Cell and Cancer Biology Branch, National Cancer Institute, NIH, Bethesda, Maryland 20892 (1)

Andrew Ellington, Department of Chemistry and Biochemistry, The University of Texas at Austin, Austin, Texas 78712 (299)

Daniel Figeys, Institute of Systems Biology, University of Ottawa, Ottawa K4A 4N2, Canada (269)

Uta Fuchs, Dr. von Haunersches Kinderspital, Ludwig Maximilians Universität München, München, Germany (75)

Brad Hall, Department of Chemistry and Biochemistry, The University of Texas at Austin, Austin, Texas 78712 (299)

Kae Ho, Department of Biostatistics, Harvard School of Public Health, Boston, Massachusetts 02115 (1)

Ning Ke, Immusol, Inc., San Diego, California 92121 (103)

Brian Korgel, Department of Chemical Engineering, The University of Texas at Austin, Austin, Texas 78712 (299)

Andrew L. Kung, Department of Pediatric Oncology, Dana-Farber Cancer Institute and Children's Hospital, Harvard Medical School, Boston, Massachusetts 02115 (191)

Alicia Lacy, Department of Biomedical Engineering, The University of Texas at Austin, Austin, Texas 78712 (299)

Jean W. Lee, Amgen Inc., Thousand Oaks, California 91320 (269)

Matthew Levy, Department of Chemistry and Biochemistry, The University of Texas at Austin, Austin, Texas 78712 (299)

Qi-Xiang Li, Immusol, Inc., San Diego, California 92121 (103)

John Marshall, Department of Chemistry and Biology, Ryerson University, Toronto, Ontario M5G1G3, Canada (23)

Da-Elene van der Merwe, Department of Pathology and Laboratory Medicine, Mount Sinai Hospital, Toronto, Ontario M5G1X5, Canada (23)

R. A. V. Milsted, Oncology Regulatory Affairs, AstraZeneca Pharmaceuticals, Alderley Park, Cheshire SK 4TF, United Kingdom (371)

Samuel C. Mok, Department of Obstetrics, Gynecology, and Reproductive Biology, Division of Gynecologic Oncology, Brigham and Women's Hospital, Harvard Medical School, Boston, Massachusetts 02115 (1)

Dawn Nida, Department of Bioengineering, Rice University, Houston, Texas 77030 (299)

Katerina Oikonomopoulou, Department of Pathology and Laboratory Medicine, Mount Sinai Hospital, Toronto, Ontario M5G1X5, Canada; Department of Laboratory Medicine and Pathobiology, University of Toronto, Toronto, Ontario M5G1L5, Canada (23)

Hardev Pandha, St. George's University of London, Cranmer, London SW17 0RE, United Kingdom (175)

Rebecca Richards-Kortum, Department of Bioengineering, Rice University, Houston, Texas 77030 (299)

Debashis Sarker, Signal Transduction and Molecular Pharmacology Team, Cancer Research UK, Centre for Cancer Therapeutics, The Institute of Cancer Research, Haddow Laboratories, Sutton, Surrey SM2 5NG, United Kingdom (213)

Christopher Sears, Division of Hematology/Oncology, Children's Hospital, Dana-Farber Cancer Institute, Harvard Medical School, Boston, Massachusetts 02115 (51)

Karol Sikora, Faculty of Medicine, Hammersmith Hospital Imperial College, London, United Kingdom; and Cancer Partners UK, London, United Kingdom (345)

Konstantin Sokolov, Department of Imaging Physics, MD Anderson Cancer Center, Houston, Texas 77030 (299)
Philip Tan, Immusol, Inc., San Diego, California 92121 (103)
Julian Vasilescu, Institute of Systems Biology, University of Ottawa, Ottawa K4A 4N2, Canada (269)
Kwong-Kwok Wong, Department of Gynecologic Oncology, M. D. Anderson Cancer Center, Houston, Texas 77030 (1)
Flossie Wong-Staal, Immusol, Inc., San Diego, California 92121 (103)
Paul Workman, Signal Transduction and Molecular Pharmacology Team, Cancer Research UK, Centre for Cancer Therapeutics, The Institute of Cancer Research, Haddow Laboratories, Sutton, Surrey SM2 5NG, United Kingdom (213)

Introduction

The development of cancer therapies over the past 50 years has largely focused on malignancy as a disease of uncontrolled proliferation. Consequently, the battery of anticancer agents in wide use today are mostly cytotoxics, tailored to kill rapidly proliferating cell populations. Chemotherapeutics, although successful in a few select settings, are associated with a high degree of toxicity, sometimes intolerable, decreasing therapeutic compliance, and overall, affording little benefit to cancer patient survival.

In the past decade, oncology drug discovery has undergone a paradigm shift, moving from broad proliferative antagonists toward "molecularly targeted" drugs, small molecules and biologics that are specifically tailored to modulate the proteins that cause or drive malignancy. The underlying rationale for these new therapies comes from our increasing understanding of the cellular traits that cells must acquire to develop a full blown malignancy (Hanahan and Weinberg, 2000), that is, self-sufficiency in growth signals, insensitivity to growth-inhibitory (antigrowth) signals, evasion of programed cell death (apoptosis), limitless replicative potential, sustained angiogenesis, and tissue invasion and metastasis. Recently marketed drugs target several of these traits such as angiogenesis [e.g., Avastin™ (bevacizumab)] and self-sufficiency in growth signaling [e.g., Gleevec™ (imatinib mesylate), Erbitux™ (cetuximab), Iressa™ (gefitinib), Tarceva™ (erlotinib), and Herceptin™ (trastuzumab)].

Although these targeted agents have clearly ushered in a new era in which cancer patients may gain considerable therapeutic benefit, it has become clear that many of these drugs are only selectively active. For example, it was initially thought that responders to EGFR-targeted therapies, such as Iressa™ or Tarceva™, would be those whose tumors expressed high levels of the EGFR protein. However, this concept could not be validated in larger series of lung, pancreatic, or colon patients. More recent evidence shows that patients whose tumors do respond particularly well harbor mutations within the EGFR gene, which engages prosurvival mechanisms within the cell (reviewed in Haber *et al.*, 2005). As subsequent studies have grown larger, the correlation between mutation and drug response is less clear-cut, underscoring the complexity of the drug–genotype relationship. Likewise, only a subset of those patients whose breast tumors are Her2 positive will respond to Herceptin^{TM}, for reasons that are not entirely clear at this time (Baselga *et al.*, 2004). The foregoing underscores that, although these drugs represent breakthroughs in cancer therapy, a major challenge will be the identification of patients that will most likely respond to specific therapies.

The shift toward developing drugs that target the molecular drivers of cancer, particularly those proteins involved in intracellular signaling, affords unique opportunities to create a "pharmacological audit trail" (Chapter 8). With the knowledge of downstream mediators of enzymatic or receptor functions within the cells, one can begin to molecularly monitor the effects of drug–target interaction, from cell-based assays through preclinical models of disease and eventually into the early phases of clinical development in man. These molecular assays enable pharmacodynamic measurements of a compound's effect on its intended target, establishing both a Proof-of-Mechanism (PoM) as well as pharmacodynamic–pharmacokinetic (PK–PD) relationships that can inform optimal drug dose and schedule. These measurements provide important guideposts for subsequent development, particularly in combination therapy settings, where the clinical effects of the new therapeutics can be accurately interpreted at a molecular level. In combination with predictive markers of response, a sufficiently rigorous test of a drug's efficacy can follow.

GENOMICS IN CANCER DRUG DISCOVERY AND DEVELOPMENT

Perhaps the single-most important driver of current oncology drug discovery is our knowledge of the cancer genome, particularly at the level of expression of encoded genes, proteins, and regulatory elements. The coalescence of genome sequencing with laboratory automation and miniaturization now enables a truly global view of the state of a cancer cell's genes and their expression. The translation of this vast genomic resource into useful insights and tools for the detection and treatment of cancer is the context in which the current volume of *Advances in Cancer Research* was conceived.

CHAPTERS 1–6: DISCOVERY OF CANCER DIAGNOSTICS AND NEW THERAPEUTIC TARGETS

By examining the signatures of gene or protein expression, cancer cells can now be readily differentiated from their normal counterparts at a molecular level, thus enabling the possibility of cancer diagnostics (Chapters 1 and 2). Expression analysis has also found significant use in aiding the selection of patients who may respond to therapy as well as delineating new therapeutic strategies for specific subtypes of cancer (Chapter 3). With improved diagnosis comes the need for new points of therapeutic intervention. Fuchs and Borkhardt (Chapter 4) and Li *et al.* (Chapter 5) illustrate how the use of small interfering RNA and ribozyme libraries, which now cover a majority of human genes, can identify those that encode proteins crucial to cell survival. These strategies provide the starting points for small molecule screening campaigns against the encoded proteins and the identification of drug leads for preclinical development. Chapter 6, by Caldwell, focuses on how the information from these screens, as well as highly parallel cell-based small molecule screens, can be used to identify new cell-active drug leads and delineate drug mechanism of action. Chapter 7 discusses the use of tumor antigens as potential targets for therapeutic intervention as well as serving as putative markers of disease.

CHAPTER 8: PRECLINICAL MODELS OF MALIGNANCY

A major challenge in the early validation of new targets is the development of well-designed, mechanism-based preclinical animal models of disease. The traditional development path for chemotherapeutic drugs has instilled the general notion that animal models of malignancy are neither realistic nor predictive of therapeutic response in man. While it is certainly true that xenografts of human tumor cells—the workhorse of preclinical oncology drug development—are hardly replicative of a human tumor, mechanism-based drug development can nonetheless benefit considerably from well-designed models that harbor specific mutations or genetic backgrounds, for which part of the "pharmacological audit trail" can be explored. As Kung points out, "We must use targeted models to ask questions about targeted therapies."

CHAPTERS 9–11: TRANSLATING NOVEL TARGETED THERAPIES FROM RODENTS TO MAN: BIOMARKERS TO MONITOR DRUG ACTION

The successful transition of targeted oncology therapeutics from rodent models to man is predicated on the transfer of preclinical knowledge of drug behavior, such as the doses and schedules required for optimal target inhibition at a molecular level, into the Phase I/II setting (Chapter 9).

Early development, rigorous validation, and transfer of these biomarker assays, both pharmacodynamic and predictive, into the clinic ensures the optimal opportunity to evaluate the relevance of the proposed mechanisms in man, and guide administration of the drug to those patients likely to respond to therapeutic intervention (Chapter 10).

Blood and tissue samples have long been used to measure surrogates of drug efficacy. Prevalent examples include monitoring the decrease of CA-125 and prostate-specific antigen (PSA) levels in patients undergoing treatment for ovarian and prostate carcinomas, respectively. Albeit well validated, these markers are nonetheless surrogates and do not enable examination of drug action in disease tissue. In contrast, optical imaging can be used to assess target engagement via labeled drug, target modulation by visualization of target substrate level, as well as early indications of drug efficacy such as tumor shrinkage or glucose uptake. Thus, imaging technologies provide a means by which the effect of novel therapies can be rapidly and effectively monitored without direct access to the site of disease (Chapter 11).

CHAPTERS 12–13: THE SOCIOECONOMIC AND REGULATORY CHALLENGES OF NOVEL TARGETED ONCOLOGY THERAPEUTICS

With new therapeutic options come considerable socioeconomic challenges (Chapter 12). Although estimates vary widely, the costs associated with bringing a new drug through approval are significant, with a mean of US$881 million (from a detailed analysis by the Biomedical Industry Advisory Group, 2005). These costs, and the need to create shareholder value for pharmaceutical companies, translate into high-priced drugs, as much as US$17,000 for Erbitux™ and US$24,000 for Zevalin™ per patient per month (http://www.slate.com/id/2102844/). Many of these drugs will be taken over long periods of time. Thus, although the benefits of these new therapies are clear, can society sustain the financial burden? Similarly, regulatory agencies face significant challenges—increasing patient advocacy, industry lobbying, pressure for more rapid approval, and the mechanisms to continue to evaluate fast-track approved drugs. The information associated with targeted agents, particularly data derived from genomic analysis of patient populations, has provoked considerable debate and has been responded to, in part, by the *Critical Path Initiative* in the US. This is likely to become a key driver for evidence-based medicine (Chapter 13).

It must be emphasized that this volume is not intended to be comprehensive, but rather to provide a snapshot of how genomic sciences are impacting the detection and treatment of cancer and how these changes are likely to affect the regulatory and socioeconomic landscape of cancer management. The chapters within should be viewed as starting points for the interested reader.

We would like to acknowledge many individuals at Elsevier who made this volume possible: Hilary Rowe who first proposed the idea to address the impact of genomics in oncology drug discovery; Melissa Turner for originally organizing the volume contributions; Phil Carpenter and Mara Conner for organizing these chapters into the *Advances in Cancer Research* volume; Drs. George van de Woude and George Klein for accepting the proposal for this volume, and Ejaz Ahmad and Prakash Kumar for rapid and excellent copy editing. Finally, we would like to sincerely thank all of the authors for their respective contributions without which this volume would not have been possible.

Garret M. Hampton
San Diego, USA

Karol Sikora
London, United Kingdom

REFERENCES

Baselga, J., Gianni, L., Geyer, C., Perez, E. A., Riva, A., and Jackisch, C. (2004). Future options with trastuzumab for primary systemic and adjuvant therapy. *Semin. Oncol.* **31,** 51–57.

Haber, D. A., Bell, D. W., Sordella, R., Kwak, E. L., Godin-Heymann, N., Sharma, S. V., Lynch, T. J., and Settleman, J. (2005). Molecular targeted therapy of lung cancer: EGFR mutations and response to EGFR inhibitors. *Cold Spring Harb. Symp. Quant. Biol.* **70,** 419–426.

Hanahan, D., and Weinberg, R. A. (2000). The hallmarks of cancer. *Cell* **100,** 57–70.

Biomarker Discovery in Epithelial Ovarian Cancer by Genomic Approaches

Samuel C. Mok,* Kevin M. Elias,[†] Kwong-Kwok Wong,[‡] Kae Ho,[§] Tomas Bonome,[†] and Michael J. Birrer[†]

**Department of Obstetrics, Gynecology, and Reproductive Biology, Division of Gynecologic Oncology, Brigham and Women's Hospital, Harvard Medical School, Boston, Massachusetts 02115;*
[†]Cell and Cancer Biology Branch, National Cancer Institute, NIH, Bethesda, Maryland 20892;
[‡]Department of Gynecologic Oncology, M. D. Anderson Cancer Center, Houston, Texas 77030;
[§]Department of Biostatistics, Harvard School of Public Health, Boston, Massachusetts 02115

I. Introduction
  A. Clinical–Pathological Features of Epithelial Ovarian Tumors
  B. Clinical and Molecular Prognostic Markers for Ovarian Cancer
II. Genomic Approaches in Biomarker Discovery
  A. Genome-wide Loss of Heterozygosity Analysis
  B. Genome-wide Comparative Genomic Hybridization Analysis
  C. Transcription Profiling
References

Ovarian cancer is the fifth most common form of cancer in women in the United States. It is a complex disease composed of different histological grades and histological types. Most of epithelial ovarian cancer cases are detected at an advanced stage. Patients usually respond to primary treatment with surgery and chemotherapy. However, the disease usually recurs and is ultimately fatal. So far, a satisfactory screening procedure and regime to treat the recurrence disease are not available. High-throughput genomic analyses have the potential to change the detection and the treatment of ovarian neoplasms. They can help diagnose subtypes of disease and predict patient survival. New diagnostic and prognostic markers for ovarian cancer are emerging. One day, profiling may influence treatment decisions, informing both which patients should receive chemotherapy and what type of chemotherapeutic agents should be employed. As greater numbers of tumor samples are analyzed, the power of these profiling studies will increase, raising the possibility that novel molecular targets and less toxic therapies will be identified. These powerful techniques hold the potential to unravel the genetic origins of ovarian cancer. Hopefully, this will translate into earlier diagnosis and better patient outcome from disease. © 2007 Elsevier Inc.

Advances in CANCER RESEARCH

0065-230X/07 $35.00
DOI: 10.1016/S0065-230X(06)96001-1

I. INTRODUCTION

Ovarian cancer is the fifth most common form of cancer in women in the United States, accounting for 4% of the total number of cancer cases and 25% of those cases occur in the female genital tract. Because of its low cure rate, it is responsible for 5% of all cancer deaths in women. It was estimated that 16,210 deaths would be caused by ovarian cancer in the year 2005 (Wu *et al.*, 2005). The vast majority of ovarian cancers are carcinomas of the surface epithelial type (Boring *et al.*, 1994). Most of epithelial ovarian cancer cases are detected at an advanced stage (where metastases are present beyond the ovaries) and are rarely curable (Boente *et al.*, 1996). This dismal prognosis has stimulated research on methods for early detection (Bast *et al.*, 1995) with the hope of improved survival. However, such work has failed thus far to yield a satisfactory screening procedure. Other than the lack of satisfactory screening procedure for detecting early stages of ovarian cancer, poor survival is also caused by the development of drug resistance in the tumor cells. Patients with early stage disease (stage I/II) have generally favorable survivals. Despite the good prognosis for early stage ovarian cancers, a subset of these patients develop aggressive recurrent disease. At this point, it is impossible for clinicians to predict which group of patients will have a poor prognosis (i.e., who should be monitored and treated differently from patients with more indolent early stage ovarian cancer).

The development of drug resistance by tumor cells contributes to the poor survival of ovarian cancer patients. While 80% of advanced ovarian cancers (stage III/IV) respond to primary treatment with surgery and chemotherapy, the disease usually recurs and is ultimately fatal. Although most patients die within 2 years of diagnosis, a subset of patients, even with clinically and morphologically indistinguishable disease, develop a more chronic form of ovarian cancer and may survive 5 years or more with treatment. It is possible that patients with indolent cancer should be monitored and treated differently from patients with rapidly progressing ovarian cancer. At this point, clinicians do not have the tools to predict the clinical course of the disease.

A. Clinical–Pathological Features of Epithelial Ovarian Tumors

The majority of ovarian neoplasms are "epithelial" and thought to originate from the single layer of cells at the surface of the ovary or the same cells lining inclusion cysts entrapped in ovarian stroma. Most of these inclusion cysts are believed to arise as a consequence of the tissue damage and repair that occur with ovulation. Epithelial ovarian tumors are classified as benign, low malignant potential (borderline), or malignant (Serov and Scullt, 1993) and further distinguished by differences in the histological type of cell. Benign ovarian tumors are lined by a single or minimally stratified layer of cells, which are columnar and often ciliated in serous tumors or contain abundant apical cytoplasmic mucin in mucinous tumors. Borderline tumors (BOTs) or tumors of low malignant potential (LMP) are characterized by atypical epitheliums with cellular proliferation and pleomorphism but without stromal invasion. Malignant epithelium demonstrates marked atypia, increased mitotic activity, and stromal invasion. Serous tumors are the most common form of ovarian neoplasm with epithelial cells resembling those of the fallopian tube. They comprise about 50–60% of primary epithelial ovarian tumors. Mucinous tumors are cystic tumors with locules lined with mucin-secreting epithelial cells resembling either endocervical or colonic epithelium. They comprise approximately 8–10% of primary epithelial ovarian tumors. Endometrioid and clear cell lesions constitute about 10% of epithelial tumors and resemble tumors that originate in the endometrium. Other tumor cell types include Brenner, mixed epithelial type, and undifferentiated (Lee *et al.*, 2003). Taken together, ovarian cancer is a heterogeneous type of disease comprising with different types and forms.

In general, women with epithelial ovarian cancer have a poor prognosis. Due to lack of symptoms with early disease, ~70% of ovarian cancer patients present with disease involving the upper abdomen (FIGO stage III/IV) (Marsden *et al.*, 2000). These patients receive platinum- and taxane-based chemotherapy after surgery, and ~70% achieve a complete clinical response initially, in which no disease can be detected by physical examination, tumor CA125 levels, and CT scan (Auersperg *et al.*, 1998). The majority of these patients will eventually relapse. Some patients relapse within 6 months and die with chemoresistant disease, while some patients have late relapses with chemosensitive disease, and thus prolonged survival. Currently, clinicians do not have tools to predict which patients with advanced stage disease will relapse with chemoresistant disease. Furthermore, new therapeutic regimens to treat these patients are needed, as all of these patients succumb to their disease.

In contrast to patients with late-stage ovarian cancer, women with early stage disease (FIGO stage I/II) have an excellent prognosis, 5-year survival rates being greater than 90%. Postoperative chemotherapy is only administered to a subgroup of "high risk" patients who have tumor involving one or both ovaries with pelvic extension and/or positive peritoneal washings/fluid (stages IC, IIA, IIB, IIC). Patients with stages IA and IB disease who have poorly differentiated or clear cell tumors are also given chemotherapy. Studies have shown that up to 60% of patients in this "high risk" early stage group will never relapse, even without chemotherapy (Trimbos *et al.*, 2003a,b). At present, however, it is impossible to predict which of these patients are most likely to recur; therefore, all patients in this group are given chemotherapy. Ability to determine likelihood of disease relapse would allow for selective treatment of those patients who are most likely to develop recurrent disease.

B. Clinical and Molecular Prognostic Markers for Ovarian Cancer

Several clinical features have been shown to have significant impact on ovarian cancer patient survival. As discussed in Section I.A, patients with early stage diseases have significantly better survival rate than those with late-stage diseases. Optimally debulking of patient surgically (ability to remove residual tumor nodules which are greater than 2 cm) is associated with improved outcomes. In addition, patients with poorly differentiated tumors and clear cell lesions have a poorer prognosis than those with low-grade tumors or BOTs, or mucinous ovarian cancer, respectively. Finally, other clinical features, such as a large volume of ascites and age at the time of diagnosis (>65 years), have also been shown to be associated with poorer clinical outcomes (Cannistra, 1993).

Recent molecular studies have identified multiple markers, which have prognostic values. Both epidermal growth factor receptor (EGFR) and HER-2/neu overexpression have been shown to be associated with poor prognosis of the disease (Kokuho *et al.*, 1993; Meden and Kuhn, 1997; Skirnisdottir *et al.*, 2001). Other oncogenes, such as *MYC* amplification (Diebold *et al.*, 1996) and *nm23* overexpression (Scambia *et al.*, 1996), have also been shown to correlate with poor survival. Clinical correlation studies have demonstrated that *TP53* mutation and p53 overexpression are associated with shorter survival in ovarian cancer (Hartmann *et al.*, 1994; Klemi *et al.*, 1995; Mayr *et al.*, 2000; Ozalp *et al.*, 2000; Reles *et al.*, 2001), and resistance to platinum-based chemotherapy (Reles *et al.*, 2001). Besides p53, studies on another tumor suppressor protein p27 have indicated that loss of p27 is a frequent event in ovarian carcinomas. This protein also served as a useful prognostic marker for predicting disease recurrence in primary ovarian carcinomas (Masciullo *et al.*, 1999). Recently, using Gynecologic Oncology Group (GOG) clinical trial specimens, we demonstrated that increased cyclin E expression is associated with amplification of the gene and a poor prognosis for patients with advanced serous ovarian cancer (Farley *et al.*, 2003). Taken together, while multiple prognostic markers have been identified in ovarian cancer, most of these markers possess limited discriminatory power owing to the small number of patients used in the studies. Furthermore, additional variables may also affect clinical outcome obscuring the true prognostic value of these markers. Genome-wide analyses may overcome some of these limitations facilitating the identification of clinically relevant diagnostic and prognostic markers associated with the disease. For example, they can evaluate multiple targets allowing the development of interrelated gene lists predicting for a clinical endpoint. In addition, expression changes and genomic alterations can be correlated on a global level.

II. GENOMIC APPROACHES IN BIOMARKER DISCOVERY

A. Genome-wide Loss of Heterozygosity Analysis

Loss of heterozygosity (LOH) analyses of solid tumors have not only enabled the delineation of specific minimally lost regions as the likely locations of critical tumor suppressor genes but also provided the molecular portrait of the pattern of accumulation of genetic alterations in a multistep progression of cancer (Fearon and Vogelstein, 1990; Lasko *et al.*, 1991; Thiagalingam *et al.*, 2001; Vogelstein *et al.*, 1988, 1989; Yokota and Ookawa, 1993). The utilization of available LOH data as markers for diagnosis and prognosis of cancer has also become generally accepted. A higher frequency of consistent LOH at defined chromosomal regions critical for specific cancers has made this a useful, reliable DNA marker for diagnosis and prognosis of cancer, regardless of whether the target gene has been identified (Dong *et al.*, 2001; Sidransky, 1997). Using multiple microsatellite markers located on different chromosomes for LOH studies, we and other investigators have identified multiple regions, which are frequently deleted or lost in both BOTs and invasive epithelial ovarian cancers (IEOC) (Auersperg *et al.*, 1998). Those loci displaying an LOH rate greater than 25% are summarized in Table I. High-grade invasive carcinomas show significantly higher LOH rates at 6q25.1-26 and 11p15 than other ovarian tumors. The identification of a significantly higher LOH rate at the *PTEN* locus in endometrioid and clear cell carcinomas versus other histotypes suggests that endometrioid and clear cell carcinomas may have distinct pathogenetic pathways. In contrast, BOTs demonstrate an LOH rate less than 25% at most of the loci. However, among BOTs, significantly higher LOH rates were identified at 3p13-14.3 and Xq11.2-q12 in comparison to IEOC, suggesting that a large subset of BOTs may not progress to IEOC.

Table I Summary of Regions Frequently Displaying LOH in Epithelial Ovarian Tumors[a]

Chromosome	Location	Loci/Flanking loci	Physical/ Genetic distance	LOH (%)	Histopathology
1p	1p36	MTHFR		43	SC
3p	3p25-pter	D3S1620		38	SC
	3p24	THRβ		33	SC
	3p13–14.3	D3S30		25–33	SC, SBOT, SBN
5q	5q13.1-21	D5S424-D5S644	22 cm	47	SC
6q	6q25.1-26	D6S473-D6S448	4 cm	44	HGSC
	6q27	D6S149	300 kb	50	SC
	6q27	D6S264-D6S297	3 cm	35–50	SC, EC, CC
7q	7q31.3	D7S655-D7S480	1300 kb	50	SC
	7q31.2	D7S486–7G14	150 kb	37	SC
	7q31.1	D7S522		73	SC
8p	8p21	D8S136		50	SC
9p	9p21	D9S171-IFNA		39	SC
	9p22-23	D9S162-D9S144		38	SC
	9q32-34	D9S16-ASS1		50	SC
10q	10q23.3	PTEN		27–42	EBN, EC, CC
11p	11p15.1	D11S1310	4 cm	47	HGSC
	11p15.3-15.5	D11S2071-D11S988	11 cm	50	HGSC
11q	11q22-23.2	D11S35-D11S925		42	SC, EC
	11q23.3-24.3	D11S934-D11S1320	8.5 mb	58–69	SC, EC
	11q23.3-qter	D11S925-D11S1336	2 mb	75	SC
		D11S912-D11S439	8 mb	67	LGSC, HGSC
12p	12p12.3-13.1	D12S89-D12S364	7 cm	26	SC
12q	12q23-qter	D12S278		30	SC
14q	14q12-13	D14S80-D14S75		45	
	14q32	D14S65-D14S267		63	
17p	17p13.3	D17S28-D17S30	15 kb	80	SC
	17q13.1	p53		35–70	SC, EC, CC, MC
17q	17q21	D17S1320-D17S1328	400 kb	65	SC
	17q25	D17S801	2 cm	60–70	SC, EC, CC, MC
18q	18q23	D18S5-D18S11		60	SC
	18q21	D18S474		36	SC
22q	22q12-q13	D22S284-CYP2D	0.5 cm	45–65	SC, EC
Xp	Xp21.1-p11.4	DXS7-DXS84		60	SC
Xq	Xq11.2-q12	DXS1161-PGK1P1	1 cm	25–60	SBOT, HGSC

[a]SC, serous carcinoma; SBOT, serous borderline ovarian tumor; SBN, serous benign tumor; HGSC, high-grade serous carcinoma; EC, endometrioid carcinoma; CC, clear cell carcinoma; EBN, endometrioid benign tumor; MC, mucinous carcinoma; LGSC, low-grade serous carcinoma.

B. Genome-wide Comparative Genomic Hybridization Analysis

Comparative genomic hybridization (CGH) is a powerful whole genome assay available for detecting gene copy number at a given genomic complement. This method involves competitive hybridization of tumor and normal reference DNA differentially labeled with distinct fluorescent molecules to normal human metaphase chromosome spreads. Based on the relative intensity of the two fluorescent colors, regions of the chromosome with a gain or loss in gene copy number can be identified in a single hybridization (Kallioniemi *et al.*, 1992). One of the advantages of chromosomal CGH is that it can make use of archival materials when frozen tissues are not available, increasing the number of cases that can be analyzed to enhance the statistical power of the analysis. This unique capability also allows for the analysis of specimens obtained from multiple centers, and the evaluation of prospective and retrospective tumor specimens to facilitate correlative clinical studies. Using chromosomal CGH, earlier studies demonstrated that more than 20% of ovarian cancers show increased copy numbers on 1p22-31, 1q25-31, 2q32-33, 3q25-26, 5p14-pter, 6p22-25, 7q22-31, 8q24, 11q14-22, 12p12, 13q31-34, 18q12-22, and 20q13; and decreased copy numbers on 8p21-23, 16q, 17p, 17q21-22, 19, and Xp (Arnold *et al.*, 1996; Iwabuchi *et al.*, 1995; Sonoda *et al.*, 1997). High-grade tumors showed significantly higher copy number abnormality (CNA) frequencies than low-grade tumors. This difference has been further confirmed by a more recent study which also showed that underrepresentation of 11p and 13q and overrepresentation of 8q and 7p correlated with high-grade tumors, while 12p underrepresentation and 18p overrepresentation were significantly more frequent in well-differentiated and moderately differentiated tumors (Kiechle *et al.*, 2001). These data suggest that high-grade and low-grade tumors may have different pathogenetic pathways. Focusing on clear cell carcinomas, a recent study demonstrated that 25–75% of cases showed increased copy numbers on 8q11-q13, 8q21-q22, 8q23, 8q24-qter, 17q25-qter, and 20q13-qter; and decreased copy number on 19p (Suehiro *et al.*, 2000). The altered frequencies of CNAs in different histological subtypes of carcinomas suggest that they may also follow distinct pathological pathways.

The correlation between disease stage and CNA frequencies has also been evaluated. One study showed that increased CNAs on 3q25-26, 5p14-pter, 8q24, 20q13 and decreased CNAs on 17p and 17q21-22 were more common in late-stage disease (Sonoda *et al.*, 1997). Using microdissection to increase the percentage of tumor cells, CGH analysis on borderline ovarian tumors has been performed. A study on 11 serous BOTs showed increased CNAs on 5p14-pter, 7q22-31, 8q24, 12p12, 13q31-34, 18q12-22, and

20q13 (Wolf *et al.*, 1999). Correlative studies have also examined CNA frequencies and chemoresistance. In a clinical study, 11 cases of ovarian cancer with high sensitivity to cisplatin-based chemotherapy and 29 cases of ovarian cancer with chemoresistance were reported. Sakamoto *et al.* (1998) identified significant decrease in copy number decrease on Xp, and increase in copy number on 19q in chemoresistant cases, suggesting that −Xp and +19q were likely to be a genetic event associated with intrinsic drug resistance.

Chromosomal CGH has a limited resolution (10–20 Mb) because of its dependence on chromosomal bands. In addition, extensive follow-up work is required to identify candidate genes after regions of gain or loss have been identified. To overcome some of these limitations, a new variation of CGH has been developed to replace normal metaphase chromosomes with a platform using long stretches of human DNA (100–200 kb) packaged in replicable units called bacterial artificial chromosomes (BACs). These BACs are selected in such a way that they are evenly distributed throughout the genome and can be arranged in an array to further resolve the loci of interest after hybridization (Pinkel *et al.*, 1998). Depending on the number of BACs selected, the entire genome can be represented in a single array.

Using high-density BAC array comparative genomic hybridization (CGH), Cheng *et al.* demonstrated amplification of chromosome 1q22 centered at the RAB25 small GTPase locus, which is implicated in apical vesicle trafficking, in approximately half of ovarian cancers. *RAB25* mRNA levels were selectively increased in stages III and IV serous epithelial ovarian cancers compared to other genes within the amplified region, implicating *RAB25* amplification as a driving event in the development of the amplicon. Furthermore, increased DNA copy number or RNA level of *RAB25* was associated with markedly decreased disease-free survival in ovarian cancer (Cheng *et al.*, 2004).

Although BAC array CGH allows the identification of DNA copy number variation at an increased mapping resolution on a locus-by-locus basis, the identification of specific genes that are involved remains tedious and challenging. In contrast to BAC array CGH, cDNA array analysis is an alternative platform, which can be used to assess DNA copy number variation. Representing over 30,000 radiation hybrid (RH)-mapped human genes, cDNA arrays provide a readily available genomic resource so that can evaluate DNA copy number changes on a gene-by-gene basis. Such an approach has been used successfully to identify DNA copy number changes in specific genes in breast cancer (Pollack *et al.*, 1999, 2002) and ovarian cancer cell lines (Bourdon *et al.*, 2002).

Using microdissected tumor tissue samples combined with a cDNA array CGH platform that we have recently developed (Tsuda *et al.*, 2004), we identified multiple genes that showed significantly higher DNA and RNA copy number simultaneously in clear cell ovarian cancer compared to serous and

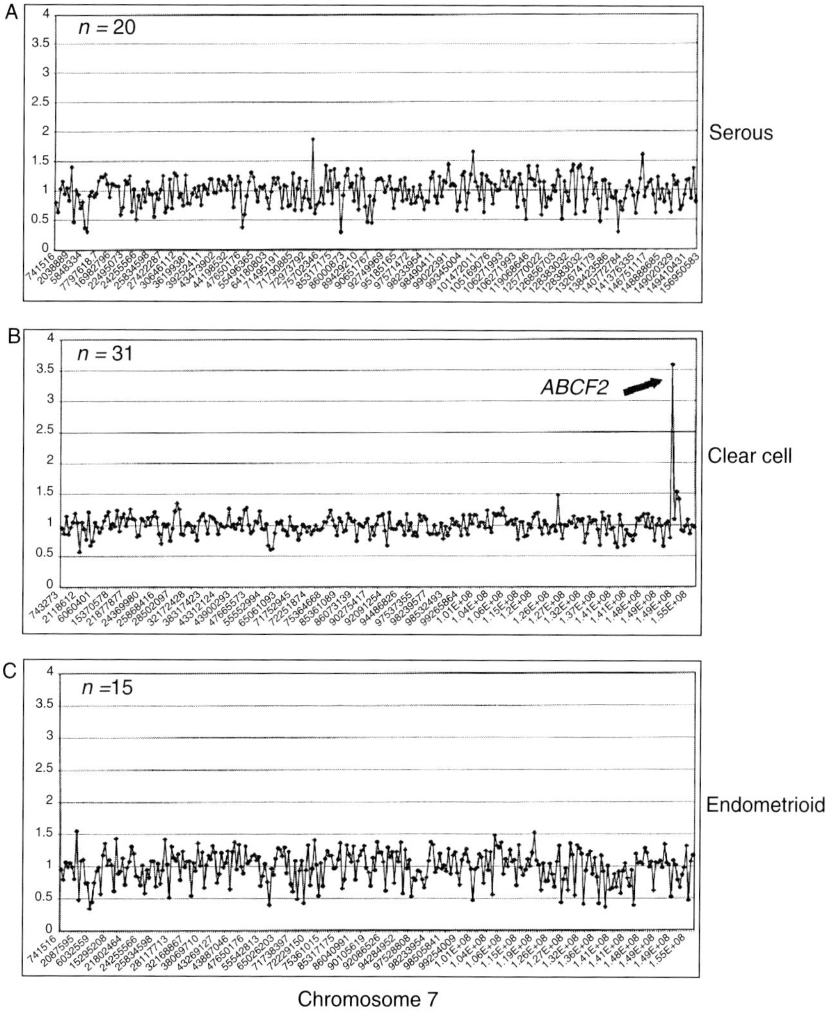

Fig. 1 cDNA array CGH profiles showing the mean DNA copy number of genes located on chromosome 7 in (A) 20 serous adenocarcinomas, (B) 31 clear cell adenocarcinomas, and (C) 15 endometrioid adenocarcinomas. The arrow indicates increase in DNA copy number of *ABCF2* located on chromosome 7q36.

endometrioid types of ovarian cancer (Tsuda *et al.*, 2005). One of them is *ABCF2* located on chromosome 7q36 (Figs. 1 and 2). Subsequent studies showed that cytoplasmic ABCF2 protein levels were significantly higher in tumor tissue samples obtained from patients who did not respond to chemotherapy compared to those who responded to the treatment (Tsuda *et al.*, 2005).

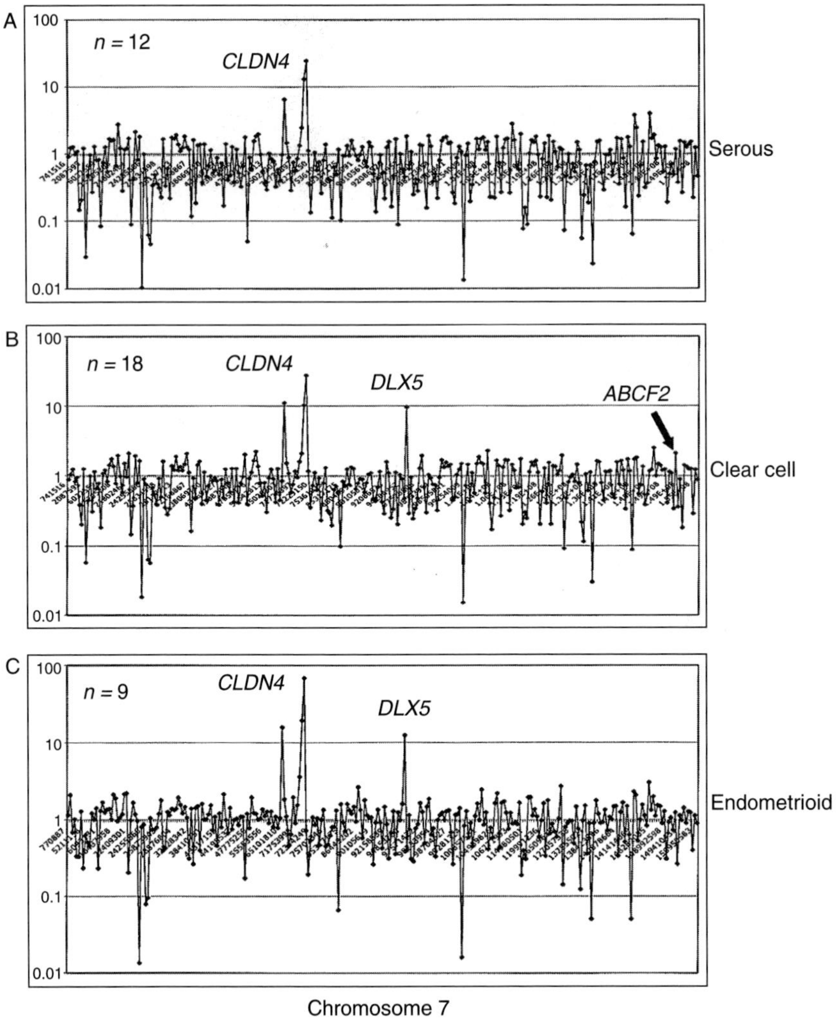

Fig. 2 Expression profiles showing the mean fold change in RNA levels of genes located on chromosome 7 in (A) 12 serous adenocarcinomas, (B) 18 clear cell adenocarcinomas, and (C) 9 endometrioid adenocarcinomas. High levels of claudin 4 (*CLDN4*) expression were identified in all histological subtypes. Significantly higher expression of *DLX5* was identified in both clear cell and endometrioid subtypes than the serous type. Significantly higher expression of *ABCF2* was identified in clear cell subtype than both the serous and the endometrioid subtypes.

Taken together, CGH analyses have identified multiple prognostic markers in ovarian cancer. However, most of these markers have limited discriminatory power due to the fact that the number of patients used in the studies is small, and clinical outcomes are influenced by multiple factors, which have not been taken into consideration, particularly in ovarian cancer, which is a heterogeneous type of disease.

C. Transcription Profiling

Transcription profiling is a large-scale approach for analyzing gene expression data, which has been widely used to identify differentially expressed genes and molecular signatures in many biological processes (Alizadeh *et al.*, 2000, p. 821; Barlund *et al.*, 2000, p. 115; DeRisi *et al.*, 1996, p. 39; DeRisi *et al.*, 1997, p. 31; Golub *et al.*, 1999, p. 6; Wang *et al.*, 2000, p. 562; Xu *et al.*, 2000, p. 566). Multiple platforms including oligonucleotide and cDNA arrays have been used. Within the last 3 years, several investigators have published intriguing results of microarray analysis on ovarian cancer specimens (Table II). Ismail *et al.* (2000) reported a study of 864 DNA elements screened against 10 ovarian cancer cells lines, and 5 normal epithelial cell lines. They identified 255 differentially expressed genes. A similar study was published in 2002 by Matei *et al.* (2002). Their study used a 12,600-element chip to screen cultures derived from 21 ovarian carcinomas and 9 normal ovaries. One hundred eleven genes were overexpressed in the cancer cells. In another noteworthy cDNA microarray study, Manderson *et al.* (2002) focused expression of genes located on chromosome 3 in ovarian cancer. They used a microchip containing 290 expression sequencing tags (ESTs) mapping to chromosome 3 to screen 4 ovarian cancer cell lines and 1 normal ovarian epithelial cell line.

Table II Select Microarray Studies of Ovarian Cancer

References	Array type	Material used	Findings
Ismail *et al.*, 2000	864-element cDNA array	10 ovarian cancer cell lines versus 5 normal HOSEs	44 HOSE-specific and 16 serous ovarian cancer-specific genes
Wong *et al.*, 2000	2400-element MICROMAX cDNA array	3 ovarian cancer cell lines versus 3 HOSEs	30 genes with expression levels greater than tenfold
Matei *et al.*, 2002	12,600-element cDNA array	21 ovarian cancer cell lines versus 9 normal ovaries	173 genes with expression levels greater than 2.5-fold
Manderson *et al.*, 2002	290-element chromosome 3 ESTs	3 ovarian cancer cell lines and 1 HOSE	25 differentially expressed ESTs
Shridhar *et al.*, 2001	18,000-element cDNA array	21 early stage versus 17 advanced stage ovarian cancer	43 genes with expression levels greater than tenfold
Sawiris *et al.*, 2002	516-element cDNA “Ovachip”	Ovarian cancer and colon cancer	2 clusters of differentially expressed genes

They found 25 differentially expressed ESTs, including 6 ESTs unique to those cancer cell lines which demonstrated significant increase in growth rate compared to that of human ovarian surface epithelial cells (HOSEs). The authors concluded that investigation of these ESTs could facilitate the identification of novel chromosome 3 genes implicated in ovarian tumorigenesis whose encoded proteins could be useful screening markers.

Using the MICROMAX^{TM} human cDNA microarray system I, which contains 2400 known human DNA elements including human genes and ESTs, we generated expression profiles from three ovarian cancer cell lines and three normal ovarian epithelial cell cultures. Thirty genes with Cy3/Cy5 signal ratios ranging from 5 to 444 were identified as overexpressed in the cancer cells (Wong *et al.*, 2000). Of particular interest were a number of secretory proteins, since these could be detectable in serum potentially serving as tumor markers. Among these, prostasin and osteopontin were chosen for further study and have been shown to be candidate markers for ovarian cancer screening (Kim *et al.*, 2002; Mok *et al.*, 2001). Furthermore, an autoantibody against another overexpressed gene, which encodes the epithelial cell antigen (Ep-CAM), has also been demonstrated to be useful in the early detection of ovarian cancer (Kim *et al.*, 2003).

To avoid potential artifacts arising from the use of cell cultures manipulated *in vitro*, other investigators have chosen to study RNA derived directly from surgically resected tumor. Shridhar *et al.* (2001) evaluated 18,000 DNA elements using RNA from 21 early stage and 17 advanced stage ovarian tumors. They concluded that further study of proteins found to be overexpressed in early stage ovarian cancer would be a good starting point for the search for markers to aid in early detection. Sawiris *et al.* (2002) produced a 516-element array termed "Ovachip" for its relevance to ovarian cancer. They found that screening with the Ovachip identified "expression patterns" that could differentiate between ovarian cancer and colon cancer specimens. This suggests that microarray technology may reveal markers able to distinguish ovarian cancer from other pelvic-abdominal tumors.

Using an oligonucleotide microarray comprising over 40,000 features, 37 advanced stage papillary serous primary carcinomas, and 6 brushings from normal ovarian surface epithelium for expression profiling analysis, Donninger *et al.* (2004) identified 22,579 sequences as informative for all of the specimens. From this list, we extracted 1191 genes differentially regulated in the tumor specimens as compared to normal hOSE at a p-value <0.001. Using gene ontology (GO) analysis, specific genes involved in cell growth, differentiation, adhesion, apoptosis, and migration were identified. In addition, through the use of PathwayAssist software, which employs a natural language-processing algorithm to identify protein interactions contained in the entire PubMed abstract database, signaling pathways associated with tumor cell proliferation, migration, spread, and invasion were characterized in the advanced stage papillary serous ovarian tumors (Fig. 3).

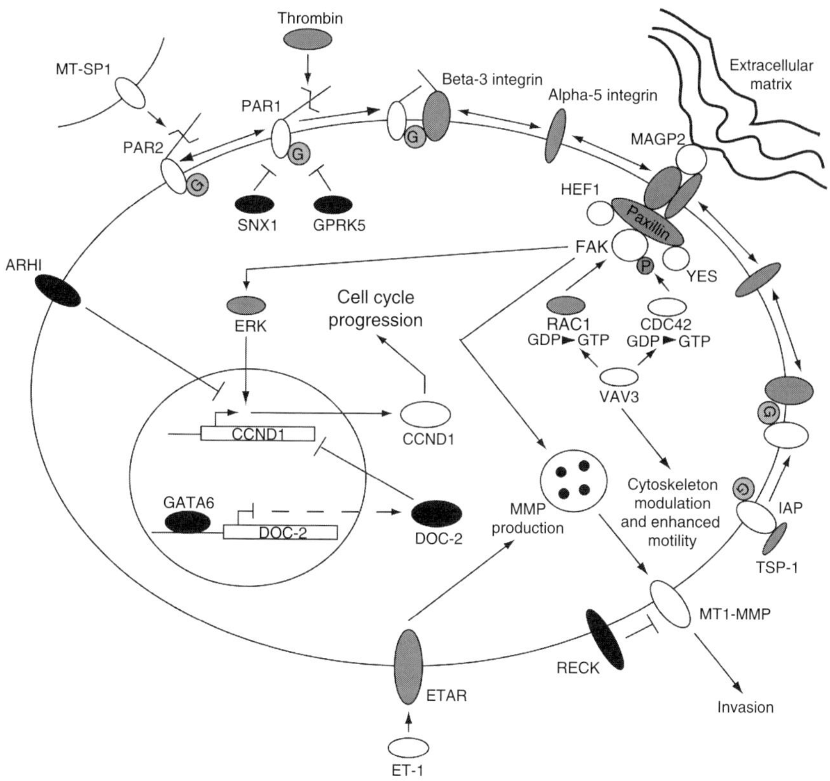

Fig. 3 Pathway analysis identifies differentially regulated genes contributing to the phenotypes associated with late-stage high-grade papillary serous ovarian cancer. PathwayAssist version 2.5 software was used to identify putative signaling pathways contributing to high-grade disease among genes identified as differentially regulated when compared to ovarian surface epithelium ($p < 0.001$). Signaling events stimulating cell cycle progression, invasion, and enhanced motility featured prominently.

Expression profiling has also been used to examine the relatedness of gynecologic cancers of different histologies and organs of origin. For instance, Shedden *et al.* (2005) looked at 103 microdissected primary ovarian and uterine carcinomas of the endometrioid or serous histotype and asked whether gene expression was predicted best by organ of origin or tumor phenotype. With a 7000-element oligonucleotide microarray, they were able to construct unique gene expression profiles for both histological subtype and organ of origin, concluding that organ and histological subtype equally contribute to gene expression profiles. Moreover, they identified subsets of ovarian and uterine endometrioid adenocarcinomas with a particular defect in Wnt signaling, and these tumors shared similar gene expression profiles not seen in tumors lacking this defect.

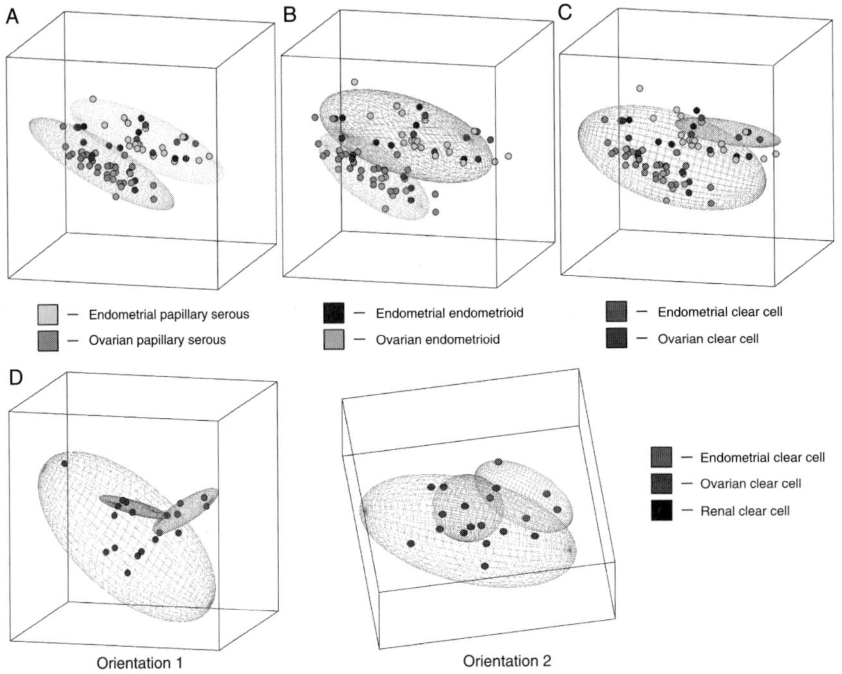

Fig. 4 Graphic depiction of principal components analysis of ovarian and endometrial cancers according to histology. Principal components analysis was performed after imputing values using Partek Pro 2000 software. These data were imported into Matlab software to allow depiction of the elliptical region where the addition of a new sample of each group would fall within a 95% confidence interval. (A) Analysis of tumors with serous histology showing two nonoverlapping elliptical regions separating endometrial (upper) from ovarian (lower) specimens. (B) Analysis of tumors with endometrioid histology showing two nonoverlapping elliptical regions separating endometrial (upper) from ovarian (lower) specimens. (C) Analysis of tumors with clear cell histology showing overlapping elliptical regions representing endometrial (upper) and ovarian (lower) specimens. (D) Analysis of tumors according to organ of origin demonstrates three overlapping elliptical regions among ovarian, endometrial, and renal clear cell specimens. Two different orientations (1 and 2) are illustrated. (See Color Insert.)

In a similar study, we used an 11,000-element cDNA array to examine the gene expression profiles of late-stage, high-grade endometrioid, serous, and clear cell ovarian and uterine carcinomas (Zorn *et al.*, 2005). Principal components analysis, which depicts the similarity among samples by their proximity within three-dimensional space, demonstrated that the gene expression profiles of endometrioid and serous subtypes of ovarian and endometrial cancers are more distinct by organ of origin than by histology (Fig. 4A–C). Interestingly, however, clear cell cancers of the ovary, endometrium, and even the kidney share such similar gene expression profiles that the organs cannot be reliably distinguished (Fig. 4D). This observation was confirmed using class prediction analysis employing leave-one-out cross-validation and permutation analysis suggesting that clear cell cancers, regardless of organ involvement, might benefit from treatment directed by histological subtype, while serous and endometrioid tumors of the ovary and endometrium should continue to be evaluated using an organ-based approach.

Expression profiling has also been used to establish the relationship among tumors of similar histology but varying grade. These types of tumors frequently have substantially different clinical courses. For instance, we sought a gene signature that could differentiate invasive papillary serous ovarian carcinomas from serous BOTs (Bonome *et al.*, 2005), using carefully microdissected specimens from the tumors of 80 previously untreated ovarian cancer patients and 10 normal (OSE) cytobrushings. Unsupervised hierarchical clustering of the genes meeting the filtering criteria generated a dendrogram depicting the relatedness among normal ovarian surface epithelium, serous BOTs, low-grade and high-grade invasive serous carcinomas (Fig. 5A). These results suggest that not only are serous BOTs clearly distinct from high-grade serous lesions, but in fact borderline serous tumors appear more closely related to OSE than to high-grade serous ovarian carcinomas. Furthermore, low-grade serous lesions more closely cluster with BOTs than with high-grade lesions, suggesting that low-grade serous ovarian carcinomas and borderline serous tumors may represent a distinct subset of tumors from high-grade disease. We validated these relationships *in silico* using binary tree prediction employing a compound covariate predictor (Fig. 5B), and with an independent set of 4 serous and 13 low-grade microdissected serous ovarian tumors.

Underlying the distinct phenotypes between high-grade invasive disease and borderline serous tumors were genes associated with cell cycle control, cell proliferation, and DNA repair (Table IIIA and B). By interrogating this list using PathwayAssist software, we identified a molecular pathway involving p21 and p53 that appears to differentiate BOTs from high-grade papillary serous ovarian carcinoma and may account for the limited invasive capacity of borderline serous lesions. This corresponds to previous studies demonstrating increased mutation of p53 in invasive, but not borderline, serous ovarian tumors (Singer *et al.*, 2005). Additional analysis involving greater numbers of tumor specimens will increase our ability to discriminate gene expression among the broad variety of ovarian tumors, and, eventually, highlight key signaling events to be targeted by novel therapies.

Recently, transcription profiles have been identified that can predict patient survival. Using bulked RNA isolated from high-grade ovarian cancer obtained from 68 patients and a 22,000-element oligonucleotide array platform, Spentzos *et al.* (2004) identified a 115-gene signature referred to

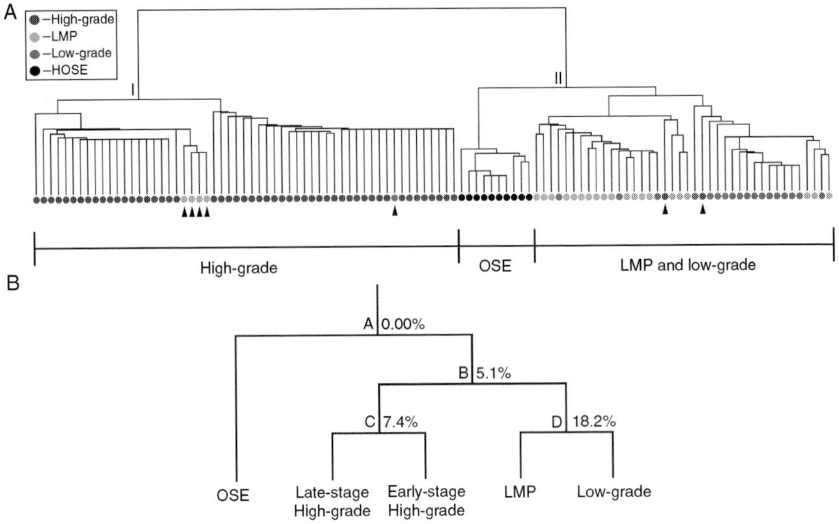

Fig. 5 Hierarchical clustering and binary tree analysis of the 14,468 probe sets passing the filtering criteria for microdissected LMP, low-grade and high-grade papillary serous tumors. (A) Clustering analysis was completed using the 1-correlation metric with centroid linkage. The close association of LMP (green) and low-grade (blue) is evidenced at node II, while high-grade (red) tumors clustered tightly at node I. Misclassified specimens are highlighted with a black arrowhead. (B) Binary tree analysis confirmed the hierarchical clustering results. Tree diagram was generated using binary tree prediction followed by leave-one-out cross-validation to estimate the error associated with the tree-building process. OSE samples were classified as basal to the ovarian cancer specimens. LMP tumors and low-grade cancers were more closely aligned to each other, as were early and late-stage high-grade tumors. Percentages indicate the misclassification error associated with each node. (See Color Insert.)

as the Ovarian Cancer Prognostic Profile (OCPP). While the signature maintained its independent prognostic value in multivariate analysis, further validation of differential gene expression by quantitative PCR has not been performed. Berchuck *et al.* (2005) also used a 22,000-element oligonucleotide array platform to examine survival in 54 patients with late-stage serous ovarian cancers. They identified an expression model that could distinguish patients who survived less than 3 years from those who lived greater than 7 years. In addition, they tested their model on the 68 patient samples from Spentzos and associates and found the same set of genes again predicted short-term versus long-term survival. They then validated the array using RT-PCR on three of the most differentially expressed genes. Interestingly, when the advanced tumors from the 24 long-term survivors in the Berchuck and associates study were compared with RNA from another group of 11 early stage cases, the predictive model showed that the cancers from the long-term survivors and the cancers from the early stage cases

Table III Signature Genes Associated with Serous Ovarian Tumors

LocusLink ID	Fold change*	Gene name	Parametric *p*-value
A. Signature genes associated with serous borderline ovarian tumors			
4246	27.28	Secretoglobin, family 2A, member 1	6e−07
80150	20.53	Asparaginase-like 1 protein	<1e−07
133690	19.92	Calcyphosine-like protein	7e−07
219670	13.26	Enkurin	<1e−07
80217	12.59	Hypothetical protein LOC80217	4e−07
84688	11.93	Ciliated bronchial epithelium 1	1.1e−06
201625	10.31	Hypothetical protein FLJ40427	<1e−07
25984	10.23	Keratin 23	4e−07
155465	9.87	Breast cancer membrane protein 11	2.7e−06
138255	9.65	Hypothetical protein LOC138255	5.4e−06
146845	9.50	Hypothetical protein LOC146845 (wdr16)	6e−07
B. Genes associated with high-grade late-stage serous ovarian carcinoma			
7849	10.36	Paired box gene 8	1.3e−06
983	9.56	Cell division cycle 2, G1 to S and G2 to M (CDC2)	<1e−07
140690	9.49	CCCTC-binding factor (zinc finger protein)-like protein	7.36e−05
10635	8.60	RAD51-associated protein 1	<1e−07
1894	8.47	Epithelial cell transforming sequence 2 oncogene	<1e−07
81610	8.31	Hypothetical protein LOC81610	<1e−07
1058	7.87	Centromere protein A, 17kDa	<1e−07
83461	7.64	Cell division cycle associated 3	<1e−07
56133	7.56	Protocadherin beta 2	8.32e−05
9133	7.43	Cyclin B2	<1e−07
55165	7.28	Hypothetical protein LOC55165	<1e−07

*Relative to OSE.

shared a similar expression pattern. This suggests a shared biology for early stage ovarian cancers and the subset of advanced tumors that follows a favorable clinical course.

Gene expression analysis has the potential to change the treatment of ovarian neoplasms. Expression profiles can help diagnose subtypes of disease and predict patient survival. One day, profiling may influence treatment decisions, informing both which patients should receive chemotherapy and what type of chemotherapeutic agents should be employed. As greater numbers of tumor samples are analyzed, the power of these profiling studies will increase, raising the possibility that novel molecular targets and less toxic therapies will be identified. These powerful techniques hold the potential to unravel the genetic origins of ovarian cancer. Hopefully, this will translate into earlier diagnosis and better patient outcome from disease. Furthermore, by integrating newly developed genotyping technology such as single nucleotide polymorphism (SNP) arrays to interrogate whole genome LOH and copy number changes with expression profiling, we will be able to get a better understanding of the underlying genetic changes in ovarian cancer.

ACKNOWLEDGMENTS

Grant Support: This work was supported in part by the DF/HCC Ovarian Cancer SPORE, and R33CA103595 from National Institute of Health, Department of Health and Human Services, Gillette Center for Women's Cancer, Adler Foundation, Inc., Edgar Astrove Fund, the Ovarian Cancer Research Fund, Inc., the Morse Family Fund, and the Natalie Pihl Fund.

REFERENCES

Alizadeh, A. A., Eisen, M. B., Davis, R. E., Ma, C., Lossos, I. S., Rosenwald, A., Boldrick, J. C., Sabet, H., Tran, T., Yu, X., Powell, J. I., Yang, L., *et al.* (2000). Distinct types of diffuse large B-cell lymphoma identified by gene expression profiling. *Nature* **403,** 503–511.

Arnold, N., Hagele, L., Walz, L., Schempp, W., Pfisterer, J., Bauknecht, T., and Kiechle, M. (1996). Overrepresentation of 3q and 8q material and loss of 18q material are recurrent findings in advanced human ovarian cancer. *Genes Chromosomes Cancer* **16,** 46–54.

Auersperg, N., Edelson, M. I., Mok, S. C., Johnson, S. W., and Hamilton, T. C. (1998). The biology of ovarian cancer. *Semin. Oncol.* **25,** 281–304.

Barlund, M., Forozan, F., Kononen, J., Bubendorf, L., Chen, Y., Bittner, M. L., Torhorst, J., Haas, P., Bucher, C., Sauter, G., Kallioniemi, O. P., and Kallioniemi, A. (2000). Detecting activation of ribosomal protein S6 kinase by complementary DNA and tissue microarray analysis. *J. Natl. Cancer Inst.* **92,** 1252–1259.

Bast, R. C., Jr., Boyer, C. M., Xu, F. J., Wiener, J., Dabel, R., Woolas, R., Jacobs, I., and Berchuck, A. (1995). Molecular approaches to prevention and detection of epithelial ovarian cancer. *J. Cell. Biochem. Suppl.* **23,** 219–222.

Berchuck, A., Iversen, E. S., Lancaster, J. M., Pittman, J., Luo, J., Lee, P., Murphy, S., Dressman, H. K., Febbo, P. G., West, M., Nevins, J. R., and Marks, J. R. (2005). Patterns of gene expression that characterize long-term survival in advanced stage serous ovarian cancers. *Clin. Cancer Res.* **11**, 3686–3696.

Boente, M. P., Hamilton, T. C., Godwin, A. K., Buetow, K., Kohler, M. F., Hogan, W. M., Berchuck, A., and Young, R. C. (1996). Early ovarian cancer: A review of its genetic and biologic factors, detection, and treatment. *Curr. Probl. Cancer* **20**, 83–137.

Bonome, T., Lee, J. Y., Park, D. C., Radonovich, M., Pise-Masison, C., Brady, J., Gardner, G. J., Hao, K., Wong, W. H., Barrett, J. C., Lu, K. H., Sood, A. K., *et al.* (2005). Expression profiling of serous low malignant potential, low-grade, and high-grade tumors of the ovary. *Cancer Res.* **65**, 10602–10612.

Boring, C. C., Squires, T. S., Tong, T., and Montgomery, S. (1994). Cancer statistics, 1994. *CA Cancer J. Clin.* **44**, 7–26.

Bourdon, V., Naef, F., Rao, P. H., Reuter, V., Mok, S. C., Bosl, G. J., Koul, S., Murty, V. V., Kucherlapati, R. S., and Chaganti, R. S. (2002). Genomic and expression analysis of the 12p11-p12 amplicon using EST arrays identifies two novel amplified and overexpressed genes. *Cancer Res.* **62**, 6218–6223.

Cannistra, S. A. (1993). Cancer of the ovary. *N. Engl. J. Med* **329**, 1550–1559.

Cheng, K. W., Lahad, J. P., Kuo, W. L., Lapuk, A., Yamada, K., Auersperg, N., Liu, J., Smith-McCune, K., Lu, K. H., Fishman, D., Gray, J. W., and Mills, G. B. (2004). The RAB25 small GTPase determines aggressiveness of ovarian and breast cancers. *Nat. Med.* **10**, 1251–1256.

DeRisi, J., Penland, L., Brown, P. O., Bittner, M. L., Meltzer, P. S., Ray, M., Chen, Y., Su, Y. A., and Trent, J. M. (1996). Use of a cDNA microarray to analyse gene expression patterns in human cancer. *Nat. Genet.* **14**, 457–460.

DeRisi, J. L., Iyer, V. R., and Brown, P. O. (1997). Exploring the metabolic and genetic control of gene expression on a genome scale. *Science* **278**, 680–686.

Diebold, J., Suchy, B., Baretton, G. B., Blasenbreu, S., Meier, W., Schmidt, M., Rabes, H., and Lohrs, U. (1996). DNA ploidy and MYC DNA amplification in ovarian carcinomas. Correlation with p53 and bcl-2 expression, proliferative activity and prognosis. *Virchows Arch.* **429**, 221–227.

Dong, S. M., Traverso, G., Johnson, C., Geng, L., Favis, R., Boynton, K., Hibi, K., Goodman, S. N., D'Allessio, M., Paty, P., Hamilton, S. R., Sidransky, D., *et al.* (2001). Detecting colorectal cancer in stool with the use of multiple genetic targets. *J. Natl. Cancer Inst.* **93**, 858–865.

Donninger, H., Bonome, T., Radonovich, M., Pise-Masison, C. A., Brady, J., Shih, J. H., Barrett, J. C., and Birrer, M. J. (2004). Whole genome expression profiling of advance stage papillary serous ovarian cancer reveals activated pathways. *Oncogene* **23**, 8065–8077.

Farley, J., Smith, L. M., Darcy, K. M., Sobel, E., O'Connor, D., Henderson, B., Morrison, L. E., and Birrer, M. J. (2003). Cyclin E expression is a significant predictor of survival in advanced, suboptimally debulked ovarian epithelial cancers: A Gynecologic Oncology Group study. *Cancer Res.* **63**, 1235–1241.

Fearon, E. R., and Vogelstein, B. (1990). A genetic model for colorectal tumorigenesis. *Cell* **61**, 759–767.

Golub, T. R., Slonim, D. K., Tamayo, P., Huard, C., Gaasenbeek, M., Mesirov, J. P., Coller, H., Loh, M. L., Downing, J. R., Caligiuri, M. A., Bloomfield, C. D., and Lander, E. S. (1999). Molecular classification of cancer: Class discovery and class prediction by gene expression monitoring. *Science* **286**, 531–537.

Hartmann, L. C., Podratz, K. C., Keeney, G. L., Kamel, N. A., Edmonson, J. H., Grill, J. P., Su, J. Q., Katzmann, J. A., and Roche, P. C. (1994). Prognostic significance of p53 immunostaining in epithelial ovarian cancer. *J. Clin. Oncol.* **12**, 64–69.

Ismail, R. S., Baldwin, R. L., Fang, J., Browning, D., Karlan, B. Y., Gasson, J. C., and Chang, D. D. (2000). Differential gene expression between normal and tumor-derived ovarian epithelial cells. *Cancer Res.* **60**, 6744–6749.

Iwabuchi, H., Sakamoto, M., Sakunaga, H., Ma, Y. Y., Carcangiu, M. L., Pinkel, D., Yang-Feng, T. L., and Gray, J. W. (1995). Genetic analysis of benign, low-grade, and high-grade ovarian tumors. *Cancer Res.* **55**, 6172–6180.

Kallioniemi, A., Kallioniemi, O. P., Sudar, D., Rutovitz, D., Gray, J. W., Waldman, F., and Pinkel, D. (1992). Comparative genomic hybridization for molecular cytogenetic analysis of solid tumors. *Science* **258,** 818–821.

Kiechle, M., Jacobsen, A., Schwarz-Boeger, U., Hedderich, J., Pfisterer, J., and Arnold, N. (2001). Comparative genomic hybridization detects genetic imbalances in primary ovarian carcinomas as correlated with grade of differentiation. *Cancer* **91,** 534–540.

Kim, J. H., Skates, S. J., Uede, T., Wong, K. K., Schorge, J. O., Feltmate, C. M., Berkowitz, R. S., Cramer, D. W., and Mok, S. C. (2002). Osteopontin as a potential diagnostic biomarker for ovarian cancer. *JAMA* **287,** 1671–1679.

Kim, J. H., Herlyn, D., Wong, K. K., Park, D. C., Schorge, J. O., Lu, K. H., Skates, S. J., Cramer, D. W., Berkowitz, R. S., and Mok, S. C. (2003). Identification of epithelial cell adhesion molecule autoantibody in patients with ovarian cancer. *Clin. Cancer Res.* **9,** 4782–4791.

Klemi, P. J., Pylkkanen, L., Kiilholma, P., Kurvinen, K., and Joensuu, H. (1995). p53 protein detected by immunohistochemistry as a prognostic factor in patients with epithelial ovarian carcinoma. *Cancer* **76,** 1201–1208.

Kokuho, M., Yoshiki, T., Hamaguchi, A., Okada, Y., Tomoyoshi, T., and Higuchi, K. (1993). Immunohistochemical study of c-erbB-2 proto-oncogene product in prostatic cancer. *Nippon Hinyokika Gakkai Zasshi* **84,** 1872–1878.

Lasko, D., Cavenee, W., and Nordenskjold, M. (1991). Loss of constitutional heterozygosity in human cancer. *Annu. Rev. Genet.* **25,** 281–314.

Lee, K. R., Tavassoli, F. A., Prat, J., Dietel, M., Gersell, D. J., Karseladze, A. I., Hauptmann, S., Rutgers, J., Russell, P., Buckley, C. H., Schwartz, P., Goldgar, D. E., *et al.* (2003). Surface epithelial-stromal tumours. *In* "World Health Organization Classification of Tumours, Pathology & Genetics" (F. A. Tavassoli and P. Devilee, Eds.), Tumor of the Breast and Female Genital Organs, pp. 117–202. IARC Press, Lyon.

Manderson, E. N., Mes-Masson, A. M., Novak, J., Lee, P. D., Provencher, D., Hudson, T. J., and Tonin, P. N. (2002). Expression profiles of 290 ESTs mapped to chromosome 3 in human epithelial ovarian cancer cell lines using DNA expression oligonucleotide microarrays. *Genome Res.* **12,** 112–121.

Marsden, D. E., Friedlander, M., and Hacker, N. F. (2000). Current management of epithelial ovarian carcinoma: A review. *Semin. Surg. Oncol.* **19,** 11–19.

Masciullo, V., Sgambato, A., Pacilio, C., Pucci, B., Ferrandina, G., Palazzo, J., Carbone, A., Cittadini, A., Mancuso, S., Scambia, G., and Giordano, A. (1999). Frequent loss of expression of the cyclin-dependent kinase inhibitor p27 in epithelial ovarian cancer. *Cancer Res.* **59,** 3790–3794.

Matei, D., Graeber, T. G., Baldwin, R. L., Karlan, B. Y., Rao, J., and Chang, D. D. (2002). Gene expression in epithelial ovarian carcinoma. *Oncogene* **21,** 6289–6298.

Mayr, D., Pannekamp, U., Baretton, G. B., Gropp, M., Meier, W., Flens, M. J., Scheper, R., and Diebold, J. (2000). Immunohistochemical analysis of drug resistance-associated proteins in ovarian carcinomas. *Pathol. Res. Pract.* **196,** 469–475.

Meden, H., and Kuhn, W. (1997). Overexpression of the oncogene c-erbB-2 (HER2/neu) in ovarian cancer: A new prognostic factor. *Eur. J. Obstet. Gynecol. Reprod. Biol.* **71,** 173–179.

Mok, S. C., Chao, J., Skates, S., Wong, K. K., Yiu, G. Y., Muto, M. G., VBerkowitz, R. S., and Cramer, D. W. (2001). Prostasin, a potential serum marker for ovarian cancer, identified through microarray technology. *J. Natl. Cancer Inst.* **93,** 1458–1464.

Ozalp, S. S., Yalcin, O. T., Basaran, G. N., Artan, S., Kabukcuoglu, S., and Minsin, T. H. (2000). Prognostic significance of deletion and over-expression of the p53 gene in epithelial ovarian cancer. *Eur. J. Gynaecol. Oncol.* **21,** 282–286.

Pinkel, D., Segraves, R., Sudar, D., Clark, S., Poole, I., Kowbel, D., Collins, C., Kuo, W. L., Chen, C., Zhai, Y., Dairkee, S. H., Ljung, B. M., *et al.* (1998). High resolution analysis of

DNA copy number variation using comparative genomic hybridization to microarrays. *Nat. Genet.* **20**, 207–211.

Pollack, J. R., Perou, C. M., Alizadeh, A. A., Eisen, M. B., Pergamenschikov, A., Williams, C. F., Jeffrey, S. S., Botstein, D., and Brown, P. O. (1999). Genome-wide analysis of DNA copy-number changes using cDNA microarrays. *Nat. Genet.* **23**, 41–46.

Pollack, J. R., Sorlie, T., Perou, C. M., Rees, C. A., Jeffrey, S. S., Lonning, P. E., Tibshirani, R., Botstein, D., Borresen-Dale, A. L., and Brown, P. O. (2002). Microarray analysis reveals a major direct role of DNA copy number alteration in the transcriptional program of human breast tumors. *Proc. Natl. Acad. Sci. USA* **99**, 12963–12968.

Reles, A., Wen, W. H., Schmider, A., Gee, C., Runnebaum, I. B., Kilian, U., Jones, L. A., El-Naggar, A., Minguillon, C., Schonborn, I., Reich, O., Kreienberg, R., *et al.* (2001). Correlation of p53 mutations with resistance to platinum-based chemotherapy and shortened survival in ovarian cancer. *Clin. Cancer Res.* **7**, 2984–2997.

Sakamoto, M., Umayahara, K., Sakamoto, H., Kawasaki, K., Suehiro, Y., Kunugi, T., Akiya, T., Iwabuchi, H., Sakunaga, H., Muroya, T., Kikuchi, Y., Sugishita, T., *et al.* (1998). Cancer-associated gene abnormalities and chemosensitivity. *Gan To Kagaku Ryoho* **25**, 1819–1831.

Sawiris, G. P., Sherman-Baust, C. A., Becker, K. G., Cheadle, C., Teichberg, D., and Morin, P. J. (2002). Development of a highly specialized cDNA array for the study and diagnosis of epithelial ovarian cancer. *Cancer Res.* **62**, 2923–2928.

Scambia, G., Ferrandina, G., Marone, M., Benedetti Panici, P., Giannitelli, C., Piantelli, M., Leone, A., and Mancuso, S. (1996). nm23 in ovarian cancer: Correlation with clinical outcome and other clinicopathologic and biochemical prognostic parameters. *J. Clin. Oncol.* **14**, 334–342.

Serov, S. F., and Scullt, R. E. (1993). Histological typing of ovarian tumors. *In* "International Histological Classification of Tumors No. 9." World Health Organization, Geneva.

Shedden, K. A., Kshirsagar, M. P., Schwartz, D. R., Wu, R., Yu, H., Misek, D. E., Hanash, S., Katabuchi, H., Ellenson, L. H., Fearon, E. R., and Cho, K. R. (2005). Histologic type, organ of origin, and Wnt pathway status: Effect on gene expression in ovarian and uterine carcinomas. *Clin. Cancer Res.* **11**, 2123–2131.

Shridhar, V., Lee, J., Pandita, A., Iturria, S., Avula, R., Staub, J., Morrissey, M., Calhoun, E., Sen, A., Kalli, K., Keeney, G., Roche, P., *et al.* (2001). Genetic analysis of early- versus late-stage ovarian tumors. *Cancer Res.* **61**, 5895–5904.

Sidransky, D. (1997). Nucleic acid-based methods for the detection of cancer. *Science* **278**, 1054–1059.

Singer, G., Stohr, R., Cope, L., Dehari, R., Hartmann, A., Cao, D. F., Wang, T. L., Kurman, R. J., and Shih Ie, M. (2005). Patterns of p53 mutations separate ovarian serous borderline tumors and low- and high-grade carcinomas and provide support for a new model of ovarian carcinogenesis: A mutational analysis with immunohistochemical correlation. *Am. J. Surg. Pathol.* **29**, 218–224.

Skirnisdottir, I., Sorbe, B., and Seidal, T. (2001). The growth factor receptors HER-2/neu and EGFR, their relationship, and their effects on the prognosis in early stage (FIGO I-II) epithelial ovarian carcinoma. *Int. J. Gynecol. Cancer* **11**, 119–129.

Sonoda, G., Palazzo, J., du Manoir, S., Godwin, A. K., Feder, M., Yakushiji, M., and Testa, J. R. (1997). Comparative genomic hybridization detects frequent overrepresentation of chromosomal material from 3q26, 8q24, and 20q13 in human ovarian carcinomas. *Genes Chromosomes Cancer* **20**, 320–328.

Spentzos, D., Levine, D. A., Ramoni, M. F., Joseph, M., Gu, X., Boyd, J., Libermann, T. A., and Cannistra, S. A. (2004). Gene expression signature with independent prognostic significance in epithelial ovarian cancer. *J. Clin. Oncol.* **22**, 4700–4710.

Suehiro, Y., Sakamoto, M., Umayahara, K., Iwabuchi, H., Sakamoto, H., Tanaka, N., Takeshima, N., Yamauchi, K., Hasumi, K., Akiya, T., Sakunaga, H., Muroya, T., *et al.* (2000). Genetic aberrations detected by comparative genomic hybridization in ovarian clear cell adenocarcinomas. *Oncology* **59,** 50–56.

Thiagalingam, S., Laken, S., Willson, J. K., Markowitz, S. D., Kinzler, K. W., Vogelstein, B., and Lengauer, C. (2001). Mechanisms underlying losses of heterozygosity in human colorectal cancers. *Proc. Natl. Acad. Sci. USA* **98,** 2698–2702.

Trimbos, J. B., Parmar, M., Vergote, I., Guthrie, D., Bolis, G., Colombo, N., Vermorken, J. B., Torri, V., Mangioni, C., Pecorelli, S., Lissoni, A., and Swart, A. M. (2003a). International Collaborative Ovarian Neoplasm trial 1 and adjuvant chemotherapy in ovarian neoplasm trial: Two parallel randomized phase III trials of adjuvant chemotherapy in patients with early-stage ovarian carcinoma. *J. Natl. Cancer Inst.* **95,** 105–112.

Trimbos, J. B., Vergote, I., Bolis, G., Vermorken, J. B., Mangioni, C., Madronal, C., Franchi, M., Tateo, S., Zanetta, G., Scarfone, G., Giurgea, L., Timmers, P., *et al.* (2003b). Impact of adjuvant chemotherapy and surgical staging in early-stage ovarian carcinoma: European organisation for research and treatment of cancer-adjuvant chemotherapy in ovarian neoplasm trial. *J. Natl. Cancer Inst.* **95,** 113–125.

Tsuda, H., Birrer, M. J., Ito, Y. M., Ohashi, Y., Lin, M., Lee, C., Wong, W. H., Rao, P. H., Lau, C. C., Berkowitz, R. S., Wong, K. K., and Mok, S. C. (2004). Identification of DNA copy number changes in microdissected serous ovarian cancer tissue using a cDNA microarray platform. *Cancer Genet. Cytogenet.* **155,** 97–107.

Tsuda, H., Ito, Y. M., Ohashi, Y., Wong, K. K., Hashiguchi, Y., Welch, W. R., Berkowitz, R. S., Birrer, M. J., and Mok, S. C. (2005). Identification of over-expression and amplification of ABCF2 in clear cell ovarian adenocarcinomas by cDNA microarray analyses. *Clin. Cancer Res.* **11,** 6880–6888.

Vogelstein, B., Fearon, E. R., Hamilton, S. R., Kern, S. E., Preisinger, A. C., Leppert, M., Nakamura, Y., White, R., Smits, A. M., and Bos, J. L. (1988). Genetic alterations during colorectal-tumor development. *N. Engl. J. Med.* **319,** 525–532.

Vogelstein, B., Fearon, E. R., Kern, S. E., Hamilton, S. R., Preisinger, A. C., Nakamura, Y., and White, R. (1989). Allelotype of colorectal carcinomas. *Science* **244,** 207–211.

Wang, T., Hopkins, D., Schmidt, C., Silva, S., Houghton, R., Takita, H., Repasky, E., and Reed, S. G. (2000). Identification of genes differentially over-expressed in lung squamous cell carcinoma using combination of cDNA subtraction and microarray analysis. *Oncogene* **19**(12), 1519–1528.

Wolf, N. G., Abdul-Karim, F. W., Farver, C., Schrock, E., du Manoir, S., and Schwartz, S. (1999). Analysis of ovarian borderline tumors using comparative genomic hybridization and fluorescence *in situ* hybridization. *Genes Chromosomes Cancer* **25,** 307–315.

Wong, K. K., Cheng, R. S., and Mok, S. C. (2000). Identification of differentially expressed genes from ovarian cancer cells by MICROMAX^{TM} cDNA microarray system. *Biotechnique* **30,** 670–675.

Wu, X., Groves, F. D., McLaughlin, C. C., Jemal, A., Martin, J., and Chen, V. W. (2005). Cancer incidence patterns among adolescents and young adults in the United States. *Cancer Causes Control* **16,** 309–320.

Xu, J., Stolk, J. A., Zhang, X., Silva, S. J., Houghton, R. L., Matsumura, M., Vedvick, T. S., Leslie, K. B., Badaro, R., and Reed, S. G. (2000). Identification of differentially expressed genes in human prostate cancer using subtraction and microarray. *Cancer Res.* **60,** 1677–1682.

Yokota, J., and Ookawa, K. (1993). Accumulation of genetic alterations during human tumor progression. *Gan To Kagaku Ryoho* **20,** 321–325.

Zorn, K. K., Bonome, T., Gangi, L., Chandramouli, G. V. R., Awtrey, C. S., Gardner, G. J., Barrett, J. C., Boyd, J., and Birrer, M. J. (2005). Gene expression profiles of serous, endometrioid, and clear cell subtypes of ovarian and endometrial cancer. *Clin. Cancer Res.* **11**(18), 6422–6430.

Mass Spectrometry: Uncovering the Cancer Proteome for Diagnostics

Da-Elene van der Merwe,[*] Katerina Oikonomopoulou,[*,†] John Marshall,[‡] and Eleftherios P. Diamandis[*,†]

[*]*Department of Pathology and Laboratory Medicine, Mount Sinai Hospital, Toronto, Ontario M5G1X5, Canada;*
[†]*Department of Laboratory Medicine and Pathobiology, University of Toronto, Toronto, Ontario M5G1L5, Canada;*
[‡]*Department of Chemistry and Biology, Ryerson University, Toronto, Ontario M5G1G3, Canada*

I. Current Cancer Biomarkers
II. Early Detection
  A. When Is an Early Detection Program Warranted?
III. The Need for New Diagnostic Strategies
IV. Mass Spectrometry
  A. Ionization Source
  B. Mass Analyzers
  C. Protein Identification
  D. Quantitation
V. Mass Spectrometry-Based Diagnostics
  A. Mass Spectrometry as a Tissue Imaging Tool
  B. Mass Spectrometry as a Biomarker Discovery Tool
  C. Mass Spectrometry as a Cancer Diagnostic Tool
VI. Current Limitations of Diagnostic Mass Spectrometry
  A. Preanalytical
  B. Analytical
  C. Postanalytical
VII. Suggestions for Future Progress
VIII. Future Direction
References

Despite impressive scientific achievements over the past few decades, cancer is still a leading cause of death. One of the major reasons is that most cancer patients are diagnosed with advanced disease. This is clearly illustrated with ovarian cancer in which the overall 5-year survival rates are only 20–30%. Conversely, when ovarian cancer is detected early (stage 1), the 5-year survival rate increases to 95%. Biomarkers, as tools for preclinical detection of cancer, have the potential to revolutionize the field of clinical diagnostics. The emerging field of clinical proteomics has found applications across a wide spectrum of cancer research. This chapter will focus on mass spectrometry as a proteomic technology implemented in three areas of cancer: diagnostics, tissue imaging, and biomarker discovery. Despite its power, it is also important to realize the preanalytical,

Advances in CANCER RESEARCH

0065-230X/07 $35.00
DOI: 10.1016/S0065-230X(06)96002-3

analytical, and postanalytical limitations currently associated with this methodology. The ultimate endpoint of clinical proteomics is individualized therapy. It is essential that research groups, the industry, and physicians collaborate to conduct large prospective, multicenter clinical trials to validate and standardize this technology, for it to have real clinical impact.

I. CURRENT CANCER BIOMARKERS

Currently, hundreds of tumor markers exist, yet most of them fall short of expectation. Clinicians expect that a marker should be beneficial to their patients in terms of improved morbidity, mortality, and quality of life. To illustrate the point, even if a biomarker is able to detect relapse a few months prior to clinical symptoms, if effective treatment does not exist, this information does not necessarily translate into improved outcome. Moreover, knowledge of tumor marker elevation may be potentially harmful since it shortens disease-free survival and adds to patient anxiety.

Despite their known shortcomings, tumor markers continue to be used in a variety of clinical settings. Some of the current applications of tumor markers and their limitations are listed in Table I.

Currently, controversy exists regarding the optimal use of tumor markers among clinicians and laboratory medicine specialists. This is reflected in practice guidelines developed by various professional societies. In 1998, the National Academy of Clinical Biochemistry (NACB)[1] sponsored a consensus conference to develop guidelines for the analytical performance and clinical utility of tumor markers (Fleisher *et al.*, 2002). The recommendations focused on pre- and postanalytical concerns, the use of reference intervals, and the manner in which tumor markers should be used clinically, with specific attention to screening, diagnosis, monitoring, or prognosis.

In the mid-1980s, a working group consisting of German scientists, physicians, and representatives of the diagnostics industry were established. In 1993, this group published a consensus statement on the criteria for use of tumor markers with respect to clinical relevance, analytical methods, and manufacturing requirements (European Group on Tumour Markers, 1999;

[1] Abbreviations: NACB, National Academy of Clinical Biochemistry; EGTM, European Group on Tumor Makers; SEER, Surveillance, epidemiology, and end results; WHO, World Health Organization; EDRN, Early Detection Research Network; ELISA, Enzyme-linked immunosorbent assay; CGAP, Cancer Genome Anatomy Project; SAGE, Serial analysis of gene expression; EST, Expressed sequence tag; SELDI-TOF, Surface-enhanced laser desorption/ ionization-time-of-flight; MUDPIT, Multidimentsional protein identification technology; HPLC, High-performance liquid chromatography; ESI, Electrospray ionization; MALDI, Matrix-assisted laser desorption; FT-ICR MS, Fourier transform ion cyclotron ionization resonance mass spectrometer; CID, Collisional-induced dissociation; SILAC, Stable-isotope labeling with amino acids in cell culture; ICAT, Isotope-coded affinity tag; LCM, Laser capture microdissection; IMAC, Immobiliszed metal affinity capture; NCI, National Cancer Institute.

Table I Current Applications of Tumor Markers

Application	Clinical value	Comments
1. Population screening	Limited	Low diagnostic sensitivity and specificity
2. Diagnosis	Limited	Low diagnostic sensitivity and specificity
3. Prognosis	Limited	Not sufficiently accurate
4. Tumor staging	Limited	Not sufficiently accurate
5. Tumor localization and targeted therapy	Limited	Low specificity, low efficiency
6. Detection of recurrence	Controversial	Short lead time, unavailable effective therapy, misleading information due to low specificity
7. Monitoring therapeutic response	Important	Biomarker usually superior to imaging modalities
8. Prediction of therapeutic response	Important	Therapy given only to those who will benefit sparing others from toxic side effects

Van Dalen, 1993). This group was formally constituted as the European Group on Tumor Markers (EGTM) in 1997. Many other clinical organizations, such as the American Society of Clinical Oncology (ASCO), formulated their own recommendations [Bast *et al.*, 2001a,b; Tumor Marker Expert Panel (ASCO), 1996]. Reviews on practice guidelines for tumor markers have been published (Duffy *et al.*, 2003; Loi *et al.*, 2004; Sturgeon, 2001, 2002) as well as strategies for their development (Oosterhuis *et al.*, 2004).

II. EARLY DETECTION

Cancer continues to be diagnosed late, when therapeutic options are limited to palliative care. In our battle against cancer, emphasis should shift from clinical diagnosis to preclinical disease detection, before cancer metastasizes and becomes incurable. In an era of evidence- and outcomes-based medicine, the following questions are relevant: (1) *Why do we need early cancer detection?* and (2) *When is an early disease detection program warranted?* (Etzioni *et al.*, 2003).

The answer to the first question is twofold: (1) Treatment of advanced disease is almost never curative. This is illustrated in the very modest gains in survival rates of patients diagnosed with advanced cancers of different organs from 1973 to 1997 (National Cancer Institute, 2002). (2) Early detection of cancer improves outcome. Ovarian cancer is a good example

Table II Projected Changes in Survival with Early Detection[a]

Cancer site	Tumors localized when detected (%)	Five-year survival rate (%)	Five-year survival rate if all tumors were localized when detected (%)
Lung	19	16	49
Colorectal	41	64	90
Breast	65	87	97
Prostate	65	90	100

[a]Based on data from SEER (National Cancer Institute, 2002) for cases diagnosed between 1990 and 1999 inclusive. Cases with *in situ* or unstaged disease have been excluded. The favorable overall 5-year survival among breast and prostate cancer patients is partly due to the prevalence of screening during the calendar years considered. Reprinted from Etzioni *et al.* (2003) with permission from copyright owners.

where early detection can have a major impact. More than two-thirds of ovarian cancer cases are detected at an advanced stage, when the cancer cells have spread away from the ovarian surface and have disseminated throughout the peritoneal cavity (Menon and Jacobs, 2002; Meyer and Rustin, 2000). The resulting 5-year survival rate is 20–30% with the best available treatment. Conversely, when the cancer is detected early (stage 1), conventional therapy leads to 95% 5-year survival (Bast *et al.*, 1983; Cohen *et al.*, 2001; Jacobs *et al.*, 1999; Menon and Jacobs, 2000). Similar figures apply to colon and other cancers. The best evidence comes from the Surveillance Epidemiology and End Results (SEER) program conducted by the National Cancer Institute (2002). Survival is excellent for the main cancers when early-stage disease is treated with existing therapies (Table II).

A. When Is an Early Detection Program Warranted?

According to the World Health Organization (WHO) the following criteria need to be fulfilled: (1) the disease must be common and associated with serious morbidity and mortality, (2) screening tests must be able to accurately detect early-stage and potentially curable disease, (3) treatment after detection through screening must show a significant advantage relative to the treatment without screening, and (4) evidence that the overall potential benefits outweigh the potential harms and costs of screening (Winawer *et al.*, 1995). For early detection to be an effective and practical approach, screening tests must satisfy four basic requirements. (1) Screening tests should distinguish healthy individuals from cancer cases with a high degree of accuracy, that is, high sensitivity and specificity and high positive and negative predictive values. (2) Detection should be possible before the disease progresses to an advanced stage, when treatment is less effective. (3) Screening tests should ideally differentiate between aggressive lesions (which require treatment) and benign tumors, avoiding the problem of overdiagnosis. (4) Tests should be inexpensive, minimally invasive, and well accepted by the targeted population.

Although screening tests are currently in use for some cancers, very few satisfy these requirements.

III. THE NEED FOR NEW DIAGNOSTIC STRATEGIES

Refinements in more conventional diagnostic strategies, such as imaging, have had a substantial benefit to patients over the last 25 years. The potential to detect early breast cancer by mammography or the ability of computed tomography, ultrasonography, and magnetic resonance imaging to reveal small masses or tumor metastasis are but a few examples. However, hybrid strategies, combining imaging with other modalities should work better. Novel biomarkers, as additional tools to detect preclinical cancers, have the potential to revolutionize the way we diagnose and manage cancer in the future.

The rapidly expanding field of cancer biomarker discovery prompted the establishment of the Early Detection Research Network (EDRN) by the National Cancer Institute (NCI) (Srivastava and Kramer, 2000). The purpose of the EDRN is to coordinate research among biomarker development laboratories, biomarker validation laboratories, clinical repositories, and population screening programs with the hope to facilitate collaboration and to promote efficiency and rigor in research. The objectives of the EDRN for biomarker development and validation can be summarized in five consecutive phases: (1) preclinical exploratory, (2) clinical assay and validation, (3) retrospective longitudinal, (4) prospective screening, and (5) cancer control (Sullivan *et al.*, 2001).

Until recently, biomarker discovery was a laborious, linear, and slow process, where each candidate biomarker is first identified and then validated for specificity and sensitivity by using mainly an enzyme-linked immunosorbent assay (ELISA). With the advent of the post-genomic era, powerful new approaches are being realized. One approach is to use bioinformatics such as digital differential display and *in silico* Northern analysis utilizing SAGE, EST, cDNA arrays, or other parallel (Brenner and Johnson, 2000) nucleic acid analysis techniques, and the databases of the Cancer Genome Anatomy Project (CGAP) to compare gene expression between healthy and cancerous tissues in order to identify overexpressed genes (Hermeking, 2003; Hess, 2003; Polyak and Riggens, 2001; Tuteja and Tuteja, 2004; Yousef *et al.*, 2003). Gene expression analysis by microarray technology is another method that identifies overexpressed genes in cancer, with the potential to develop cancer biomarkers (Hampton and Frierson, 2003; Hellstrom *et al.*, 2003; Lu *et al.*, 2004; Welsh *et al.*, 2001, 2003; Zarrinkar *et al.*, 2001). However, some of the best cancer biomarkers (such as PSA) are not overexpressed in cancer (Magklara *et al.*, 2000).

The emerging field of clinical proteomics is not only well suited to the discovery and implementation of new biomarkers, but it could also be applied across the spectrum of cancer research (Fig. 1). Proteomics refers to the systematic study of the total protein complement (proteome) encoded and expressed by a genome or by a particular cell, tissue, or organism (Pusch *et al.*, 2003). Many researchers have hypothesized that the best cancer biomarkers will likely be secreted proteins (Welsh *et al.*, 2003). Approximately 20–25% of all cell proteins are secreted. Proteins, or their fragments, originating from cancer cells or their microenvironment, may eventually enter the circulation. The patterns of expression of these proteins could be analyzed by mass spectrometry in combination with mathematical algorithms. Proteomic pattern diagnostics include proteomic

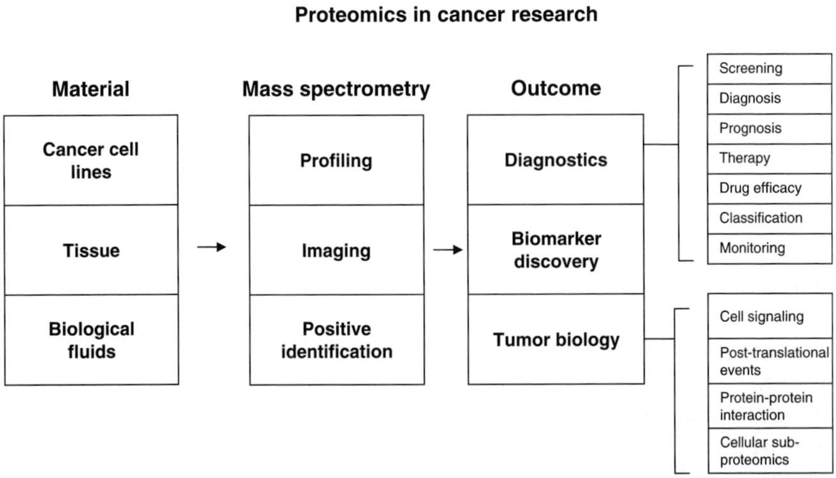

Fig. 1 Application of clinical proteomics in cancer research. Clinical material (cell lines, tissues, or biological fluids) is analyzed by mass spectroscopy with or without chromatographic separation for either imaging, proteomic profiling, or for identification of putative biomarkers. This analysis can lead to development of novel diagnostics or for understanding tumor biology.

pattern profiling of serum by surface-enhanced laser desorption/ionization-time-of-flight (SELDI-TOF) mass spectrometry combined with bioinformatic tools (Petricoin *et al.*, 2002c). The rest of this chapter will focus on mass spectrometry as a tool for biomarker discovery and as a diagnostic platform for cancer.

IV. MASS SPECTROMETRY

The concept of global protein analysis as a complete inventory of human proteins was proposed 20 years ago (Anderson and Anderson, 2002; Anderson *et al.*, 2004), and proteomic research was driven in the mid-1990s by the development in three areas: two-dimensional gel electrophoresis, mass spectrometry, and bioinformatic databases. In "top-down" proteomics, intact proteins are analyzed. In "bottom-up" proteomics, the proteins are proteolytically cleaved intentionally, using enzymes. In contrast, the endogenous peptides of serum or plasma presumably result from physiological proteolysis *in vivo* or *in vitro*.

Commonly available mass spectrometers are sensitive to the hundreds of atto mols and zeptomolar sensitivity has been demonstrated (Dick Smith). However, in practice sensitivity is overwhelmingly dependent on sample preparations.

Mass spectrometry-based proteomics has become the method of choice for the analysis of complex protein samples. Mass spectrometers have been used for many decades as diagnostic tools in clinical laboratories and have enjoyed many successes in the area of identification and quantification of relatively small molecules (molecular mass $<$ 1000 Da). Recent interest in this technology for studying larger molecules, such as nucleic acids and proteins, has escalated significantly (Aebersold and Mann, 2003; Fenn *et al.*, 1989; Pedrioli *et al.*, 2004; Tyers and Mann, 2003). This has been made possible not only by the availability of genome sequence databases, but particularly by the discovery and development of novel protein ionization methods recognized by the 2002 Nobel Prize in chemistry.

A mass spectrometer consists of an ion source, a mass analyzer that measures the mass-to-charge ratio (m/z) of the ionized analytes and a detector that registers the number of ions at each m/z value (Aebersold and Mann, 2003). A typical proteomic experiment usually consists of five stages. (1) The proteins to be analyzed, present in cell lysates, tissues, or fluids are separated by various fractionation or affinity selection techniques (Lim and Elenitoba-Johnson, 2004). This defines the "subproteome" to be analyzed. The most powerful recent strategy integrates different separation methods as multidimensional combinations (MUDPIT) such as ion-exchange and reverse-phase HPLC. (2) Enzymatic protein degradation to peptides (usually by trypsin). MS of whole proteins is less sensitive than peptide MS and the intact protein by itself may not be as easily detected, although methods for examining large proteins are rapidly advancing. (3) Peptides are routinely separated by high-performance liquid chromatography in very fine capillaries and eluted into an electrospray ion source where they are nebulized in small, highly charged droplets. After evaporation, multiple protonated peptides enter the mass spectrometer. (4) A mass spectrum of the peptides eluting at each time point is taken. (5) These peptides are prioritized for fragmentation and a series of tandem mass spectrometric (MS/MS) experiments ensues to obtain sequence information. Identified peptides are matched against protein sequence databases to eventually identify the proteins of interest.

Essential to proteomic studies is the simplification of a complex mixture of proteins into less complex components. In general, measurement of peptide masses by MS is experimentally and mathematically (Mann *et al.*, 2001) simpler than the calculation of intact protein masses. The ability to accurately determine the mass of a unique peptide that originates from a particular protein greatly facilitates the identification of that protein (Hunt and Shabanowitz, 1987; Smith and Anderson, 2002).

A. Ionization Source

Mass spectrometric measurements are carried out in the gas phase on ionized analytes. Two techniques are most commonly used to volatize or ionize the proteins or peptides, namely electrospray ionization (ESI) and matrix-assisted laser desorption/ionization (MALDI) (Karas and Hillenkamp, 1988; Nakanishi *et al.*, 1994). ESI ionizes analytes out of a solution and is therefore readily coupled to liquid-based separation tools (e.g., chromatographic and electrophoretic). MALDI ionizes the samples out of dry, crystalline matrix via laser pulses. MALDI-MS is normally applied to relatively simple peptide mixtures, compared to ESI-MS combined with liquid-chromatography (LC-MS), which is preferred for the analysis of complex samples. A variant MALDI technology, which has been used extensively in diagnostics, is surface-enhanced laser desorption ionization (SELDI) (Merchant and Weinberger, 2000). In this technology, a surface (ProteinChip^{TM}) functions as a solid phase extraction tool. The objective is to overcome the requirement for purification and separation of proteins prior to MS analysis (Aebersold and Goodlett, 2001).

B. Mass Analyzers

The mass analyzer separates ions according to *m/z* ratio. In terms of proteomics, its key parameters are sensitivity, resolution, mass accuracy, and the ability to generate information-rich mass spectra from peptide fragments (Mann *et al.*, 2001; Pandey and Mann, 2000; Wilkins *et al.*, 1998). Four basic types of mass analyzers are commonly used in proteomic research: the ion trap, time-of-flight (TOF), quadruple, and Fourier transform ion cyclotron resonance mass spectrometer (FT-ICR MS) analyzer. They all have different characteristics and can be used on their own or in combination with each other to optimize results (Lim and Elenitoba-Johnson, 2004).

C. Protein Identification

1. PEPTIDE MASS FINGERPRINTING

This is the simplest method for protein identification which combines enzymatic digestion, mass spectrometry, and data analysis. The peptides generated are analyzed by MS and the masses are compared with theoretical mass spectra of proteins listed in databases. Software algorithms for peptide mass mapping include PeptIdent/MultiIdent and ProFound (MacCoss *et al.*, 2002; Zhang and Chait, 2000).

2. PEPTIDE SEQUENCING BY TANDEM MASS SPECTROMETRY

This technique is based on collisional-induced dissociation (CID) that randomly cleaves peptide bonds between adjacent amino acid residues. This yields ion series that eventually reveal the amino acid sequence of a peptide.

D. Quantitation

Small molecules are routinely quantified on triple stage quadrupole mass spectrometers and this may one day be extended to peptides. A quantitative dimension has been added to MS experiments by stable-isotope dilution (SILAC), which is based on the principle that pairs of chemically identical analytes of different stable-isotope composition can be differentiated by

MS owing to their mass difference, and that the ratio of signal intensities for such analyte pairs accurately indicates the abundance ratio for the two analytes (Conrads *et al.*, 2002; Mirgorodskaya *et al.*, 2000; Yao *et al.*, 2001). Another technology, isotope-coded affinity tags (ICAT), relies on stable isotope labeling of cysteine residues (Gygi *et al.*, 1999; Von Haller *et al.*, 2003a, b). The advantage of this method is that it allows evaluation of low-abundance proteins and proteins at both extremes of molecular weight and isoelectric point. Absolute quantitation requires prior identification of the analyte and the use of external or internal standards.

V. MASS SPECTROMETRY-BASED DIAGNOSTICS

Mass spectrometry has been used in two different settings in the area of cancer diagnostics, first for the discovery of novel cancer biomarkers and second as a cancer diagnostic and imaging tool. The discovery of biomarkers and their use as early detectors of cancer is based on the hypothesis that a complex interplay exists between a tumor and its host microenvironment (Liotta and Kohn, 2001). As blood perfuses through a diseased organ, the serum protein profile is altered as a result of ongoing physiological and pathological events. This may include proteins being overexpressed and/or abnormally shed, clipped, modified, or removed due to abnormal activation of the proteolytic degradation pathway, generating a unique signature in blood (Fig. 2). As a consequence, the expressed serum protein profile is different between normal and diseased states. This creates a unique opportunity to exploit accessible body fluids, such as serum, urine, saliva, seminal plasma, malignant ascites, or cerebrospinal fluid, for the discovery of novel biomarkers.

A. Mass Spectrometry as a Tissue Imaging Tool

A recent advance, laser capture microdissection (LCM) provides a means of rapidly procuring pure cell populations from the surrounding heterogeneous tissue, allowing the use of tissue as an additional medium to discover novel biomarkers (Banks, 1999; Emmert-Buck *et al.*, 1996). The concept of MALDI-MS-based imaging mass spectrometry was introduced in 1997 by Caprioli *et al.* (1997). MS is used to map the distribution of peptides and proteins directly from thin tissue sections and allows visualization of 500–1000 individual protein signals in the molecular weight range from 2000 to 200,000. Matrix is deposited uniformly over the section and

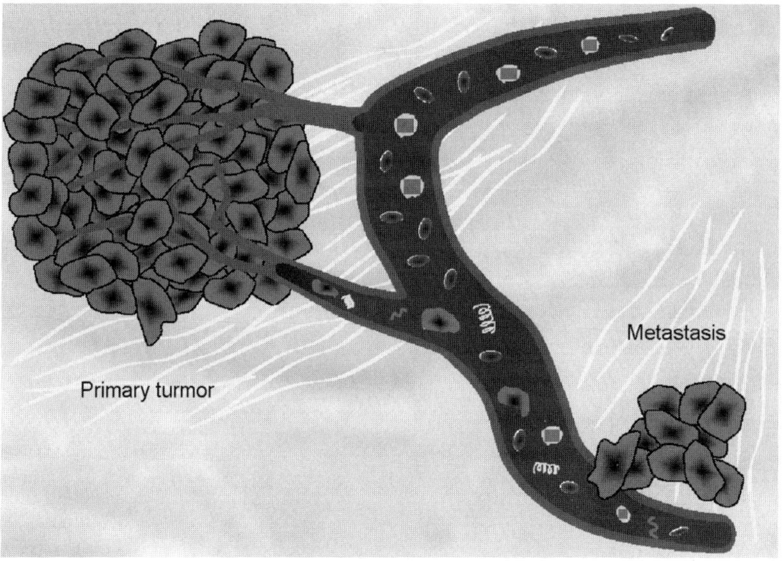

Fig. 2 Schematic representation of blood composition after contribution of molecules or cells by tumor due to angiogenesis or tissue destruction. The enrichment of blood with tumor- or microenvironment-derived components can be used for diagnostics.

analysis of the tissue is performed over a predetermined two-dimensional array or grid generating a full mass spectrum at each grid coordinate. Each spectrum is generated with an average of 15–50 consecutive laser shots at each coordinate. From the intensity of a given *m/z* value, a density map or image can be constructed. It is essential to maintain three conditions: (1) the deposition process must not disperse or translocate proteins within the section, (2) the matrix solution must wet the tissue surface to form crystals which contain cocrystallized proteins, and (3) the crystal dimensions must be smaller than the image resolution (Chaurand *et al.*, 2002). Imaging mass spectrometry is still in an early developmental stage and many improvements in sample preparation, handling, and instrumentation can be expected in future. However, this technique yields a wealth of information about the protein pattern trends within a tissue sample, and differentially expressed protein profiling between healthy and cancerous tissues has already been explored for novel cancer biomarker identification (Schwartz *et al.*, 2004; Yanagisawa *et al.*, 2003). More recently, this method has been used to predict tumor response to molecular therapeutics (Reyzer *et al.*, 2004). This may become an important means to delineate surgical margins in real time during surgery.

B. Mass Spectrometry as a Biomarker Discovery Tool

The use of MS as a biomarker discovery technique is conceptually straightforward. Fluids or tissue extracts from a diseased group, as well as a control group, are analyzed by MS and the differentially expressed peaks are identified. These peaks potentially represent molecules that could be measured with simpler and cheaper techniques, such as ELISA, for the purpose of diagnosis and management of cancer. A list of candidate biomarkers identified by MS is shown in Table III (Cho *et al.*, 2002; Johnson *et al.*, 1994; Koomen *et al.*, 2005; Zhang *et al.*, 2004). So far, MS discovery efforts have focused on three subsets of the proteome: (1) polypeptides and whole

Table III Serum Concentration of Some Abundant Proteins, Classical Cancer Biomarkers, and Putative New Cancer Biomarkers Identified by Mass Spectrometry[a]

Compound	Approximate concentration (pmol/liter)	Biomarker for cancer type	References
Serum proteins			
Albumin	600,000,000	–	Johnson *et al.*, 1994
Immunoglobulins	30,000,000	–	Johnson *et al.*, 1994
C-reactive protein	40,000	–	Johnson *et al.*, 1994
Classical tumor markers			
α-Fetoprotein	150	Hepatoma, testicular	Johnson *et al.*, 1994
Prostate-specific antigen	140	Prostate	Johnson *et al.*, 1994
Carcinoembryonic antigen	30	Colon, pancreas, lung, breast	Johnson *et al.*, 1994
Choriogonadotropin	20	Testicular, choriocarcinoma	Johnson *et al.*, 1994
β-Subunit of choriogonadotropin	2	Testicular, choriocarcinoma	Johnson *et al.*, 1994
Mass spectrometry-identified proteins			
Apolipoprotein A1	40,000,000	Ovarian, pancreatic	Liotta *et al.*, 2003; Zhang *et al.*, 2004
Transthyretin fragment	6,000,000	Ovarian	Zhang *et al.*, 2004
Inter-α-trypsin inhibitor fragment	4,000,000	Ovarian, pancreatic	Koomen *et al.*, 2005; Zhang *et al.*, 2004
Haptoglobin-α-subunit	1,000,000	Ovarian, pancreatic	Koomen *et al.*, 2005
Vitamin D-binding protein	10,000,000	Prostate	Zhang *et al.*, 2004
Serum amyloid A	20,000,000	Nasopharyngeal, pancreatic	Koomen *et al.*, 2005; Cho *et al.*, 2002
α_1-Antitrypsin	10,000,000	Pancreatic	Koomen *et al.*, 2005
α_1-Antichymotrypsin	5,000,000	Pancreatic	Koomen *et al.*, 2005

[a]Reproduced from Diamandis and van der Merwe (2005) with permission from copyright owners.

proteins that can be analyzed by electrophoresis with or without prior fractionation, (2) enzymatic peptide fragments separated by liquid chromatography and analyzed with either ESI or MALDI, typically after one or more chromatographic or other fractionation steps, and (3) naturally occurring peptides (the peptidome), which provide a complementary picture of many events at the low mass end of the plasma proteome (Liotta *et al.*, 2003; Loboda and Krutchinsky, 2000; Marshall and Jankowski, 2004; Villanueva *et al.*, 2004). The biggest challenge in uncovering potential biomarkers present in serum lies in the complexity and dynamic range of the proteome. Various prefractionation steps have been applied to mine into the subproteome in order to reach the low-abundance and likely the most informative molecules (Fig. 3).

C. Mass Spectrometry as a Cancer Diagnostic Tool

The concept and utility of multivariate protein markers as opposed to a single indicator to diagnose disease has been established for some time. More than 20 years ago, it became clear that different tumor cell types could be distinguished based on patterns of metabolites analyzed by GC-MS (Jellum *et al.*, 1981). Investigators are currently using two types of proteomic technologies to mine the proteomic signature in order to differentiate between normal and diseased states: protein microarrays and mass spectrometry. We will concentrate on the latter for the purpose of this chapter.

Mass spectrometry of endogenous human serum peptides using the Ciphergen Biosystems TOF in the MALDI or SELDI mode (Weinberger *et al.*, 2000) as a diagnostic tool and their identification by MALDI-Qq-TOF was successfully demonstrated by Jackowski and coworkers (Takahashi *et al.*, 2001). Later Petricoin and coworkers proposed using only the SELDI pattern of the unidentified peaks as a diagnostic tool (Petricoin *et al.*, 2002a). Biovision (BioVisioN AG, Hannover, Germany) proposed the examination of the MALDI profile of endogenous peptides prepared by reversed phase chromatography against a previously established library of analytes. Their approach is based on identifying patterns of differentially expressed proteins analyzed by computer algorithms, between samples from diseased and nondiseased subjects, without requiring knowledge of the identity of the individual discriminating molecules (Tammen *et al.*, 2003). Since then, many papers have been published on using protein pattern profiling in diagnosing various types of cancer (Table IV) (Adam *et al.*, 2002; Dolios *et al.*, 2003; Ferrari *et al.*, 2000; Koopmann *et al.*, 2004; Kozak *et al.*, 2002; Langridge *et al.*, 1993; Lehrer *et al.*, 2003; Li *et al.*, 2002; Liotta and Petricoin, 2000; Petricoin *et al.*, 2002a,b; Poon *et al.*, 2003; Qu *et al.*, 2002; Rosty *et al.*, 2002; Sasaki *et al.*, 2002; Sauter *et al.*, 2002; Stegner *et al.*, 2004;

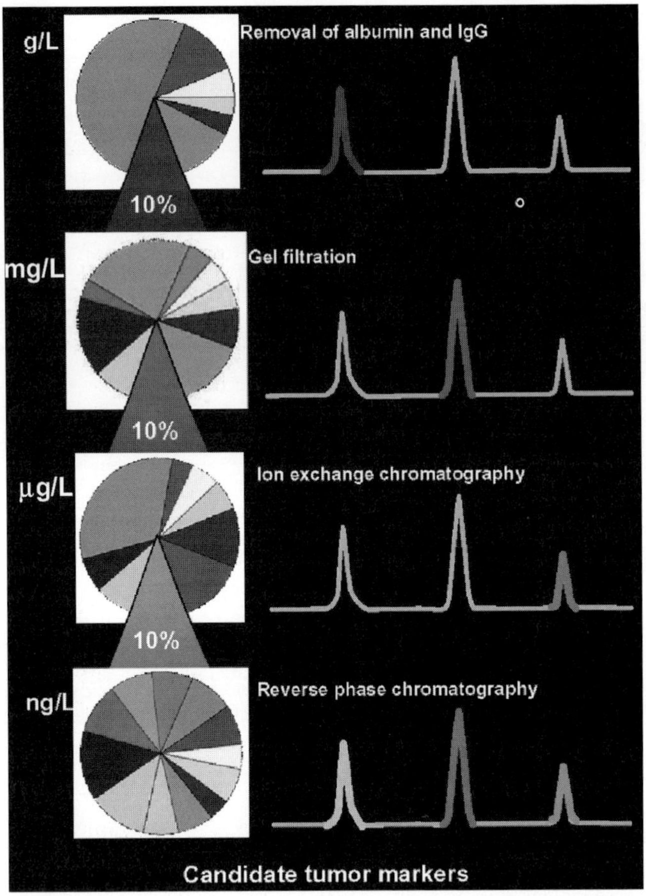

Fig. 3 Schematic representation of successive chromatographic separations for enrichment of fractions with low-abundance proteins. For further discussion see text. (See Color Insert.)

Tammen *et al.*, 2003; Vlahou *et al.*, 2001; von Eggeling *et al.*, 2000; Wadsworth *et al.*, 2004; Wright *et al.*, 1999; Zhukov *et al.*, 2003). The vast majority of the data were generated using SELDI-TOF technology (Ciphergen Biosystems, Fremont, CA). In general, it has been reported that this technology can achieve much higher diagnostic sensitivities and specificities (nearly 100%) compared to classical single biomarkers (Conrads *et al.*, 2004; Powell, 2003). If these findings are reproduced and validated, they could have immediate clinical impact. However, it is important to highlight some limitations of this technique as well and discuss a number of controversial issues surrounding its implementation into clinical practice.

Table IV Protein Pattern Profiling for Cancer Diagnosis

Cancer type	References
Ovarian	Petricoin *et al.*, 2002a; Kozak *et al.*, 2002
Breast	Liotta and Petricoin, 2000; Li *et al.*, 2002; Sauter *et al.*, 2002; Stegner *et al.*, 2004
Prostate	Adam *et al.*, 2002; Lehrer *et al.*, 2003; Petricoin *et al.*, 2002; Qu *et al.*, 2002; Wright *et al.*, 1999
Bladder	Langridge *et al.*, 1993; Vlahou *et al.*, 2001
Pancreatic	Koopmann *et al.*, 2004; Rosty *et al.*, 2002; Sasaki *et al.*, 2002
Head and neck	von Eggeling *et al.*, 2000; Wadsworth *et al.*, 2004
Lung	Zhukov *et al.*, 2003
Colon	Dolios *et al.*, 2003
Melanoma	Ferrari *et al.*, 2000
Hepatocellular	Poon *et al.*, 2003

VI. CURRENT LIMITATIONS OF DIAGNOSTIC MASS SPECTROMETRY

The greatest challenge for proteomic technologies is the inherent complexity of cellular proteomes. Contrary to the genome, the proteome is a dynamic entity, constantly changing in response to cellular or environmental stimuli. Different cells have different proteomes and the proteins within a proteome are structurally quite diverse.

Most of the discussion below will focus on SELDI-TOF mass spectrometry. The possible limitations mentioned here are not unique to this technology, but are relevant to other platforms as well (Diamandis, 2003, 2004a,b; Diamandis and van der Merwe, 2005). The controversy surrounding the method has raised questions as to whether mass spectrometry can meet the standards of reproducibility and performance expected of established clinical tests (Coombes, 2005; Hortin, 2005). A commendable report by Semmes *et al.* (2005) examined the reproducibility between different laboratories and highlighted that this technology does not as yet meet the desired standards to be applied in clinical laboratory practice.

A few common steps are involved in SELDI-TOF procedures. The biological fluid is fractionated with a protein chip, enabling the analysis of subgroups of proteins based on their affinity for a given surface (e.g., normal phase, reverse phase, immobilized metal affinity capture (IMAC), ion-exchange or ligand-binding affinity chromatography) to capture proteins from complex biological samples. After washing, the immobilized proteins are analyzed by SELDI-TOF MS (Fig. 4). The associated shortcomings of the method can be divided into preanalytical, analytical, and postanalytical.

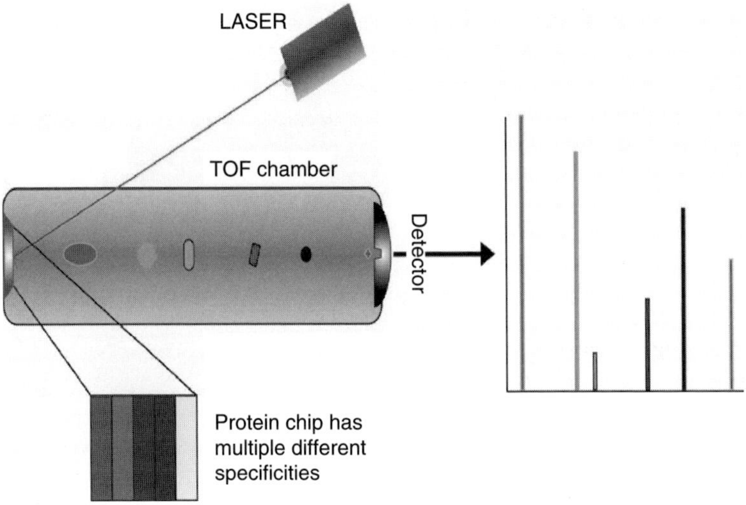

Fig. 4 Schematic representation of SELDI-TOF mass spectrometry. Unfractionated sample is applied to a protein chip, which is coated with various functional groups (1–5) to enable the analysis of a subset of proteins based on affinity for a given surface. Unbound proteins are washed away. A laser beam desorbs and ionizes the proteins, which are cocrystallized with matrix. The mass-to-charge (*m/z*) ratio is determined by the TOF detector and proteomic patterns are analyzed by suitable software. (See Color Insert.)

A. Preanalytical

Many factors influence the concentration of proteins in plasma besides disease. When these are not considered, the detection of medically meaningful changes becomes dubious. The effects of sample storage and processing, sample type (plasma versus serum), patient selection, and different biological variables (gender, age, ethnicity, exercise, menopause, nutrition, drugs, and so on) have as yet not been established yet for mass spectrometric analyses of this type.

B. Analytical

1. DYNAMIC RANGE

The dynamic range of established techniques, such as ELISA, encompasses molecules like albumin in the very high-abundance end (35–50 mg/ml) as well as in the very low-abundance end, for example interleukin-6 (5 pg/ml). The abundance of these two molecules in plasma differs by a factor of 10^{10}. Unbiased protein identification by techniques such as LC/MS/MS have typical dynamic ranges of only 10^2–10^4, falling short of the requirement for comprehensive proteome mapping by at least 6–8 orders of magnitude. Various fractionation methods (chromatography, immunoaffinity subtraction, preparative isoelectric focusing, or precipitation) improve, but still fall short of the desired dynamic range.

2. ANALYTICAL SENSITIVITY

An important question is whether SELDI-TOF or other associated technologies are sensitive enough to capture putative proteomic changes caused by early stage tumors (Diamandis, 2003, 2004a,b). As currently used, these techniques are unlikely to detect any serum component at concentrations of <1 μg/ml (Diamandis, 2003). This concentration is ~1000-fold higher than the concentrations of known tumor markers in the circulation (see Table III for quantitative comparisons).

3. IDENTIFICATION OF ESTABLISHED CANCER BIOMARKERS

PSA is an established biomarker that can reasonably distinguish cancer from noncancer patients. Free and complexed PSA have molecular masses of ~30 and 100 kDa, respectively, which are well within the current capabilities of mass spectrometry. None of the published studies with breast, prostate, or ovarian cancer identified any of the classical cancer biomarkers as distinguishing molecules. This is likely due to the inadequate sensitivity of currently used protocols, as exemplified in detail elsewhere (Diamandis, 2003).

4. BIAS TOWARD HIGH-ABUNDANCE MOLECULES

Serum contains a wide range of molecules as mentioned earlier (Anderson and Anderson, 2002; Anderson *et al.*, 2004), therefore, competition between high-abundance and low-abundance molecules for immobilization on the protein chip will take place once the sample is applied. For example, PSA concentration in serum of healthy males is in the order of 1 ng/ml compared to a total protein of 80×10^6 ng/ml. When proteins are exposed to the chip, each PSA molecule (or other molecules of similar abundance) will encounter competition for binding to the same matrix by millions of irrelevant molecules of high abundance. Therefore, low-abundance molecules will likely escape binding and detection. Also, the relative amplitudes of peaks in MS spectra will not accurately represent their abundance compared to pure standards. The theoretical sensitivity of MS could be very high (e.g., in the zeptomolar range; Smith and Anderson, 2002), but whether this is achievable in a complex mixture of high-abundance, as well as low-abundance proteins, remains to be seen. From the available experimental data, current protocols detect primarily or exclusively high-abundance molecules in the concentration range of milligrams per milliliters (Table III).

5. IONIZATION EFFICIENCY

It is not well established whether the same concentration of an informative molecule on the protein chip produces a peak of the same amplitude if it is surrounded by variable amounts of irrelevant proteins that are simultaneously ionized during laser desorption, therefore causing ionization suppression of the relevant molecule. This issue needs to be experimentally examined for each analyte of interest.

6. IDENTITY AND ORIGIN OF DISCRIMINATORY PEAKS

Two different opinions exist in the literature: (1) the identity of the discriminatory peaks produced by MS is not essential and that the diagnostic endpoint is a differentially expressed proteomic profile containing a multitude of molecules reflecting tumor–host interaction, (2) the identity of these peaks is essential, the reason being threefold: to relate their biological connection to cancer, to exclude artifacts originating *ex vivo* during sample handling, and to examine if the findings represent cancer epiphenomena. Most of the discriminatory molecules identified thus far are acute-phase proteins (Table III), released by the liver likely in response to malignancy-associated inflammation (Diamandis, 2003).

7. REPRODUCIBILITY

Some questions and concerns regarding the reproducibility of protein patterns by SELDI-TOF have been raised. There are no systematic studies showing that similar data can be obtained by using different batches of SELDI chips, different technologists, different instrumentation, or at different time points. One hypothesis for the published data is that the differences in serum proteomic patterns between controls and cases are due to the presence of cancer in the latter group. Alternatively, these differences could be due to an unrelated effect, that is, analytical variables or mass spectrometric, bioinformatics, and statistical biases. To date several groups have reported good reproducibility by offline preseparation by C18 partition chromatography prior to MALDI-TOF analysis (Marshall *et al.*, 2003).

Three recent publications dealt with the issue of reproducibility of the serum proteomic test for ovarian cancer (Baggerly *et al.*, 2005; Liotta *et al.*, 2005; Ransohoff, 2005). Baggerly *et al.* (2005) concluded, after analyzing sets of data produced by Liotta *et al.*, that the discriminatory peaks do not represent biologically important changes in cancer patients and the resulting classification may have arisen by chance. On the other hand, Liotta *et al.* (2005) suggested, as we have proposed previously, to characterize discriminatory peaks so that future classifications are more reproducible and robust. Ransohoff draws attention to biases of experimental designs and suggested that future clinical trials should avoid biological, analytical, statistical, and epidemiological biases (Ransohoff, 2005).

8. ROBUSTNESS

The long-term robustness of this technology needs to be established. Rogers *et al.* (2003) reported that the diagnostic sensitivity in renal cell carcinoma fell from 98 to 41%, tested on two different occasions, 10 months apart. This kind of variability is unacceptable for tests destined to reach the clinic.

C. Postanalytical

1. BIOINFORMATIC ARTIFACTS

SELDI-TOF or associated experiments use a fraction of the clinical samples as a "training set" to derive the interpretation algorithm, while the remaining samples are used as a "test set." Qu *et al.* (2002) pointed out that one of the concerns about learning algorithms is the potential to overfit the data. It is unknown if these algorithms will remain stable over time or when different sets of clinical samples are used. The data of Rogers *et al.* (2003) cast doubt on algorithm stability over time (Poon *et al.*, 2003).

Furthermore, many discriminatory peaks identified to differentiate cancer often have an *m/z* ratio <2000, discarded by many as noise due to matrix effects. Reanalysis of the data of Petricoin *et al.* (2002a) by Sorace and Zhan (2003) and Baggerly *et al.* (2004) revealed that many peaks within the *m/z* range <2000 had powerful discriminatory ability, concluding that a nonbiological bias may best explain the published data and not the presence of cancer.

2. EXTERNAL VALIDATION

The real value of new biomarkers and discovery approaches will ultimately be decided at validation (Ransohoff, 2003, 2004). Efforts to standardize the methodology and test the reproducibility in various laboratories under different clinical settings are underway (Banez *et al.*, 2003; Semmes, 2004;

Table V Some Open Questions Related to Diagnostic SELDI-TOF Technology[a]

1. Identity and serum concentrations of discriminating molecules mostly unknown. These molecules may represent artifacts or cancer epiphenomena
2. Mass spectrometry is a largely qualitative technique
3. Discriminating peaks identified by different investigators for the same disease are different
4. Data are not easily reproducible between laboratories, making validation difficult
5. Optimal sample preparation for the same disease differs between investigators
6. Validated serum cancer markers (PSA, CA125, and so on) that could serve as positive controls are not identified by this technology due to low sensitivity
7. Nonspecific absorption matrices favor extraction of high-abundance proteins and loss of low-abundance proteins
8. Analytical sensitivity of mass spectrometry in a complex mixture (e.g., serum) is unknown
9. No known relationship has been demonstrated between discriminatory molecules and cancer biology

[a]This table was modified from Diamandis (2003) and published with permission from the copyright owners.

Semmes *et al.*, 2005). A summary of the limitations of current MS diagnostic protocols is presented in Table V (Diamandis, 2003).

VII. SUGGESTIONS FOR FUTURE PROGRESS

1. Future investigations should report, whenever possible, the identity of the discriminatory peaks and attempts should be made to link them to cancer biology.
2. Internal controls should be included to correct for peak amplitudes in different experiments.
3. Standardized statistical algorithms that will not vary over time, such as ANOVA, should be used to compare samples and populations (Marshall *et al.*, 2003). Bioinformatic algorithms should be tested periodically to validate their robustness over time.
4. Different bioinformatic algorithms should be compared on the same set of data to determine whether discriminatory peaks and similar diagnostic sensitivity and specificity can be obtained.
5. The analytical sensitivity of mass spectrometry as applied to unfractionated serum samples needs to be determined.
6. Establish whether certain discriminatory peaks originated *ex vivo* (after sample collection) or *in vivo*. Samples from the same patients should be collected with or without proteinase inhibitors and processed in various ways, as described by Marshall *et al.* (2003). In general, serum contains many protein fragments generated during the coagulation cascade.
7. Studies should be performed to establish the effect of pre- and postanalytical variables on proteomic patterns generated, as described earlier. The possibility of bias needs to identified and excluded during each step.

VIII. FUTURE DIRECTION

The complete sequence of the human genome, the development of novel bioinformatic tools, and recent advances in biological mass spectrometry and microarrays sparked optimism that the time has come to discover novel cancer biomarkers. Over the last 3 years, we witnessed an exponential growth of mass spectrometry-based diagnostics with claims of unprecedented clinical sensitivities and specificities. However, various analytical and clinical shortcomings have been recognized. The controversy can be resolved with well-designed validation studies, which are currently underway by investigators, diagnostic companies, and organizations such as the EDRN. Despite these difficulties, it is clear that the opportunities are enormous. For example, mass spectrometry and protein microarrays offer a unique way to simultaneously monitor hundreds to thousands of proteins at the same time. Newer developments may improve the analytical sensitivity of mass spectrometry, allowing measurement of molecules present in biological fluids at very low concentrations. A new discipline called "peptidomics" deals with small peptides in biological fluids which may carry unique information on proteolysis around the cancer microenvironment. Mass spectrometry is ideally suited for high-throughput analysis of a large number of different peptides.

We conclude that mass spectrometry-based diagnostics will continue to grow in the future, with multiparametric analysis of high- and low molecular weight proteins/peptides present in biological fluids in low abundance. Such analysis combined with bioinformatics will eventually lead to novel ways of diagnosing and monitoring cancer. This approach may eventually replace the traditional use of single biomarkers for diagnosis and monitoring of cancer.

REFERENCES

Adam, B.-L., Qu, Y., Davies, J. W., Ward, M. D., Clements, M. A., Cazares, L. H., Semmes, O. J., Schellhammer, P. F., Yassui, Y. M., Feng, Z., and Wright, G. L., Jr. (2002). Serum protein fingerprinting coupled with a pattern-matching algorithm distinguishes prostate cancer from benign prostate hyperplasia and healthy men. *Cancer Res.* **62**, 3609–3614.

Aebersold, R., and Goodlett, D. R. (2001). Mass spectrometry in proteomics. *Chem. Rev.* **101**, 269–295.

Aebersold, R., and Mann, M. (2003). Mass spectrometry-based proteomics. *Nature* **422**, 198–207.

Anderson, N. L., and Anderson, N. G. (2002). The human plasma proteome: History, character, and diagnostic prospects. *Mol. Cell. Proteomics* **1**, 845–867.

Anderson, N. L., Polanski, M., Pieper, R., Gatlin, T., Tirumalai, R. S., Conrads, T. P., Veenstra, T. D., Adkins, J. N., Pounds, J. G., Fagan, R., and Loley, A. (2004). The human plasma proteome: A nonredundant list developed by combination of four separate sources. *Mol. Cell. Proteomics* **3**, 311–326.

Baggerly, K. A., Morris, J. S., and Coombes, K. R. (2004). Reproducibility of SELDI-TOF protein patterns in serum: Comparing data sets from different experiments. *In* "Bioinformatics Advanced Access," pp. 1–9. Oxford, UK.

Baggerly, K. A., Morris, J. S., Edmonson, S. R., and Coombes, K. R. (2005). Signal in noise: Evaluating reported reproducibility of serum proteomic tests for ovarian cancer. *J. Natl. Cancer Inst.* **94,** 307–309.

Banez, L. L., Prasanna, P., Sun, L., Alo, A., Zou, Z., Adam, B. L., McLeod, D. J., Moul, J. W., and Srivastava, S. (2003). Diagnostic potential of serum proteomic patterns in prostate. *J. Urol.* 3(170), 442–446.

Banks, R. E. (1999). The potential use of laser capture microdissection to selectively obtain distinct populations of cells for proteomic analysis: Preliminary findings. *Electrophoresis* **20,** 689–700.

Bast, R. C., Klug, T. L., St. John, E., Jenison, E., Niloff, J. M., Lazarus, H., Berkowitz, H. S., Leavitt, T., Griffiths, C. T., Parker, L., Zurawski, V. R., and Knapp, R. C. (1983). A radioimmunoassay using a monoclonal antibody to monitor the course of epithelial ovarian cancer. *N. Engl. J. Med.* **309,** 883–887.

Bast, R. C., Radvin, P., Hayes, D. F., Bates, S., Fritshce, H., Jessup, J. M., Kemeny, N., Locker, G. Y., Mennel, R. G., and Somerfield, M. R. (2001a). 2000 update of recommendations for the use of tumor markers in breast and colorectal cancer: Clinical practice guidelines of the American Society of Clinical Oncology. *J. Clin. Oncol.* **19,** 1865–1878.

Bast, R. C., Radvin, P., Hayes, D. F., Bates, S., Fritshce, H., Jessup, J. M., Kemeny, N., Locker, G. Y., Mennel, R. G., and Somerfield, M. R. (2001b). 2000 update of recommendations for the use of tumor markers in breast and colorectal cancer: Clinical practice guidelines of the American Society of Clinical Oncology. *J. Clin. Oncol.* **19,** 4185–4188.

Brenner, S., and Johnson, M. (2000). Gene expression analysis by massively parallel signature sequencing (MPSS) on microbead arrays. *Nat. Biotechnol.* **18,** 630–634.

Caprioli, R. M., Farmer, T. B., and Gile, J. (1997). Molecular imaging of biological samples: Localization of peptides and proteins using MALDI-TOF MS. *Anal. Chem.* **69,** 4751–4760.

Chaurand, P., Schwartz, S. A., and Caprioli, R. M. (2002). Imaging mass spectrometry: A new tool to investigate the spatial organization of peptides and proteins in mammalian tissue sections. *Curr. Opin. Chem. Biol.* **6,** 676–681.

Cohen, L. S., Escobar, P. F., Scharm, C., Glimco, B., and Fishman, D. A. (2001). Three-dimensional power Doppler ultrasound improves the diagnostic accuracy for ovarian cancer prediction. *Gynecol. Oncol.* **82,** 40–48.

Conrads, T. P., Issaq, H. J., and Veenstra, T. D. (2002). New tools for quantitative phosphoproteome analysis. *Biochem. Biophys. Res. Commun.* **290,** 885–890.

Conrads, T. P., Fusaro, V. A., Ross, S., Johann, D., Vinodh, R., Hitt, B. A., Steinberg, S. M., Kohn, E. C., Fishman, D. A., Whiteley, G., Barrett, J. C., Liotta, L. A., *et al.* (2004). High-resolution serum proteomic features for ovarian cancer detection. *Endocr. Relat. Cancer* **11,** 163–178.

Coombes, K. R. (2005). Analysis of mass spectrometry profiles of the serum proteome. *Clin. Chem.* **51,** 1–2.

Diamandis, E. P. (2003). Point: Proteomic patterns in biological fluids: Do they represent the future of cancer diagnostics? *Clin. Chem.* **49,** 1272–1275.

Diamandis, E. P. (2004a). Analysis of serum proteomic patterns for early cancer diagnosis: Drawing attention to potential problems. *J. Natl. Cancer Inst.* **96,** 353–356.

Diamandis, E. P. (2004b). Mass spectrometry as a diagnostic and a cancer biomarker discovery tool. Opportunities and potential limitations. *Mol. Cell. Proteomics* **3,** 367–378.

Diamandis, E. P., and van der Merwe, D. (2005). Plasma protein profiling by mass spectrometry for cancer diagnosis: Opportunities and limitations. *Clin. Cancer Res.* **11,** 963–965.

Dolios, G., Roboz, J., and Wang, R. (2003). Identification of colon cancer associated protein in plasma using MALDI-TOF mass spectrometry. 51st ASMS Conference on Mass Spectrometry and Allied Topics.

Duffy, M. J., van Dalen, A., Haglund, C., Jansson, L., Klapdor, R., Lamerz, R., Nilsson, O., Sturgeon, C., and Topolcan, O. (2003). Clinical utility of biochemical markers in colorectal cancer: European Group on Tumour Markers (EGTM) guidelines. *Eur. J. Cancer* **39,** 718–727.

Emmert-Buck, M. R., Bonner, R. F., Smith, P. D., Chuaqui, R. F., Zhuang, Z., Goldstein, S. R., Weiss, R. A., and Liotta, L. A. (1996). Laser capture microdissection. *Science* **20,** 998–1001.

Etzioni, R., Urban, M., Ramsey, S., McIntosh, M., Schwarz, S., Reid, B., Radich, J., Anderson, G., and Hartwell, L. (2003). The case for early detection. *Nat. Rev. Cancer* **3,** 243–252.

European Group on Tumour Markers (1999). Tumor markers in gastrointestinal cancers-EGTM Recommendations. *Anticancer Res.* **19,** 2785–2820.

Fenn, J. B., Mann, M., Meng, C. K., Eong, S. F., and Whitehouse, C. M. (1989). Electrospray ionization for the mass spectrometry of large biomolecules. *Science* **2456,** 64–71.

Ferrari, L., Seraglia, R., Rossi, C. R., Bertazzo, A., Lise, M., Allegri, G., and Traldi, P. (2000). Protein profiles in sera of patients with malignant cutaneous melanoma. *Rapid Commun. Mass Spectrom.* **14,** 1149–1154.

Fleisher, M., Dnistrian, A. M., Sturgeon, C. M., Lamerz, R., and Wittliff, J. L. (2002). *In* "Tumour Markers: Physiology, Pathobiology, Technology and Clinical Applications" (E. P. Diamandis, H. A. Fritsche, H. Lilja, D. W. Chan, and M. K. Schwartz, Eds.), pp. 33–63. AACC Press, Washington, DC.

Gygi, S. P., Rist, B., Gerber, S. A., Turecek, F., Gelb, M. H., and Aebersold, R. (1999). Quantitative analysis of complex protein mixtures using isotope-coded affinity tags. *Nat. Biotechnol.* **17,** 994–999.

Hampton, G. M., and Frierson, H. F. (2003). Classifying human cancer by analysis of gene expression. *Trend. Mol. Med.* **9,** 5–10.

Hellstrom, I., Raycraft, J., Hayden-Ledbetter, M., Ledbetter, J. A., Schummer, M., McIntosh, M., Drescher, C., Urban, N., and Hellstrom, K. E. (2003). The HE4 (WFDC2) protein is a biomarker for ovarian carcinoma. *Cancer Res.* **63,** 3695–3700.

Hermeking, H. (2003). Serial analysis of gene expression and cancer. *Curr. Opin. Oncol.* **15,** 44–49.

Hess, J. L. (2003). The Cancer Genome Anatomy Project: Power tools for cancer biologists. *Cancer Invest.* **21,** 325–326.

Hortin, G. L. (2005). Can mass spectrometric protein profiling meet desired standards of clinical laboratory practice. *Clin. Chem.* **51,** 3–5.

Hunt, D. F., and Shabanowitz, J. (1987). Tandem quadrupole Fourier-transform mass spectrometry of oligopeptides and small proteins. *Proc. Natl. Acad. Sci. USA* **84,** 620–623.

Jacobs, I. J., Skates, S. J., MacDonald, N., Menon, U., Rosenthal, A. N., Davies, A. P., Woolas, R., Jeyarajah, A. R., Sibley, K., Lowe, D. G., and Oram, D. H. (1999). Screening for ovarian cancer: A pilot randomized controlled trial. *Lancet* **353,** 1207–1210.

Jellum, E., Bjornson, I., Nesbakken, R., Johansson, E., and Wold, S. (1981). Classification of human cancer cells by means of capillary gas chromatography and pattern recognition analysis. *J. Chromatogr.* **217,** 231–237.

Johnson, M. A., Rohlfs, E. M., and Silverman, L. M. (1994). Proteins. *In* "Tietz Textbook of Clinical Chemistry" (C. A. Burtis and E. R. Ashwood, Eds.), pp. 477–540. WB Saunders Co., Philadelphia.

Karas, M., and Hillenkamp, F. (1988). Laser desorption ionization of proteins with molecular mass exceeding 10,000 daltons. *Anal. Chem.* **60,** 2299–2301.

Koomen, J. M., Shih, L. N., Coombes, K. R., Li, D., Xiao, L.-C., Fidler, I. J., Abbruzzese, J. L., and Kobayashi, R. (2005). Plasma protein profiling for diagnosis of pancreatic cancer reveals the presence of host response proteins. *Clin. Cancer Res.* **11,** 1110–1118.

Koopmann, J., Zhang, Z., Rosenzweig, J., Fedarko, N., Jagannah, S., Canto, M. I., Yeo, C. J., Chan, D. W., and Goggins, M. (2004). Serum diagnosis of pancreatic adencarcinoma using surface-enhance laser desorption and ionization mass spectrometry. *Clin. Cancer Res.* **10,** 860–868.

Kozak, K. R., Amneus, Z., Rosenzweig, J., Shih, L., Pham, T., Fung, E. T., Sokoll, L. J., and Chan, D. W. (2002). Proteomic approaches to tumor marker discovery. *Arch. Pathol. Lab. Med.* **126,** 1518–1526.

Langridge, J., McClure, T., El-Shakawi, S., Fielding, A. S. K., and Newton, R. (1993). Gas chromatography/mass spectrometric analysis of urinary nucleosides in cancer patients. Potential of modified nucleosides as tumour markers. *Rapid Commun. Mass Spectrom.* **7,** 427–434.

Lehrer, S., Roboz, J., Ding, H., Zhao, S., Diamon, E. J., Holland, J. F., Stone, N. N., Droller, M. J., and Stock, R. G. (2003). Putative protein markers in the sera of men with prostatic neoplasms. *Br. J. U. Intl.* **92,** 223–225.

Li, J., Zhang, Z., Rosenzweig, J., Wang, Y. Y., and Chan, D. W. (2002). Proteomics and bioinformatics approach for identification of serum biomarkers to detect breast cancer. *Clin. Chem.* **48,** 1296–1304.

Lim, M. S., and Elenitoba-Johnson, K. S. J. (2004). Proteomics in pathology research. *Lab. Invest.* **84,** 1227–1244.

Liotta, L., and Petricoin, E. F. (2000). Molecular profiling of human cancer. *Nat. Rev. Genet.* **1,** 48–165.

Liotta, L. A., and Kohn, E. C. (2001). The microenvironment of the tumour-host interface. *Nature* **411,** 375–379.

Liotta, L. A., Ferrari, M., and Petricoin, E. (2003). Clinical proteomics: Written in blood. *Nature* **425,** 905.

Liotta, L. A., Lowenthal, M., Mehta, A., Conrads, T. P., Veenstra, T. D., Fishman, D. A., and Petricoin, E. F. (2005). Importance of communication between producers and consumers of publicly available experimental data. *J. Natl. Cancer Inst.* **97,** 310–314.

Loboda, A. V., and Krutchinsky, A. N. (2000). A tandem quadrupole/time-of-flight mass spectrometer with a matrix-assisted laser desorption/ionization source: Design and performance. *Rapid Commun. Mass Spectrom.* **14,** 1047–1057.

Loi, S., Haydon, A. M., Shapiro, J., Schwarz, M. A., and Schneider, H. G. (2004). Towards evidence-based use of serum tumour marker requests: An audit of use in a tertiary hospital. *Int. Med. J.* **34,** 545–550.

Lu, K. H., Patterson, A. P., Wang, L., Marquez, R. T., Atkinson, E. N., Baggerly, K. A., Ramoth, L. R., Rosen, D. G., Liu, J., Hellstrom, I., Smith, D., Harmann, L., *et al.* (2004). Selection of potential markers for epithelial ovarian cancer with gene expression arrays and recursive descent partition analysis. *Clin. Cancer Res.* **10,** 3291–3300.

MacCoss, M. J., Wu, C. C., and Yates, J. R., III (2002). Probability-based validation of protein identifications using a modified SEQUEST algorithm. *Anal. Chem.* **74,** 5593–5599.

Magklara, A., Scorilas, A., Stephen, C., Kristiansen, G. O., Hauptmann, S., Jung, K., and Diamandis, E. P. (2000). Decreased concentrations of prostate-specific antigen and human glandular kallikrein 2 in malignant versus nonmalignant prostatic tissue. *Urology* **56,** 527–532.

Mann, M., Hendrickson, R. C., and Pandey, A. (2001). Analysis of proteins and proteomics by mass spectrometry. *Ann. Rev. Biochem.* **70,** 437–473.

Marshall, J., and Jankowski, A. (2004). Human serum proteins preseparated by electrophoresis or chromatography followed by tandem mass spectrometry. *J. Proteome Res.* **3,** 364–382.

Marshall, J., Kupchak, P., Zhu, W., Yantha, J., Vrees, T., Furesz, S., Jacks, K., Smith, C., Kireeva, I., Zhang, R., Takahashi, M., Stanton, E., *et al.* (2003). Processing of serum proteins underlies the mass spectral fingerprinting of myocardial infarction. *J. Proteome Res.* **2,** 361–372.

Menon, U., and Jacobs, I. J. (2000). Recent developments in ovarian cancer screening. *Curr. Opin. Obstet. Gynecol.* **12,** 39–42.

Menon, U., and Jacobs, I. J. (2002). Tumor markers. *In* "Principles and Practice of Gynecologic Oncology" (W. J. Hoskins, C. A. Perez, and R. C. Young, Eds.), 3rd edn. pp. 165–182. Lippincott, Williams & Wilkins, Philadelphia, PA.

Merchant, M., and Weinberger, S. R. (2000). Recent advancements in surface-enhanced laser desertion/ionization-time of flight mass spectrometry. *Electrophoresis* **21,** 1164–1177.

Meyer, T., and Rustin, G. J. (2000). Role of tumour markers in monitoring epithelial ovarian cancer. *Br. J. Cancer* **82,** 1535–1538.

Mirgorodskaya, O. A., Kozmin, Y. P., Titov, M. I., Korner, R., Sonksen, C. P., and Roepstorff, P. (2000). Quantitation of peptides and proteins by matrix-assisted laser desorption/ionization mass spectrometry using (18)O-labeled internal standards. *Rapid Commun. Mass Spectrom.* **14,** 1226–1232.

Nakanishi, T., Okamoto, N., Tanaka, K., and Shimizu, A. (1994). Laser desorption time-of-flight mass spectrometric analysis of transferring precipitated with antiserum: A unique simple method to identify molecular weight variants. *Biol. Mass Spectrom.* **23,** 220–223.

National Cancer Institute. (2002). Surveillance epidemiology and end results program (online), (cited November 19, 2002), http://seer.cancer.gov/2002.

Oosterhuis, W. P., Bruns, D. E., Watine, J., Sandberg, S., and Horvath, A. R. (2004). Evidence-based guidelines in laboratory medicine: Principles and methods. *Clin. Chem.* **50,** 806–818.

Pandey, A., and Mann, M. (2000). Proteomics to study genes and genomes. *Nature* **405,** 837–846.

Pedrioli, P. G., Eng, J. K., Hubley, R., Vogelzang, M., Deutsch, E. W., Raught, B., Pratt, B., Nilsson, E., Angeletti, R. H., Apweiler, R., Cheing, K., Costello, C. E., *et al.* (2004). A common open representation of mass spectrometry data and its application to proteomics research. *Nat. Biotechnol.* **22,** 1459–1466.

Petricoin, E. F., III, Ardekani, A. M., Hitt, B. A., Levine, P., Fusaro, V. A., Steinberg, S., Mills, G. B., Simcoe, C., Fishman, D. A., Kohn, D. C., and Liotta, L. A. (2002a). Use of proteomic patterns in serum to identify ovarian cancer. *Lancet* **359,** 572–575.

Petricoin, E. F., III, Ornstein, D. K., Paweletz, C. P., Ardekani, A., Hackett, P. S., and Hitt, B. A. (2002b). Serum proteomic patterns for detection if prostate cancer. *J. Natl. Cancer Inst.* **94,** 1576–1578.

Petricoin, E. F., III, Zoon, K. C., Kohn, E. C., Barret, J. C., and Liotta, L. A. (2002c). Clinical proteomics: Translating benchside promise into bedside reality. *Nat. Rev. Drug Disc.* **1,** 683–695.

Polyak, K., and Riggens, G. J. (2001). Gene discovery using the serial analysis of gene expression technique: Implications for cancer research. *J. Clin. Oncol.* **19,** 2948–2958.

Poon, T. C. W., Yip, T., Chan, A. T. C., Yip, C., Yip, V., Mok, T. S. K., Lee, C. C. Y., Leung, T. W. T., Ho, S. K. W., and Johnson, P. J. (2003). Comprehensive proteomic profiling identifies serum proteomic signatures for detection of hepatocellular carcinoma and its subtypes. *Clin. Chem.* **49,** 752–760.

Powell, K. (2003). Proteomics delivers on promise of cancer biomarkers. *Nat. Med.* **9,** 980.

Pusch, W., Floccoo, M. T., Leung, S., Thiele, H., and Kostrzewa, M. (2003). Mass spectrometry-based clinical proteomics. *Pharmagenomics* **4,** 463–476.

Qu, Y., Adam, B.-L., Yasui, Y., Ward, M. D., Cazares, L. H., Schellhammer, P. F., Feng, Z., Semmes, O. J., and Wright, G. L., Jr. (2002). Boosted decision tree analysis of surface-enhanced laser desorption/ionization mass spectral serum profiles disciminated prostate cancer from non-prostate patients. *Clin. Chem.* **48,** 1835–1843.

Ransohoff, D. F. (2003). Cancer. Developing molecular biomarkers for cancer. *Science* **299,** 1679–1680.

Ransohoff, D. F. (2004). Rules of evidence for cancer molecular-marker discovery and validation. *Nat. Rev. Cancer* **4,** 309–314.

Ransohoff, D. F. (2005). Lessons from controversy: Ovarian cancer screening and serum proteomics. *J. Natl. Cancer Inst.* **97,** 315–319.

Reyzer, M. L., Cladwell, R. L., Dugger, T. C., Forbes, J. T., Ritter, C. A., Guix, M., Arteaga, C. L., and Caprioli, R. M. (2004). Early changes in protein expression detected by mass spectrometry predict tumor response to molecular therapeutics. *Cancer Res.* **64,** 9093–9100.

Rogers, M. A., Clarke, P., Noble, J., Munro, N. P., Paul, A., and Selby, P. J. (2003). Proteomic profiling of urinary proteins in renal cancer by surface enhanced laser desorption ionization and neural network analysis: Identification of key issues affecting potential clinical utility. *Cancer Res.* **63,** 6971–6983.

Rosty, C., Christa, L. M., Kuzdzal, S., Baldwin, W. M., Zahurak, M. L., Carnot, F., Chan, D. W., Canto, M., Lillemoe, K. D., Cameron, J. L., Yeo, C. J., Hruban, R. H., *et al.* (2002). Identification of hepatocarcinoma-intestine-pancreas/pancreatitis-associated protein 1 as a biomarker for pancreatic ductal adenocarcinoma by protein biochip technology. *Cancer Res.* **62,** 1868–1875.

Sasaki, K., Sato, K., Akiyama, Y., Yanagihara, K., Masaaki, O., and Yamauchi, K. (2002). Peptidomics-based approach reveals the secretion of the 29-residue COOH-terminal fragment of the putative tumor suppressor protein DMBT 1 from pancreatic adenocarcinoma cell lines. *Cancer Res.* **62,** 4894–4898.

Sauter, E. R., Zhu, W., Wassell, R. P., Chervoneva, I., and Du Bois, G. C. (2002). Proteomic analysis of nipple aspirate fluid to detect biologic markers of breast cancer. *Br. J. Cancer* **86,** 1440–1443.

Schwartz, S. A., Weil, R. J., Johnson, M. D., Toms, S. A., and Caprioli, M. R. (2004). Protein profiling in brain tumors using mass spectrometry: Feasibility of a new technique for the analysis of protein expression. *Clin. Cancer Res.* **10,** 981–987.

Semmes, O. J. (2004). Defining the role of mass spectrometry in cancer diagnostics. (Editorial). *Cancer Epidemiol. Biomarkers Prev.* **13,** 1555–1557.

Semmes, O. J., Ziding, F., Dam, B.-L., Banez, L. L., Bigbee, W. L., Campos, D., Cazares, L. H., Chan, D. W., Grizzle, W. E., Izbicka, E., Kagan, J., Malik, G., *et al.* (2005). Evaluation of serum protein profiling by surface-enhanced laser desorption/ionization time-of-flight mass spectrometry for the detection of prostate cancer: I. Assessment of platform reproducibility. *Clin. Chem.* **51,** 102–112.

Smith, R. D., and Anderson, G. A. (2002). An accurate mass tag strategy for quantitative and high-throughput proteome measurements. *Proteomics* **2,** 513–523.

Sorace, J. M., and Zhan, M. A. (2003). Data review and re-assessment of ovarian cancer serum proteomic profiling. *BMC Bioinformatics* **4,** 24.

Srivastava, S., and Kramer, B. S. (2000). Early detection cancer research network [Editorial]. *Lab. Invest.* **80,** 1147–1148.

Stegner, A. H., Wagner-Mann, C., Du Bois, G. C., and Sauters, E. R. (2004). Proteomic analysis to identify breast cancer biomarkers in nipple aspirate fluid. *Clin. Cancer Res.* **10,** 7500–7510.

Sturgeon, C. M. (2001). Tumor markers in the laboratory: Closing the guideline-practice gap. *Clin. Biochem.* **34,** 353–359.

Sturgeon, C. M. (2002). Practice guidelines for tumor marker use in the clinic. *Clin. Chem.* **48,** 1151–1159.

Sullivan, P. M., Etzioni, R., Feng, Z., Potter, J. D., Thompson, M. L., Thornquist, M., Winget, M., and Yasui, Y. (2001). Phases of biomarker development for early detection of cancer. *J. Natl. Cancer Inst.* **93,** 1054–1061.

Takahashi, M., Stanton, E., Moreno, J. I., and Jackowski, G. (2001). Delivering on the promise of proteomics. (cited April 2001), www.synxpharma.com

Tammen, H., Kreipe, H., Hess, R., Kellmann, M., Lehmann, U., Pich, A., Lamping, N., Schulz-Knappe, P., Zucht, H. D., and Lilischkis, R. (2003). Expression profiling of breast cancer cell by differential peptide display. *Breast Cancer Res. Treat.* **79,** 83–93.

Tumor Marker Expert Panel (ASCO). (1996). Clinical practice guidelines for the use of tumor markers in breast and colorectal cancer. *J. Clin. Oncol.* **14**, 2843–2877.

Tuteja, R., and Tuteja, N. (2004). Serial analysis of gene expression (SAGE): Unraveling the bioinformatics tools. *Bioessays* **26**, 916–922.

Tyers, M., and Mann, M. (2003). From genomics to proteomics. *Nature* **422**, 193–197.

Van Dalen, A. (1993). Quality control and standardization of tumor marker tests. *Tumor Biol.* **14**, 131–135.

Villanueva, J., Philip, J., Entenberg, D., Chaparro, C. A., Tanwar, M. K., Holland, E. C., and Tempst, P. (2004). Serum peptide profiling by magnetic particle-assisted, automated sample processing and MALDI-TOF mass spectrometry. *Anal. Chem.* **76**, 1560–1570.

Vlahou, A., Schellhammer, P. F., Mendinos, S., Patel, K., Kondylis, F. I., Gong, L., Nasim, S., and Wright, G. L. J. (2001). Development of a novel proteomic approach for the detection of transitional cell carcinoma of the bladder in urine. *Am. J. Pathol.* **158**, 1491–1502.

von Eggeling, F., Davies, H., Lomas, L., Fiedler, W., Junker, K., Classen, U., and Ernst, G. (2000). Tissue-specific microdissection coupled with protein chip array technologies: Applications in cancer research. *Biotechniques* **29**, 1066–1069.

Von Haller, P. D., Yi, E., Donohoe, S., Vaughn, K., Keller, A., Nesvizhskii, A. I., Eng, J., Li, X. J., Goodlett, D. R., Aebersold, R., and Watts, J. D. (2003a). The application of new software tools to quantitative protein profiling via isotope-coded affinity tag (ICAT) and tandem mass spectrometry: I. Statistically annotated datasets for peptide sequences and proteins identified via the application of ICAT and tandem mass spectrometry to proteins copurifying with T cell lipid rafts. *Mol. Cell. Proteomics* **2**, 426–427.

Von Haller, P. D., Yi, E., Donohoe, S., Vaughn, K., Keller, A., Nesvizhskii, A. I., Eng, J., Li, X. J., Goodlett, D. R., Aebersold, R., and Watts, J. D. (2003b). The application of new software tools to quantitative protein profiling via isotope-coded affinity tag (ICAT) and tandem mass spectrometry: II. Evaluation of tandem mass spectrometry methodologies for large-scale protein analysis, and the application of statistical tools for data analysis and interpretation. *Mol. Cell. Proteomics* **2**, 428–442.

Wadsworth, J. T., Somers, K. D., Stack, B. C. J., Cazares, L., Malik, G., Adam, B. L., Wright, G. L., and Semmes, O. J. (2004). Identification of patients with head and neck cancer using serum protein profiles. *Arch. Otolaryngol. Head Neck Surg.* **130**, 98–104.

Weinberger, S.R, Morris, T. S., and Pawlak, M. (2000). Recent trends in protein biochip technology. *Pharmacogenomics* **1**, 395–416.

Welsh, J. B., Sapinoso, L. M., Su, A. I., Kern, S. G., Wang-Rodriquez, J., Moskaluk, C. A., Frierson, H. F., Jr., and Hampton, G. M. (2001). Analysis of gene expression identifies candidate markers and pharmacological targets in prostate cancer. *Cancer Res.* **61**, 5974–5978.

Welsh, J. B., Sapinoso, L. M., Kern, S. G., Brown, D. A., Liu, T., Bauskin, A. R., Ward, R. L., Hawkins, N.J, Quinn, D. I., Russell, P. J., Sutherland, R. L., Breit, S. N., *et al.* (2003). Large scale delineation of secreted protein biomarkers overexpressed in cancer tissue and serum. *Proc. Natl. Acad. Sci. USA* **100**, 3410–3415.

Wilkins, M. R., Gasteiger, E., Wheeler, C. H., Lindskog, I., Sanchez, J. C., Bairoch, A., Appel, R. D., Dunn, M. J., and Hochstrasser, D. F. (1998). Multiple parameter cross-species protein identification using Multi-indent: A world-wide web accessible tool. *Electrophoresis* **19**, 3199–3206.

Winawer, S. J., St. John, D. J., Bond, J. H., Rozen, P., Burt, R. W., Waye, J. D., Kronborg, O., O'Brien, M. J., Bishop, D. T., Kurtz, R. C., *et al.* (1995). Prevention of colorectal cancer: Guidelines based on new data. WHO Collaborating centre for the Prevention of Colorectal. *Cancer Bull. World Health Organ.* **73**, 7–10.

Wright, G. L., Cazares, L. H., Leung, S., Nasim, S., Adam, B., Yip, T., Schellhammer, P. F., Gong, L., and Vlahou, A. (1999). Protein chip surface enhanced laser desorption/ionization

(SELDI) mass spectrometry: A novel biochip technology for detection of prostate cancer biomarkers in complex protein mixtures. *Prost. Cancer Prost. Dis.* **2,** 264–276.

Yanagisawa, K., Shyr, Y., Xu, B. J., Massion, P. P., Larsen, P. H., White, B. C., Roberts, J. R., Edgerton, M., Gonzalez, A., Nadaf, S., Moore, J. H., Caprioli, R. M., *et al.* (2003). Proteomic patterns of tumor subsets of non-small-cell lung carcinoma. *Lancet* **362,** 433–439.

Yao, X., Freas, A., Ramirez, J., Demirev, P. A., and Fenselau, C. (2001). Proteolytic 18-O labeling for comparative proteomics: Model studies with two serotype of adenovirus. *Anal. Chem.* **73,** 2836–2842.

Yousef, G. M., Polymeris, M. E., Yaccoub, G. M., Scorilas, A., Soosaipillai, A., Popalis, C., Fracchioli, S., Katsaros, D., and Diamandis, E. P. (2003). Parallel overexpression of seven kallikrein genes in ovarian cancer. *Cancer Res.* **63,** 2223–2227.

Zarrinkar, P. P., Mainquist, J. K., Zamora, M., Stern, D., Welsh, J. B., Sapinoso, L. M., Hampton, G. M., and Lockhart, D. J. (2001). Arrays of arrays for high-throughput gene expression profiling. *Genome Res.* **11,** 1256–1261.

Zhang, W., and Chait, B. T. (2000). ProFound: An expert system for protein identification using mass spectrometric peptide mapping information. *Anal. Chem.* **72,** 2482–2489.

Zhang, Z., Bast, R. C., Jr., Yu, Y., Li, J., Sokoll, L. J., Rai, A. J., Rosenzweig, J. M., Cameron, B., Wang, Y. Y., Meng, X. Y., Berchuck, A., van Haaften-Day, C., *et al.* (2004). Three biomarkers identified from serum proteomic analysis for the detection of early stage ovarian cancer. *Cancer Res.* **64,** 5882–5890.

Zhukov, T. A., Johanson, R. A., Cantor, A. B., Clark, R. A., and Tockman, M. S. (2003). Discovery of distinct protein profiles specific for lung tumors and premalignant lung lesions by SELDI mass spectrometry. *Lung Cancer* **40,** 267–279.

Microarrays to Identify New Therapeutic Strategies for Cancer

Christopher Sears* and Scott A. Armstrong†

**Division of Hematology/Oncology, Children's Hospital, Dana-Farber Cancer Institute, Harvard Medical School, Boston, Massachusetts 02115;*
†Department of Pediatric Oncology, Dana-Farber Cancer Institute, Harvard Medical School, Boston, Massachusetts 02115

I. Introduction
II. Microarray Technologies
  A. cDNA Microarrays
  B. Tissue Microarrays
  C. Microarray Comparative Genomic Hybridization
  D. Nonsense-Mediated mRNA Decay
  E. Small Molecule Microarrays
III. Microarrays in Drug Development
  A. Pharmacogenomics
  B. Toxicity and Pharmacogenomic Evaluation in Clinical Trials
IV. Microarrays to Direct the Use of Cancer Therapeutics
  A. Identification of Patients for Adjuvant Therapy
  B. Imatinib Mesylate (Gleevec)
  C. Farnesyltransferase Inhibitor, R115777
  D. Histone Deacetylase Inhibitors
  E. DNA Methylases, Epigenetic Silencing, and Tumor Suppressors
V. Identification of Therapeutic Targets in Distinct Disease Subtypes
  A. Leukemias and Lymphomas
  B. Breast Cancer
VI. Conclusions
  References

Over the past decade, microarrays have emerged as an important tool for the characterization of cancer cells. Numerous studies have demonstrated that cDNA arrays can help delineate biological subsets of disease that have prognostic relevance. Such studies provide hope that introduction of this information into clinical trials will lead to more biologically based stratification schemes such that appropriately tailored therapies can be developed. While the identification of unique subsets of cancer promises to improve our ability to predict which cancers are unlikely to have a significant response to therapy, new therapeutic approaches are needed in most cases. The wealth of information that comes from microarray analysis of cancer likely holds the information necessary to develop such approaches. This chapter will provide examples where

Advances in CANCER RESEARCH

0065-230X/07 $35.00
DOI: 10.1016/S0065-230X(06)96003-5

microarray analysis has been used in an attempt to either direct the use of current therapies or identify new potential therapeutic avenues.

I. INTRODUCTION

As a result of the public and private human genome sequencing efforts, the scientific community is developing an unprecedented inventory of the blueprints for disease. The information accumulated in the post-genome era will provide scientists/clinicians with detailed information on the ~35,000 genes that comprise the human genome, the proteins they encode, and their variation in human populations that place individuals at risk for disease. This variation is increasingly viewed as playing a key role in the etiology of disease and determining the response to therapy. Since the genomes of roughly 1000 other organisms have also been sequenced in addition to humans, it should be possible to develop model organisms for disease states and relate the similarities and dissimilarities to humans at the most basic level.

In conjunction with the inventory of genetic building blocks and their variation, the development of microarray technologies has led to new ways in which to understand global regulatory networks. Microarrays allow one to study the behavior of a large complement of genes or proteins in parallel. The profiling of cancer using microarrays promises to revolutionize the field by identifying tumor subclasses and target-specific genes for diagnosis and therapy. In this chapter, we will briefly outline the types of microarrays currently available, followed by examples of their use in drug development and target identification.

II. MICROARRAY TECHNOLOGIES

A. cDNA Microarrays

In the genomic era, cDNA microarray (DNA chip) technology has evolved into an important and powerful tool for high-throughput comprehensive analysis of gene expression, genotyping, and resequencing applications in almost every field of biomedical research (Fig. 1). Large-scale transcriptional profiling analyses with microarrays are frequently used to explore gene expression patterns in order to better understand the molecular mechanisms of physiology and pathogenesis. New diagnostic and therapeutic strategies can be formulated on characterization of a specific cell or population of cells and their subsets of biomolecules.

Laser microdissection and laser pressure catapulting (LMPC) have been used to obtain pure and relatively homogenous samples of cells or subcellular components that can be analyzed with single-cell messenger RNA (mRNA)

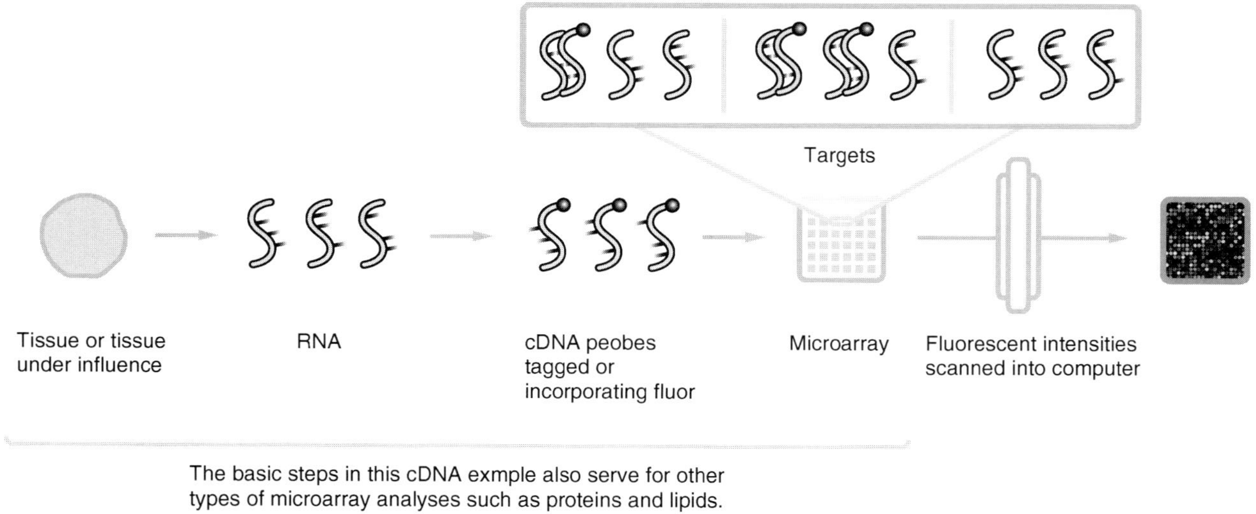

Fig. 1 Schematic representation of microarray technology. The general principles of array technology involve isolating biomolecules from particular biological samples, tagging the molecules for future detection, and hybridizing to a finely spaced grid of complementary molecules. The input biomolecules are then detected by laser scanning. Sophisticated informatics subsequently process the raw image data and generate numerical representations of expression levels as well as false color maps of the array and differential gene expression. (See Color Insert.)

extraction, PCR, and microarray techniques. The improved methods of sample generation and handling increases the potential that such approaches can provide insight for possible early detection of disease, therapeutic targeting, and development of custom therapies for patients (Niyaz *et al.*, 2005).

cDNA microarrays successfully measure global cellular transcriptional states, which are then related to the underlying biology in order to elucidate biological pathways. Determining the number of patterns or clusters for use in data interpretation is critical for proper separation of gene groups to occur. Methods which rely on gene annotation linked to decompositional analysis of global gene expression data are proving increasingly valuable. Specific activity on strongly coupled signaling pathways can be estimated, and in some cases, estimation of the activity of specific signaling proteins can be obtained (Bidaut *et al.*, 2006; Yang *et al.*, 2005). These studies demonstrate that when gene ontology or transcription factor databases are utilized with microarray data, downstream indicators of pathway activity can be provided. This information can be used to better understand signaling activity in normal and disease processes and allow investigation of the specificity and success of targeted therapeutics (Bidaut *et al.*, 2006).

Microarrays enable scientists to resolve one or more genes from the global assembly and to highlight their function in some experimental condition or disease state. A global protein interaction network (interactome) analysis is another effective tool to understand the relationships between genes in a microarray study. When the topological features in the interactions of differentially expressed lung squamous cancer genes were assessed, it was observed from microarray gene profiling data that the differentially elevated genes were well connected (Wachi *et al.*, 2005). The suppressed and randomly selected genes were not as well connected. This may indicate that when a topological analysis of cancer genes is carried out using protein interaction data, it is possible to place the gene list, often of a disparate nature, into the global context of the cell.

While the cDNA microarray remains the predominant form in current practice, there are additional useful array technologies for chemicals and biomolecules. The numerous variants of microarray technology are also complementary to one another. As described in the following sections, sophisticated studies are now employing multiple high-throughput array technologies in order to address different aspects of the same problem (Fig. 2).

B. Tissue Microarrays

Tissue microarray (TMA) technology enables one to quickly analyze a large number of clinical specimens in one experiment. The advantage of TMAs is that they maximize returns in cellular pathology while using a minimum of

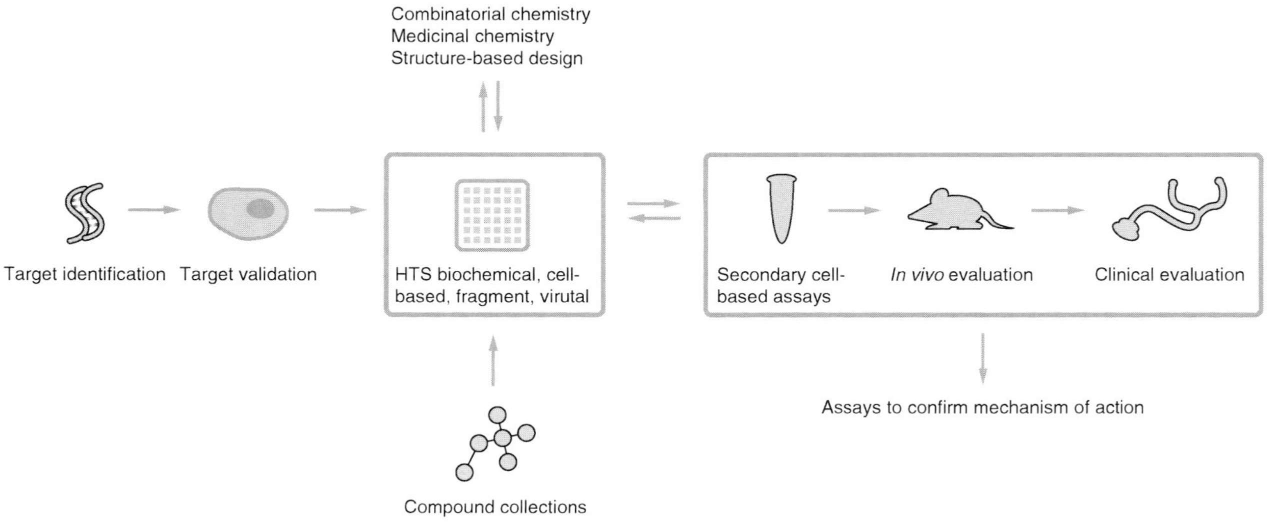

Fig. 2 Schematic representation of the drug discovery process with emphasis on stages at which microarray technology are currently employed. Microarrays are involved across the drug discovery process: from early research stages of target identification and validation, through screening of small molecule compounds and biomolecules, to toxicology and pharmacodynamic analyses in cell-based and *in vivo* systems (after Clarke *et al.*, 2004).

cells and tissues. Cores of tissue are taken from multiple donors and from this a single recipient block is constructed. Several hundred tissues may be represented on one TMA that is immunohistochemically stained to provide protein expression data across multiple samples. This technique may be used with cDNA microarray analysis to assist with identification of disease markers and potential therapeutic targets. Cellular pathology is consequently being revolutionized by this high-throughput technology, which yields large amounts of information from previously very low-throughput samples (Warford, 2004).

This powerful, high-throughput method permits the parallel analysis of any molecular alterations in DNA, RNA, or proteins in thousands of tissue specimens, thereby enabling large, cost- and time-effective, genome-scale molecular pathology studies. Adequate computing power and storage is necessary to accommodate the vast quantities of data that can be generated by one TMA experiment. Databases may contain various aspects of each specimen, including pathological information, TMA construction protocol, experimental protocol, results from various immunocytological and histochemical staining experiments, and any scanned images from TMA cores. Many immunostaining results may be compared for each TMA core so that gene products of clinical importance, such as therapeutic or prognostic markers, may be quickly identified (Sharma-Oates *et al.*, 2005; Tzankov *et al.*, 2005).

TMA and chromogenic *in situ* hybridization techniques were utilized to study which colorectal carcinoma patients would be most suitable for cetuximab therapy, an antiepidermal growth factor receptor drug. The amplification of the epidermal growth factor receptor gene in colorectal cancer and its relationship with protein expression were evaluated via immunohistochemistry (IHC). It was found that epidermal growth factor receptor gene amplification may not be useful in predicting response to cetuximab-based therapy because only a small fraction of EGF receptor-positive colorectal carcinomas detected by ICH were associated with gene amplification (Shia *et al.*, 2005). The combination of cDNA microarrays and TMAs promises to provide much greater insight than either technique when used alone.

C. Microarray Comparative Genomic Hybridization

Microarray comparative genomic hybridization (mCGH) is a technique that can be used to detect gene dosage alterations and breakpoints in various diseases including cancer. High-density BAC, fosmid, or oligo arrays span the entire nonrepetitive regions of the human genome, tiling the full genome in a single array. Spontaneously transformed mouse ovarian surface epithelial (MOSE) cell lines were analyzed using cDNA microarrays to identify any imbalances in their genomes. Two related genomic variants were identified among six cell lines studied. This approach has proved suitable for studying the initiation and progression of human ovarian cancer and for evaluation of therapies (Urzua *et al.*, 2005).

mCGH analysis has been used to identify genes of prognostic value in breast cancer. Tumors were screened by array-CGH and were divided into two groups depending on if they were isolated from long-term survivors or early disease-related deaths associated with a subset of genes. The genes EMS1, TOP2A, CCNE1, and ERBB2 were amplified and a fluorescent *in situ* hybridization using a TMA was carried out to evaluate the prognostic significance of the four genes. It was determined that TOP2A and ERBB2 were associated with adverse disease-related outcome; the combined amplification of TOP2A, ERBB2, and EMSI also resulted in adverse disease-related outcome. Amplification of EMS1 alone had no prognostic significance, nor did CCNE1. Array-CGH and TMAs were thus a useful combination in the identification and validation of molecular markers and together were able to classify breast cancers into prognostic groups (Callagy *et al.*, 2005).

Testicular germ-cell tumors (TGCTs) were analyzed using mCGH analysis to gather more information about the genetic basis of cisplatin resistance. Three cisplatin-sensitive and three cisplatin-resistant cell lines were analyzed by mCGH to obtain information about genomic and expression difference. While no consistent genomic region changes were found in the three cell lines, additional information was uncovered related to oncogenes, tumor-suppressor genes, and drug resistance-related genes. When the latter results were examined further via comparative expressed sequence hybridization, a technique for gene expression profiling along chromosomes, the scientists discovered a consistently overexpressed chromosomal region in all three resistant lines compared with their parent lines (Wilson *et al.*, 2005).

Another experiment utilizing mCGH involved analysis of primary lung adenocarcinoma bacterial artificial chromosome clones in order to gather data related to genome-wide copy number changes in tumors. mCGH of 800 chromosomal loci resulted in the identification of large numbers of chromosomal alterations in cancer-related genes. Three subgroups of lung adenocarcinoma were revealed by an unsupervised hierarchical clustering analysis of multiple alterations. Distinct genetic alterations, smoking history, and gender were associated with these abnormalities. Such information may be beneficial for the discovery of novel cancer-related genes, estimating patient prognosis and choosing therapeutic targets (Shibata *et al.*, 2005).

D. Nonsense-Mediated mRNA Decay

The multistep process of cancer pathogenesis relies on the evolution and selection of deleterious mutations. Identifying gene mutations is therefore important in cancer studies and may be applied to diagnostics and development of therapeutic targets, but has been an extremely slow and laborious process. In order to increase the rate at which genes undergoing mutations are found, a new approach was developed to prioritize and focus the search. Nonsense-mediated mRNA decay (NMD) inhibition and microarray analysis (NMD microarrays) were used to identify transcripts that contained nonsense mutations (Losson and Lacroute, 1979; Maquat, 2005). This technology relies on the fact that mRNAs containing nonsense mutations are selectively and rapidly degraded by the NMD pathway during translation. By inhibiting this pathway, and analyzing the global transcriptional state, one can obtain an inventory of genes that may have mutations. Identification of tumor-suppressor genes that were inactivated in cancer was accomplished by combining NMD microarrays with array-based CGH. When prostate cancer cell lines were screened with such a "mutatomics" approach, inactivating mutations in the EPHB2 gene were identified. The screening of metastatic uncultured prostate cancers also uncovered mutations of this gene, which may be responsible for loss of tissue architecture. While the utility of this approach is still under investigation, identification of novel mutated genes in cancer cell lines could be accomplished quickly with NMD microarray analysis, and thus may help speed further investigation (Wolf *et al.*, 2005).

E. Small Molecule Microarrays

Small molecule microarrays (SMMs) can robustly screen for the identification and validation of potential drug development targets. They utilize the capability of combinatorial chemistry to produce myriad compounds along with the throughput of microarrays, with the resulting tool possessing the characteristics of versatility and rapid analysis and discovery. This technology can assist with the identification of biologically significant molecules, both natural and synthetic, and the subsequent exploration of medicinal and diagnostic applications (Uttamchandani *et al.*, 2005).

A chemical genomics experiment was performed in which microarray transcriptional signatures of engineered candidate kinase targets were directly compared. Some of the signatures were elicited with an inhibitor whose specificity was not known and some were elicited by pharmacological inhibitors that were highly specific. Two cyclin-dependent kinases, Cdk1 and Pho85, were identified as the targets of the inhibitor GW400426. When Cdk1 and Pho85 are simultaneously inhibited, but not either alone by GW400426, expression of specific transcripts involved in cell growth is controlled, thus revealing a cellular process that is uniquely sensitive to the multiplex inhibition of two cellular kinases. These results illustrate how classes of genes can be used as targets for therapeutic intervention, and how a greater understanding of multiplex protein kinase inhibitors can be obtained than would have been possible if only the sum of individual inhibitor–kinase interactions had been considered (Kung *et al.*, 2005).

III. MICROARRAYS IN DRUG DEVELOPMENT

A. Pharmacogenomics

Microarrays are proving useful in predicting physiological reactions to pharmaceuticals. The cytochrome P450 (CYP) pathway and its variants, particularly CYP2D6 and CYP2C19, are involved in the metabolism of ~25% of all prescription drugs, with antidepressants, antipsychotics, immunosuppressive, and anticancer drugs being most heavily assessed in this area. If the metabolic status of an individual taking the drugs can be ascertained, many ill effects caused by drug-metabolizing enzymes might be anticipated. Pharmacogenetics might prove useful to help develop such personalized medicine (Evans and Relling, 1999). Microarrays can be utilized to test for CYP polymorphisms, and the data can be used both in the study of disease pathophysiology and drug metabolism and toxicology.

The AmpliChip CYP450 from Roche Molecular Diagnostics (Alameda, CA) is a combination of Roche's PCR technology with the GeneChip microarray system from Affymetrix (Santa Clara, CA). It is the first Food and Drug Administration (FDA)-approved microarray molecular diagnostic test for analyzing 29 polymorphisms and mutations of the CYP2D6 gene, and two polymorphisms of the CYP2C19 gene. Such testing is not meant to be a solitary tool that can determine optimum drug dosage. Rather, it is meant to be used along with clinical evaluation and other methods to select the treatment that is best suited for a patient. The results obtained from AmpliChip CYP450, when combined with pharmacotherapy knowledge, could greatly assist with selecting the optimal drug and dosage in individual cases, and may facilitate the development of personalized medicine (Jain, 2005).

B. Toxicity and Pharmacogenomic Evaluation in Clinical Trials

There have been encouraging correlations seen when microarray-based expression profiling studies in the field of oncology have been used to compare tumor transcriptional profiles and eventual patient outcomes. Several studies over the past few years based on the profiling of archived tumor tissues imply that prognosis, and in some cases, even a predicted response to specific therapies, may be obtained on transcriptional analysis. Because of this, there has been great interest in trying to apply transcriptional profiling to real-time clinical trial samples, and increasingly, clinical trial study designs utilize transcriptional profiling strategies to meet clinical pharmacogenomic objectives. The FDA recently released voluntary guidelines for genomic data to help evaluate any potential benefit that may be obtained by implementing expression profiling studies during the preclinical and clinical phases of drug development, which can be used by both regulatory agencies and pharmaceutical companies. There is great promise afforded by this technology but there are still a number of practical impediments in the way that must be solved in order to reap the ultimate benefit of applying transcriptional profiling to personalized treatment strategies. These impediments include: (1) ensuring that the legal integrity of informed consents and patient confidentiality are maintained, (2) ensuring that chain-of-custody and specimen traceability are maintained, and (3) ensuring the integrity (i.e., accuracy, completeness, and reliability) of data reported (Burczynski *et al.*, 2005). Protocols designed to address the data integrity issues have recently been developed by the regulatory administrations.[1]

IV. MICROARRAYS TO DIRECT THE USE OF CANCER THERAPEUTICS

Microarrays are being heavily utilized in cancer research and are proving to be useful in all aspects of study, including the classification of cancer, the study of biochemical pathways, and the identification of potential targets for novel therapeutics. Gene expression technologies are also being used to distinguish on-target versus off-target effects of cancer therapeutics, mechanisms of resistance to treatment, mechanisms of therapeutic function, and prediction of drug response.

Resistance to chemotherapy drugs is a major barrier to the successful long-term treatment of cancer. In order to better understand the mechanisms involved, gene families were identified that appear to contribute to the evolution of drug resistance and may be regulated through a multiple pathway gene expression program. Microarray analysis of tumor samples will make feasible the identification of critical genes that are most relevant to clinical drug resistance, and the data can be used to develop strategies for

[1] The Food and Drug Administration (FDA) Good Laboratory Practice (GLP) regulation [21 Code of Federal Regulations (CFR) Part 58].

circumvention of resistance (Holleman *et al.*, 2004; Richardson and Kaye, 2005).

When cancer patients are given adjuvant chemotherapy, there may not be any benefit because either locoregional treatment alone has cured their cancer or the patient may be resistant to the regimens employed. If prognostic factors could be improved, then selection of patients for adjuvant therapy would be facilitated. Likewise, if the appropriate predictive factors could be identified, they would contribute to the ease of selecting an optimal therapeutic strategy. There are several ongoing studies exploring genomic prognostic factors for the purpose of optimizing the indications for adjuvant chemotherapy. A large randomized trial utilizing *m*icroarray *i*n *n*ode-negative *d*isease may *a*void *c*hemo*t*herapy (MINDACT) is being conducted to discern genomic signatures of good prognosis breast cancer from breast cancers with a worse prognosis. This will be done by comparing the information obtained with genomic profiling to the classical clinicopathologic index. Other trials are being conducted to assess if neoadjuvant chemotherapy with docetaxel is more effective than an anthracycline-containing regimen for treating p53-mutated tumors. Within this context an additional study will evaluate the ability of gene profiles to predict p53 status (Mauriac *et al.*, 2005).

The anthracycline antibiotic doxorubicin is a cancer chemotherapy agent used in multiple cancers, but resistant cells often emerge from the treated population. Using cDNA microarray and RNA interference (RNAi) analysis, genes were screened that regulate doxorubicin susceptibility in highly tumorigenic breast cancer cells. Genes associated with both proliferation and cell cycle arrest after treatment with doxorubicin were identified. A model in which a distinct transcriptional response to doxorubicin is induced in highly tumorigenic breast cancer cells that differs from less malignant cells was supported with these results. It may be possible to target the induced genes, which regulate drug susceptibility positively and negatively, for therapeutic intervention (Mallory *et al.*, 2005).

Drug resistance in colon cancer has also been studied utilizing microarray technology. After being treated with 5-fluorouracil (5-FU) or oxaliplatin, HCT115 colorectal cancer cells that exhibited resistance to these agents were selected and a DNA microarray was used to analyze their transcriptional profile. On bioinformatic analysis, it was found that the drug resistant cells contained sets of genes that were constitutively dysregulated and then transiently altered after they were exposed to the chemotherapy drugs. The molecular signatures of sensitivity to 5-FU and oxaliplatin may be represented by these genes (Boyer *et al.*, 2006).

Microarray data can be used to predict drug response. A combination of chemotherapies called M-VAC is a neoadjuvant therapy used for invasive bladder cancer, and consists of administering a regimen of methotrexate, vinblastine, doxorubicin, and cisplatin. Some patients experience tumor shrinkage and improved prognosis, while others suffer from severe adverse drug reactions to this treatment without any obvious benefit. There is no existing method that can assist with the prediction of how an individual patient will respond to chemotherapy. Using cDNA microarrays, gene expression profiles of biopsy tissue from 27 invasive bladder cancers were analyzed in order to attempt to predict a response to M-VAC therapy. Laser capture microdissection (LCM) was used to purify the populations of cancer cells for this analysis. There were 14 genes shown to be predictive, and after devising a numerical prediction scoring system to clearly delineate responder tumors from nonresponder tumors, this system could accurately predict drug responses in 8 out of 9 test cases that were taken from the original 27 cases. The RT-PCR data for the 14 genes were highly concordant with the cDNA microarray data. A feasible prediction system for bladder cancer sensitivity to M-VAC neoadjuvant chemotherapy might be developed based on RT-PCR which could potentially be used in the clinic to personalize therapy (Takata *et al.*, 2005).

A. Identification of Patients for Adjuvant Therapy

After several years of profiling breast cancer gene expression at the Netherlands Cancer Institute using a microarray platform containing 25,000 genes, a prognostic classifier was identified that consists of 70 genes. This information can help distinguish those patients who are either at a high or low risk for developing distant metastases, and could indicate who might require adjuvant therapy and their response to it. Furthermore, this analysis has identified the estrogen receptor, HER-2 status, c-kit mutation, and the epidermal growth factor (EGF) mutation as predictors for response to systemic therapy. There are several neoadjuvant studies either planned or underway to test the ability of gene expression profiling as a successful means of predicting the therapeutic response to chemotherapy drugs in breast cancer patients. Clinical trials assessing genomic analysis and docetaxel, paclitaxel, and doxorubicin response are currently underway, with an increasing number of such trials expected in the coming years (van de Vijver, 2005).

B. Imatinib Mesylate (Gleevec)

Imatinib is a 2-phenylaminopyrimidine derivative that functions as a potent inhibitor of a number of tyrosine kinase enzymes, with particularly high affinity for kinases involved in leukemia. Chronic myeloid leukemia (CML) has recently been successfully treated in the early stages of the disease (Druker *et al.*, 2001), but the blastic phase (BP), which is characterized by rapid expansion of therapy-refractory and differentiation-arrested blast cells, still remains a challenge therapeutically. In order to develop better therapeutic strategies with an increased understanding of the molecular mechanisms that govern the disease progression, comparative gene expression profiling may prove particularly useful. If transcriptional signatures could be obtained that would explain the pathological characteristics and aggressive behavior of BP blasts, new insights for treatment might be uncovered. Comparative gene expression profiling using Affymetrix oligonucleotide arrays was carried out with CD34+ Ph+ cells that were purified from untreated newly diagnosed chronic phase CML patients, and from patients who were in BP. Signatures were identified that were correlated with progression from the chronic to BP. Further characterization of these signatures may identify programs important for this progression and thus identify potential pathways for intervention (Zheng *et al.*, 2006).

Microarrays were assessed for feasibility to measure molecular end points before and after neoadjuvant treatment of prostate cancer patients in the intermediate/high-risk category with imatinib mesylate. Biopsy tissue was obtained either from ultrasound-guided biopsies taken before treatment or posttreatment radical prostatectomy specimens. High-quality microarray data was generated via this study using LCM and RNA amplification (Febbo *et al.*, 2006). Results after gene set enrichment analysis suggest that imatinib mesylate therapy causes apoptosis of microvascular endothelial cells, an observation also obtained anecdotally by IHC.

In order to gather more information regarding the molecular mechanism of action of imatinib mesylate, anaplastic thyroid carcinomas (ATCs) were studied utilizing TMAs. Out of 12 tumors that were histologically proven to be ATCs, and had been treated with imatinib, 6 of them expressed at least one imatinib-sensitive tyrosine kinase. Imatinib reduced the metastatic potential, and halved the proliferation of thyroid cells without affecting apoptosis. Imatinib's antitumor activity may be a therapeutic option for some ATC patients (Rao *et al.*, 2006).

C. Farnesyltransferase Inhibitor, R115777

The farnesyltransferase inhibitors (FTIs) are a class of experimental cancer drugs that target a posttranslational modification important for the activity of a number of signaling proteins in cancer cells (RAS and others). A microarray study was used to assess the FTI R115777 in rat mammary tumors that were positive or negative for the Ha-ras mutation. All tumors responded to the FTI inhibitor but the tumors with Ha-ras mutations were extremely sensitive. The authors assessed whether the gene expression profile before FTI treatment could be used to identify highly sensitive tumors (Ha-ras positive) and tumors that had variable sensitivity (Ha-ras negative). Both untreated and FT-treated tumors

(Ha-ras positive or negative) were examined using an oligonucleotide array, and a large number of genes were differentially expressed in control rat mammary tumors which either contained or lacked the activated Ha-ras mutation. This suggested that a microarray analysis might be useful for differentiating both highly sensitive and variably sensitive tumors. Further study of gene expression changes in FTI-treated or untreated rat mammary adenocarcinomas are necessary to identify any potential pharmacodynamic markers of FTI treatment in addition to any potential molecular targets of FTIs (Yao *et al.*, 2006).

D. Histone Deacetylase Inhibitors

Histone deacetylases (HDACs) are a class of enzymes that remove acetyl groups from an ϵ-*N*-acetyl lysine amino acid on histones or other proteins. Deacetylation alters the electric charge on the histone, which increases the affinity for DNA, and down-regulates transcription by blocking the access of transcription factors. Impaired histone acetylation occurs in the process of carcinogenesis. In addition, breast cancer cell differentiation is induced and tumor growth restrained when HDAC is inhibited. The protein expression of the genes HDAC-1 and -3 in breast tumors was analyzed immunohistochemically using a TMA. Using 600 core biopsies from 200 patients, the expression of HDAC-1 and -3 was correlated to the steroid hormone receptor Her-2/neu, to whether the tumors proliferated or not, and to disease-free and overall survival. It was discovered that the expression of HDAC-1 and -3 significantly correlated with the receptor expression of estrogen and progesterone. HDAC-1 was shown to be an independent prognostic marker after multivariate analysis was carried out, and its expression predicted which women would have significantly better chances of disease-free survival (DFS), especially in those who had small tumors of all differentiation types. The evaluation of HDAC-1 protein expression could be valuable for breast cancer patients because it enables a more precise assessment of their prognosis, and it may be clinically useful in assisting with the selection of tailor-made adjuvant systemic therapy (Krusche *et al.*, 2005).

E. DNA Methylases, Epigenetic Silencing, and Tumor Suppressors

Epigenetic mechanisms silence many tumor-suppressor genes, and this realization has spurred the assessment of novel tumor-suppressor genes. In one example, microarray analysis, among other techniques, was used to search for genes that are epigenetically silenced in human endometrial cancers. The endometrial cancer cell line Ishikawa exhibited changes in global gene expression that suggested the tazarotene-induced gene-1 (Tig1) and

CCAAT/enhancer-binding protein-α (C/ebpα) might function as tumor suppressor proteins. Forced expression of either Tig1 or C/ebpα resulted in significant reduction of the growth of Ishikawa cells, highlighting a potential role for these genes in endometrial cancer (Takai *et al.*, 2005).

Gene expression silencing via methylation may play a role in gastric cancers. The gene RUNX3 appears to contribute to human gastric carcinogenesis, but its role in progression and metastastis is not clear. Clinical samples of peritoneal metastases from gastric cancers were characterized for RUNX3 expression. Stable RUNX3 transfectants of gastric cancer cells were used in animal experiments so that changes in metastatic potential could be assessed. A cDNA microarray was used to analyze changes in global expression. Primary tumors and peritoneal metastases of gastric cancers exhibited significant downregulation of RUNX3 through methylation of the promoter region. Cell proliferation was inhibited slightly when RUNX3 was transfected, with modest antiproliferative and apoptotic effects observed when induced with transforming growth factor-β (TGF-β). In an animal model, RUNX3 strongly inhibited peritoneal metastases of gastric cancers. A cDNA microarray was used to globally analyze expression profiles of $\sim$21,000 genes in cells stably transfected with RUNX3, and 28 genes were identified that could possibly be under the downstream control of RUNX3 and thus possibly involved in various aspects of peritoneal metastases (Sakakura *et al.*, 2005).

In order to understand the genetic basis of Wilms tumor, the most frequent renal neoplasm in children, cDNA microarray experiments were performed using 63 primary Wilms tumors. It was hoped that new candidate genes would be detected that are associated with malignancy and tumor progression. In this study, all tumors were treated with preoperative chemotherapy as mandated by the SIOP protocol. Clear differences in expression were seen when relapse-free tumors were compared to relapsed tumors, and also when tumors with intermediate risk were compared to high-risk tumors. Microarray data uncovered several differentially expressed genes that were associated with the progression of Wilms tumor. The observation was made that in advanced tumors, the retinoic acid pathway was deregulated at different levels, leading to the hypothesis that the retinoic acid pathway might be perturbed in aggressive Wilms tumors (Zirn *et al.*, 2006).

V. IDENTIFICATION OF THERAPEUTIC TARGETS IN DISTINCT DISEASE SUBTYPES

A. Leukemias and Lymphomas

Diffuse large B-cell lymphoma (DLBCL) is the most common subtype of aggressive non-Hodgkin's lymphoma. There are biologically distinct subgroups within DLBCL, as shown by microarray gene expression studies, and different subtypes have been correlated with outcome of the disease (Alizadeh *et al.*, 2000; Shipp *et al.*, 2002). The 5-year overall survival rates range from 26 to 73% and those with limited-stage disease are treated with brief chemotherapy and radiation, while those with advanced disease are treated with extended chemotherapy. Gene expression profiling should enable additional identification of lymphoma-specific therapeutic targets and allow for novel agents to be developed that can be specifically tailored to the patient and disease (Sehn and Connors, 2005). An early example of such an approach includes the identification of the NF-κB pathway in a subset of DLBCL through the use of cDNA microarray analysis. In this study, multiple genes activated by the NF-κB pathway were identified as highly expressed in a subtype of DLBCL with subsequent demonstration that this pathway was indeed highly active DLBCL cells (Davis *et al.*, 2001). A similar approach was used to identify phosphodiesterase 4B as a potential therapeutic target in DLBCL (Smith *et al.*, 2005). Clinical trials are underway to assess agents that inhibit these pathways.

In the study of acute myeloid leukemia (AML), research has been conducted to attempt to define biologically and clinically relevant entities. There are well-defined cytogenetic subgroups, which display considerable heterogeneity, and many subtypes exist for which the pathogenic event is still not known. Using DNA microarray technology to survey the expression levels of thousands of genes in parallel has proven valuable to diagnose different cytogenetic subtypes, ascertain novel AML subclasses, and attempt to predict clinical outcome (Bullinger *et al.*, 2004; Ross *et al.*, 2004; Valk *et al.*, 2004). The AML gene expression profiling studies done so far exhibit a remarkable level of concordance which may enable an increasingly refined molecular taxonomy to be developed. Further insights into the pathobiologic nature of AML may be possible if gene expression profiling is utilized with other microarray-based applications, high-throughput mutational analyses, and proteomic strategies (Bullinger and Valk, 2005).

FLT3 mutations are one of the most common genetic abnormalities in AML. The presence of these mutations has been shown to be an independent prognostic indicator of poor outcome and response to standard chemotherapeutic interventions (Gilliland and Griffin, 2002). Thus, when FLT3 was identified by cDNA microarray analysis as highly expressed in specific subtypes of childhood leukemias, this prompted further assessment of FLT3 as a potential therapeutic target in those diseases (Armstrong *et al.*, 2002) (Fig. 3). Subsequent studies have demonstrated the presence of activating FLT3 mutations in the subsets of childhood leukemias with particularly high-level expression of FLT3, and FLT3 inhibitors effectively treat the leukemias *in vitro* and in model systems (Armstrong *et al.*, 2003; Brown *et al.*, 2005; Stam *et al.*, 2005) (Fig. 4). In a subsequent study, nine adult lymphoid leukemia patients were evaluated using oligonucleotide microarray analysis to further elucidate the

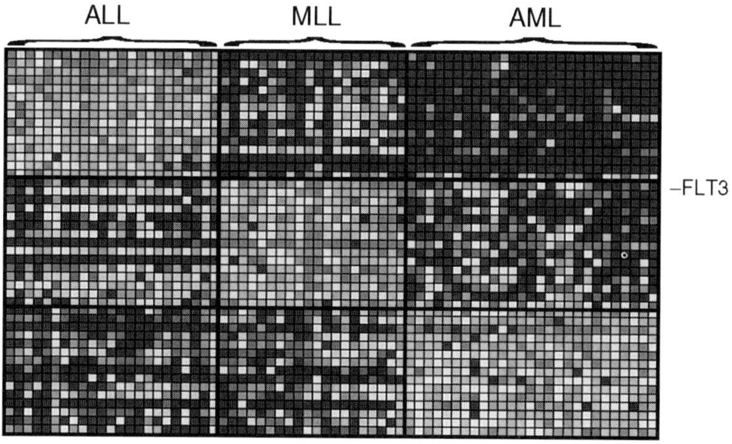

Fig. 3 Gene expression profiling identifies the receptor tyrosine kinase FLT3 as highly expressed in *MLL*-rearranged ALL. Gene expression profiles of acute lymphoblastic leukemias with rearrangement of the *MLL* gene on chromosome 11 were compared to other acute leukemias. The receptor tyrosine kinase FLT3 was identified as consistently highly expressed. Image taken from Armstrong *et al.* (2002). (See Color Insert.)

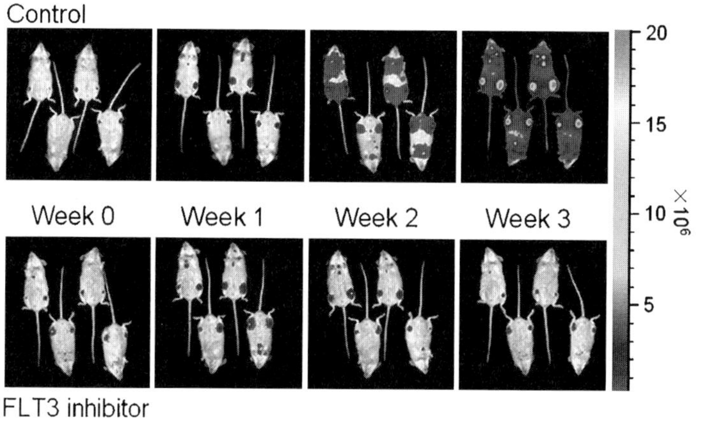

Fig. 4 A FLT3 inhibitor is effective in a model of *MLL*-rearranged ALL. Based on the high-level expression of FLT3 in this subset of acute leukemias, the FLT3 inhibitor was tested in a luminescent model of leukemia progression. The inhibitor showed significant activity against the human leukemia cells in this model, thus leading to the development of clinical trials that will assess FLT3 inhibitors in this poor prognosis leukemia. Image taken from Amstrong *et al.* (2003). (See Color Insert.)

pathogenetic role of the FLT3 protein. Increased FLT3 mRNA levels were measured, which was corroborated by increased protein expression seen on Western blots. When the effect of FLT3 inhibition on signal transduction and survival was investigated, it was seen that the FLT3 inhibitor decreased the survival of leukemia cells when compared to untreated cells (Torelli *et al.*, 2005). The FLT3 protein may play an important role in those acute lymphoblastic leukemia (ALL) patients with a particularly poor prognosis and specific inhibition of FLT3 may prove to be an innovative therapeutic tool for treating certain subsets of ALL. Thus, the presence of high-level expression of a receptor tyrosine kinase in specific subsets of leukemia can direct one to a potential therapeutic target for the disease. Based on these findings, an FLT3 diagnostic assay tests for two mutations, the internal tandem duplication (ITD) and D835 mutations, was developed by Genzyme (Cambridge, MA). Somatic mutations of FLT3 consisting of ITD occur in approximately 20–30% of patients with AML and a missense mutation at aspartic acid residue 835 occurs ~7% of AML (Gilliland and Griffin, 2002). Detection of these mutations is clinically significant because patients with these mutations have a poor prognosis, and they may benefit from a proactive, more aggressive intervention treatment. While advances in the treatment of AML have resulted in improved remission rates (60–70% of adults achieve a complete response), more than 50% of these patients will relapse and the long-term survival is less than 40%.

In the study of ALL, microarray analysis has permitted further investigation into the specific patterns which are connected with known molecular abnormalities, phenotypic characteristics, and prognostic features (Yeoh *et al.*, 2002). In the study of chronic lymphocytic leukemia (CLL), this approach has illustrated that there are distinct variants of this disease characterized by different mutations in IgVH (Rosenwald *et al.*, 2001). Much like the studies referenced in previous sections for ALL, further information about expression patterns may improve our ability to accurately predict patient outcome, and identify new methods for treatment (Chiaretti *et al.*, 2005).

A unique approach combined cDNA microarrays, 2D gel electrophoresis and computational biology to study the effects of retinoic acid (RA) and arsenic trioxide (ATO) on acute promyelocytic leukemia (APL) cells. The synergy between these two agents can result in DFS for patients whose leukemia was refractory to conventional chemotherapy. The experimental results indicated that relevant molecular networks, including granulocytic differentiation and apoptosis, are coordinately regulated. Combined approach such as this may provide more detailed insight into drug response and resistance (Zheng *et al.*, 2005).

B. Breast Cancer

In breast cancer patients, overexpression of the HER2 gene is linked to tamoxifen resistance, and patients overexpressing HER1, HER2, and HER3 have reduced survival rates. TMA analysis has shown that the status of HER1–HER3 and progesterone receptor (PR) can be used to predict which estrogen receptor (ER)-positive tamoxifen-treated patients will have an early relapse. When HER1–HER3 (but not HER4) were overexpressed, there was an early relapse for those being treated with tamoxifen. Those cases that were PR-negative were also more likely to relapse while being treated with tamoxifen, and both of these treatment groups were considered high risk. It was concluded that this information applies only to early relapse while on tamoxifen; after 3 years of tamoxifen treatment, any disease relapse was not related to HER/PR status (Tovey *et al.*, 2005).

The presence of tumor-specific vascular endothelial growth factor A (VEGF-A) may be a prognostic indicator for tamoxifen response in breast cancer patients, although it is presently difficult to avoid conflicting results. Tumors were analyzed from postmenopausal breast cancer patients who had been treated for 2 years with tamoxifen or had received no treatment. A microarray system was utilized to analyze VEGF-A and its receptor, vascular endothelial growth receptor 2 (VEGFR2), in parallel with angiogenic factors and hormone receptor status. The expression of VEGF-1 and PR negativity was strongly correlated with tumor-specific expression of VEGFR2; VEGF-A was not shown to be associated with hormone receptor status. For those patients who were ER positive, there was a statistically significant response to tamoxifen in terms of DFS at 10 and 18 years if their tumors were VEGF-A negative, but in women whose tumors were VEGF-A positive, tamoxifen had no beneficial effect. For those women with ER fraction greater than 10%, the status of VEGFR2 did not foretell any information regarding tamoxifen response. However, in women whose ER fraction was greater than 90%, the status of both VEGF-A and VEGFR2 was associated with the augmented response to tamoxifen treatment (Ryden *et al.*, 2005).

Breast cancer patients who are steroid hormone receptor-positive exhibit resistance to endocrine therapy, and resistance may be associated with DNA methylation of gene promoters, a gene silencing mechanism that shuts off tumor suppressors in cancer (see preceding sections). To investigate any relationship between promoter DNA methylation and resistance to endocrine therapy in recurrent breast cancer, a microarray-based technology was used to study steroid hormone receptor-positive tumors in patients who were treated with tamoxifen. Of the 117 genes analyzed, it was seen that the promoter DNA methylation status of 10 of them was meaningfully associated with tamoxifen therapy clinical outcome. There was favorable clinical outcome associated with the promoter hypermethylation of the strongest marker, phosphoserine aminotransferase (PSAT1), and the more PSAT1 was methylated, the less it was expressed at the mRNA level. Thus, if promoter hypermethylation and low mRNA expression occurs with PSAT1, this could be an indicator as to the success of tamoxifen therapy when it is administered to steroid hormone receptor-positive breast cancer patients. While all such studies as those described here require confirmation, there is hope that detailed assessment of breast cancer cells will not only allow identification of those that will/would not respond to therapy, but open up new avenues for therapy (Martens *et al.*, 2005).

VI. CONCLUSIONS

Microarray technology has become a valuable tool for screening genetic material in order to quickly identify global transcriptional states. Analysis of gene expression patterns will likely facilitate a better understanding of the molecular mechanisms of cancer pathogenesis. Histological analysis may be implemented on TMAs, gene dosage alterations associated with disease may be detected using mCGH, and SMMs are used to screen for potential drug development candidates.

Microarray technology has made substantial contributions to cancer research in recent years. It is particularly suited for identifying the genes, and gene expression profiles associated tumor subtypes and tumor progression. All pathways involved in tumorigenesis may be studied simultaneously, including apoptosis, senescence, cell adhesion, and proliferation. The role that genes play in metastasis may be explored at the molecular level, and receptor and protein-signaling data may be quickly gathered.

Microarrays allow screening for possible toxic reactions to pharmaceuticals, antibody drugs, chemotherapy drugs, or radiation. Drug resistance in numerous cancers has been studied at the cellular level utilizing microarray technology in order to identify possible therapeutic gene targets. Microarray-based gene expression analysis may be useful for the prognostic identification of which adjuvant therapy should be recommended for cancer patients, and the potential response. Further detailed studies of newer drugs, such as imatinib mesylate, erlotinib, and FTIs, have been implemented with microarrays and may help guide their use.

To date, microarray analysis has had its greatest impact in the characterization of leukemias and lymphomas. This is due in large part to the ease with which a pure population of tumor cells can be obtained. Also, detailed study of hematological malignancy over the past 30 years provided a strong foundation on which microarray studies could be developed. It is now clear that the different recurrent genetic abnormalities found in leukemias and lymphomas can be correlated with specific gene expression profiles, and in some cases, those profiles can be used to identify new therapeutic targets such as FLT3 in mixed lineage leukemia (MLL)-rearranged leukemias. However, for these gene expression profiles to be fully understood, much work will need to be done in model systems of disease.

REFERENCES

Alizadeh, A. A., Eisen, M. B., Davis, R. E., Ma, C., Lossos, I. S., Rosenwald, A., Boldrick, J. C., Sabet, H., Tran, T., Yu, X., Powell, J. I., Yang, L., *et al.* (2000). Distinct types of diffuse large B-cell lymphoma identified by gene expression profiling. *Nature* **403**, 503–511.

Armstrong, S. A., Staunton, J. E., Silverman, L. B., Pieters, R., den Boer, M. L., Minden, M. D., Sallan, S. E., Lander, E. S., Golub, T. R., and Korsmeyer, S. J. (2002). MLL translocations specify a distinct gene expression profile that distinguishes a unique leukemia. *Nat. Genet.* **30**, 41–47.

Armstrong, S. A., Kung, A. L., Mabon, M. E., Silverman, L. B., Stam, R. W., Den Boer, M. L., Pieters, R., Kersey, J. H., Sallan, S. E., Fletcher, J. A., Golub, T. R., Griffin, J. D., *et al.* (2003). Validation of a therapeutic target identified by gene expression based classification. *Cancer Cell* **3**, 173–183.

Bidaut, G., Suhre, K., Claverie, J. M., and Ochs, M. F. (2006). Determination of strongly overlapping signaling activity from microarray data. *BMC Bioinform.* **7**, 99.

Boyer, J., Allen, W. L., McLean, E. G., Wilson, P. M., McCulla, A., Moore, S., Longley, D. B., Caldas, C., and Johnston, P. G. (2006). Pharmacogenomic identification of novel determinants of response to chemotherapy in colon cancer. *Cancer Res.* **66**, 2765–2777.

Brown, P., Levis, M., Shurtleff, S., Campana, D., Downing, J., and Small, D. (2005). FLT3 inhibition selectively kills childhood acute lymphoblastic leukemia cells with high levels of FLT3 expression. *Blood* **105**, 812–820.

Bullinger, L., and Valk, P. J. (2005). Gene expression profiling in acute myeloid leukemia. *J. Clin. Oncol.* **23**, 6296–6305.

Bullinger, L., Dohner, K., Bair, E., Frohling, S., Schlenk, R. F., Tibshirani, R., Dohner, H., and Pollack, J. R. (2004). Use of gene-expression profiling to identify prognostic subclasses in adult acute myeloid leukemia. *N. Engl. J. Med.* **350**, 1605–1616.

Burczynski, M. E., Oestreicher, J. L., Cahilly, M. J., Mounts, D. P., Whitley, M. Z., Speicher, L. A., and Trepicchio, W. L. (2005). Clinical pharmacogenomics and transcriptional profiling in early phase oncology clinical trials. *Curr. Mol. Med.* **5**, 83–102.

Callagy, G., Pharoah, P., Chin, S. F., Sangan, T., Daigo, Y., Jackson, L., and Caldas, C. (2005). Identification and validation of prognostic markers in breast cancer with the complementary use of array-CGH and tissue microarrays. *J. Pathol.* **205**, 388–396.

Chiaretti, S., Ritz, J., and Foa, R. (2005). Genomic analysis in lymphoid leukemias. *Rev. Clin. Exp. Hematol.* **9**, E3.

Clarke, P. A., te Poele, R., and Workman, P. (2004). Gene expression microarray technologies in the development of new therapeutic agents. *Eur. J. Cancer* **40**, 2560–2591.

Davis, R. E., Brown, K. D., Siebenlist, U., and Staudt, L. M. (2001). Constitutive nuclear factor kappaB activity is required for survival of activated B cell-like diffuse large B cell lymphoma cells. *J. Exp. Med.* **194**, 1861–1874.

Druker, B. J., Talpaz, M., Resta, D. J., Peng, B., Buchdunger, E., Ford, J. M., Lydon, N. B., Kantarjian, H., Capdeville, R., Ohno-Jones, S., and Sawyers, C. L. (2001). Efficacy and safety of a specific inhibitor of the BCR-ABL tyrosine kinase in chronic myeloid leukemia. *N. Engl. J. Med.* **344**, 1031–1037.

Evans, W. E., and Relling, M. V. (1999). Pharmacogenomics: Translating functional genomics into rational therapeutics. *Science* **286**, 487–491.

Febbo, P. G., Thorner, A., Rubin, M. A., Loda, M., Kantoff, P. W., Oh, W. K., Golub, T., and George, D. (2006). Application of oligonucleotide microarrays to assess the biological effects of neoadjuvant imatinib mesylate treatment for localized prostate cancer. *Clin. Cancer Res.* **12,** 152–158.

Gilliland, D. G., and Griffin, J. D. (2002). The roles of FLT3 in hematopoiesis and leukemia. *Blood* **100,** 1532–1542.

Holleman, A., Cheok, M. H., den Boer, M. L., Yang, W., Veerman, A. J., Kazemier, K. M., Pei, D., Cheng, C., Pui, C. H., Relling, M. V., Janka-Schaub, G. E., Pieters, R., *et al.* (2004). Gene-expression patterns in drug-resistant acute lymphoblastic leukemia cells and response to treatment. *N. Engl. J. Med.* **351,** 533–542.

Jain, K. K. (2005). Applications of AmpliChip CYP450. *Mol. Diagn.* **9,** 119–127.

Krusche, C. A., Wulfing, P., Kersting, C., Vloet, A., Bocker, W., Kiesel, L., Beier, H. M., and Alfer, J. (2005). Histone deacetylase-1 and -3 protein expression in human breast cancer: A tissue microarray analysis. *Breast Cancer Res. Treat.* **90,** 15–23.

Kung, C., Kenski, D. M., Dickerson, S. H., Howson, R. W., Kuyper, L. F., Madhani, H. D., and Shokat, K. M. (2005). Chemical genomic profiling to identify intracellular targets of a multiplex kinase inhibitor. *Proc. Natl. Acad. Sci. USA* **102,** 3587–3592.

Losson, R., and Lacroute, F. (1979). Interference of nonsense mutations with eukaryotic messenger RNA stability. *Proc. Natl. Acad. Sci. USA* **76,** 5134–5137.

Mallory, J. C., Crudden, G., Oliva, A., Saunders, C., Stromberg, A., and Craven, R. J. (2005). A novel group of genes regulates susceptibility to antineoplastic drugs in highly tumorigenic breast cancer cells. *Mol. Pharmacol.* **68,** 1747–1756.

Maquat, L. E. (2005). Nonsense-mediated mRNA decay in mammals. *J. Cell Sci.* **118,** 1773–1776.

Martens, J. W., Nimmrich, I., Koenig, T., Look, M. P., Harbeck, N., Model, F., Kluth, A., Bolt-de Vries, J., Sieuwerts, A. M., Portengen, H., Meijer-Van Gelder, M. E., Piepenbrock, C., *et al.* (2005). Association of DNA methylation of phosphoserine aminotransferase with response to endocrine therapy in patients with recurrent breast cancer. *Cancer Res.* **65,** 4101–4117.

Mauriac, L., Debled, M., and MacGrogan, G. (2005). When will more useful predictive factors be ready for use? *Breast* **14,** 617–623.

Niyaz, Y., Stich, M., Sagmuller, B., Burgemeister, R., Friedemann, G., Sauer, U., Gangnus, R., and Schutze, K. (2005). Noncontact laser microdissection and pressure catapulting: Sample preparation for genomic, transcriptomic, and proteomic analysis. *Methods Mol. Med.* **114,** 1–24.

Rao, A. S., Kremenevskaja, N., von Wasielewski, R., Jakubcakova, V., Kant, S., Resch, J., and Brabant, G. (2006). Wnt/beta-catenin signaling mediates antineoplastic effects of imatinibmesylate (gleevec) in anaplastic thyroid cancer. *J. Clin. Endocrinol. Metab.* **91,** 159–168.

Richardson, A., and Kaye, S. B. (2005). Drug resistance in ovarian cancer: The emerging importance of gene transcription and spatio-temporal regulation of resistance. *Drug Resist. Updat.* **8,** 311–321.

Rosenwald, A., Alizadeh, A. A., Widhopf, G., Simon, R., Davis, R. E., Yu, X., Yang, L., Pickeral, O. K., Rassenti, L. Z., Powell, J., Botstein, D., Byrd, J. C., *et al.* (2001). Relation of gene expression phenotype to immunoglobulin mutation genotype in B cell chronic lymphocytic leukemia. *J. Exp. Med.* **194,** 1639–1647.

Ross, M. E., Mahfouz, R., Onciu, M., Liu, H. C., Zhou, X., Song, G., Shurtleff, S. A., Pounds, S., Cheng, C., Ma, J., Ribeiro, R. C., Rubnitz, J. E., *et al.* (2004). Gene expression profiling of pediatric acute myelogenous leukemia. *Blood* **104,** 3679–3687.

Ryden, L., Stendahl, M., Jonsson, H., Emdin, S., Bengtsson, N. O., and Landberg, G. (2005). Tumor-specific VEGF-A and VEGFR2 in postmenopausal breast cancer patients with

long-term follow-up. Implication of a link between VEGF pathway and tamoxifen response. *Breast Cancer Res. Treat.* **89,** 135–143.
Sakakura, C., Hasegawa, K., Miyagawa, K., Nakashima, S., Yoshikawa, T., Kin, S., Nakase, Y., Yazumi, S., Yamagishi, H., Okanoue, T., Chiba, T., and Hagiwara, A. (2005). Possible involvement of RUNX3 silencing in the peritoneal metastases of gastric cancers. *Clin. Cancer Res.* **11,** 6479–6488.
Sehn, L. H., and Connors, J. M. (2005). Treatment of aggressive non-Hodgkin's lymphoma: A North American perspective. *Oncology (Williston Park)* **19,** 26–34.
Sharma-Oates, A., Quirke, P., and Westhead, D. R. (2005). TmaDB: A repository for tissue microarray data. *BMC Bioinform.* **6,** 218.
Shia, J., Klimstra, D. S., Li, A. R., Qin, J., Saltz, L., Teruya-Feldstein, J., Akram, M., Chung, K. Y., Yao, D., Paty, P. B., Gerald, W., and Chen, B. (2005). Epidermal growth factor receptor expression and gene amplification in colorectal carcinoma: An immunohistochemical and chromogenic *in situ* hybridization study. *Mod. Pathol.* **18,** 1350–1356.
Shibata, T., Uryu, S., Kokubu, A., Hosoda, F., Ohki, M., Sakiyama, T., Matsuno, Y., Tsuchiya, R., Kanai, Y., Kondo, T., Imoto, I., Inazawa, J., *et al.* (2005). Genetic classification of lung adenocarcinoma based on array-based comparative genomic hybridization analysis: Its association with clinicopathologic features. *Clin. Cancer Res.* **11,** 6177–6185.
Shipp, M. A., Ross, K. N., Tamayo, P., Weng, A. P., Kutok, J. L., Aguiar, R. C., Gaasenbeek, M., Angelo, M., Reich, M., Pinkus, G. S., Ray, T. S., Koval, M. A., *et al.* (2002). Diffuse large B-cell lymphoma outcome prediction by gene-expression profiling and supervised machine learning. *Nat. Med.* **8,** 68–74.
Smith, P. G., Wang, F., Wilkinson, K. N., Savage, K. J., Klein, U., Neuberg, D. S., Bollag, G., Shipp, M. A., and Aguiar, R. C. (2005). The phosphodiesterase PDE4B limits cAMP-associated PI3K/AKT-dependent apoptosis in diffuse large B-cell lymphoma. *Blood* **105,** 308–316.
Stam, R. W., den Boer, M. L., Schneider, P., Nollau, P., Horstmann, M., Beverloo, H. B., van der Voort, E., Valsecchi, M. G., de Lorenzo, P., Sallan, S. E., Armstrong, S. A., and Pieters, R. (2005). Targeting FLT3 in primary MLL-gene-rearranged infant acute lymphoblastic leukemia. *Blood* **106,** 2484–2490.
Takai, N., Kawamata, N., Walsh, C. S., Gery, S., Desmond, J. C., Whittaker, S., Said, J. W., Popoviciu, L. M., Jones, P. A., Miyakawa, I., and Koeffler, H. P. (2005). Discovery of epigenetically masked tumor suppressor genes in endometrial cancer. *Mol. Cancer Res.* **3,** 261–269.
Takata, R., Katagiri, T., Kanehira, M., Tsunoda, T., Shuin, T., Miki, T., Namiki, M., Kohri, K., Matsushita, Y., Fujioka, T., and Nakamura, Y. (2005). Predicting response to methotrexate, vinblastine, doxorubicin, and cisplatin neoadjuvant chemotherapy for bladder cancers through genome-wide gene expression profiling. *Clin. Cancer Res.* **11,** 2625–2636.
Torelli, G. F., Guarini, A., Porzia, A., Chiaretti, S., Tatarelli, C., Diverio, D., Maggio, R., Vitale, A., Ritz, J., and Foa, R. (2005). FLT3 inhibition in t(4; 11)+ adult acute lymphoid leukaemia. *Br. J. Haematol.* **130,** 43–50.
Tovey, S., Dunne, B., Witton, C. J., Forsyth, A., Cooke, T. G., and Bartlett, J. M. (2005). Can molecular markers predict when to implement treatment with aromatase inhibitors in invasive breast cancer? *Clin. Cancer Res.* **11,** 4835–4842.
Tzankov, A., Went, P., Zimpfer, A., and Dirnhofer, S. (2005). Tissue microarray technology: Principles, pitfalls and perspectives: Lessons learned from hematological malignancies. *Exp. Gerontol.* **40,** 737–744.
Urzua, U., Frankenberger, C., Gangi, L., Mayer, S., Burkett, S., and Munroe, D. J. (2005). Microarray comparative genomic hybridization profile of a murine model for epithelial ovarian cancer reveals genomic imbalances resembling human ovarian carcinomas. *Tumour Biol.* **26,** 236–244.

Uttamchandani, M., Walsh, D. P., Yao, S. Q., and Chang, Y. T. (2005). Small molecule microarrays: Recent advances and applications. *Curr. Opin. Chem. Biol.* **9,** 4–13.

Valk, P. J., Verhaak, R. G., Beijen, M. A., Erpelinck, C. A., Barjesteh van Waalwijk van Doorn-Khosrovani, S., Boer, J. M., Beverloo, H. B., Moorhouse, M. J., van der Spek, P. J., Lowenberg, B., and Delwel, R. (2004). Prognostically useful gene-expression profiles in acute myeloid leukemia. *N. Engl. J. Med.* **350,** 1617–1628.

van de Vijver, M. (2005). Gene-expression profiling and the future of adjuvant therapy. *Oncologist* **10**(Suppl. 2), 30–34.

Wachi, S., Yoneda, K., and Wu, R. (2005). Interactome-transcriptome analysis reveals the high centrality of genes differentially expressed in lung cancer tissues. *Bioinformatics* **21,** 4205–4208.

Warford, A. (2004). Tissue microarrays: Fast-tracking protein expression at the cellular level. *Expert Rev. Proteomics* **1,** 283–292.

Wilson, C., Yang, J., Strefford, J. C., Summersgill, B., Young, B. D., Shipley, J., Oliver, T., and Lu, Y. J. (2005). Overexpression of genes on 16q associated with cisplatin resistance of testicular germ cell tumor cell lines. *Genes Chromosomes Cancer* **43,** 211–216.

Wolf, M., Edgren, H., Muggerud, A., Kilpinen, S., Huusko, P., Sorlie, T., Mousses, S., and Kallioniemi, O. (2005). NMD microarray analysis for rapid genome-wide screen of mutated genes in cancer. *Cell Oncol.* **27,** 169–173.

Yang, S. X., Simon, R. M., Tan, A. R., Nguyen, D., and Swain, S. M. (2005). Gene expression patterns and profile changes pre- and post-erlotinib treatment in patients with metastatic breast cancer. *Clin. Cancer Res.* **11,** 6226–6232.

Yao, R., Wang, Y., Lu, Y., Lemon, W. J., End, D. W., Grubbs, C. J., Lubet, R. A., and You, M. (2006). Efficacy of the farnesyltransferase inhibitor R115777 in a rat mammary tumor model: Role of Ha-ras mutations and use of microarray analysis to identify potential targets. *Carcinogenesis* **27**(7), 1420–1431.

Yeoh, E. J., Ross, M. E., Shurtleff, S. A., Williams, W. K., Patel, D., Mahfouz, R., Behm, F. G., Raimondi, S. C., Relling, M. V., Patel, A., Cheng, C., Campana, D., *et al.* (2002). Classification, subtype discovery, and prediction of outcome in pediatric acute lymphoblastic leukemia by gene expression profiling. *Cancer Cell* **1,** 133–143.

Zheng, C., Li, L., Haak, M., Brors, B., Frank, O., Giehl, M., Fabarius, A., Schatz, M., Weisser, A., Lorentz, C., Gretz, N., Hehlmann, R., *et al.* (2006). Gene expression profiling of CD34+ cells identifies a molecular signature of chronic myeloid leukemia blast crisis. *Leukemia* **20**(6), 1028–1034.

Zheng, P. Z., Wang, K. K., Zhang, Q. Y., Huang, Q. H., Du, Y. Z., Zhang, Q. H., Xiao, D. K., Shen, S. H., Imbeaud, S., Eveno, E., Zhao, C. J., Chen, Y. L., *et al.* (2005). Systems analysis of transcriptome and proteome in retinoic acid/arsenic trioxide-induced cell differentiation/apoptosis of promyelocytic leukemia. *Proc. Natl. Acad. Sci. USA* **102,** 7653–7658.

Zirn, B., Hartmann, O., Samans, B., Krause, M., Wittmann, S., Mertens, F., Graf, N., Eilers, M., and Gessler, M. (2006). Expression profiling of Wilms tumors reveals new candidate genes for different clinical parameters. *Int. J. Cancer* **118,** 1954–1962.

The Application of siRNA Technology to Cancer Biology Discovery

Uta Fuchs and Arndt Borkhardt

Dr. von Haunersches Kinderspital, Ludwig Maximilians Universität München, München, Germany

I. Introduction
II. The Mechanism of RNAi
III. Transcriptional Gene Silencing by siRNAs
IV. siRNAs Delivery: Strategies and Difficulties
A. Synthetic siRNAs
B. DNA Expression Vectors
C. Viral Vectors
V. RNAi as Discovery Tool in Cancer Biology
VI. RNAi Screens
A. By Synthetic siRNAs
B. By Transfected Cell Assays
C. By siRNA Expression Vectors
VII. Limitations of siRNAs as Cancer Therapeutics: Not Related to Delivery Problems
VIII. Summary
References

RNA interference (RNAi) is a naturally occurring cellular defense mechanism against viral infections and transposon invasion. Short double-stranded RNA molecules, so-called small-interfering (si)RNAs, bind their complementary mRNA leading to the mRNA's degradation. During the past few years, RNAi has become a valuable tool for transient as well as stable repression of gene expression rendering the time-consuming production of knockout animals superfluous. In this chapter the usability of the RNAi technology in cancer research will be described, focusing on the application of large-scale screens for identification of new components in cancer-relevant signal pathways (e.g., p53, RAS). The screens are especially helpful in the detection of potential anticancer drug targets or siRNAs with therapeutic potential. © 2007 Elsevier Inc.

I. INTRODUCTION

In the past, homologous recombination was the method of choice to determine the function of a certain gene in mammals, for example, a putative oncogene or tumor suppressor gene. The production of such knockout

Advances in CANCER RESEARCH

0065-230X/07 $35.00
DOI: 10.1016/S0065-230X(06)96004-7

animals was often time consuming as well as expensive and in some cases the function of the targeted gene could not be determined due to unexpected early embryonic lethality of the knockout or due to redundant phenotypes. Alternatively, gene function studies have been performed using antisense technology or ribozymes which also have been examined for their therapeutic potential (Dykxhoorn *et al.*, 2003).

Since 1998, a completely new field of gene regulation has been unraveled involving small double-stranded RNA (dsRNA) molecules. The discovery that the injection of dsRNA into *Caenorhabditis elegans* triggered the silencing of the related gene started the investigation of gene silencing by RNA (Fire *et al.*, 1998). To date, various forms of silencing mediated by small RNAs have been reported, affecting transcription, mRNA turnover, or protein synthesis. These mechanisms are ancient as indicated by the degree of conservation between species and even kingdoms. The first hints that this additional layer of regulation exists were observed in plants and fungi (reviewed in Matzke and Matzke, 2004; Pickford *et al.*, 2002) but the mechanisms are now known to be functional in almost all eukaryotes.

One form of dsRNA-mediated gene silencing is known as RNA interference (RNAi), since it was first found to operate in cultured mammalian cells (Elbashir *et al.*, 2001a) it has become widely used as a research tool for gene-inactivation studies.

Especially in cancer research, RNAi holds great promises. Inactivation of oncogenes by RNAi allows for the induction of differentiation or apoptosis, whereas the inactivation of tumor suppressor genes contributes to a better understanding of tumorigenesis. The prospect of a therapeutic application of small-interfering RNAs (siRNAs) incited many investigators to validate this approach. Finally, RNAi screens for the identification of novel tumor suppressors or drug targets have been performed successfully, demonstrating the enormous potential of the RNAi technology in cancer research.

II. THE MECHANISM OF RNAi

In the RNAi process, long dsRNA is cleaved by a ribonuclease into fragments of 21–23 nucleotides in length exhibiting a two-nucleotide 3′-overhang, a characteristic feature of ribonuclease III enzymes. Soon the responsible enzyme was identified and termed Dicer (Bernstein *et al.*, 2001). Dicer is a member of an enzyme family containing a PAZ domain responsible for the nucleic acid binding by ribonucleases. This domain can also be found in a second group of RNAi-related proteins, the Ago protein family (Sontheimer, 2005). The small interfering RNAs (siRNAs) produced by Dicer are then channeled into RNA-induced silencing complex (RISC).

In this multienzyme complex, the duplex siRNA is unwound and the sense strand is abolished. The remaining antisense strand targets this activated RISC to the complementary mRNA by base pairing leading to an endonucleolytic cleavage at a single site, 10 nucleotides from the 5′ end of the siRNA/mRNA-duplex region (Elbashir *et al.*, 2001b) (Fig. 1). The presence of ATP accelerates RISC turnover probably by promoting siRNA unwinding (Haley and Zamore, 2004). It has been speculated that the selection of the antisense strand of the siRNA is dependent on the direction of Dicer processing (Sontheimer, 2005), an important consideration for the production of siRNAs from long dsRNAs *in vitro*. In extracts from human cells, target mRNA cleavage is mediated by the argonaute protein Ago2 (Liu *et al.*, 2004; Meister *et al.*, 2004), which tightly binds the small RNA (Martinez and Tuschl, 2004). Argonaute proteins are characterized by their possession of a PAZ as well as a PIWI domain. While PAZ domains are responsible for RNA binding, the PIWI domain mediates the interaction with Dicer. Different organisms possess different numbers of Ago proteins, although these proteins do not seem to be redundant (reviewed in Meister and Tuschl, 2004).

It is thought that RNAi evolved as a cellular defense mechanism against virus infection and it has been hypothesized that these mechanisms act as a kind of immune system of the genome, protecting against viruses and transposons. This proposal has been supported by the observation that several viruses have developed defense mechanisms against RNAi (Li *et al.*, 2002; Voinnet, 2001). As the immune system has to recognize an intruder as nonself, in the case of the genome's immune system the nonself feature could simply be dsRNA which does not occur in the cell naturally (Plasterk, 2002). In *C. elegans*, RNAi has been proposed as a protective mechanism against endogenous transposon activity, therefore contributing to the maintenance of genome stability (Tabara *et al.*, 1999).

A cell-to-cell spreading of the gene-silencing effect after application of siRNAs targeting a certain mRNA has been observed in several RNAi models (*C. elegans*, *Arabidopsis thaliana*) but has not yet been demonstrated in mammals. However, the molecule responsible for the spreading of the RNAi effect has been identified in *C. elegans* as the transmembrane protein SID-1. The structure of SID-1 is similar to those of the ABC family of transmembrane transporters for macromolecules (Winston *et al.*, 2002). Recently, the enhancement of siRNA uptake by expression of the human SID-1 homologue has been demonstrated in human pancreatic ductal adenocarcinoma cells (Duxbury *et al.*, 2005). If the fundamental pathways of RNAi spreading are conserved in mammals, ways might be found to induce systemic RNAi in higher eukaryotes in the near future. Clearly, this would dramatically increase the direct therapeutic potential of RNAi in the field of gene therapy.

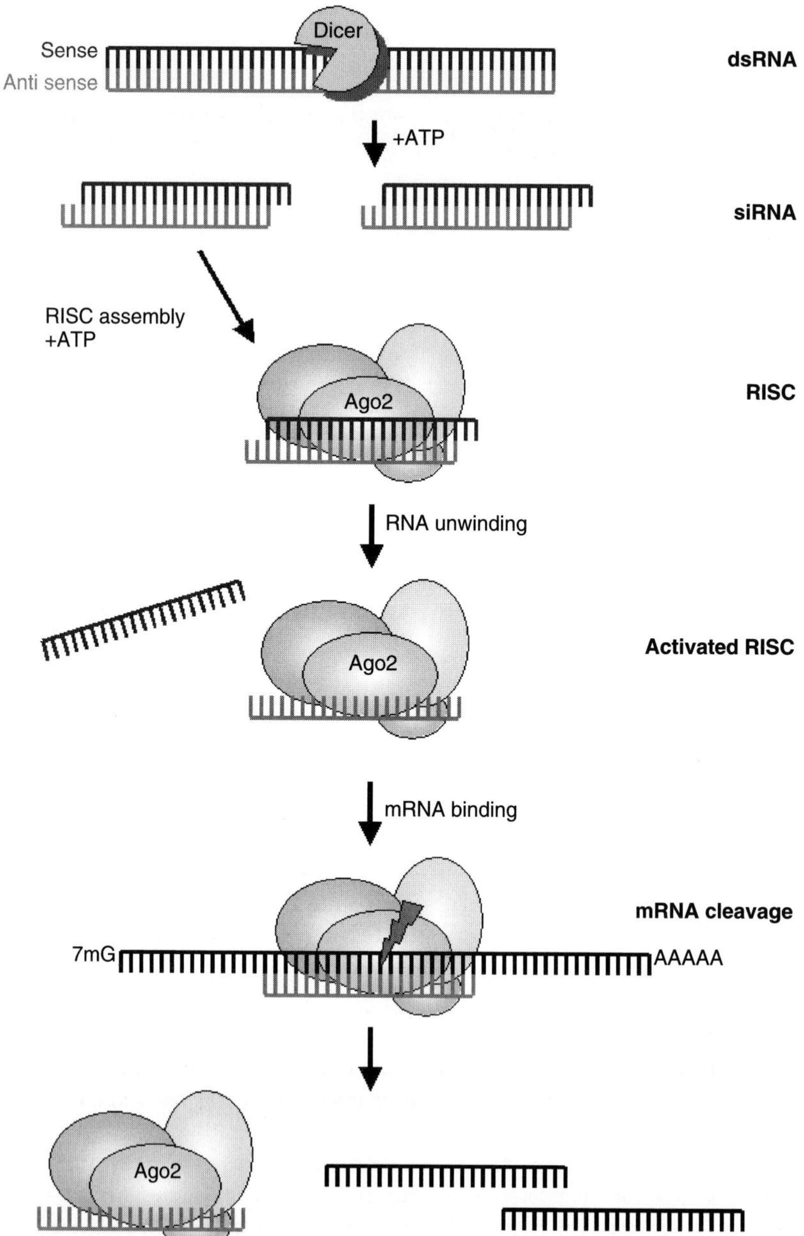

Fig. 1 Mechanism of RNA interference. Double-stranded RNAs are cleaved by Dicer into fragments of 21 nucleotides with two-nucleotide 3′-overhangs. These siRNAs are incorporated into RNA-induced silencing complex (RISC) and unwound. The antisense strand (red) guides the activated RISC complex to the target mRNA (black). Cleavage of the mRNA is performed by a not yet molecularly characterized endonuclease (slicer), the mRNA fragments are then exonucleatically digested. (See Color Insert.)

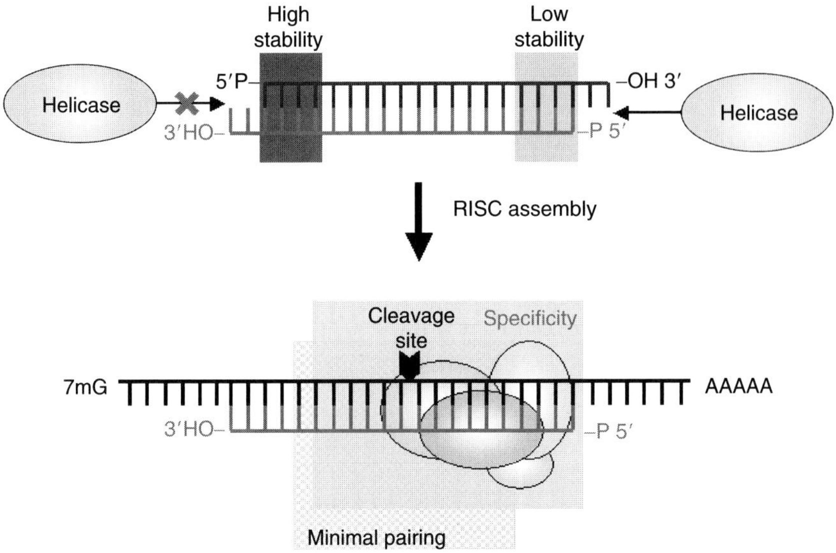

Fig. 2 Characteristics of efficient siRNAs. The thermodynamic stability of the first base pairs of either siRNA strand guides incorporation into RISC. The unwinding of the double strand starts at the end with low stability (light gray box). To ensure an efficient incorporation of the antisense strand into the RISC complex, the 5′ end of the antisense strand should lay in this region. The 5′ half of the siRNA has a greater importance for target recognition than the 3′ half of the molecule (red box). The central 13 base pairs are the minimal recognition area sufficient for mRNA cleavage (blue box) (Dorsett and Tuschl, 2004). (See Color Insert.)

RNAi in mammals is complicated by the cellular interferon response to dsRNA longer than ~30 bps (Stark *et al.*, 1998). Therefore, short dsRNAs have to be used to circumvent this mechanism. Since these individual siRNAs vary extremely in their efficiency to induce down-regulation of the target, several siRNAs have to be designed for each target gene and evaluated for their efficiency and the lack of side effects. Although rules have been established to design effective siRNAs (Fig. 2), secondary structures of the target mRNA or unknown factors render many siRNAs unable to initiate RNAi. The possible determinants of efficient silencing by siRNAs involve optimal incorporation of the siRNA into RISC and its stability in RISC, the perfect base pairing with the target mRNA, since mismatches can abolish target degradation (Wilda *et al.*, 2002), efficient cleavage of the mRNA, and the turnover rate of the complex after cleavage. With regard to the mRNA, the position of the siRNA chosen seems to be important since secondary and tertiary structures as well as mRNA-associated proteins can render an otherwise perfect target site inaccessible for siRNA binding (Vickers *et al.*, 2003). The abundance of an mRNA might influence the effectiveness of an siRNA by saturating the system and the targets' localization within the cell might be meaningful for efficient down-regulation. The rate of mRNA translation has been proposed to affect the gene silencing since the binding of multiple ribosomes to a single mRNA might block the binding of RISC (Dykxhoorn *et al.*, 2003).

Apart from that, the stimulation of the innate immune system by siRNAs has to be considered. Synthetic siRNAs formulated in a nonviral delivery system have been shown to potently induce interferons and inflammatory cytokines both *in vivo* in mice and *in vitro* in human blood. This immunostimulatory activity and the derived toxicity seemed to be dependent on the nucleotide sequence (Judge *et al.*, 2005). The induction of interferon-α (INF-α) in plasmacytoid dendritic cells (specialized antigen-presenting cells in the human body) by siRNAs was mediated by Toll-like receptor (TLR) 7. Interestingly, locked nucleic acid modifications at the 3′ but not at the 5′ end of the siRNA's sense strand strongly inhibited the induction of IFN-α (Hornung *et al.*, 2005). Sioud demonstrated that both the double-stranded molecule and the sense or antisense strand of a given siRNA had the ability to induce an inflammatory response in blood mononuclear cells, whereas in other cases only the single strands activated the response. Remarkably, both groups did not observe any immune-stimulating effects after delivery of the siRNAs by electroporation, indicating a direct link to the lipofection reagents used.

III. TRANSCRIPTIONAL GENE SILENCING BY siRNAs

Systemic suppression involves at least two distinct components, a post-transcriptional silencing mechanism due to mRNA degradation and a related transcriptional gene-silencing (TGS) mechanism. TGS seems to involve epigenetic modifications such as have been initially observed in plants. Here, targeting a promoter was shown to trigger RNA-directed DNA methylation and initiate transcriptional silencing accompanied by the production of siRNAs (Mette *et al.*, 1999).

siRNA-induced transcriptional silencing in human cells has been observed. The transcription of an integrated proviral GFP reporter gene construct as well as the transcription derived from the endogenous EF1A promoter has been silenced efficiently. The silencing was reversed by treating the cells with the inhibitor of histone deacetylases Trichostatin A and with 5-azacytidine, an inhibitor of DNA methyltransferases indicating that the siRNA induced epigenetic changes in the chromatin. Essential for these effects was the efficient delivery of the siRNA into the nucleus, in this case achieved by a peptide containing a nuclear targeting signal. In comparison, cytoplasmic delivery of the siRNA had no effect on promoter activity (Morris *et al.*, 2004). Others demonstrated in human cells that synthetic- as well as vector-based siRNAs were able to induce sequence-specific RNA-mediated DNA methylation. The methylation was dependent on methyltransferases DNMT1 and DNMT3B. In addition, methylation of histone H3 lysine 9 has been observed (Kawasaki and Taira, 2004). The authors used this methodology to target the erbB2 (HER-2/neu) promoter in a human breast adenocarcinoma cell line thereby inducing erbB2 promoter methylation and reduction of erbB2 transcription. This led to a significant reduction of cell proliferation rates. Similar observations have been made in human breast cancer cells after silencing erbB2 posttranscriptionally by siRNAs (Faltus *et al.*, 2004). This approach of targeting regulatory rather than coding regions might also provide a method for gene silencing in cancer biology research as well as in cancer therapy.

IV. siRNAs DELIVERY: STRATEGIES AND DIFFICULTIES

A. Synthetic siRNAs

The ideal delivery technique should be able to bind siRNAs in a reversible manner to permit the efficient release of the siRNA in the targeted cell, should protect the siRNA molecules from nucleases during transit, should be neither toxic nor immunogenic, and should be degradable but avoid fast clearance with regard to therapeutic applications (Rossi, 2005).

Chemically synthesized siRNAs have been the most broadly used method for induction of RNAi especially in cell culture models. Apart from that, *in vitro* transcribed expression cassettes as well as siRNAs isolated from *Drosophila* embryo protein extracts have been used successfully (Lipardi *et al.*, 2001; Yu *et al.*, 2002). Chemically synthesized siRNAs are convenient but of course nonrenewable and still expensive, both limiting factors for their application in genome-wide screens. Apart from that, the effect of these siRNAs is temporally limited. An additional problem to be faced is the delivery of these siRNAs to the target cells. Several specialized transfection reagents for small RNAs have been developed, but these reagents are expensive and the transfection efficiencies are often poor, especially when dealing with primary cells.

Covalent modifications of siRNAs have been tested to enhance the uptake into the target cell. For example, the attachment of cholesterol to one of the siRNA strands has facilitated the direct intravenous, low pressure injection of the molecules into mice. These siRNAs showed improved pharmacological properties *in vitro* and *in vivo* (Lorenz *et al.*, 2004). Conjugation of the cholesterol to the 3′ end of the siRNA's sense strand did not result in a significant loss of gene-silencing activity in cell culture as has been demonstrated by targeting the apoB mRNA *in vitro* and *in vivo* (Soutschek *et al.*, 2004). Recently, receptor-targeted siRNAs have also been proposed to overcome *in vivo* delivery problems. Heavy-chain antibody fragment (Fab) fusions that deliver noncovalently bound siRNAs via surface receptors were employed. The binding of the siRNA was achieved by protamine, a highly basic cellular protein fused to the Fab fragment. The small protamine (51 amino acids) molecule possesses a positive charge and binds the negatively charged siRNA readily. The Fab used was directed against extracellularly displayed HIV-1 envelope glycoprotein. The mixing of the conjugate with siRNAs and incubation of HIV-1 envelope protein expressing cells led to the efficient uptake of fluorescently labeled siRNAs by these cells. siRNAs targeting c-MYC, MDM2, and vascular endothelial growth factor (VEGF) were mixed with the protamine-Fab conjugate and applied intratumorally. A reduction of tumor volume was achieved, whereas the application of naked siRNA yielded no effect on tumor growth. Also many other ligands could be employed for receptor-mediated uptake of siRNAs using this protamine-fusion method. Modifications in the backbone of the siRNA could improve the binding by protamine and could therefore reduce the relative concentration of siRNAs necessary (Rossi, 2005).

In addition, a set of small molecules has been identified which enhance the transdermal penetration of macromolecules including oligonucleotides. These molecules could be potentially used for the transdermal delivery of siRNA (Dorsett and Tuschl, 2004; Karande *et al.*, 2004). Also electroporation has been applied in transfection of cells with siRNAs as an alternative to viral transduction and high-throughput electroporation devices have been developed to facilitate large-scale RNAi screens (Ovcharenko *et al.*, 2005; Thomas *et al.*, 2005; Weil *et al.*, 2002; Wilson *et al.*, 2003).

B. DNA Expression Vectors

The use of DNA vectors for endogenous production of siRNAs is helpful in solving at least some of the aforementioned problems. Vectors containing a U6 or H1 promoters have been applied for the intracellular expression of siRNAs. These promoters are bound by RNA polymerase III (pol III), the polymerase usually responsible for the synthesis of tRNAs, 5S rRNA, and certain snRNAs. The advantage of pol III promoters is the fact that they do not need any additional transcriptional elements. The termination signal for pol III consists of four to five thymidine residues, allowing for transcript length determination but also limiting these vectors to inserts without naturally occurring poly-T stretches. The transcripts terminate after the second thymidine resulting in a two nucleotide overhang identical to the overhang found in effective siRNAs. The design of the inserts for these vectors has to be performed carefully, since the transcripts have to form a stem-loop structure to be efficiently processed by Dicer into functional siRNAs. In addition, the specialization of pol III for short transcripts limits the length of the insert for efficient transcription (Brummelkamp *et al.*, 2002b; Miyagishi and Taira, 2002; Myslinski *et al.*, 2001).

Also, RNA pol II (usually responsible for the generation of mRNAs) promoters have been employed for the production of short hairpin RNAs. Stable pol II-mediated expression of a 500 nt hairpin structure induced RNAi in a murine embryonic cell line (Paddison *et al.*, 2002). Careful design of such a hairpin might facilitate the targeting of more than one mRNA and could therefore overcome the problem of resistance to silencing mediated by point mutations (Shuey *et al.*, 2002). An advantage of pol II promoters is the inducible, tissue- or cell-type-specific RNA expression as has been demonstrated by Kennerdell and Carthew (2000), applying a Gal4-inducible system for the expression of a hairpin RNA. In this system, the expression of the Gal4 transactivator was controlled by the heat shock protein 70 that allowed the induction of the system by a simple temperature shift (Kennerdell and Carthew, 2000). Hutvagner and Zamore (2002) demonstrated that the introduction of 100 nM of a hairpin construct into HeLa extracts resulted in the generation of $\sim$5 nM of Dicer-produced small RNA. This RNA in turn targeted the complementary mRNA as efficiently as 100 nM of a corresponding synthetic siRNA indicating a more efficient incorporation of Dicer-produced small RNAs into the downstream complex, RISC (Hutvagner and Zamore, 2002).

C. Viral Vectors

Retroviral vectors have been used to introduce siRNA expression constructs into cells. The vectors are based, in general, on oncoretroviruses as Moloney murine stem cell virus (MoMuLV) (Paddison and Hannon, 2002) or the murine stem cell virus (MSCV) (Brummelkamp *et al.*, 2002a). Also another group of retroviruses has been applied, namely lentiviruses. These lentiviral vectors have the advantage to infect both dividing and nondividing cells (Naldini *et al.*, 1996; Qin *et al.*, 2003). Unlike oncoretroviruses, lentiviral vectors do not undergo proviral silencing and can therefore be used for the generation of transgenic animals (Svoboda *et al.*, 2000). Lentiviral vectors have been used to generate transgene-based RNAi in mice, indicating that siRNA-mediated gene silencing functions in all cell types and tissues tested from early embryos to adult mice (Tiscornia *et al.*, 2003).

Obstacle of the application of viral vectors is the unpredictability of the virus integration site into the genome of the targeted cell. In the Paris' gene therapy trial for X-linked severe combined immunodeficiency, 3 out of 14 patients developed a treatment-related T-cell leukemia due to the accidental activation of the oncogenes like LMO2 and others, indicating that a tight control of viral integration has to be achieved before these viruses can be safely used for siRNA delivery in humans (Hacein-Bey-Abina *et al.*, 2003; Marshall, 2003) and unpublished own observation.

V. RNAi AS DISCOVERY TOOL IN CANCER BIOLOGY

Since its discovery, RNAi was widely used as a valuable tool in the examination of cancer biology. Several knockdown studies have been facilitated to determine the function of a gene in cancer biology without the need to generate knockout animals [e.g., DP97 in Rajendran *et al.* (2003), PLK1 in Spankuch-Schmitt *et al.* (2002), or DNMT1 in Robert *et al.* (2003)]. Many groups have focused on targets exhibiting a certain therapeutic potential of which some examples shall be mentioned here.

Since mutations of the p53 gene *TP53* are key events during tumorigenesis, particularly in solid tumors, and result in genomic instability, deregulation of the cell cycle and resistance to certain chemotherapies (Vousden and Lu, 2002), several groups have used the RNAi approach to target p53. Small hairpin RNAs targeting different sites of the p53 gene resulted in different levels of p53 knockdown in hematopoietic stem cells derived from mice aberrantly expressing the *myc* oncogene in their lymphocytes. Reconstitution of the immune system of lethally irradiated mice with these so-called p53-hypomorphs led to Myc-induced lymphomagenesis whose severity correlated with the degree of p53 silencing (Hemann *et al.*, 2003). Brummelkamp *et al.* (2002a) examined another key regulator in oncogenesis, the *RAS* oncogene. They used a viral expression vector for stable expression of an siRNA targeting the constitutively active form of RAS (RAS^{V12}) which differs from the wild type in a single nucleotide exchange. This construct was able to target RAS^{V12} without altering the levels of wild-type RAS and decreased the oncogenic potential of a pancreatic carcinoma cell line (Brummelkamp *et al.*, 2002a). Using siRNAs, the inhibition of oncogenic fusion genes in leukemias was one of the first applications of RNAi in the field of cancer. The *BCR/ABL* fusion gene, generated by translocation t(9;22) is consistently found in chronic myeloid leukemias (CML) and a smaller fraction of cases with acute lymphoblastic leukemia. The formed *BCR/ABL* hybrid gene directly triggers leukemogenesis and when inhibited the cells underwent apoptosis. These results were achieved by transient transfection as well as stable transduction of a BCR-ABL expressing cell line (Scherr *et al.*, 2003, 2005; Wilda *et al.*, 2002; Wohlbold *et al.*, 2003). Down-regulating the *TEL/PDGFRβ* fusion gene in transformed hematopoietic cells demonstrated the possibility to sensitize these cells to small molecule inhibitors by stable expression of the siRNA giving another encouragement for the development of a combinatorial therapy strategy (Chen *et al.*, 2004). The *RUNX1/CBFA2 T1* and *MLL/AF4* fusions also have successfully been targeted by siRNAs (Heidenreich *et al.*, 2003). The transient inhibition of *MLL/AF4* resulted in reduced proliferation and clonogenicity due to induction of apoptosis associated with caspase-3 activation and diminished BCL-X_L expression. The transfection of the leukemic cells with MLL/AF4 siRNAs reduced leukemia-associated mortality in a xenotransplant model (Thomas *et al.*, 2005). A draw back for the targeting of oncogenic fusion genes is the fact that the fusion sites, although recurring, differ slightly from patient to patient making a generalized siRNA therapy difficult (Damm-Welk *et al.*, 2003). In these cases, molecules downstream of the fusion products have to be targeted or, like *ABL* in *BCR/ABL*, a fraction of the fusion whose wild-type molecule is lacking in the cell (Lieberman *et al.*, 2003). Such a downstream approach was performed by targeting LYN kinase in drug resistant and *BCR/ABL* positive CML blast crisis cells. The exposition of K562 cells to anti-LYN siRNA resulted in an inhibition of cell proliferation and LYN-siRNA-treated lymphoid CML blasts underwent a rapid and massive induction of apoptosis (Ptasznik *et al.*, 2004).

Having in mind a generalized RNAi-based therapy, genes overexpressed in cancer are probably easier to target. The cell cycle regulator polo-like kinase 1 (PLK1) is overexpressed in various tumors therefore being an interesting target for the application of RNAi. Mice with human xenograft tumors were treated with plasmids expressing PLK1-shRNAs. The plasmids were treated with aurintricarboxylic acid to prevent nucleolytic degradation in the murine circulation. Indeed, these plasmids reduced tumor burden to 18% in comparison to mice treated with control plasmids. Plasmids without acid treatment showed a lower efficiency, indicating their degradation in the blood (Spankuch *et al.*, 2004). These experiments are of special interest since the mice were treated with small volumes of plasmid solutions without addition of transfection reagents administered by low pressure vein injection, a treatment strategy potentially transferable to the human system.

Overexpression of the multidrug resistance gene *MDR1* confers resistance to chemotherapeutics to cancer cells. Conversely, down-regulation of MDR1 reversed the *MDR* phenotype, for example, in the doxorubicin-resistant cell lines (Stege *et al.*, 2004; Yague *et al.*, 2004).

Migration and invasion of adjacent tissues is characteristic for the malignant phenotype of a cell, therefore proteins involved in cell migration

Table I Oncogenes Recently Targeted by RNAi

Oncogene	Cancer	Cell lines/animal models used	Mode of delivery	References
E6, E7	Cervix carcinoma	HeLa	Viral	(Putral *et al.*, 2005)
B-RAF	Melanoma	Melan-a	Lipofection	(Wellbrock *et al.*, 2004; Wellbrock and Marais, 2005)
K-RAS	Colorectal cancer	Murine C26	Viral	(Smakman *et al.*, 2005)
$p28^{GANK}$	Hepatocellular carcinoma	HepG2, Hep3B, HuH-7	Viral	(Li *et al.*, 2005)
CRD-BP/IMP1	Breast cancer	MCF-7	Lipofection	(Ioannidis *et al.*, 2005)
c-MYC	Breast cancer	MCF-7	Lipofection	(Wang *et al.*, 2005)
TPR-MET	Gastric cancer	Mouse embryonal fibroblasts and epithelial cells	Viral	(Taulli *et al.*, 2005)
N-RAS	Melanoma	224, BL, A375, 397	Lipofection	(Eskandarpour *et al.*, 2005)
CEACAM6	Pancreatic adenocarcinoma	BxPC3	Viral	(Duxbury *et al.*, 2004)
JPO2	Medulloblastoma	UW228	Viral	(Huang *et al.*, 2005)
PARP-1	Colorectal cancer	SW480, HCT116	Lipofection	(Idogawa *et al.*, 2005)
SKP2	Lung cancer	A549, H1792	Lipofection	(Jiang *et al.*, 2005)
FAS	Breast cancer	MCF-7/AdrR	Lipofection	(Menendez *et al.*, 2005)
VAV1	Pancreatic adenocarcinoma	AsPc1, BxPc3, Capan2, CFPAC1, PANC1	Electroporation	(Menendez *et al.*, 2005)
EWS/FLI-1	Ewing tumors	A673	Lipofection	(Prieur *et al.*, 2004)

processes have been targeted by RNAi. Knockdown of serine protease urokinase-type plasminogen activator (u-PA) resulted in a reduction of cell migration, invasion, and proliferation in human hepatocellular carcinoma-derived cell lines (Salvi *et al.*, 2004). Additional examples for oncogene deactivation by RNAi are listed in Table I.

VI. RNAi SCREENS

A. By Synthetic siRNAs

High-throughput gene expression studies have been performed primarily by the use of cDNA or oligonucleotide microarrays in which the expression of almost the complete genome can be analyzed simultaneously. The information produced by microarray analyses focus on the transcriptome and therefore does not provide any insights into the expression of proteins or their posttranslational modifications, often importantly altering the functional effects of gene expression. In contrast, RNAi screens can be used in two ways: either for the identification of the most potent siRNA sequence targeting a gene of interest or for the identification of a gene product responsible for certain aspects in cancer biology. The latter is accomplished by transfection of siRNA libraries followed by cell biological screens. Cells reacting strikingly can be analyzed for the corresponding siRNA and thereby for the gene/protein responsible for the reaction.

The analysis of protein function is especially important in cancer biology research since the majority of targets for drug development are proteins (Vanhecke and Janitz, 2005). Therefore, the identification of key factors in oncogenesis is the main area for application of RNAi screens. High-throughput assays to identify siRNAs specifically reducing mRNA and protein levels have been performed in 6- to 384-well-plate formats. The parallel measurement of the endogenous mRNA levels and protein levels of an exogenously expressed tagged gene rendered the possibilities to correctly predict siRNA molecules which efficiently reduce mRNA and protein levels and identify proteins exhibiting a long half-life and therefore being suboptimal for targeting by RNAi (Vanhecke and Janitz, 2005; Wu *et al.*, 2004).

A similar transfection assay employing well plates was developed to identify modulators of TRAIL-induced apoptosis. TRAIL has the ability to selectively kill tumor cells, although the precise molecular mechanism is not yet fully understood. Therefore, an siRNA library directed against 510 genes including most kinases was screened by measurement of viability and induction of apoptosis. Several modulators of cell growth were identified, capable of either enhancing or repressing cell viability in TRAIL-induced apoptosis. This led to the identification of two new members of TRAIL signaling, namely DOBI and MIRSA. DOBI is placed downstream of BID transducing the apoptotic signal to the mitochondria prior to cytochrome *c*, whereas MIRSA seemed to have a rather broad antiapoptotic function since its inhibition induced a low level of caspase activation in the absence of TRAIL (Aza-Blanc *et al.*, 2003). An opposed approach focused on human kinases and phosphatases in antiapoptotic survival pathways. The role of these molecules in apoptosis is not well understood and a functional screen could serve to identify new cell survival regulators and drug targets for chemotherapy. By large-scale RNAi kinases as well as phosphatases were identified essential for survival, including a new group of phosphatases with tumor suppressor characteristics. HeLa cervical carcinoma cells were transfected with two siRNAs targeting each of 650 known or putative kinases. Seventy-three survival kinases were found of which CDK6, RPS6KL1, ROR1, and NLK were the most potent with RPS6KL1 and ROR1 being novel with unknown function. In addition, 222 known phosphatases were screened indicating a set of survival phosphatases including serine/threonine protein phosphatases as well as protein tyrosine phosphatases. To identify phosphatases which normally sensitize cells to apoptosis (cell-death phosphatases) and whose loss of function would therefore lead to resistance to induced apoptosis, HeLa cells were transfected with a human phosphatase siRNA library and then treated with cisplatin, Taxol, or etoposide to induce cell death, thereby identifying MK-STYX as the most potent cell-death phosphatase. These kind of phosphatases are especially important due to their role as possible tumor suppressors, and a total of 12 phosphatases were shown to act in this way. Applying the same methodology in a breast carcinoma cell line showing additional survival pathways the findings could be confirmed, indicating a broad meaning of these results in cancer biology. In addition, it demonstrated synergistic effects achieved by the application of siRNAs along with chemotherapeutics, an approach especially interesting as a therapeutic strategy (MacKeigan *et al.*, 2005).

An elegant combination of microarray analysis and RNAi screen has been developed. Colon cancer cell lines were segregated phenotypically based on their c-SRC protein kinase activity, because an SRC-specific transformation expression profile has been observed in SRC-transformed rodent cells as well as in human colon tumors. Gene expression patterns were identified that statistically associated with highly or weakly transformed phenotypes (phenotypic anchored gene expression profiles) and up-regulated genes were subsequently validated for their role in facilitating invasiveness by an RNAi screen. By this strategy, the first two levels of gene transcription responsible for invasiveness were identified, the first level directly controlled by SRC activity and the second level controlled by the first level, giving a nice example for an RNAi screen providing a better understanding of the biology underlying the SRC-mediated transformation (Irby *et al.*, 2005).

B. By Transfected Cell Assays

An alternative for these conservative transfection assays are so-called transfected cell arrays (TCAs), a method based on reverse transfection (Fig. 3). Expression vectors or siRNAs in gelatin solution are printed at high density on a modified glass slide along with a lipid transfection reagent. The array is then placed in tissue culture dishes, and cultured cells in medium are added to the array in a sufficient number to achieve near confluence at the end of the experiment. The array becomes covered with a monolayer of recipient cells and transfection efficiency is measured by either cotransfection of reporter plasmids or fluorochrome conjugated siRNAs (Mousses *et al.*, 2003; Silva *et al.*, 2004; Wu *et al.*, 2004). This technique has been used to perform large-scale loss-of-function studies in which synthetic siRNAs as well as shRNA expression vectors have been applied successfully (Kumar *et al.*, 2003; Mousses *et al.*, 2003; Silva *et al.*, 2004; Wu *et al.*, 2004). Gene silencing can be monitored by fluorescently labeled antibodies leading to the detection of dark patches in the cell layer corresponding to certain siRNA spots. Alternatively, phosphorylation-sensitive antibodies for transcription factors or receptors can be applied to monitor the activation status of these molecules with respect to a certain siRNA. This is especially interesting when searching for previously unknown members of cell-signaling cascades or protein kinases or phosphatases as drug targets (Gschwind *et al.*, 2004; van Huijsduijnen *et al.*, 2002). Induction of apoptosis by certain siRNAs can be monitored on the slides by reagents like Annexin V, whereas changes in cell morphology can be inspected directly on the array.

The TCA technique is still at the beginning and certain improvements have to be realized to render this technology applicable for the majority of researchers. Since up to 6000 distinct spots of siRNAs or expression vectors have been reported to be spotted on one slide, and depending on the cell line used, each spot can be covered by 30–500 transfected cells, high resolution imaging with automated image analysis is necessary to examine the slides. To date, no adequate cell-array specialized devices to perform these analysis is commercially available. Researchers therefore have to adjust automated microscopy and even cDNA-microarray readers to acquire the cell-array information. Applying molecules as small as siRNAs, an accurate spotting is also essential for the detection of effects. Nevertheless, the technology has the ability to reduce the amount of effort necessary for rapid cell-based RNAi screens (Wheeler *et al.*, 2005).

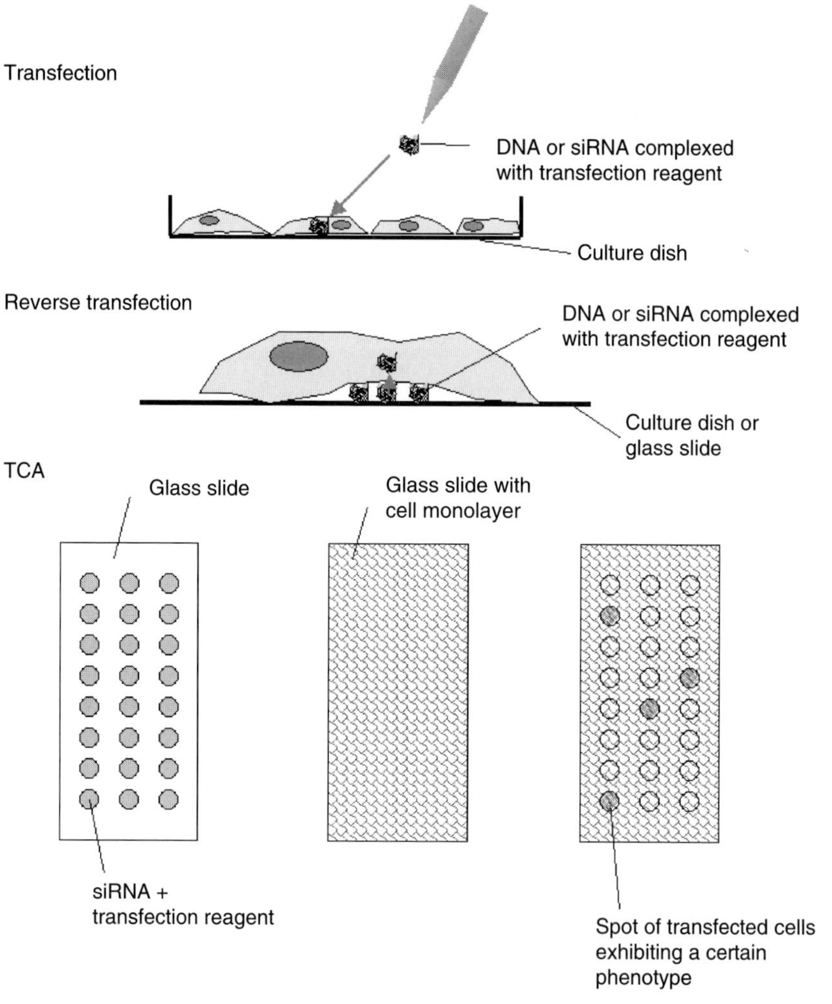

Fig. 3 Transfected cell array (TCA) technology. In normal transfection (upper panel), cultured cells are overlaid with a solution containing DNA or RNA complexed with a transfection reagent. In reverse transfection (middle panel), cells are grown on a solid support covered with nucleic acids complexed with the transfection reagent. TCAs (lower panel) base on reverse transfection. Plasmid DNA or siRNA are spotted along with a transfection reagent on glass slides. Cells are grown on these slides until confluence and examined for a certain phenotype (e.g., surface marker expression, apoptosis, and so on) (Vanhecke and Janitz, 2005).

Although the majority of reports using TCAs focused on adherent cells, the method may be applicable also to nonadherent cells, for example, for cells derived from hematological malignancies. For this purposes, glass slides were modified with a cell anchoring reagent which immobilized the human erythroleukemic cell line K562 on the slide. A transfection of these cells by cDNA expression vectors as well as siRNAs was demonstrated. Since the modification of the slides is easy to perform and even a manual spotting of the transfection mixture was demonstrated this approach seems to be easily transferable to any oncological research laboratory (Kato *et al.*, 2004). It is possible to adjust the TCA technology for simultaneous transfection of siRNAs targeting two different genes for identification of synthetic phenotypes. Similar approaches will be useful for synthetic lethal experiments in which mutation/knockdown of either gene alone is compatible with viability but mutation/knockdown of both leads to death. Targeting a gene that is synthetic lethal to a cancer-relevant mutation by this method should kill only cancer cells and spare normal cells therefore being especially interesting for the development of cancer-specific cytotoxic agents (Kaelin, 2005). The treatment of cells with a drug or RNAi reagent prior to adding them to the cell microarray will be useful for epistatic analysis. Screening of gene—small molecule drug combinations or pairs of genes that confer lethality to the cell when knocked down in concert could be helpful in the search for new cancer drug targets. By cotransfection of cDNA expression vectors and siRNAs, genes could be identified whose knockdown yields lethality in cells overexpressing a certain oncogene therefore providing an insight into new possibilities for cancer therapies (Wheeler *et al.*, 2005).

C. By siRNA Expression Vectors

As an alternative to siRNA, screens in array format pooled siRNA libraries can be used. In array screens, each siRNA sequence is tested individually whereas in pooled library screens, the effect of silencing multiple genes in one cell can be examined, reducing the number of assays which have to be performed. In this type of screen shRNA vectors have been applied for their ability to be identified by a unique DNA barcode provided by the siRNA sequence within the vector insert. The barcodes are detected through hybridization with microarrays containing oligonucleotide probes complementary to these barcodes, allowing for fast and convenient identification of the siRNAs involved (Berns *et al.*, 2004; Paddison *et al.*, 2004) (Fig. 4). One of the most promising application of expression vector-mediated RNAi screens in cancer biology discovery is the identification of novel components of signal transduction pathways frequently involved in tumorigenesis and/or the detection of potential drug targets in these pathways. Some examples of already performed screens targeting these questions shall be mentioned in the following paragraphs.

The p53 tumor suppressor pathway is crucial for genome integrity and transmits antiproliferative as well as proapoptotic signals in response to

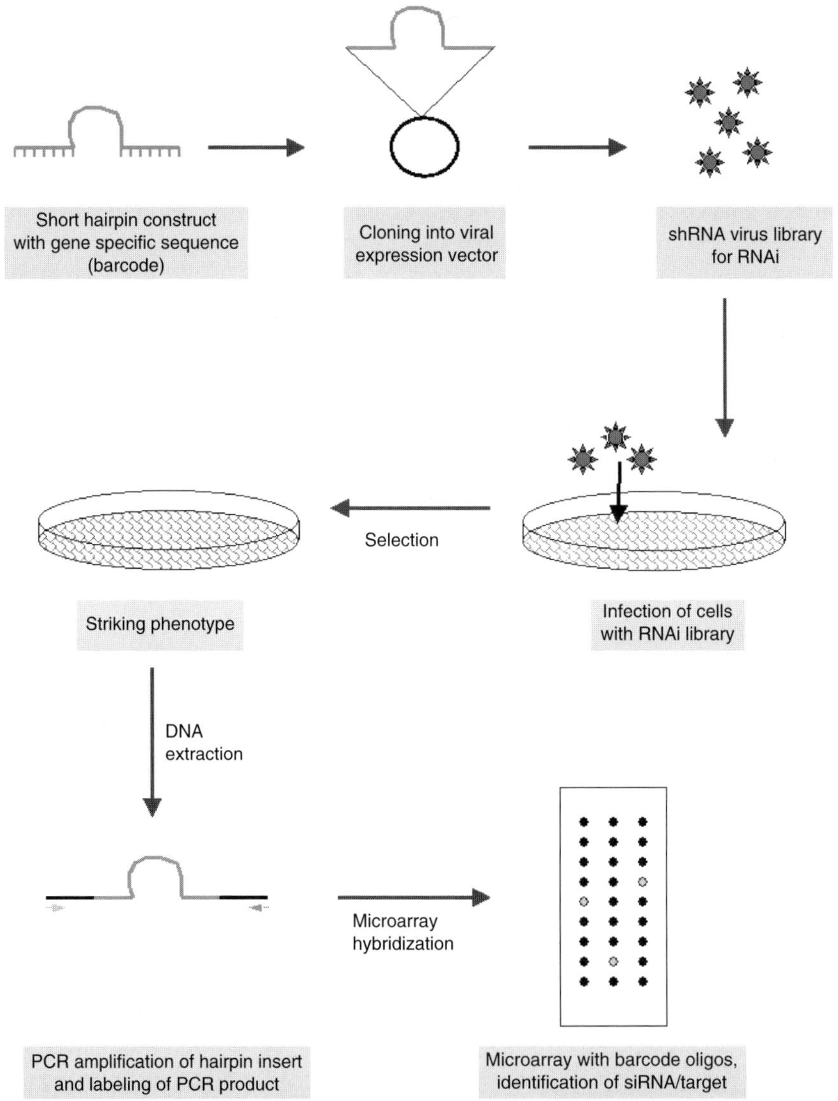

Fig. 4 RNAi screens using barcode identification. RNAi screens by shRNA expressing viral vector libraries use a mixture of vectors for transduction. After observation of an interesting phenotype the siRNAs and targets knocked down in these cells have to be identified. To simplify this identification process, a so-called "barcode screen" is performed. The short hairpin inserts of the viral vectors are gene specific and therefore individual for each vector. This molecular "barcode" is PCR amplified and labeled. The PCR products are then hybridized to a microarray containing oligonucleotides complementary to the siRNAs used in the screen. The methodology accelerates the target identification and allows a screening for synthetic phenotypes caused by the simultaneous down-regulation of more than one target gene (Brummelkamp and Bernards, 2003).

cellular stress. Mutations of the p53 locus can be found in a variety of cancers resulting in a decreased expression of growth inhibitory genes (Steele and Lane, 2005). To identify novel components of the p53 tumor suppressor pathway, a retroviral vector library was produced encoding 23,742 distinct shRNAs targeting 7914 different human genes and engineered human primary fibroblasts were transduced with this library. In this particularly well-understood genetic background, the screen enabled the identification of five novel modulators of p53-dependent proliferation arrest. Suppression of these genes conferred resistance to p53 as well as to $p19^{ARF}$-dependent proliferation arrest and abolished damage-induced G1 cell cycle arrest (Berns *et al.*, 2004).

In contrast to the widely used short hairpin RNA vectors, specific siRNA sequences can be inserted between two opposing promoters, for example, mouse U6 and human H1, leading to the expression of dsRNAs. This strategy was employed in a high-throughput screen for genes involved in the NF-κB signaling pathway. NF-κB is involved in protecting cells from undergoing apoptosis in response to DNA damage and modulation of the NF-κB pathway can potentiate the effects of chemotherapy and radiation therapy in the treatment of cancer (Yamamoto and Gaynor, 2001). The library constructed for the NF-κB screen contained two siRNAs per gene for more than 8000 genes and resulted in the identification of 12 previously unreported genes contributing to NF-κB signaling (Zheng *et al.*, 2004).

Although the use of two opposing promoters can obviously be applied for the production of siRNA libraries, shRNA expression vectors may yield significantly higher suppressive activity than tandem-type siRNA expression vectors, as has been reported in an siRNA expression library evaluation by screening for apoptosis-related genes. Induction of apoptosis was performed by dsRNA transfection with subsequent identification of expression vectors inhibiting dsRNA-dependent apoptosis (Miyagishi *et al.*, 2004).

An impressive application of an RNAi screen applied in cancer biology has been reported by Brummelkamp *et al.* (2003). The gene family that mediates ubiquitin conjugation was studied by a screen with RNAi vectors suppressing 50 human deubiquitinating enzymes in cancer-relevant pathways. The meaning of the deubiquitinating enzymes in cancer development is emphasized by the fact that this family contains oncogenes as well as tumor suppressor genes (Hershko and Ciechanover, 1998). The inhibition of one of these enzymes, CYLD, involved in familiar cylindromatosis but having no known function, enhanced activation of the transcription factor NF-κB. Inhibition of CYLD increased the cells resistance to apoptosis, indicating a mechanism through which loss of CYLD function contributes to oncogenesis. This effect could be relieved by application of aspirin derivatives known to inhibit NF-κB activity thereby leading to the direct test of this drug in a clinical trial on cylindromas (Brummelkamp *et al.*, 2003).

Since RAS is one of the most frequently involved oncogenes in human tumors, a retroviral transduction of cells with siRNA-expressing vectors was used by Kolfschoten *et al.* (2005) for a screen of knockdown constructs that transform human primary cells in the absence of ectopically introduced oncogenic RAS. The screen was designed to identify proteins whose inhibition induced the RAS pathway and tumorigenicity. Colonies were picked and the shRNA inserts identified, indicating that the knockdown of Krüppel-like transcription factor KLF4, proapoptotic calcium-binding protein PDCD6, and homeodomain pituitary transcription factor PITX1 yields a transforming activity comparable to the overexpression of oncogenic RAS^{V12}. KLF4 and PDCD6 were already attributed with a putative tumor suppressive activity, but it was the first report for PITX1 to act as tumor suppressor in the RAS pathway whose knockdown led to elevated levels of activated RAS. The PITX1 knockdown cells showed a similar phenotype as cells displaying an oncogenic activation of the RAS pathway. In support of these findings, PITX1 expression is reduced in prostate and bladder cancers relative to their normal tissue (Kolfschoten *et al.*, 2005).

Taken together, the ideal concept of such an identification attempt can be summarized as follows: (1) screen of an siRNA library; (2) validation of the siRNA effect by additional siRNAs targeting the same gene product; (3) identification of the physiological role of the gene product identified; (4) recapitulation of the *in vitro* findings *in vivo*; (5) optimization of the delivery rate to the target cells, for example, by RNA modifications like cholesterol conjugation; and (6) avoiding of unspecific and undesired siRNA effects *in vivo*.

The identification of candidate tumor suppressor genes in human mammary epithelial cells was recently addressed by Westbrook *et al.* (2005). For the understanding of the tumorigenic process it is necessary to identify genes providing tumor cells with the capabilities for their initiation and progression. In most cancer cells these genes are veiled by dozens of additional mutations and chromosomal alterations without any influence on tumorigenesis. Therefore, the authors used for their screen immortalized human mammary epithelial cells expressing hTERT and SV40 LT instead of a breast cancer cell line. Each shRNA was linked to a 60 nucleotide DNA barcode to facilitate fast identification of the shRNA by microarray technology. Readout of the screen was anchorage-independent growth in a semisolid medium, a hallmark of malignant transformation. The majority of colonies growing in this screen contained shRNAs directed against eight genes. One of the genes identified, *TGFBR2*, was validated by the transduction of the cells with a known dominant negative mutant form and with an SMAD7 construct, a negative regulator of the transforming growth factor-β (TGF-β) receptor signaling. Both constructs conferred growth in semisolid medium to the cells giving an example for appropriate controls in RNAi screens (Westbrook *et al.*, 2005).

The execution of adequate controls [as nicely summarized by Huppi *et al.* (2005)] has been emphasized repeatedly in this chapter, since it has been reported that silencing by siRNAs is not entirely sequence specific (Jackson *et al.*, 2003; Semizarov *et al.*, 2003). Sequence independent off-target effects due to induction of the interferon response can result in false positive results (Bridge *et al.*, 2003; Sledz *et al.*, 2003), but also false negative results are possible, owing to redundancy of the target (e.g., gene function might be taken over by another protein of the same family) or simply by inefficient knock-down of the target mRNA or by the long half-life of the analyzed protein.

VII. LIMITATIONS OF siRNAs AS CANCER THERAPEUTICS: NOT RELATED TO DELIVERY PROBLEMS

RNAi strategies activate a normal cellular process resulting in a highly specific posttranscriptional degradation of the targeted mRNA. siRNAs have been shown to be far more potent and longer lasting than various types of DNA oligonucleotides (ODNs) (Kretschmer-Kazemi and Sczakiel, 2003; Miyagishi *et al.*, 2003) and experimental evidence indicates that they are also more efficient than ribozymes (Yokota *et al.*, 2004). In the light of these findings, it is not surprising that many companies have started to develop an siRNA-based drugs. A proof of the therapeutic principle is already being established in trials with siRNAs against VEGF in the treatment of macular degeneration (Rye and Stigbrand, 2004).

However, beside their aforementioned immune-stimulatory role via activation of TLRs, siRNAs may activate protein kinase R (PKR), of a dsRNA-dependent protein kinase. Exposure of cells to dsRNA can activate the type I interferon response leading to PKR expression mediated by STAT. In addition to this expression activation, PKR is directly activated by binding to dsRNAs. PKR inhibition leads to a shut down of translation via phosphorylation of the small subunit of the eukaryotic initiation factor 2 (eIF2) α and induces apoptosis (Stark *et al.*, 1998). Thus, induction of apotosis may also hit nontumor cells causing significant side effects of such a therapeutic approach. In addition, cells confronted with dsRNAs synthesize 2′–5′ polyadenylic acid, which activates nonspecific RNase L leading to altered metabolism and activation of apoptosis (Gil and Esteban, 2000). Although the use of small RNAs evades these cellular reactions in general, the induction of an interferon response seems to be at least in part dependent on the composition of the applied siRNAs (Bridge *et al.*, 2003).

Another aspect to be considered is the possibility that siRNAs with partial complementarity to an mRNA could act like an endogenous miRNA and unintentionally repress the translation of this mRNA (Doench *et al.*, 2003; Saxena *et al.*, 2003; Zeng *et al.*, 2003). This underscores the necessity for a careful design of the siRNA used.

VIII. SUMMARY

RNAi has proved to be an extremely helpful tool in cancer biology discovery. RNAi screens have the potential to identify new components of cancer-relevant signaling pathways like the p53, RB, or NF-κB pathway. In combination with gene expression profiling and functional genomics, siRNAs or shRNAs permit the identification of new therapeutically promising targets. In addition, an "siRNA drug" may allow direct targeting of cancer-causing RNAs and proteins that are not "druggable" by other means. Additional progress in the field of delivery of siRNA is needed to implement their great therapeutic potential into successful clinical trials. In this respect, efficient *in vivo* delivery of cholesterol-linked siRNAs may help to show the way. Whether the recently discovered immune-modulatory effect of many siRNAs may be therapeutically beneficial or should rather be regarded as undesired side effect is one of the major open questions.

REFERENCES

Aza-Blanc, P., Cooper, C. L., Wagner, K., Batalov, S., Deveraux, Q. L., and Cooke, M. P. (2003). Identification of modulators of TRAIL-induced apoptosis via RNAi-based phenotypic screening. *Mol. Cell* **12**, 627–637.

Berns, K., Hijmans, E. M., Mullenders, J., Brummelkamp, T. R., Velds, A., Heimerikx, M., Kerkhoven, R. M., Madiredjo, M., Nijkamp, W., Weigelt, B., Agami, R., Ge, W., *et al.* (2004). A large-scale RNAi screen in human cells identifies new components of the p53 pathway. *Nature* **428**, 431–437.

Bernstein, E., Caudy, A. A., Hammond, S. M., and Hannon, G. J. (2001). Role for a bidentate ribonuclease in the initiation step of RNA interference. *Nature* **409**, 363–366.

Bridge, A. J., Pebernard, S., Ducraux, A., Nicoulaz, A. L., and Iggo, R. (2003). Induction of an interferon response by RNAi vectors in mammalian cells. *Nat. Genet.* **34**, 263–264.

Brummelkamp, T. R., and Bernards, R. (2003). New tools for functional mammalian cancer genetics. *Nat. Rev. Cancer* **3**, 781–789.

Brummelkamp, T. R., Bernards, R., and Agami, R. (2002a). Stable suppression of tumorigenicity by virus-mediated RNA interference. *Cancer Cell* **2**, 243–247.

Brummelkamp, T. R., Bernards, R., and Agami, R. (2002b). A system for stable expression of short interfering RNAs in mammalian cells. *Science* **296**, 550–553.

Brummelkamp, T. R., Nijman, S. M., Dirac, A. M., and Bernards, R. (2003). Loss of the cylindromatosis tumour suppressor inhibits apoptosis by activating NF-kappaB. *Nature* **424**, 797–801.

Chen, J., Wall, N. R., Kocher, K., Duclos, N., Fabbro, D., Neuberg, D., Griffin, J. D., Shi, Y., and Gilliland, D. G. (2004). Stable expression of small interfering RNA sensitizes TEL-PDGFbetaR to inhibition with imatinib or rapamycin. *J. Clin. Invest.* **113,** 1784–1791.

Damm-Welk, C., Fuchs, U., Wössmann, W., and Borkhardt, A. (2003). Targeting oncogenic fusion genes in leukemias and lymphomas by RNA interference. *Semin. Cancer Biol.* **13,** 283–292.

Doench, J. G., Petersen, C. P., and Sharp, P. A. (2003). siRNAs can function as miRNAs. *Genes Dev.* **17,** 438–442.

Dorsett, Y., and Tuschl, T. (2004). siRNAs: Applications in functional genomics and potential as therapeutics. *Nat. Rev. Drug Discov.* **3,** 318–329.

Duxbury, M. S., Ito, H., Benoit, E., Ashley, S. W., and Whang, E. E. (2004). CEACAM6 is a determinant of pancreatic adenocarcinoma cellular invasiveness. *Br. J. Cancer* **91,** 1384–1390.

Duxbury, M. S., Ashley, S. W., and Whang, E. E. (2005). RNA interference: A mammalian SID-1 homologue enhances siRNA uptake and gene silencing efficacy in human cells. *Biochem. Biophys. Res. Commun.* **331,** 459–463.

Dykxhoorn, D. M., Novina, C. D., and Sharp, P. A. (2003). Killing the messenger: Short RNAs that silence gene expression. *Nat. Rev. Mol. Cell Biol.* **4,** 457–467.

Elbashir, S. M., Harborth, J., Lendeckel, W., Yalcin, A., Weber, K., and Tuschl, T. (2001a). Duplexes of 21-nucleotide RNAs mediate RNA interference in cultured mammalian cells. *Nature* **411,** 494–498.

Elbashir, S. M., Lendeckel, W., and Tuschl, T. (2001b). RNA interference is mediated by 21- and 22-nucleotide RNAs. *Genes Dev.* **15,** 188–200.

Eskandarpour, M., Kiaii, S., Zhu, C., Castro, J., Sakko, A. J., and Hansson, J. (2005). Suppression of oncogenic NRAS by RNA interference induces apoptosis of human melanoma cells. *Int. J. Cancer* **115,** 65–73.

Faltus, T., Yuan, J., Zimmer, B., Kramer, A., Loibl, S., Kaufmann, M., and Strebhardt, K. (2004). Silencing of the HER2/neu gene by siRNA inhibits proliferation and induces apoptosis in HER2/neu-overexpressing breast cancer cells. *Neoplasia* **6,** 786–795.

Fire, A., Xu, S., Montgomery, M. K., Kostas, S. A., Driver, S. E., and Mello, C. C. (1998). Potent and specific genetic interference by double-stranded RNA in *Caenorhabditis elegans. Nature* **391,** 806–811.

Gil, J., and Esteban, M. (2000). Induction of apoptosis by the dsRNA-dependent protein kinase (PKR): Mechanism of action. *Apoptosis* **5,** 107–114.

Gschwind, A., Fischer, O. M., and Ullrich, A. (2004). The discovery of receptor tyrosine kinases: Targets for cancer therapy. *Nat. Rev. Cancer* **4,** 361–370.

Hacein-Bey-Abina, S., Von Kalle, C., Schmidt, M., Le Deist, F., Wulffraat, N., McIntyre, E., Radford, I., Villeval, J. L., Fraser, C. C., Cavazzana-Calvo, M., and Fischer, A. (2003). A serious adverse event after successful gene therapy for X-linked severe combined immunodeficiency. *N. Engl. J. Med.* **348,** 255–256.

Haley, B., and Zamore, P. D. (2004). Kinetic analysis of the RNAi enzyme complex. *Nat. Struct. Mol. Biol.* **11,** 599–606.

Heidenreich, O., Krauter, J., Riehle, H., Hadwiger, P., John, M., Heil, G., Vornlocher, H. P., and Nordheim, A. (2003). AML1/MTG8 oncogene suppression by small interfering RNAs supports myeloid differentiation of t(8;21)-positive leukemic cells. *Blood* **101,** 3157–3163.

Hemann, M. T., Fridman, J. S., Zilfou, J. T., Hernando, E., Paddison, P. J., Cordon-Cardo, C., Hannon, G. J., and Lowe, S. W. (2003). An epi-allelic series of p53 hypomorphs created by stable RNAi produces distinct tumor phenotypes *in vivo*. *Nat. Genet.* **33,** 396–400.

Hershko, A., and Ciechanover, A. (1998). The ubiquitin system. *Annu. Rev. Biochem.* **67,** 425–479.

Hornung, V., Guenthner-Biller, M., Bourquin, C., Ablasser, A., Schlee, M., Uematsu, S., Noronha, A., Manoharan, M., Akira, S., de, F. A., Endres, S., and Hartmann, G. (2005). Sequence-specific potent induction of IFN-alpha by short interfering RNA in plasmacytoid dendritic cells through TLR7. *Nat. Med.* **11,** 263–270.

Huang, A., Ho, C. S., Ponzielli, R., Barsyte-Lovejoy, D., Bouffet, E., Picard, D., Hawkins, C. E., and Penn, L. Z. (2005). Identification of a novel c-Myc protein interactor, JPO2, with transforming activity in medulloblastoma cells. *Cancer Res.* **65**, 5607–5619.

Huppi, K., Martin, S. E., and Caplen, N. J. (2005). Defining and assaying RNAi in mammalian cells. *Mol. Cell* **17**, 1–10.

Hutvagner, G., and Zamore, P. D. (2002). A microRNA in a multiple-turnover RNAi enzyme complex. *Science* **297**, 2056–2060.

Idogawa, M., Yamada, T., Honda, K., Sato, S., Imai, K., and Hirohashi, S. (2005). Poly (ADP-ribose) polymerase-1 is a component of the oncogenic T-cell factor-4/beta-catenin complex. *Gastroenterology* **128**, 1919–1936.

Ioannidis, P., Mahaira, L. G., Perez, S. A., Gritzapis, A. D., Sotiropoulou, P. A., Kavalakis, G. J., Antsaklis, A. I., Baxevanis, C. N., and Papamichail, M. (2005). CRD-BP/IMP1 expression characterizes cord blood CD34+ stem cells and affects c-myc and IGF-II expression in MCF-7 cancer cells. *J. Biol. Chem.* **280**, 20086–20093.

Irby, R. B., Malek, R. L., Bloom, G., Tsai, J., Letwin, N., Frank, B. C., Verratti, K., Yeatman, T. J., and Lee, N. H. (2005). Iterative microarray and RNA interference-based interrogation of the SRC-induced invasive phenotype. *Cancer Res.* **65**, 1814–1821.

Jackson, A. L., Bartz, S. R., Schelter, J., Kobayashi, S. V., Burchard, J., Mao, M., Li, B., Cavet, G., and Linsley, P. S. (2003). Expression profiling reveals off-target gene regulation by RNAi. *Nat. Biotechnol.* **21**, 635–637.

Jiang, F., Caraway, N. P., Li, R., and Katz, R. L. (2005). RNA silencing of S-phase kinase-interacting protein 2 inhibits proliferation and centrosome amplification in lung cancer cells. *Oncogene* **24**, 3409–3418.

Judge, A. D., Sood, V., Shaw, J. R., Fang, D., McClintock, K., and MacLachlan, I. (2005). Sequence-dependent stimulation of the mammalian innate immune response by synthetic siRNA. *Nat. Biotechnol.* **23**, 457–462.

Kaelin, W. G. (2005). The concept of synthetic lethality in the context of anticancer therapy. *Nat. Rev. Cancer* **5**, 689–698.

Karande, P., Jain, A., and Mitragotri, S. (2004). Discovery of transdermal penetration enhancers by high-throughput screening. *Nat. Biotechnol.* **22**, 192–197.

Kato, K., Umezawa, K., Miyake, M., Miyake, J., and Nagamune, T. (2004). Transfection microarray of nonadherent cells on an oleyl poly(ethylene glycol) ether-modified glass slide. *Biotechniques* **37**, 444–448, 450, 452.

Kawasaki, H., and Taira, K. (2004). Induction of DNA methylation and gene silencing by short interfering RNAs in human cells. *Nature* **431**, 211–217.

Kennerdell, J. R., and Carthew, R. W. (2000). Heritable gene silencing in *Drosophila* using double-stranded RNA. *Nat. Biotechnol.* **18**, 896–898.

Kolfschoten, I. G., van, L. B., Berns, K., Mullenders, J., Beijersbergen, R. L., Bernards, R., Voorhoeve, P. M., and Agami, R. (2005). A genetic screen identifies PITX1 as a suppressor of RAS activity and tumorigenicity. *Cell* **121**, 849–858.

Kretschmer-Kazemi, F. R., and Sczakiel, G. (2003). The activity of siRNA in mammalian cells is related to structural target accessibility: A comparison with antisense oligonucleotides. *Nucleic Acids Res.* **31**, 4417–4424.

Kumar, R., Conklin, D. S., and Mittal, V. (2003). High-throughput selection of effective RNAi probes for gene silencing. *Genome Res.* **13**, 2333–2340.

Li, H., Li, W. X., and Ding, S. W. (2002). Induction and suppression of RNA silencing by an animal virus. *Science* **296**, 1319–1321.

Li, H., Fu, X., Chen, Y., Hong, Y., Tan, Y., Cao, H., Wu, M., and Wang, H. (2005). Use of adenovirus-delivered siRNA to target oncoprotein p28GANK in hepatocellular carcinoma. *Gastroenterology* **128**, 2029–2041.

Lieberman, J., Song, E., Lee, S. K., and Shankar, P. (2003). Interfering with disease: Opportunities and roadblocks to harnessing RNA interference. *Trends Mol. Med.* **9**, 397–403.

Lipardi, C., Wei, Q., and Paterson, B. M. (2001). RNAi as random degradative PCR: siRNA primers convert mRNA into dsRNAs that are degraded to generate new siRNAs. *Cell* **107**, 297–307.

Liu, J., Carmell, M. A., Rivas, F. V., Marsden, C. G., Thomson, J. M., Song, J. J., Hammond, S. M., Joshua-Tor, L., and Hannon, G. J. (2004). Argonaute2 is the catalytic engine of mammalian RNAi. *Science* **305**, 1437–1441.

Lorenz, C., Hadwiger, P., John, M., Vornlocher, H. P., and Unverzagt, C. (2004). Steroid and lipid conjugates of siRNAs to enhance cellular uptake and gene silencing in liver cells. *Bioorg. Med. Chem. Lett.* **14**, 4975–4977.

MacKeigan, J. P., Murphy, L. O., and Blenis, J. (2005). Sensitized RNAi screen of human kinases and phosphatases identifies new regulators of apoptosis and chemoresistance. *Nat. Cell Biol.* **7**, 591–600.

Marshall, E. (2003). Gene therapy. Second child in French trial is found to have leukemia. *Science* **299**, 320.

Martinez, J., and Tuschl, T. (2004). RISC is a 5′ phosphomonoester-producing RNA endonuclease. *Genes Dev.* **18**, 975–980.

Matzke, M. A., and Matzke, A. J. (2004). Planting the seeds of a new paradigm. *PLoS Biol.* **2**, E133.

Meister, G., and Tuschl, T. (2004). Mechanisms of gene silencing by double-stranded RNA. *Nature* **431**, 343–349.

Meister, G., Landthaler, M., Patkaniowska, A., Dorsett, Y., Teng, G., and Tuschl, T. (2004). Human Argonaute2 mediates RNA cleavage targeted by miRNAs and siRNAs. *Mol. Cell* **15**, 185–197.

Menendez, J. A., Vellon, L., Colomer, R., and Lupu, R. (2005). Pharmacological and small interference RNA-mediated inhibition of breast cancer-associated fatty acid synthase (oncogenic antigen-519) synergistically enhances Taxol (paclitaxel)-induced cytotoxicity. *Int. J. Cancer* **115**, 19–35.

Mette, M. F., van der, W. J., Matzke, M. A., and Matzke, A. J. (1999). Production of aberrant promoter transcripts contributes to methylation and silencing of unlinked homologous promoters in trans. *EMBO J.* **18**, 241–248.

Miyagishi, M., and Taira, K. (2002). U6 promoter-driven siRNAs with four uridine 3′ overhangs efficiently suppress targeted gene expression in mammalian cells. *Nat. Biotechnol.* **20**, 497–500.

Miyagishi, M., Hayashi, M., and Taira, K. (2003). Comparison of the suppressive effects of antisense oligonucleotides and siRNAs directed against the same targets in mammalian cells. *Antisense Nucleic Acid Drug Dev.* **13**, 1–7.

Miyagishi, M., Matsumoto, S., and Taira, K. (2004). Generation of an shRNAi expression library against the whole human transcripts. *Virus Res.* **102**, 117–124.

Morris, K. V., Chan, S. W., Jacobsen, S. E., and Looney, D. J. (2004). Small interfering RNA-induced transcriptional gene silencing in human cells. *Science* **305**, 1289–1292.

Mousses, S., Caplen, N. J., Cornelison, R., Weaver, D., Basik, M., Hautaniemi, S., Elkahloun, A. G., Lotufo, R. A., Choudary, A., Dougherty, E. R., Suh, E., and Kallioniemi, O. (2003). RNAi microarray analysis in cultured mammalian cells. *Genome Res.* **13**, 2341–2347.

Myslinski, E., Ame, J. C., Krol, A., and Carbon, P. (2001). An unusually compact external promoter for RNA polymerase III transcription of the human H1RNA gene. *Nucleic Acids Res.* **29**, 2502–2509.

Naldini, L., Blomer, U., Gallay, P., Ory, D., Mulligan, R., Gage, F. H., Verma, I. M., and Trono, D. (1996). *In vivo* gene delivery and stable transduction of nondividing cells by a lentiviral vector. *Science* **272,** 263–267.

Ovcharenko, D., Jarvis, R., Hunicke-Smith, S., Kelnar, K., and Brown, D. (2005). High-throughput RNAi screening *in vitro*: From cell lines to primary cells. *RNA* **11,** 985–993.

Paddison, P. J., and Hannon, G. J. (2002). RNA interference: The new somatic cell genetics? *Cancer Cell* **2,** 17–23.

Paddison, P. J., Caudy, A. A., and Hannon, G. J. (2002). Stable suppression of gene expression by RNAi in mammalian cells. *Proc. Natl. Acad. Sci. USA* **99,** 1443–1448.

Paddison, P. J., Silva, J. M., Conklin, D. S., Schlabach, M., Li, M., Aruleba, S., Balija, V., O'Shaughnessy, A., Gnoj, L., Scobie, K., Chang, K., Westbrook, T., *et al.* (2004). A resource for large-scale RNA-interference-based screens in mammals. *Nature* **428,** 427–431.

Pickford, A. S., Catalanotto, C., Cogoni, C., and Macino, G. (2002). Quelling in Neurospora crassa. *Adv. Genet.* **46,** 277–303.

Plasterk, R. H. (2002). RNA silencing: The genome's immune system. *Science* **296,** 1263–1265.

Prieur, A., Tirode, F., Cohen, P., and Delattre, O. (2004). EWS/FLI-1 silencing and gene profiling of Ewing cells reveal downstream oncogenic pathways and a crucial role for repression of insulin-like growth factor binding protein 3. *Mol. Cell. Biol.* **24,** 7275–7283.

Ptasznik, A., Nakata, Y., Kalota, A., Emerson, S. G., and Gewirtz, A. M. (2004). Short interfering RNA (siRNA) targeting the Lyn kinase induces apoptosis in primary, and drug-resistant, BCR-ABL1(+) leukemia cells. *Nat. Med.* **10,** 1187–1189.

Putral, L. N., Bywater, M. J., Gu, W., Saunders, N. A., Gabrielli, B. G., Leggatt, G. R., and McMillan, N. A. (2005). RNAi against HPV oncogenes in cervical cancer cells results in increased sensitivity to cisplatin. *Mol. Pharmacol.* **68,** 1311–1319.

Qin, X. F., An, D. S., Chen, I. S., and Baltimore, D. (2003). Inhibiting HIV-1 infection in human T cells by lentiviral-mediated delivery of small interfering RNA against CCR5. *Proc. Natl. Acad. Sci. USA* **100,** 183–188.

Rajendran, R. R., Nye, A. C., Frasor, J., Balsara, R. D., Martini, P. G., and Katzenellenbogen, B. S. (2003). Regulation of nuclear receptor transcriptional activity by a novel DEAD box RNA helicase (DP97). *J. Biol. Chem.* **278,** 4628–4638.

Robert, M. F., Morin, S., Beaulieu, N., Gauthier, F., Chute, I. C., Barsalou, A., and MacLeod, A. R. (2003). DNMT1 is required to maintain CpG methylation and aberrant gene silencing in human cancer cells. *Nat. Genet.* **33,** 61–65.

Rossi, J. J. (2005). Receptor-targeted siRNAs. *Nat. Biotechnol.* **23,** 682–684.

Rye, P. D., and Stigbrand, T. (2004). Interfering with cancer: A brief outline of advances in RNA interference in oncology. *Tumour Biol.* **25,** 329–336.

Salvi, A., Arici, B., De, P. G., and Barlati, S. (2004). Small interfering RNA urokinase silencing inhibits invasion and migration of human hepatocellular carcinoma cells. *Mol. Cancer Ther.* **3,** 671–678.

Saxena, S., Jonsson, Z. O., and Dutta, A. (2003). Small RNAs with imperfect match to endogenous mRNA repress translation. Implications for off-target activity of small inhibitory RNA in mammalian cells. *J. Biol. Chem.* **278,** 44312–44319.

Scherr, M., Battmer, K., Winkler, T., Heidenreich, O., Ganser, A., and Eder, M. (2003). Specific inhibition of bcr-abl gene expression by small interfering RNA. *Blood* **101,** 1566–1569.

Scherr, M., Battmer, K., Schultheis, B., Ganser, A., and Eder, M. (2005). Stable RNA interference (RNAi) as an option for anti-bcr-abl therapy. *Gene Ther.* **12,** 12–21.

Semizarov, D., Frost, L., Sarthy, A., Kroeger, P., Halbert, D. N., and Fesik, S. W. (2003). Specificity of short interfering RNA determined through gene expression signatures. *Proc. Natl. Acad. Sci. USA* **100,** 6347–6352.

Shuey, D. J., McCallus, D. E., and Giordano, T. (2002). RNAi: Gene-silencing in therapeutic intervention. *Drug Discov. Today* **7,** 1040–1046.

Silva, J. M., Mizuno, H., Brady, A., Lucito, R., and Hannon, G. J. (2004). RNA interference microarrays: High-throughput loss-of-function genetics in mammalian cells. *Proc. Natl. Acad. Sci. USA* **101,** 6548–6552.

Sledz, C. A., Holko, M., De Veer, M. J., Silverman, R. H., and Williams, B. R. (2003). Activation of the interferon system by short-interfering RNAs. *Nat. Cell Biol.* **5,** 834–839.

Smakman, N., Veenendaal, L. M., van, D. P., Bos, R., Offringa, R., Borel, R.I, and Kranenburg, O. (2005). Dual effect of Kras(D12) knockdown on tumorigenesis: Increased immune-mediated tumor clearance and abrogation of tumor malignancy. *Oncogene* **24**(56), 8338–8342.

Sontheimer, E. J. (2005). Assembly and function of RNA silencing complexes. *Nat. Rev. Mol. Cell Biol.* **6,** 127–138.

Soutschek, J., Akinc, A., Bramlage, B., Charisse, K., Constien, R., Donoghue, M., Elbashir, S., Geick, A., Hadwiger, P., Harborth, J., John, M., Kesavan, V., *et al.* (2004). Therapeutic silencing of an endogenous gene by systemic administration of modified siRNAs. *Nature* **432,** 173–178.

Spankuch, B., Matthess, Y., Knecht, R., Zimmer, B., Kaufmann, M., and Strebhardt, K. (2004). Cancer inhibition in nude mice after systemic application of U6 promoter-driven short hairpin RNAs against PLK1. *J. Natl. Cancer Inst.* **96,** 862–872.

Spankuch-Schmitt, B., Bereiter-Hahn, J., Kaufmann, M., and Strebhardt, K. (2002). Effect of RNA silencing of polo-like kinase-1 (PLK1) on apoptosis and spindle formation in human cancer cells. *J. Natl. Cancer Inst.* **94,** 1863–1877.

Stark, G. R., Kerr, I. M., Williams, B. R., Silverman, R. H., and Schreiber, R. D. (1998). How cells respond to interferons. *Annu. Rev. Biochem.* **67,** 227–264.

Steele, R. J., and Lane, D. P. (2005). P53 in cancer: A paradigm for modern management of cancer. *Surgeon* **3,** 197–205.

Stege, A., Priebsch, A., Nieth, C., and Lage, H. (2004). Stable and complete overcoming of MDR1/P-glycoprotein-mediated multidrug resistance in human gastric carcinoma cells by RNA interference. *Cancer Gene Ther.* **11,** 699–706.

Svoboda, J., Hejnar, J., Geryk, J., Elleder, D., and Vernerova, Z. (2000). Retroviruses in foreign species and the problem of provirus silencing. *Gene* **261,** 181–188.

Tabara, H., Sarkissian, M., Kelly, W. G., Fleenor, J., Grishok, A., Timmons, L., Fire, A., and Mello, C. C. (1999). The rde-1 gene, RNA interference, and transposon silencing in *C. elegans*. *Cell* **99,** 123–132.

Taulli, R., Accornero, P., Follenzi, A., Mangano, T., Morotti, A., Scuoppo, C., Forni, P. E., Bersani, F., Crepaldi, T., Chiarle, R., Naldini, L., and Ponzetto, C. (2005). RNAi technology and lentiviral delivery as a powerful tool to suppress Tpr-Met-mediated tumorigenesis. *Cancer Gene Ther.* **12,** 456–463.

Thomas, M., Gessner, A., Vornlocher, H. P., Hadwiger, P., Greil, J., and Heidenreich, O. (2005). Targeting MLL-AF4 with short interfering RNAs inhibits clonogenicity and engraftment of t(4;11)-positive human leukemic cells. *Blood* **106**(10), 3559–3566.

Tiscornia, G., Singer, O., Ikawa, M., and Verma, I. M. (2003). A general method for gene knockdown in mice by using lentiviral vectors expressing small interfering RNA. *Proc. Natl. Acad. Sci. USA* **100,** 1844–1848.

van Huijsduijnen, R. H., Bombrun, A., and Swinnen, D. (2002). Selecting protein tyrosine phosphatases as drug targets. *Drug Discov. Today* **7,** 1013–1019.

Vanhecke, D., and Janitz, M. (2005). Functional genomics using high-throughput RNA interference. *Drug Discov. Today* **10,** 205–212.

Vickers, T. A., Koo, S., Bennett, C. F., Crooke, S. T., Dean, N. M., and Baker, B. F. (2003). Efficient reduction of target RNAs by small interfering RNA and RNase H-dependent antisense agents. A comparative analysis. *J. Biol. Chem.* **278,** 7108–7118.

Voinnet, O. (2001). RNA silencing as a plant immune system against viruses. *Trends Genet.* **17,** 449–459.

Vousden, K. H., and Lu, X. (2002). Live or let die: The cell's response to p53. *Nat. Rev. Cancer* **2,** 594–604.

Wang, Y. H., Liu, S., Zhang, G., Zhou, C. Q., Zhu, H. X., Zhou, X. B., Quan, L. P., Bai, J. F., and Xu, N. Z. (2005). Knockdown of c-Myc expression by RNAi inhibits MCF-7 breast tumor cells growth *in vitro* and *in vivo*. *Breast Cancer Res.* **7,** R220–R228.

Weil, D., Garcon, L., Harper, M., Dumenil, D., Dautry, F., and Kress, M. (2002). Targeting the kinesin Eg5 to monitor siRNA transfection in mammalian cells. *Biotechniques* **33,** 1244–1248.

Wellbrock, C., and Marais, R. (2005). Elevated expression of MITF counteracts B-RAF-stimulated melanocyte and melanoma cell proliferation. *J. Cell Biol.* **170,** 703–708.

Wellbrock, C., Ogilvie, L., Hedley, D., Karasarides, M., Martin, J., Niculescu-Duvaz, D., Springer, C. J., and Marais, R. (2004). V599 EB-RAF is an oncogene in melanocytes. *Cancer Res.* **64,** 2338–2342.

Westbrook, T. F., Martin, E. S., Schlabach, M. R., Leng, Y., Liang, A. C., Feng, B., Zhao, J. J., Roberts, T. M., Mandel, G., Hannon, G. J., Depinho, R. A., Chin, L., *et al.* (2005). A genetic screen for candidate tumor suppressors identifies rest. *Cell* **121,** 837–848.

Wheeler, D. B., Carpenter, A. E., and Sabatini, D. M. (2005). Cell microarrays and RNA interference chip away at gene function. *Nat. Genet.* **37**(Suppl.), S25–S30.

Wilda, M., Fuchs, U., Wossmann, W., and Borkhardt, A. (2002). Killing of leukemic cells with a BCR/ABL fusion gene by RNA interference (RNAi). *Oncogene* **21,** 5716–5724.

Wilson, J. A., Jayasena, S., Khvorova, A., Sabatinos, S., Rodrigue-Gervais, I. G., Arya, S., Sarangi, F., Harris-Brandts, M., Beaulieu, S., and Richardson, C. D. (2003). RNA interference blocks gene expression and RNA synthesis from hepatitis C replicons propagated in human liver cells. *Proc. Natl. Acad. Sci. USA* **100,** 2783–2788.

Winston, W. M., Molodowitch, C., and Hunter, C. P. (2002). Systemic RNAi in *C. elegans* requires the putative transmembrane protein SID-1. *Science* **295,** 2456–2459.

Wohlbold, L., Van Der, K. H., Miething, C., Vornlocher, H. P., Knabbe, C., Duyster, J., and Aulitzky, W. E. (2003). Inhibition of bcr-abl gene expression by small interfering RNA sensitizes for imatinib mesylate (STI571). *Blood* **102,** 2236–2239.

Wu, W., Hodges, E., Redelius, J., and Hoog, C. (2004). A novel approach for evaluating the efficiency of siRNAs on protein levels in cultured cells. *Nucleic Acids Res.* **32,** e17.

Yague, E., Higgins, C. F., and Raguz, S. (2004). Complete reversal of multidrug resistance by stable expression of small interfering RNAs targeting MDR1. *Gene Ther.* **11,** 1170–1174.

Yamamoto, Y., and Gaynor, R. B. (2001). Therapeutic potential of inhibition of the NF-kappaB pathway in the treatment of inflammation and cancer. *J. Clin. Invest.* **107,** 135–142.

Yokota, T., Miyagishi, M., Hino, T., Matsumura, R., Tasinato, A., Urushitani, M., Rao, R. V., Takahashi, R., Bredesen, D. E., Taira, K., and Mizusawa, H. (2004). siRNA-based inhibition specific for mutant SOD1 with single nucleotide alternation in familial ALS, compared with ribozyme and DNA enzyme. *Biochem. Biophys. Res. Commun.* **314,** 283–291.

Yu, J. Y., DeRuiter, S. L., and Turner, D. L. (2002). RNA interference by expression of short-interfering RNAs and hairpin RNAs in mammalian cells. *Proc. Natl. Acad. Sci. USA* **99,** 6047–6052.

Zeng, Y., Yi, R., and Cullen, B. R. (2003). MicroRNAs and small interfering RNAs can inhibit mRNA expression by similar mechanisms. *Proc. Natl. Acad. Sci. USA* **100,** 9779–9784.

Ribozyme Technology for Cancer Gene Target Identification and Validation

Qi-Xiang Li, Philip Tan, Ning Ke, and Flossie Wong-Staal

Immusol, Inc., San Diego, California 92121

I. Brief Biology of Ribozymes
II. Ribozymes as Tools for Gene Inactivation
  A. Design of Ribozymes for Targeting a Specific mRNA of Interest
  B. Delivery of Ribozymes into Cells
  C. Factors That Influence the Effectiveness of Ribozyme-Mediated Gene Inactivation
III. Ribozymes as Tools in Gene Target Discovery and Validation
  A. Principles of Ribozyme-Based Gene Target Validation
  B. Drug Target Discovery Using a Combinatorial Ribozyme Gene Library
IV. Ribozyme-Based Genomic Technology in Cancer Gene Target Discovery and Validation
  A. HeLa/HeLaHF Cervical Cancer Cell System and Anchorage-Independent Growth
  B. Activation of Apoptosis Induced by External Stimuli
  C. Identification of Genes Involved in Cell Invasion Using an *In Vitro* Invasion Assay
  D. Other Inverse Genomics® Screenings for Cancer Targets
V. Summary
  References

Ribozymes are naturally occurring RNAs with catalytic activities including *cis*- or *trans*- cleavage of RNA at predefined sequence sites. This activity has been exploited for specific gene inactivation in cells during the last two decades, and ribozymes have been important functional genomics tools, especially in the pre-RNAi era. It has also been broadly applied in drug target identification and validation in pharmaceutical R&D. This chapter covers many application principles and case studies of ribozyme technology in the areas of cancer research. We also described RNAi applications in some of the same studies for comparison. Although RNAi may be more effective than ribozymes in many respects, they are nonetheless built on many of the same principles.

I. BRIEF BIOLOGY OF RIBOZYMES

The discovery of ribozymes, or catalytic RNA, in the early 1980s (Cech *et al.*, 1981; Guerrier-Takada *et al.*, 1983; Kruger *et al.*, 1982) not only dispelled the then existing view that enzymatic activities were exclusive to proteins, but it also led to broad and practical applications. The enzymatic activities of ribozymes include cleavage, ligation, and *trans*-splicing of RNA

Advances in CANCER RESEARCH

0065-230X/07 $35.00
DOI: 10.1016/S0065-230X(06)96005-9

molecules. There are several classes of ribozymes found in nature: self-splicing group I and group II intron RNAs (responsible for RNA processing/splicing), ribonuclease P RNA subunit (responsible for processing the 5′ end of precursor tRNA and some rRNA in *Escherichia coli*), and self-cleaving RNAs including hammerhead and hairpin ribozymes. The most commonly used ribozymes in molecular biology research are the small hammerhead and hairpin ribozymes, as they can be engineered to cleave heterologous RNA molecules *in trans* in a sequence-specific manner.

Hairpin ribozymes (Fig. 1A) are naturally occurring RNA molecules derived from the negative strand of the satellite RNA of tobacco ringspot virus (Hampel and Tritz, 1989; Hampel *et al.*, 1990). Each ribozyme is composed of a 50-nucleotide long sequence which folds into a two-dimensional "hairpin" structure containing several helical and loop regions. Helices I and II result from the binding of the two target recognition sequences to the substrate RNA in a complementary Watson–Crick base pairing manner (Fig. 1A). Cleavage occurs 5′ to a GUC triplet in a loop region of the target sequence between helices I and II.

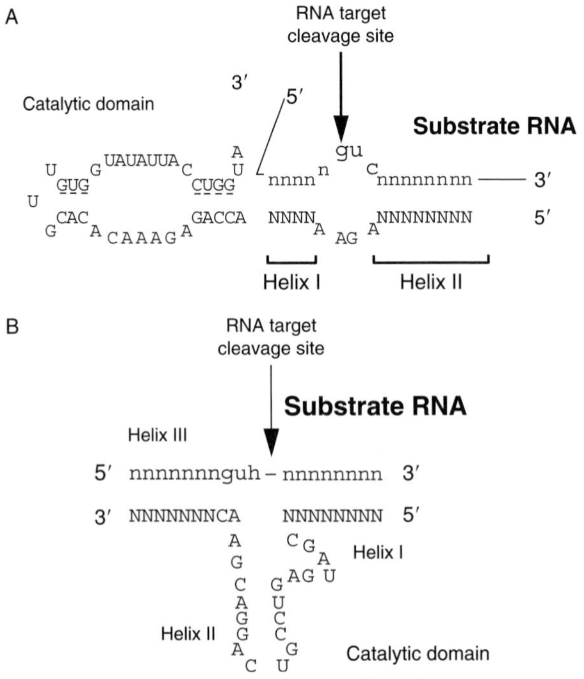

Fig. 1 The structures of small ribozymes. (Panel A) Hammerhead ribozyme; (panel B) hairpin ribozyme.

Hammerhead ribozymes, named after their secondary structure, are also naturally occurring RNA molecules found in plant viroid pathogens (Fig. 1B). The hammerhead structure formed after binding to the substrate consists of three base-paired stems (helices I, II, and III). Helices I and III flank the susceptible phosphodiester bond in the target (Fig. 1B). Cleavage of RNA occurs at the position immediately 3′ of the NUX target sequence.

The activity of both hammerhead and hairpin ribozyme activities can be assessed in an *in vitro* cleavage reaction (test tube) to determine the kinetic parameters K_m (nM) and k_{cat} (min^{-1}). The k_{cat}/K_m ratio defines the efficiency of the ribozyme. The k_{cat}/K_m of the native hairpin ribozyme is around 0.07 $nM^{-1}min^{-1}$(Hendry *et al.*, 1997; Yu and Burke, 1997).

II. RIBOZYMES AS TOOLS FOR GENE INACTIVATION

Both hammerhead and hairpin ribozymes are self-cleaving in nature, but can be designed to target any specific RNA in *trans*. The cleavage can also occur intracellularly (*in vivo*) or in the test tube (*in vitro*). All the engineered ribozymes share conserved sequences, including the catalytic and structural determinants, but contain variable regions that specify target sequence recognition. Ribozymes have been used for the identification of genes and the determination of gene function.

A. Design of Ribozymes for Targeting a Specific mRNA of Interest

Engineered hairpin ribozymes are usually 50–70 bases in length. There are two target recognition sequences on the ribozyme, which bind to complementary sequences on the target RNA, forming helices I (with variable length of 6–8 nucleotides) and II (fixed length of 4 nucleotides). The ribozyme-binding sequences on the target RNA flank an obligatory GUC triplet, which is cleaved by the ribozyme at the 5′ end (Fig. 1A). The sequences flanking the GUC site are flexible. Since most mRNA molecules contain multiple GUC sites (statistically, this should occur every 64 nucleotides), they all likely contain one or more ribozyme-cleavable-binding sequences, although not every potential GUC site is accessible to ribozyme cleavage in practice (see following section).

The engineered hammerhead ribozyme (Fig. 1B) functions similarly to the hairpin ribozyme. It also has two target recognition sequences, helix I (3–10 nucleotides) and helix III (6–10 nucleotides), complementary to the ribozyme-binding sites on the target mRNA. The binding sites on the target flanked an obligatory NUH triplet (where N represents any ribonucleotide,

H represents anyone of A, U, or C), but are otherwise quite flexible. Thus, by varying the sequence of the recognition sequences, the ribozyme can be engineered to target any NUH site. Almost any mRNA molecule will contain multiple NUH sites (statistically, this should occur every 12 nucleotides) and therefore multiple hammerhead ribozymes can be designed for each target.

The typical procedure for ribozyme design is: (1) identify and select the obligatory NUH/GUC triplet cleavage sites on the intended target; (2) identify the flanking target-binding sequences, and then determine the complementary target recognition sequences on the ribozyme; and (3) assemble the variable target recognition sequences and the conserved sequences of ribozyme into unique full-length ribozyme sequences (Fig. 1). Additional considerations include various modifications to increase the stability and activity of the ribozymes *in vivo* (see following section). Control ribozymes can be designed which contain the target recognition sequences but lack catalytic activity. In the hairpin ribozyme, this is usually achieved by mutating the AAA at positions 22–24 in the loop II catalytic domain to CGU (Fig. 1A). These "disabled" negative control ribozymes should have little or no effect on the target mRNA levels, demonstrating that knockdown obtained with the active ribozymes is indeed due to their specific catalytic activity.

B. Delivery of Ribozymes into Cells

Ribozyme RNAs can be prepared by chemical synthesis or *in vitro* transcription, but these are rarely used for cell biology studies due to the labile nature of RNA. More commonly, ribozyme genes are engineered into expression vectors that are introduced into cells for transcription (Fig. 2A). The intracellularly transcribed ribozymes (Fig. 2B and C) then function to cleave

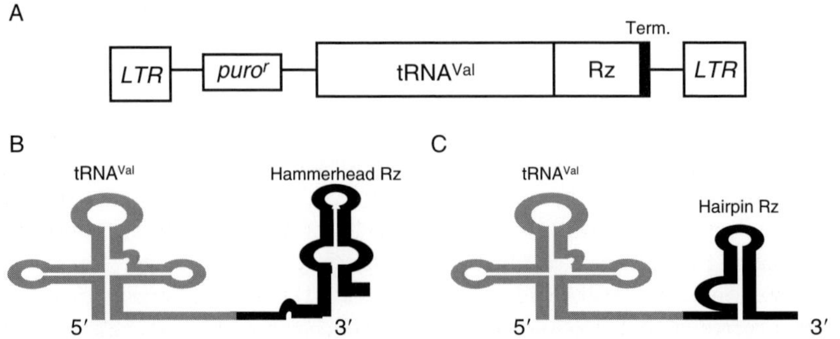

Fig. 2 The typical structures of ribozyme gene vector and ribozyme gene products. (Panel A) Retroviral ribozyme vector containing $tRNA^{Val}$ promoter driven ribozyme expression cassette; (panel B) $tRNA^{Val}$-hammerhead ribozyme hybrid; (panel C) $tRNA^{Val}$-hairpin ribozyme hybrid.

their intended targets within the transfected or transduced cells. Should the ribozyme be stably expressed in cells, the intended target may be inactivated permanently.

Various RNA polymerase II promoters (pol II promoters) and pol III promoters have been tested for driving ribozyme gene expression in a broad range of cell types (Tritz *et al.*, 1999). The pol II promoters tested include cytomegalovirus (CMV)-, respiratory syncytial virus (RSV)- (Bertrand *et al.*, 1997), MMTV (mouse mammary tumor virus)-, and HIV long terminal repeats (HIV-LTR) (L'Huillier *et al.*, 1996) promoters. Pol II promoters can readily be tailored for inducible or cell type-specific expression.

Pol III promoters, typically used by RNA polymerase III to transcribe tRNA and some other small RNAs, are much more efficient than Pol II promoters, and therefore are the most commonly used promoters for ribozyme expression (Tritz *et al.*, 1999). In particular, the tRNA (e.g., $tRNA^{Val}$ promoter) or U6 small nuclear RNA (snRNA) pol III promoters are favored for their high transcription levels in most cells and simple termination signals. A typical tRNA promoter ribozyme expression cassette is shown in Fig. 2A. Unlike the U6 promoter, tRNA promoters are embedded within the tRNA structural genes. Therefore, the transcripts of tRNA promoter-ribozyme expression cassettes produce chimeric RNA molecules with the tRNA at its 5′ end and ribozyme at the 3′ end (Fig. 2B and C). The $tRNA^{Val}$-ribozyme chimera was shown to be efficiently exported from the nucleus by virtue of its tRNA component (Kuwabara *et al.*, 2001), and exhibited a fivefold greater cleavage of cytoplasmic target mRNAs than a $tRNA^{Met}$-ribozyme chimera that remained in the nucleus (Kuwabara *et al.*, 2001). Another pol III promoter construct, juxtaposing the dual extragenic U6 promoter and $tRNA^{Lys3}$ gene/promoter, produces a greater transcriptional activity that resulted in a threefold increase in expression of the $tRNA^{Lys3}$-ribozyme chimera relative to the $tRNA^{Lys3}$ promoter alone (Chang *et al.*, 2002). In addition, the $tRNA^{Lys3}$-ribozyme chimera was distributed evenly between the cytoplasm and nucleus, unlike normal U6 promoter driven ribozymes that remain primarily in the nucleus, which correlated well with increased cleavage of the cytoplasmic target (Chang *et al.*, 2002).

The ribozyme gene expression cassettes can be introduced into cells via DNA transfection or packaged into a viral vector for transduction (Fig. 2A). For transient expression, transfection of DNA vectors is convenient and efficient for delivery of ribozymes for cells that are highly transfectable, since each transfected cell can acquire many copies of vectors to ensure inactivation of the target gene. However, if the transfection efficiency is low, a viral vector such as adenovirus vector would be a better alternative due to their high transduction efficiency and high vector number introduced per transduced cell. Retroviral or lentiviral vectors are commonly used for stable ribozyme gene delivery and thus for stable inactivation of targeted gene functions. A schematic representation of a retroviral vector carrying a tRNA promoter driven ribozyme gene is depicted in Fig. 2A. The vector usually also contains a selectable marker (e.g., drug resistance gene) driven by a pol II promoter. We have previously demonstrated the effectiveness of the hairpin ribozymes expressed from such vectors for gene inactivation in several systems (Feng *et al.*, 2000; Li *et al.*, 2000b).

C. Factors That Influence the Effectiveness of Ribozyme-Mediated Gene Inactivation

Many factors influence the effectiveness of the ribozyme-mediated gene inactivation. For *cis*-cleavage by native ribozymes, the ribozyme and target are automatically colocalized and the cleavage reaction occurs stoichiometrically (one ribozyme to one target), resulting in maximal cleavage efficiency. In contrast, the efficiency of *trans*-cleavage is dependent on the ability to localize the ribozyme to the same subcellular location as the target. Subcellular localization of the ribozyme is largely determined by the promoter and the addition of appended accessory sequences. For example, the packaging signal from MoMLV tethered to a ribozyme significantly increased the cleavage of target sequences that are on the MoMLV viral vector (Tritz *et al.*, 1999). *trans*-Cleavage ribyzome reactions are also dependent on the level of ribozyme gene expression, particularly if the target is expressed at high levels, even though the reaction is catalytic, and some turnover of the ribozyme occurs.

Another factor that affects the cleavage efficiency is the accessibility of the ribozyme-binding site on the target RNA, which is influenced by RNA secondary structure or bound proteins that sterically hinder ribozyme binding. Many techniques have been described to increase the accessibility of ribozyme-binding sites (Chatterton *et al.*, 2004). One approach is to identify the accessible sites prior to ribozyme design. Computer programs such as M-fold have been used to predict target RNA secondary structure prior to ribozyme design (Mathews *et al.*, 1999), but such analyses are generally not predictive. In an experimental approach, Rossi and coworkers described utilizing semirandom oligonucleotide libraries to identify potentially accessible ribozyme-binding sites (Scherr and Rossi, 1998; Scherr *et al.*, 2000, 2001) in which oligonucleotide-directed RNase H-sensitive sites of target mRNA were determined in cell extracts that approximates the intracellular environment. These identified RNase H-sensitive sites correlated highly with efficient cleavage by subsequently engineered ribozymes. Others have used a random ribozyme library to directly identify ribozymes that cleave the target mRNA efficiently *in vitro* (Pan *et al.*, 2003). Although these empirical determinations are more predictive, they are also quite laborious, thus limiting their use for high-throughput target validation.

Another strategy is to increase target site accessibility by physically disrupting the secondary structure of mRNA target to expose the binding sites. For example, hammerhead ribozyme activity was significantly enhanced when the ribozyme was appended to RNA helicase-binding elements such as the poly (A) motif (Fig. 2C, Rz-A60) or a constitutive transport element derived from a retroviral RNA, thereby linking the ribozymes to the unwinding activity of an endogenous RNA helicase (Chatterton *et al.*, 2004; Kawasaki and Taira, 2002; Kawasaki *et al.*, 2002a,b; Warashina *et al.*, 2001). These hybrid ribozymes were shown to cleave target RNA irrespective of its secondary structure. In addition, Kawasaki and Taira (2002) also created heterodimeric ribozymes (maxizymes) that cleave the mRNA target at two separate locations. Binding of the maxizyme to an accessible site (the "anchor") dramatically increased cleavage at a second inaccessible site, presumably by creating more energetically favorable binding conditions for the second site. Moreover, maxizymes also cleaved more efficiently than conventional ribozymes at accessible sites (80% versus 50% reduction in reporter gene activity, respectively). It was suggested that a maxizyme can be designed with a "universal anchor," such as the poly (A) tail, to make this approach generally applicable (Kawasaki and Taira, 2002). Shahi and Banerjea (2002) also constructed multitarget ribozymes by linking two hammerhead ribozymes in tandem in a single transcript. These tandem ribozymes cleaved their target more efficiently than either of the individual ribozymes.

III. RIBOZYMES AS TOOLS IN GENE TARGET DISCOVERY AND VALIDATION

A. Principles of Ribozyme-Based Gene Target Validation

The gene inactivation ability of ribozymes has been exploited for pharmaceutical drug target validation. A typical drug target is a gene or gene product (protein) whose inhibition results in a therapeutic effect. However, prior to the expensive process of drug discovery and development, putative drug target genes are often screened or validated by functional genetics approaches. If manipulation of gene expression leads to a desired therapeutic phenotype, then the gene is validated and may be then selected as a candidate drug target. Traditional approaches for revealing gene function include transgene and gene knockout experiments. In transgene experiments, the candidate gene is overexpressed in cells or in animals to increase the amount and activity of the gene product. This "gain-of-function" can lead to an acquired phenotype, which will help to elucidate function of the gene. A limitation to this approach is that the artificially created transgene may not be subject to the native regulation and thus produce an abnormal physiology. Also, the gain-of-function does not reflect the typical antagonistic actions of drugs.

Somatic cell gene knockout procedures, in contrast, produce a "loss-of-function" phenotype that more accurately mimics drug action. The phenotype that results from a gene knockout can reveal gene function. However, the procedure is usually achieved via DNA recombination, which is complex and not always successful. In addition, many animal knockouts are developmentally lethal, yielding no clues about gene function. Furthermore, the complete and permanent removal of a gene in a gene knockout procedure is not an ideal mimic of the activity of drugs, which only partially and temporarily inhibit their targets. Ribozymes, by contrast, usually only partially inhibit target gene expression and thus better resemble the partial inhibitory activity of drugs, thereby providing an alternative and perhaps more accurate method for target gene validation.

The typical steps for target validation using ribozymes include: (1) determining or developing the functional cell-based assays for disease phenotype, (2) designing ribozymes against the intended target and determining whether the designed ribozymes target only the gene of interest through database searches, (3) construct and prepare ribozyme vectors, (4) transduce cells and perform the functional phenotypic assay for gene validation, and (5) biochemical analysis for mRNA and protein levels in the ribozyme-treated cells to correlate knockdown of the gene to the phenotype alteration. If an expected and desirable phenotype is correlated with gene knockdown, then the gene target is considered functionally validated.

B. Drug Target Discovery Using a Combinatorial Ribozyme Gene Library

1. RATIONALE AND PROCEDURES

We initially described a combinatorial approach based on the use of a randomized ribozyme gene library, and demonstrated that it can be used for screening to identify functionally relevant genes, without prior sequence information (Kruger *et al.*, 2000; Li *et al.*, 2000b; Welch *et al.*, 2000). In this approach, the target recognition sequences encoded by the ribozyme genes are randomized so that theoretically every mRNA in the human transcriptome can be targeted (Fig. 1A). This unbiased phenotype-driven approach is similar to a classical "forward genetics" approach in principle. We called this randomized ribozyme library gene identification approach Inverse Genomics®. Subsequently, Taira and colleagues utilized a similar strategy to identify functional genes using hammerhead ribozyme libraries (Kawasaki and Taira, 2002; Kawasaki *et al.*, 2002a,b).

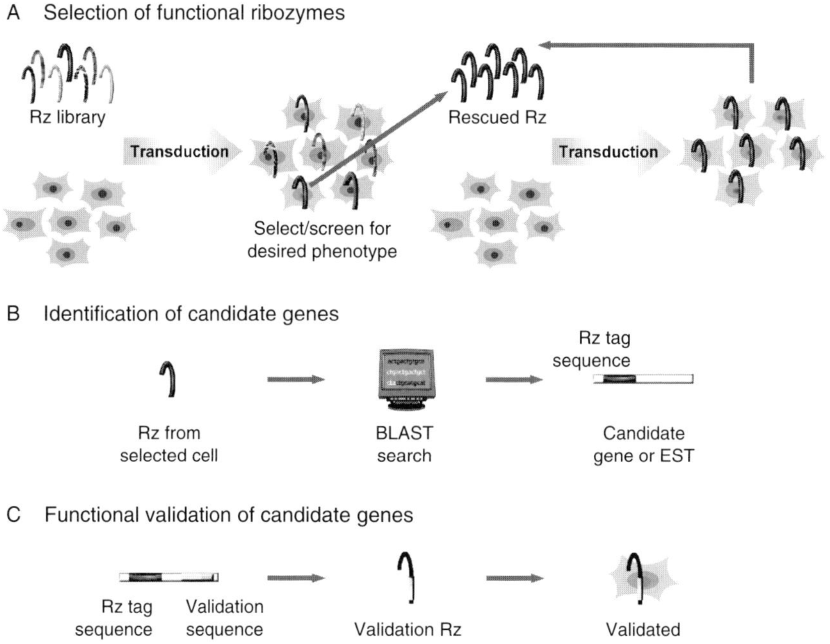

Fig. 3 Target discovery process of Inverse Genomics®.

The typical process of the Inverse Genomics® for gene discovery is depicted in Fig. 3. First, the library is introduced into experimental model cells, usually via retroviral vectors. Thus, a single gene is presumably inactivated in each transduced cell. The population of stably transduced cells is then subjected to a phenotypic screen to isolate the cells that have acquired a phenotype of interest. The ribozymes are then recovered from the isolated cells by polymerase chain reaction (PCR). Due to the high complexity of the ribozyme library, multiple cycles of selections and enrichment of the functional ribozymes may be needed. The enriched ribozymes are then confirmed individually. Next, the confirmed riboyzmes are sequenced to identify the ribozyme sequence tags (RSTs) and the targets (Fig. 3B). Finally, the candidate gene sequences are used to produce additional validation ribozymes that target separate regions of the gene (Fig. 3C). If these validation ribozymes also produce the same phenotype, then the identified gene is validated.

2. HAIRPIN RIBOZYME GENE LIBRARY

The first described combinatorial ribozyme library is the hairpin ribozyme library (Kruger *et al.*, 2000; Li *et al.*, 2000b; Welch *et al.*, 2000).

The retroviral vector construct containing a hairpin ribozyme gene library of randomized target recognition sequences is shown in Fig. 1A, where eight nucleotides in helix I and four nucleotides in helix II are completely randomized. The maximal complexity of the resulting library should be 1.7×10^7 ($=4^{8+4}$). The detailed procedure for ribozyme gene library in a retroviral vector construction (Fig. 2A) was described previously (Kruger *et al.*, 2000; Li *et al.*, 2000b; Welch *et al.*, 2000). A total of 5×10^7 independent transformed colonies were generated, demonstrating that the resultant ribozyme library should have complexity of 1.7×10^7 with 95% confidence. The randomness of the library was also evaluated by sequencing 64 independent colonies from the original transformation. All 64 sequences are unique, suggesting randomization of the sequences. The frequency of the four nucleotides were found to deviate somewhat from the expected 25% (A, 29.0′4.5%; T, 29.0′4.5%; C, 24.2′4.2%; G, 17.7′3.9%), which likely resulted from some inherent bias during chemical synthesis of the library oligos. The actual complexity of the library is difficult to assess, and it may be lower than the maximal complexity due to these variations.

3. HAMMERHEAD RIBOZYME GENE LIBRARY

Taira and colleagues created a randomized hammerhead ribozyme gene library (Kawasaki and Taira, 2002; Kawasaki *et al.*, 2002a,b) conceptually similar to the hairpin ribozyme gene library described in preceding section. This ribozyme gene library has a total of 20 nucleotides of randomized target recognition sequences. It also has a poly A(60) attached to its 3′ end, which is expected to interact with cellular RNA helicase and thus increase the efficiency and specificity of the ribozymes (Fig. 2C). The authors found that the hybrid ribozyme library (Rz-A60 library) significantly increased the positive rate (fivefold) of the screening over the randomized ribozyme library that lacks the poly (A) tail.

4. SUMMARY OF ADVANTAGES OF INVERSE GENOMICS® FOR GENE IDENTIFICATION

Inverse Genomics® has several advantages over many other functional genomics approaches in identifying potential disease drug targets. First, it identifies genes using a phenotype-driven forward genetics approach in which no bias or presumptions are introduced. It also establishes a more direct causal relationship between the identified genes and disease in basically one step. In contrast, other genomics approaches such as comparative expression profiling merely correlate gene expression with disease, and additional gene function studies such as gene knockdown or transgene expression must still be performed to validate the identified genes. Second, ribozyme sequences can be directly used to identify target genes, making gene identification extremely rapid and convenient. Third, ribozymes identified through functional genomics screens can be further coupled to other genomics methods, such as comparative expression profiling, to identify additional gene targets and to elucidate relevant disease pathways. Finally, Inverse Genomics® can be applied to many therapeutic areas, provided that cell-based phenotypic screens or selections are available.

IV. RIBOZYME-BASED GENOMIC TECHNOLOGY IN CANCER GENE TARGET DISCOVERY AND VALIDATION

Cancers are genetic diseases caused by multiple alterations of three types of genes: oncogenes, tumor suppressors, and DNA stability genes (Vogelstein and Kinzler, 2004). The cancer-related alterations include mutations leading to constitutive activation of oncogenes, inactivation of tumor suppressor and DNA repair genes, and aberrant expression of these gene products. Identification of these etiological genetic factors and understanding their functions are the foundation for future cancer therapeutics.

The key to discover and validate cancer-related genes using Inverse Genomics® is to identify relevant cancer cell model systems and develop robust genetic screens or selections for desired therapeutic phenotypes. Ideal assays can positively select for a disease-relevant phenotype with high reproducibility and signal-to-noise ratio, either based on an inherently measurable disease phenotype or on a surrogate reporter gene for the disease phenotype. Cancer phenotypes include loss of cell cycle control, elevated cell proliferation rate, resistance to apoptosis induced by various stress or external death signals, capacity for indefinite proliferation, and increased migratory and invasion potential. Most of these attributes can be examined and even utilized to separate transformed cells from nontransformed cells (Table I).

There are two cell-based functional genomic selections for cancer-related phenotypes. One selects live cells gaining proliferation from gene-silencing agents such as ribozyme (positive selection). However, this method yields gene targets that negatively regulate cell growth ("tumor suppressor" type of targets). While positive selection is technically easier to achieve, tumor suppressors are not ideal drug targets since the majority of drugs are antagonists. However, this selection can be coupled with gene expression profiling, comparing the parental and gene-silenced cells to identify oncogene candidates (see following section). Another method selects cells that either cease to grow or undergo apoptosis, which is considered negative selection and technically more challenging. However, it yields oncogene

Table 1 Oncology Assays

Phenotypes	Assays	Treatments
Transformation		
• *In vitro*	Proliferation[a]	Anchorage dependent (liquid culture (LC)) Anchorage independent (soft agar (SA))
	Apoptosis/survival[b]	10% serum culture (no stress) 1% serum for growth factor deprivation stress DNA-damage agents: etoposide Protein unfolding stress: Brefeldin A Extrinsic stimuli including TRAIL and Fas Ab Cell detachment (anoikis)
	Clonogenicity[c]	Anchorage dependent/independent
• *In vivo*	Tumorigenesis	
Metastasis/angiogenesis		
• *In vitro*	Migration[d] Invasion[e]	
• *In vivo*	Metastasis	

[a] Can usually easily measured by colorimetric assay.
[b] Colony formation assays are used.
[c] Apoptosis can be measured via survival or apoptosis-specific assays, for example, caspase 3/7 activity, DNA fragmentation (TUNEL), annexin V, and so on.
[d] Migration can be assessed by a Boyden chamber chemotaxis assay and wound migration assay.
[e] Invasion can be assessed by the Chemicon Cell Invasion Assay Kit, where cell culture inserts with 8-μm pore membranes are coated with a solid gel of basement proteins prepared from the Engelbreth–Holm–Swarm (EHS) mouse tumor.

targets that are therapeutically relevant. Once targets are identified from either library screening or expression profiling, phenotypic validation becomes the next important step. In addition to functional validation, candidate cancer gene targets must also fulfill additional criteria. First, the molecular epidemiology should ideally show an up-regulation of the identified genes in human cancers relative to normal tissues. This is important for drug-target specificity and because most drugs antagonize or inhibit overly expressed or overly active proteins. Second, the gene products should be readily "druggable," meaning that they are either enzymes containing active sites that can be readily targeted and inactivated by small molecule drugs or cell surface proteins/receptors and secreted proteins that can be targeted by either small molecule drugs or protein-based therapeutics such as antibodies. One should bear in mind, however, that the definition of "druggability" is everchanging, as technology to identify and develop drugs that target other interactions (e.g., protein–protein, protein–nucleic acids) becomes more facile. Finally, it is important to understand the biological pathway and thus the potential drug actions for each target, which is not only critical toward predicting the efficacy but also the potential toxicology profile.

However, for this chapter, we will discuss only the identification and validation of cancer gene targets, using gene inactivation via ribozyme (or siRNA). The following section describes several cancer experimental systems in which cancer-related genes have been identified and validated.

A. HeLa/HeLaHF Cervical Cancer Cell System and Anchorage-Independent Growth

Cancers of cervical origin are the second most common gynecological cancers and HeLa is the most well-characterized experimental cell line of cervical cancer (adenocarcinoma) origin (Masters, 2002). HeLa cells show a strong transformed phenotype including anchorage-independent growth *in vitro* and tumorigenesis *in vivo*. HeLaHF is a revertant variant isolated from HeLa cells following exposure to the mutagen ethylmethane sulphonate (EMS) (Boylan *et al.*, 1996). HeLaHF demonstrates a "nontransformed" phenotype exhibited by the appearance of flat and nonrefractile morphology, significantly decreased growth in soft agar, and loss of *in vivo* tumorigenicity (Boylan *et al.*, 1996) (Fig. 4A and B). Cell fusion studies indicated that the transformation reversion likely resulted from activation of tumor suppressor(s). Since HeLa and HeLaHF constitute an isogenic cell pair with distinct transformation phenotypes, the pair is an ideal system for identification of genes involved in cell transformation and/or transformation reversion. Many of these genes could be interesting candidate targets for cancer therapeutics.

Anchorage-independent growth is a hallmark of cell transformation, and this *in vitro* assay most closely mimics *in vivo* tumor growth. Growth in soft agar, a semisolid culture medium, measures cell anchorage-independent growth potential. The difference in soft-agar growth between HeLa and the HeLaHF is easily visible and measurable (Fig. 4A and B), reflecting the reduced transformation potential of the HeLaHF revertant cell line. We recently developed a 1-week 96-well soft-agar assay, which makes it possible to screen and validate multiple identified targets (Ke *et al.*, 2004a) (Fig. 4B). This format, together with the traditional larger format soft-agar assay (Claassen *et al.*, 2004) (Fig. 4A and 9C), can be used to characterize a large number of candidate genes derived from the HeLa/HeLaHF cell pair rapidly and reliably. In addition, this HTS format can readily be coupled with other 96-well format cell-based oncology assays (Table I) (Fig. 4B), for example, apoptosis, liquid culture growth/survival, and so on, to comprehensively analyze a large number of cancer gene targets. This integrated and multiplexed cell-based approach for gene identification using ribozyme and siRNA technology (Fig. 4C) greatly enhances the productivity of cancer target validation.

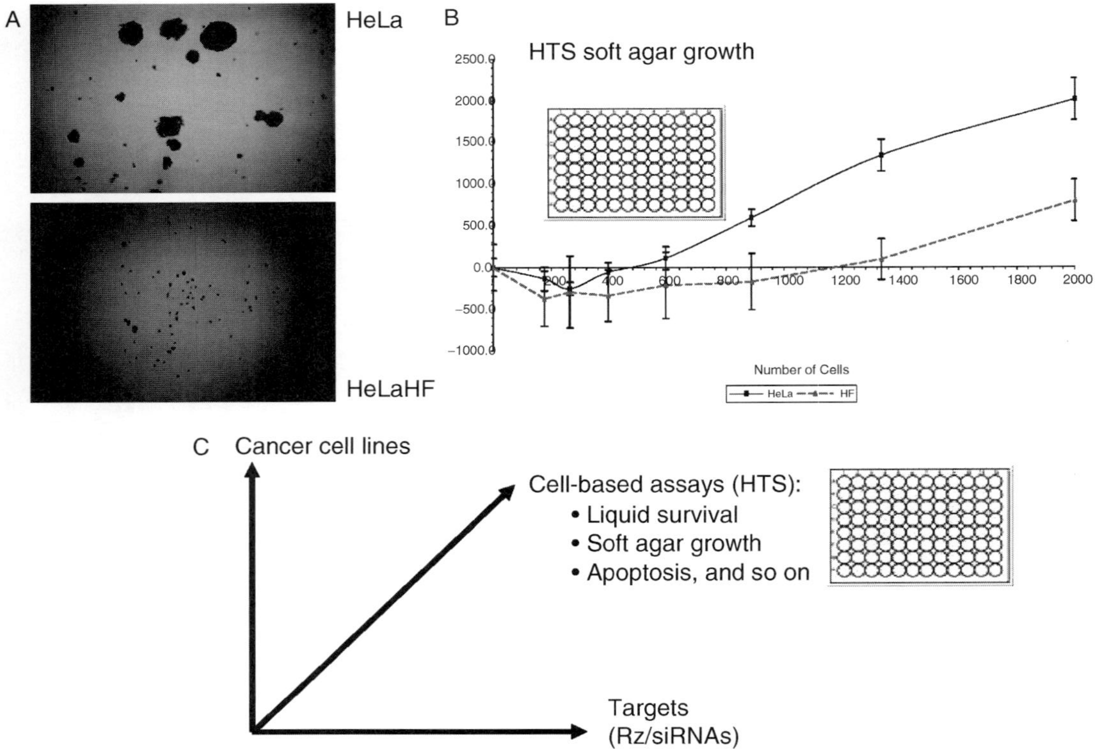

Fig. 4 Cell-based oncology assays. (A) Soft-agar colony formation assay measuring anchorage-independent cell cloning efficiency. A desired number of HeLa or HeLaHF cells are plated in the soft-agar media in a 10-cm dish as described (Claassen *et al.*, 2004; Ke *et al.*, 2004a). Colonies are allowed to grow for 3 weeks at 37 °C with 5% CO_2. Microscopic morphology of HeLa cell colonies is shown. Macroscopic overview of colonies can also be visualized and recorded using Q-count (Claassen *et al.*, 2004) (Fig. 9C). (B) One-week 96-well soft-agar assay measuring overall anchorage-independent growth. Different numbers of HeLa or HeLaHF cells were plated into the 96-well soft-agar media as described (Ke *et al.*, 2004a). Cells were allowed to

1. TUMOR SUPPRESSORS IDENTIFIED THROUGH DIRECT INVERSE GENOMICS®

We initially used our hairpin ribozyme library in nontransformed HeLaHF cells to select for ribozyme-induced transformed cells (Welch *et al.*, 2000). The stably transduced cells were seeded into soft agar and incubated for 3 weeks. A modest increase in colony number was observed in the library-transduced cells compared to the negative control ribozyme-transduced CNR3 or TL1 cells. After several rounds of enrichment, several ribozymes from the library-transduced cells were recovered and functionally confirmed. Rz568 targets the human orthologue of *Drosophila ppan* gene, which was known to be involved in cell cycle and its overexpression could lead to cell death, consistent with our observation and thereby verifying the utility of the Inverse Genomics® approach for gene identification. Rz619, even though its target is unknown, displays stronger transformation potential than that of Rz568 in soft-agar assay. The transformation effect of Rz619 was also confirmed by its ability to mediate tumor growth of HF cells in a xenograft tumor model (Fig. 5). This result also suggests *in vivo* tumor model can potentially be used for target identification and validation via ribozyme-mediated gene inactivation (Fig. 5, see following section).

The validation data for the third ribozyme, Rz-HFSC1, are presented in following section. This ribozyme, when reintroduced into HeLaHF cells, increased cell growth in 96-well soft-agar culture (Fig. 6B). A BLAST search revealed BRI/ITM2B (integral membrane protein 2B, accession number: NM_021999) as a candidate gene target. This hit was a 15/16 match with a mismatch located at the last nucleotide of helix I, which is considered tolerable (Li *et al.*, 2000b; Welch *et al.*, 2000). siRNAs against ITM2B also increased HeLaHF cell soft-agar growth (Fig. 6C), validating ITM2B gene as a putative tumor suppressor gene.

ITM2B is a 266-amino acid type II membrane protein. It contains an ATPase and a BRICHOS domain (137–231AA). BRICHOS domains are found in a variety of proteins implicated in dementia, respiratory distress, and cancer. ITM2B is widely expressed in a variety of tissues. Although its biological function is unknown, the mouse ITM2B orthologue has been implicated as a proapoptotic gene (Fleischer *et al.*, 2002a,b). ITM2B is a cytoplasmic and mitochondrial protein that also contains a BH3 domain similar to that of the Bcl-2 family members, which are key regulators of apoptosis. Expression of

grow at 37 °C for 1 week and quantitated AlamarBlue staining. The relationship of AlamarBlue staining to cell number is depicted. (C) Concept of multiplexed cell-based phenotype assays for comprehensive phenotype analysis. Cells containing either ribozymes or siRNAs can be assayed in 96-well plates under various treatments and subjected to a variety of phenotype readouts (growth/survival/apoptosis). (See Color Insert.)

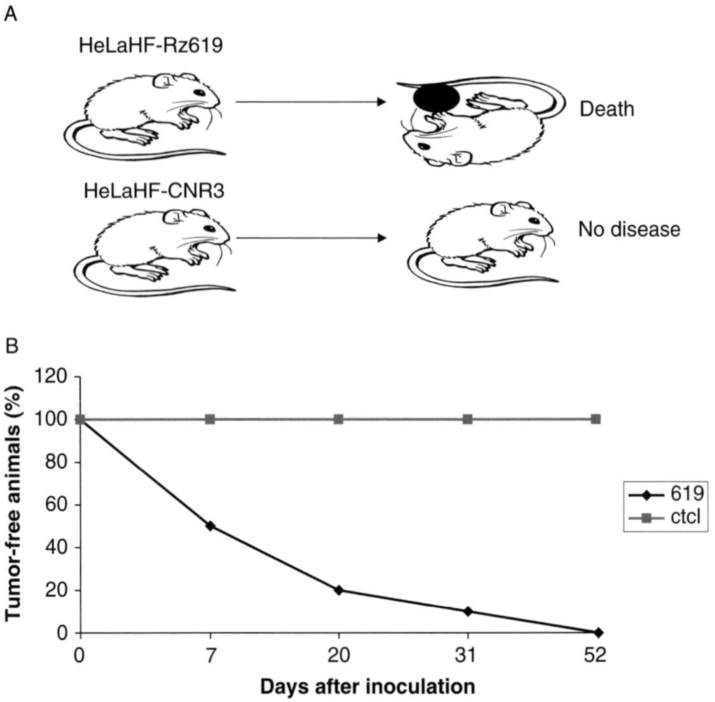

Fig. 5 Tumor growth of HeLaHF cells transformed by Rz619. (A) HeLaHF cells with either the control ribozyme (CNR3) or Rz619 are injected subcutaneously into athymic nude mice and allowed to form tumors. (B) The survival of animals (percentage of tumor-free animals) is presented.

ITM2B induces apoptosis in IL-2-stimulated cells, where it interacts with the antiapoptotic protein Bcl-2, but not with the proapoptotic protein, Bad. Mutation of the critical L and D residues within the BH3 domain abolished the ability of ITM2B to promote apoptosis. Our results verify the importance of ITM2B in cancer, and further suggest that it functions as a tumor suppressor. This function is consistent with molecular epidemiology data demonstrating low expression levels in several cancer tissues, including colon, pancreatic, breast, and prostate cancers (NCBI virtual Northern).

2. ONCOGENES IDENTIFIED THROUGH GENE EXPRESSION PROFILING

As discussed in the preceding section, using the screen for potential tumor suppressors in HeLaHF cells, we identified several tumor suppressor genes, validating the Inverse Genomics® approach. However, since our goal is to

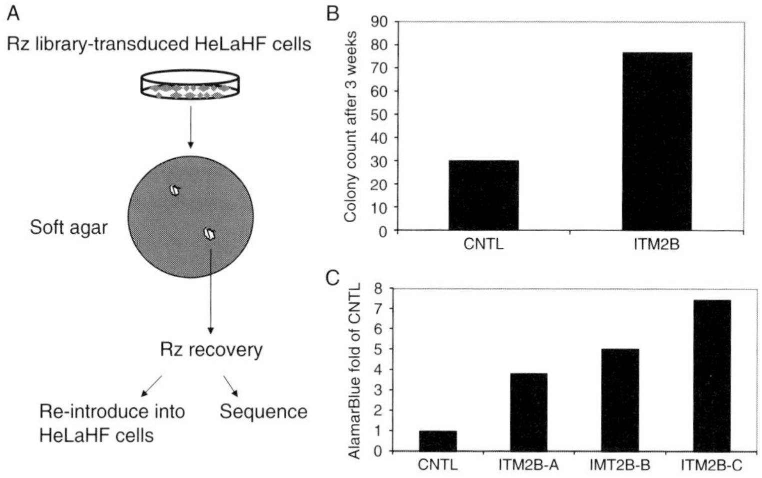

Fig. 6 Inverse Genomics® screening for tumor suppressors by soft-agar growth. (Panel A) Scheme of Inverse Genomics® soft-agar growth selection for ribozymes inducing anchorage-independent growth. HeLaHF cells containing the ribozyme library were plated onto the soft-agar plates. Soft-agar-forming colonies were then picked, expanded, and ribozymes recovered. Ribozymes were reintroduced into HeLaHF cells to confirm the transformation phenotype; at the meantime, ribozymes were also subjected to sequencing analysis to identify potential targets. (Panel B) Soft-agar growth of HeLaHF containing IMT2B ribozyme. The soft-agar growth using the 96-well soft-agar assay was shown for HeLaHF cells containing either the control or ITM2B ribozyme. (Panel C) Soft-agar growth of HeLaHF cells transfected with ITM2B siRNAs. *In vitro* transcribed siRNAs against IMT2B or the control siRNA were transiently transfected into HeLaHF cells in 96-well plates. The cells were then harvested and the equivalent number of cells were seeded in the 96-well plates of liquid and soft-agar cultures.

identify oncogenes rather than tumor suppressor genes, we performed gene expression profiles to identify potential oncogenes that may be up-regulated when the cells become transformed. We compared HeLa cells and the ribozyme transformed revertant HeLaHF cells to the parental nontransformed HeLaHF cells using Affymetrix GeneChip® analysis and cDNA-based microarrays (Fig. 7). Using this approach, we identified genes with significantly altered expression levels in the revertant compared to parental cells that are likely associated with the transformation phenotype (Table II). By definition, the genes up-regulated in the transformed HeLa or HeLaHF/Rzs cells relative to the nontransformed HeLaHF cells have possible oncogenic properties, while the down-regulated ones have possible tumor suppressor properties. A similar comparison was also made with the *in vivo* xenograft tumors derived from Rz619-transformed HeLaHF cells and HeLaHF cells (Fig. 7) with regards to its potential role in *in vivo* tumor formation. The genes that were found to be associated with the transformation phenotype both *in vitro* and *in vivo* were prioritized for downstream characterization. The comparison of isogenic

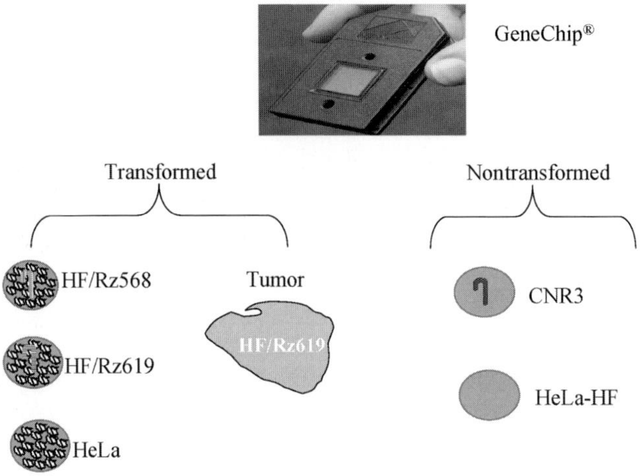

Fig. 7 Expression profiling analysis of the HeLa/HeLaHF isogenic cell pairs and HeLaHF/HeLaHF-Rz tumors to identify genes associated with HeLa cell transformation. RNAs from either transformed cells (HF/Rz568, HF/Rz619, or HeLa) or tumors (HF-Rz619) were prepared and compared to RNAs from nontransformed HF/RzCNR3 or parental HF cells using the Affymatrix GeneChip®. The common changes were selected for phenotype analysis.

cell pairs generated by ribozymes has significant advantages over just comparing HeLa and HeLaHF cells because of potentially nonrelated changes among the comparing pairs. The validation data for some of the candidate targets derived from this pathway analysis are presented in the following section.

EphB4, or HTK (hepatoma transmembrane tyrosine kinase, accession number: M_004444), is one of the up-regulated genes in HeLa as compared to that in HeLaHF, suggesting possible oncogenic properties. For functional validation of EphB4, ribozyme HTK-317, designed to target the EphB4 message, was introduced into HeLa cells via retroviral transduction and the stably transduced cells (pooled) were plated in 100-mm dish for soft-agar colony formation. Cloning efficiency (colonies >50 cells) was determined after 3-week growth. The results showed that ribozyme HTK-317 caused 50% reduction in soft-agar cloning efficiency, as compared to that of the control ribozyme (dTL3) (Fig. 8A), which correlated to ribozyme-mediated EphB4 down-regulation (Fig. 8A) as assessed by real-time RT-PCR. In contrast, no difference was detected in anchorage-dependent growth (data not shown). This result was also further confirmed by transiently transfecting chemically synthesized siRNA targeting EphB4 into HeLa cells in the described 96-well format soft-agar culture (Ke *et al.*, 2004a) (Fig. 8B). Since EphB4 was shown to be significantly up-regulated in breast cancers as compared to the corresponding normal tissues (see following section), we also tested the siRNA effect on a breast cancer line T47D. A combination of EphB4 A/B

Table II Summary of Genes Involved in HeLa Cell Transformation

Gene	Expression in HF	Phenotypes in HeLa	Phenotypes in non-HeLa cells	Pathway	References
P53	No change at mRNA levels, but stabilized in protein	Reduced anchorage-independent growth, via anoikis (partial)	Tumor suppressor	P53 pathway	Yu *et al.* (2006)
PHTS	Up-regulated (3×)	Reduced anchorage-independent growth, not via anoikis (partial)	Tumor suppressor	Not p53 pathway	Yu *et al.* (2006)
Receptor-X	Down-regulated > 10×	Increase resistance to apoptosis, anchorage-independent growth	Oncogene	Cell adhesion pathway/integrin/Fyn, and so on	Ke *et al.* (unpublished)
Kinase X	Down-regulated	Increase resistance to apoptosis	Oncogene	Apoptosis	Ke *et al.* (unpublished)
Ppan	No significant change at mRNA levels	Block cell cycle or cause apoptosis		Cell cycle	Welch *et al.* (2000)
NR4A1, 2, and 3	Down-regulated	Prosurvival, resistance to apoptosis	Sensitize to apoptosis	Apoptosis	Ke *et al.* (2004b)
IGFBP3	Up-regulated	Tumor suppressor, reduced	Tumor suppressor	P53	Athanassiou *et al.* (1999)
ITM2B	No change at mRNA levels	Reduces apoptosis	Tumor suppressor/proapoptosis	Mitochondrial/Apoptosis	
EphB4	Down-regulated	Increases soft-agar growth	Proepithelial cell growth		
CXCR4	Down-regulated	Increase soft-agar growth	Oncogene	Reduced survival/migration	Ke *et al.* (unpublished)
PLK	Down-regulated	Inactivation cause cell-growth arrest	Oncogene	Cell cycle	Ke *et al.* (2004a)
Fyn	Down-regulated		Oncogene	Adhesion	Ke *et al.* (unpublished)

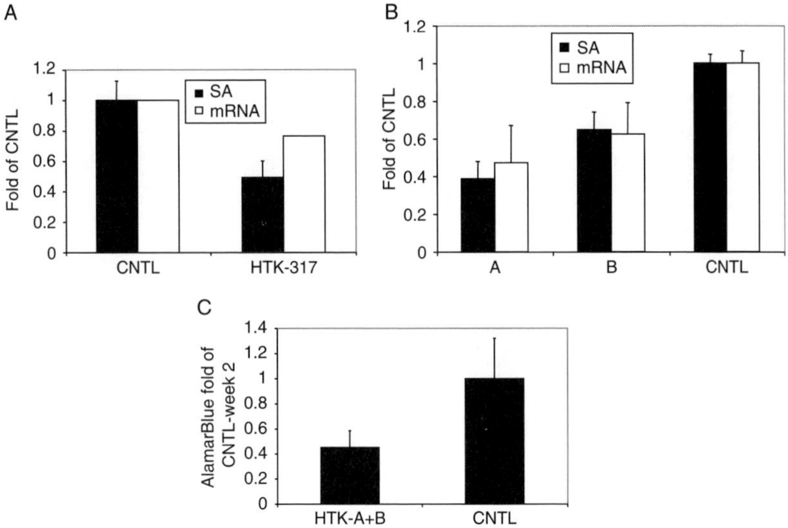

Fig. 8 Oncogene properties of EphB4. (Panel A) Soft-agar growth of HeLa cells transduced with ribozyme-317 targeting EphB4 (312–326 on accession number: NM_004444, cleavage position: 317) (RST): 5′-GTT GGG TCC CAC GGC G-3′). The stably transduced HeLa cells with either control or HTK-317 ribozyme were plated onto either liquid or soft-agar culture in the 10-cm dishes. Colony formation in liquid culture was quantitated 1 week later, while colony formation in soft agar were quantitated 3 weeks later. The soft-agar colony numbers were normalized against the liquid culture colony number, and the ratio [colony-forming efficiency (CFE)] is displayed. Each sample was performed in triplicate. EphB4 mRNA level is also determined by real-time RT-PCR for the HeLa cells containing either the CNTL or the HTK-317 ribozymes. The average of two experiments is presented (p-value: < 0.03). (Panel B) Soft-agar growth of HeLa cells transfected with siRNAs. HeLa cells were transfected with chemically synthesized siRNAs against EphB4: (A: aatgtcaagacgctgcgtctg; B: aatgtcaccactgaccgagag). Cells were plated into the 96-well plates in either liquid culture or soft-agar culture 24 h posttransfection. The cell growth was measured by AlamarBlue staining on day 1 for liquid culture and day 7 for soft-agar culture. The soft-agar readings were normalized against the day 1 readings of the liquid cultures. Each sample was performed in triplicate. p-values for: siRNA A, <0.002; siRNA B, <0.02. At the meantime, real-time RT-PCR analysis of EphB4 mRNA in HeLa cells transfected with siRNAs against EphB4 was also conducted 48 h after transfection (samples are in duplicates and the average is presented). (Panel C) Soft-agar growth of T47D cells transfected with siRNAs (mixture of A+B) against EphB4. siRNA were transfected into T47D cells and then seeded in 96-well plate of liquid and soft-agar cultures. The cells growth was assayed as in (B) [the samples are triplicates (p-value: 0.005)].

siRNAs also suppressed T47D soft-agar growth (Fig. 8C), while no effect on cell proliferation (anchorage-dependent growth) was observed (data not shown). Therefore, knockdown of EphB4 in multiple transformed cell lines leads to reduced transformed phenotype.

EphB4 is a member of erythropoietin-producing hepatocellular (Eph) receptor tyrosine kinases (RTKs), the largest subfamily of RTKs (Andres *et al.*, 1994; Bennett *et al.*, 1994; Stapleton *et al.*, 1999). Like other Ephs, EphB4 is a large multidomain type I receptor membrane protein, containing an extracellular domain (ECD), a membrane-spanning region and an intracellular region, which contains a conserved tyrosine kinase domain and an sterile alpha motif (SAM) domain. Its functional ligand, ephrin B2, binds to the EphB4 ligand-binding domain located on its ECD. These functional domains together are responsible for the activation of receptor kinase activities and downstream signal transduction events. A ras/MAPK-signaling cascades have been implicated in EphB4 signaling (Kim *et al.*, 2002), and an antibody recognizing the EphB4 ECD was found to be agonistic for its tyrosine phosphorylation (Bennett *et al.*, 1994).

EphB4 is widely expressed in fetal and adult tissues as well as many malignant cell lines (Bennett *et al.*, 1994). Ephrin B2 ligand and EphB4 are believed to play important roles in various biological processes during embryonic development, including the targeting behavior of migratory neurons, erythropoiesis (Inada *et al.*, 1997), angiogenesis, and vascular network assembly (Helbling *et al.*, 2000; Suenobu *et al.*, 2002). In addition, EphB4 is also implicated in cell growth and cancer development. Overexpression of EphB4 has been observed in several cancers, including colon cancer (Stephenson *et al.*, 2001), small cell lung cancer (SCLC) (Gerety *et al.*, 1999), leukemia-lymphoma cells (Steube *et al.*, 1999), glioma, ovarian cancers, and breast cancer (Berclaz *et al.*, 1996). Coexpression of the receptor and ligand has been observed in colon carcinoma (Liu *et al.*, 2002) and SCLC (Gerety *et al.*, 1999). Interestingly, it was also observed that expression of the ligand, which is normally present, can be lost during the progression of carcinogenesis (Nikolova *et al.*, 1998). In addition, transgenic expression in mammary tissue of mice produced an invasive phenotype in mammary tumors associated with increased cell proliferation and reduced apoptosis (Munarini *et al.*, 2002). Thus, these data, in general, support the role of EphB4 as an oncogene.

NR4A1, 2, and *3*, three members of the Nur subfamily of nuclear receptors, are all up-regulated in transformed HeLa cells as well as in tumors derived from the Rz619-retransformed HeLaHF cells, as compared to non-transformed HeLaHF cells (Ke *et al.*, 2004a) (Fig. 6), suggesting potential oncogenic properties. NR4A1, also named nerve growth factor-induced clone B (NGFI-B), Nur77, TR3, and NAK-1, is a transcription factor closely related to NR4A2 (Nurr1, TINUR, HZF-3, RNR-1) and NR4A3 (NOR1, MINOR) (Law *et al.*, 1992; Ohkura *et al.*, 1994). NR4A1 silencing mediated by siRNA transient transfection of HeLa cells led to a drastic reduction in both anchorage-dependent growth (not shown) and anchorage-independent growth, which correlated with mRNA down-regulation (Fig. 9A). The same effect has been observed in many other cancer cells including PC3 prostate cancer cell line, DLD1 colon cancer cell line, and so on (Table III). The reduced growth seems largely due to increased apoptosis, which was observed when transient transduction was conducted with lentiviral siRNA vector in HeLa and M14 cells (Ke *et al.*, 2004b) (Fig. 9B, Table III). However, attempts to generate stable NR4A1-silenced HeLa cells using the same lentiviral vector failed, most likely due to counter selection (not shown). This is not surprising considering NR4A1 siRNA induces apoptosis and plays a prosurvival role.

Similar antiapoptotic properties have also been observed for NR4A2 using the same transient siRNA transduction approach (Ke *et al.*, 2004b) (Fig. 9B, Table III). Interestingly, the lentiviral NR4A2-siRNA vector yields stably silenced HeLa cells, although transient transduction of the same vector induces apoptosis (Fig. 9B). Apparently, the cells eventually adapt to the reduced NR4A2 expression. However, these stably transduced cells

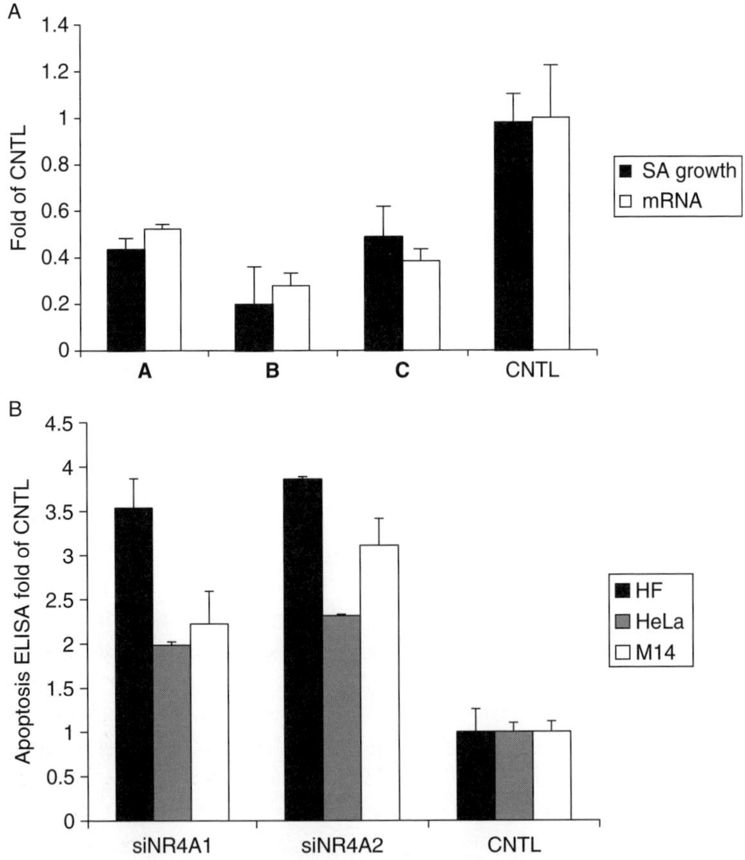

Fig. 9 (*continued*)

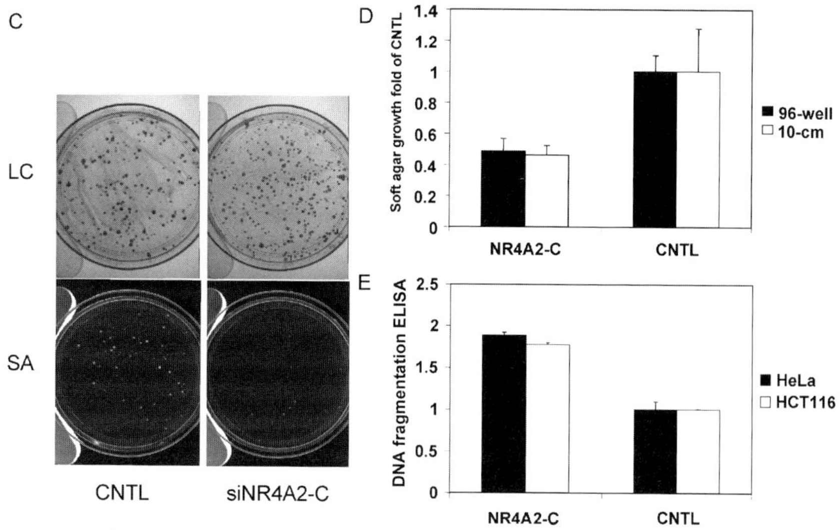

Fig. 9 Prosurvival effect of NR4A1 and NR4A2. (Panel A) Transient transfection with NR4A1 siRNAs. HeLa cells were transfected with siRNAs against NR4A1 [(A) sense strand siRNA: GUUCGAGGACUUCCAGGUGtt and antisense strand siRNA: CACCUGGAAGUC-CUCGAACtt; (B) sense strand siRNA: GGAAGUUGUCCGAACAGACtt and antisense strand siRNA: GUCUGUUCGGACAACUUCCtt; (C) sense strand siRNA: GAUCUUCAUGGA-CACGCUGtt and antisense strand siRNA: CAGCGUGUCCAUGAAGAUCtt]. The transfected cells were plated into the 96-well liquid and soft-agar culture plates as described in Fig. 6. (Panel B) TUNEL-based apoptosis assay for cells after transient transduction by lentiviral vector expressing siRNA (siNR4A1B) against NR4A1 and NR4A2 (siNR4A2). Transduction was conducted with HF, HeLa, and M14 cells. Apoptosis was measured 4 days after transduction using the DNA-fragmentation ELISA kit (Roche). (Panel C/D) Cloning efficiency (both liquid culture and soft-agar culture) of stable NR4A2-silenced HeLa cells. Stable HeLa cells containing either the control siRNA or NR4A2C siRNA vectors were generated through transduction and stable selection. Five hundred cells were plated in the liquid culture media and allowed to grow for 7–10 days before the colonies were stained with Coomassie Blue and counted by Q-count. For soft-agar culture, 5000 cells were plated and allowed to grow for 3–4 weeks before the colonies were counted by Q-count. The colony image in liquid culture and soft-agar culture were shown in (C), and the soft-agar CFE was shown in (D) along with soft-agar growth quantitated by 96-well soft-agar method. (Panel E) Anoikis assay. Stable HeLa or HCT116 cells with either the CNTL or NR4A2 siRNA were seeded into methylcellulose media and apoptosis was detected using the DNA-fragmentation ELISA assay 18 h later as described.

still display greatly reduced anchorage-independent growth, with minimal reduction on anchorage-dependent growth, as measured by cloning efficiency assays (Fig. 9C and D). Further experiments indicate that this reduced anchorage-independent growth was largely due to increased anoikis (Ke *et al.*, 2004b). In addition, our preliminary result also points to the similar prosurvival effect for NR4A3.

Table III Properties of NR4A1, 2, and 3

Phenotypes	NR4A1	NR4A2	NR4A3
Soft-agar (SA) growth	HeLa, PC3, DLD1, AsPC1, A2058, U87	HeLa, PC3, DLD1, HCT116, AsPC1	HeLa
Liquid culture (LC)	HeLa, PC3	HeLa	
Apoptosis	HeLa, M14, DLD1, HCT116, U87	HeLa, M14	

Note: HeLa, cervical cancer; PC3, prostate cancer; DLD1 and HCT116, colon carcinoma; A2058 and M14, melanoma; U87, glomoa; AsPC1, pancreatic cancer.

NR4A1, 2, and 3 have been previously implicated in cell growth/survival/apoptosis. In self-reactive immature thymocytes, NR4A1 expression was induced and increased apoptosis was observed following stimulation of the T-cell receptor (TCR) (Cheng *et al.*, 1997; Liu *et al.*, 1994; Woronicz *et al.*, 1994). Recent studies also indicate that NR4A1 translocates to the mitochondria and induces cytochrome *c* release and apoptosis in LNCaP human prostate cancer cells in response to apoptotic stimuli (Li *et al.*, 2000a). In other circumstances, NR4A1 was reported to play antiapoptotic functions (Bras *et al.*, 2000; Suzuki *et al.*, 2003). NR4A3 plays redundant functions with NR4A1 in T-cell apoptosis (Cheng *et al.*, 1997). In contrast, for NR4A2, only prosurvival effects have been observed in neuronal cells (Le *et al.*, 1999). Therefore, the cumulative information suggests that the Nur77/NGFIB family of nuclear hormone receptors seems to share similar prosurvival effects at least in certain cancer cells, and thus they could potentially serve as anticancer therapeutic targets (Table III).

Our gene profiling analysis has identified other well known oncogenes that are up-regulated in HeLa cells. We have confirmed the causal effect of many of these genes using RNAi, including polo-like kinase (PLK) (Ke *et al.*, 2004a,b; Yu *et al.*, 2005), chemokine receptor CXCR4, and protooncogene *c*-Fyn (Ke *et al.*, unpublished observation). We have also validated several down-regulated genes in HeLaHF cells as potential tumor suppressors, including a novel, putative HeLa tumor suppressor PHTS (Yu *et al.*, 2006). Other investigators also used similar expression profiling of HeLaHF transformation reversion coupled with transgene experiments to identify IGFBP3 and Dickkopf-1 (DKK-1) as tumor suppressors in the p53 pathway (Athanassiou *et al.*, 1999; Mikheev *et al.*, 2004). All these results demonstrate the power of coupling expression profiling with phenotype validation.

In summary, integrating Inverse Genomics®, expression profiling, and multiplexed cell-based assays for cellular transformation, we were able to effectively identify and validate relevant cancer genes from the HeLa/

HeLaHF isogenic cell line pair as well as other cancer cell lines. In addition, this approach enables us to reveal the pathways that are involved in HeLa cell transformation including anoikis (p53) pathway, apoptosis pathways (NR4A1, 2, and 3, kinase-X), adhesion pathways (Fyn/integrins), and so on (Table II). The genes identified include novel genes not previously known to be involved in cancer development, demonstrating the utilities of the integrated cell-based validation approach.

B. Activation of Apoptosis Induced by External Stimuli

Another hallmark of cancer development is resistance to apoptosis. Apoptosis is normally triggered either intrinsically by stress conditions (DNA damage, protein misfolding, and so on) through the activation of caspase 9 or extrinsically by extracellular death ligands (FasL, TRAIL, TNF-α) through activation of caspase 8. These two pathways converge downstream leading to the activation of caspases 3 and 7 to irreversibly commit cells to apoptosis. These pathways are regulated at many steps by a large number of repressors, such as inhibitors of apoptosis (IAPs), to precisely control the induction of and commitment to apoptosis. Cancers feature mutations in genes that enhance these antiapoptotic repressors or that inactivate proapoptotic members of the apoptosis machinery. The apoptosis repressors are thus potential therapeutic targets.

For the identification of therapeutic targets, induction of the extrinsic pathway by the addition of death ligand is the preferred experimental method because apoptosis induction is rapid and synchronized, leading to assay readouts with a high signal-to-noise ratio. In contrast, induction of the intrinsic apoptosis pathway through ultraviolet irradiation or chemotherapeutic agents is slower and unsynchronized within a cell population. Apoptosis can be directly measured by detection of apoptotic markers, such as exposure of phosphatidylserine on the plasma membrane outer leaflet, where apoptotic cells can be labeled and separated from nonapoptotic cells by fluorescence-activated cell sorting (FACS). For the Inverse Genomics® technology, the ribozymes from isolated apoptotic cells can then be used to identify therapeutic genes. Alternatively, apoptosis can be indirectly measured by selection of cells that survive an apoptotic stimulus. This approach is readily achievable technically since it is a positive selection. However, targeting proapoptotic genes enhance survival, while targeting antiapoptotic genes promotes apoptosis and is unlikely to be identified.

At Immusol, we developed assays that directly measure apoptosis to identify potentially therapeutic genes using our Inverse Genomics® technology. We first identified cell lines that resist apoptosis after addition of exogenous death ligands but become sensitized to extrinsic pathway-induced apoptosis

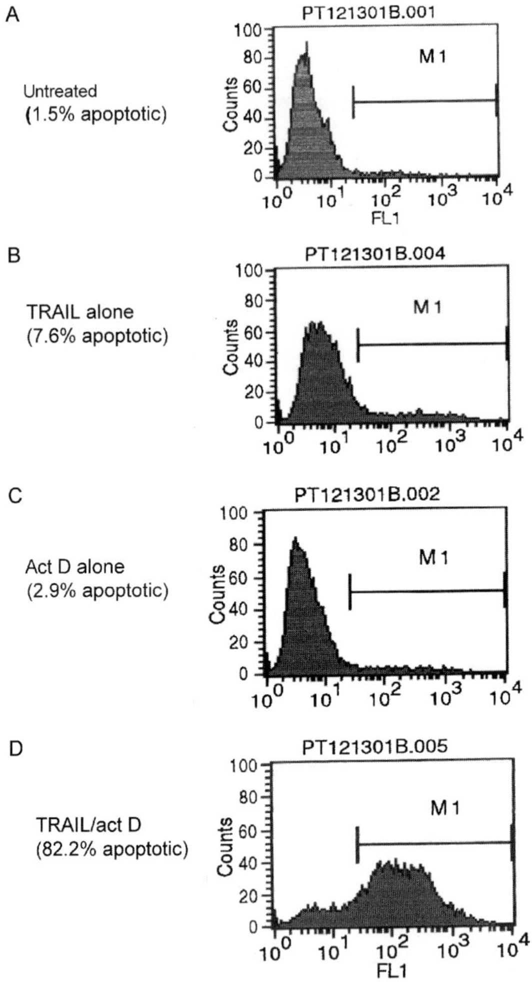

Fig. 10 Death ligand-induced apoptosis project scheme. AsPC-1 pancreatic cancer cells were either left untreated (panel A) or treated with the death ligand TRAIL (panel B), actinomycin D (act D, panel C), or the combination of TRAIL and act D (panel D) for 16 h prior to harvest and staining of cells using the fluorescein isothiocyanate (FITC) *in situ* cell death detection kit (Roche), which fluoresces apoptotic cells. Cells were then analyzed by FACS to determine the percentage of fluorescent or apoptotic cells. This figure illustrates a cell line with a desired death ligand phenotype for ribozyme library screening. AsPC-1 cells are generally resistant to the TRAIL (panel B) but are sensitive to TRAIL when it is combined with the sensitizer act D (panel D), showing that they are resistant yet responsive to the death ligand. Other cell lines with this phenotype include DLD1 (with the Fas death ligand), and A2058 and DU-145 cells (with TRAIL and FasL). After identifying cells with this phenotype, cells are stably transduced with the ribozyme library and then treated with death ligand alone (as in panel B). The apoptotic cells are then rendered fluorescent and recovered by FACS to identify ribozymes that sensitize cells to death ligand-induced apoptosis.

in the presence of another added compound or protein, termed a sensitizer (Fig. 10). Although the mechanism of the sensitizers is unknown, they presumably act by countering the activity of repressors of apoptosis. Nonetheless, the apoptotic induction by the sensitizer indicates that the apoptosis machinery is intact and can be activated, and thus such cell lines are useful for the identification of novel therapeutic targets. Genome-wide ribozyme library was introduced to identify ribozymes that activate apoptosis on addition of death ligands in the absence of the additional chemical or protein sensitizer (Fig. 10). The apoptosis-promoting ribozymes, therefore, function as death ligand sensitizers by presumably down-regulating gene targets that inhibit apoptosis. Thus, the targets of the identified ribozymes are potential drug targets for cancer.

We describe several model systems that use this approach in the next section. Survival assays performed by other groups are also described in the following section.

1. IDENTIFICATION OF RIBOZYMES THAT ENHANCE EXTRINSIC APOPTOTIC PATHWAY

a. Activation of Fas-Induced Apoptosis in Colon Carcinoma DLD1 Cells

We first used the hairpin ribozyme-based library approach to identify genes that enhance Fas-induced apoptosis pathways in colon carcinoma DLD1 cells. Fas is a member of the tumor necrosis factor receptor (TNFR) family and it induces apoptosis on binding to Fas ligand. In our studies, a commonly used agonistic Fas antibody was used to mimic the function of Fas ligand. The DLD1 colon carcinoma cell line was chosen for three reasons: (1) it is resistant to FaS-induced apoptosis, (2) it expresses a high level of Fas receptor, and (3) it can be activated to undergo Fas-induced apoptosis by pretreatment with the sensitizer interferon-γ. As described earlier, this resistant yet responsive phenotype shows that the cell line exhibits an intact machinery for extrinsic pathway apoptosis that somehow is inhibited. This cell line expresses a mutant p53 protein and has been reported to express several known inhibitors of apoptosis such as FAP-1 (an inhibitor of Fas-induced apoptosis), SAG and survivin (both generalized inhibitors of apoptosis). We chose annexin V or tunnel-based FACS staining for quantifying the level of apoptosis in a cell population and for the isolation of apoptotic cells.

We first stably transduced the hairpin ribozyme library into DLD1 cells. The transduced cells were then treated with the agonistic α-Fas antibody to induce extrinsic pathway apoptosis, and the apoptotic cells were labeled with annexin V and collected by FACS sorting. The apoptosis-inducing ribozymes were then recovered from the apoptotic cells using PCR.

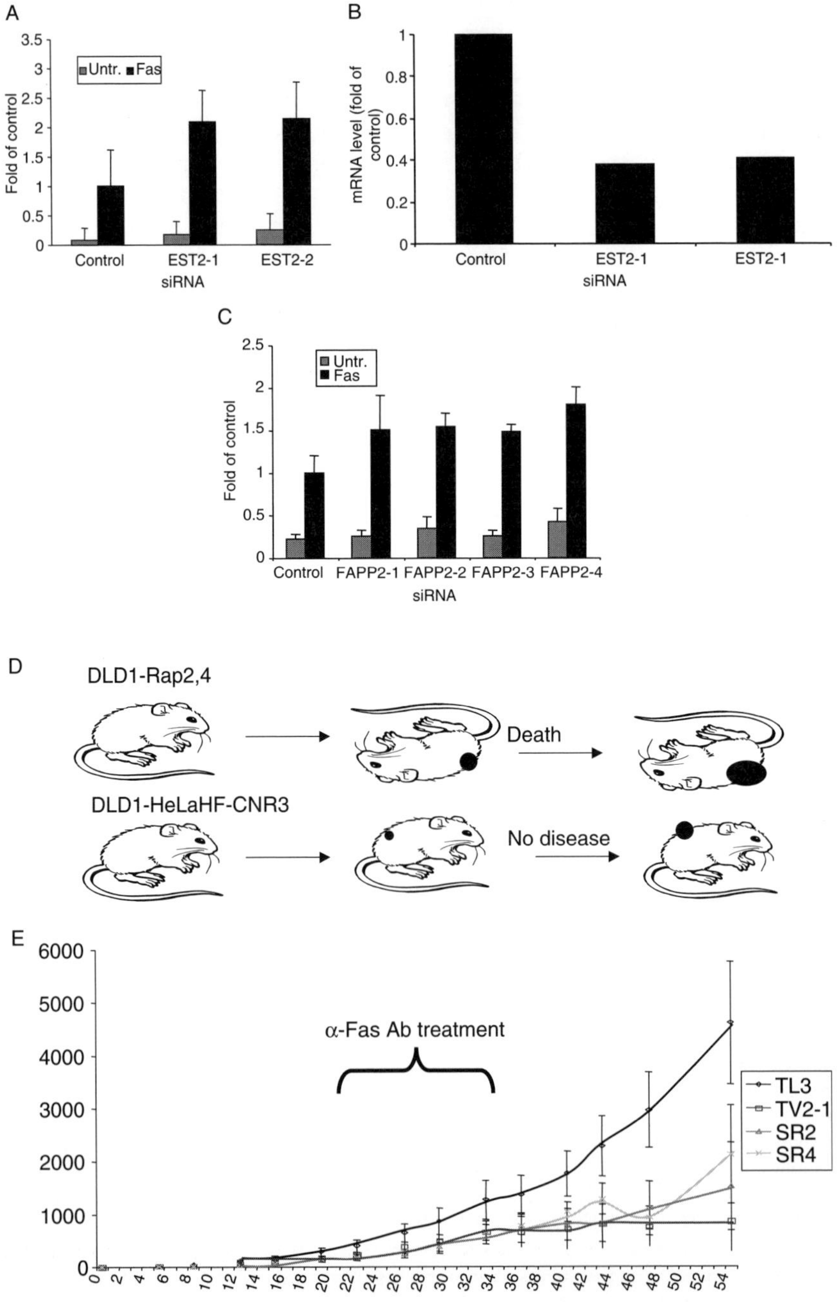

Fig. 11 Genes involved in Fas apoptosis pathways. (A) siRNA against Rap2 increases Fas-induced apoptosis in DLD1 cells. Control siRNA or two indicated siRNAs against Rap2 (EST2–1 and EST2–2) were transfected into DLD1 cells and 48 h later the cells were either left untreated

These rescued ribozymes were reintroduced into DLD1 cells and the entire procedure of phenotypic screening and ribozyme recovery was repeated five times to enrich for the active ribozymes. We ultimately identified five ribozymes which were able to activate DLD1 cells to undergo apoptosis after Fas induction. The ribozymes were confirmed by transducing them individually into the DLD1 cells and assaying with annexin V staining relative to a negative control ribozyme vector. The negative control vector contained a ribozyme against the hepatitis C virus core protein and was designated as LPR-TL3 or LPR-dTL3 for the disabled version of the ribozyme (data not shown). This negative control ribozyme does not target any known cellular genes.

The corresponding RSTs from the apoptosis-inducing ribozymes were used to identify potential target genes using BLAST queries of the public databases. The RST for a ribozyme we named RAp2 identified an EST sequence that we used to identify a full-length gene on screening a brain cDNA library. This gene has an ORF of $\sim$1000 nt and codes for a protein of $\sim$300 aa. The protein has significant identity with the amino terminal Fer-CIP4 homology (FCH) domains of ARHGAP14/SRGAP2 and FNBP2, ARHGAP family proteins are involved in actin–myosin cytoskeletal interactions and the Rho-GTPase signaling pathway and recently shown to be overexpressed in several human cancer cell lines (Katoh, 2003, 2004). We validated this gene using additional ribozymes that target distinct sites and also by using two siRNA molecules named Est2-1 and Est2-2 (Fig. 11).

To further validate the RAp2 gene target *in vivo* using the xenograft model (ARHGAP family member), we subcutaneously inoculated RAp2 ribozyme- and RAp2 validation ribozyme 1-transduced DLD1 cells into severe combined immune deficiency (SCID) mice and monitored DLD1 tumor cell growth after administration of agonistic Fas antibody (Fig. 11). Compared to control ribozyme-transduced DLD1 cells, cells transduced

(left columns) or treated for 16 h with agonistic Fas antibody (right columns). Cells were then assayed for apoptosis using the DNA-fragmentation ELISA assay (Roche). Data are plotted as a fold relative to the Fas-treated control transfection, and represent the average and standard deviation of four separate experiments. "Control" refers to transfection with an siRNA with no known target. (B) DLD1 cells were stably transfected with a control empty plasmid or plasmids encoding Rap2 siRNA est2-1 or est2-2. Stably transfected cells were then subjected to Taqman analysis to determine est2 mRNA levels. Data are from one experiment and a repeat produced similar results. The knockdown of Rap2 by the est2-1 and est2-2 siRNA correlates with the increased sensitivity to Fas-induced apoptosis seen in panel A. (C) siRNA against FAPP2 increases Fas-induced apoptosis in DLD1 cells. Cells were transfected with the indicated siRNA and were then examined for apoptosis as in panel A. (D and E) DLD1 cells containing either control (CNR3) ribozyme or Rap ribozymes (Rap2,4) were subcutaneously inoculated into SCID mice, and tumor growth were monitored by volume measurement after administration of agonistic Fas antibody every 2 days.

with the RAp2 ribozymes showed a markedly reduced tumor growth *in vivo*. This experiment further demonstrates the importance of the RAp2 gene in cancer *in vivo*.

The RST for another ribozyme recovered from the DLD1/Fas screen, ribozyme RAp594, identified four-phosphate-adaptor protein 2 (FAPP2) from the BLAST of the NCBI nucleotide database. FAPP2 contains a plekstrin homology (PH) domain and was recently shown to be involved in membrane trafficking from the Golgi to the plasma membrane (Godi *et al.*, 2004). The FAPP2 gene RST/Rap594 ribozyme match contains a G:U wobble pair in the middle of the long helix for substrate/ribozyme binding. Nevertheless, this gene was validated in our apoptotic assays using target validation ribozymes and RNAi (Fig. 11). The identification of FAPP2 and an ARHGAP family member protein (the RAp2 ribozyme gene target) in this screen for apoptosis suggests roles for the secretory pathway and the actin cytoskeleton, respectively, in the development and/or maintenance of cancer.

b. Trail Activation of Apoptosis in AsPC-1 and A2058 Cells

TNF-α-related apoptosis-inducing ligand (TRAIL) is another death ligand that is related to Fas ligand in structure and function. However, unlike the Fas ligand, TRAIL exhibits little or no liver toxicity and is therefore a potential cancer therapeutic agent. We sought to identify genes that enhance TRAIL-induced apoptosis in cancer cells. Such target genes could then be used to produce new therapies for cancer that can be used either individually or in combination with TRAIL therapy.

We first identified cell lines that resist apoptosis with TRAIL treatment alone yet undergo apoptosis on the addition of the sensitizer actinomycin D (Fig. 10). Two cell lines were selected for this study: the pancreatic carcinoma cell line AsPC-1 and the malignant melanoma line A2058. Both are resistant to TRAIL-induced apoptosis yet express high levels of TRAIL receptor. The AsPC-1 cells have been reported to express a mutant p53 protein and FLICE-inhibitory proteins (FLIP), while A2058 cells express a wild-type p53.

A2058 and AsPC cells were stably transduced with the ribozyme library and treated with TRAIL for 16 h. The cells were then subjected to an *in situ* TUNEL procedure to label the apoptotic cells (Fig. 10), which were then isolated by FACS. The ribozyme genes were then recovered by PCR and reintroduced into the cells for another round of screening. Each cell line was screened for a total of seven rounds. A total of 18 ribozyme sequences were enriched from the selection. Ten of these ribozymes were confirmed individually, that is, they yielded a significant increase (p-value < 0.05) in TUNEL staining of at least twofold versus the CNR3 ribozyme control (data not shown).

BLAST queries of the RSTs from the 10 confirmed ribozymes yielded several interesting candidate genes. In particular, the A2058-1 ribozyme matched perfectly with an RST in the gene for insulin-like growth factor binding protein 2, or IGFBP2 (also called IBP2, IGF-BP53), accession number: NM_000597. IGFBP2 is one of seven IGFBPs that bind to the insulin-like growth factors (IGFs) I and II in the plasma to affect the bioavailability of these mitogens. This IGF system performs many functions including the regulation of normal and malignant growth. Enhanced IGF expression has been found in various tumors, while enhanced IGFBP2 expression has been found in gliomas and prostate cancers. Changes in IGFBP2 expression are linked to nutrition status, insulin secretion, fetal development, and malignant transformation. The gene is expressed as a 1.4-kb transcript with no known isoforms. CGAP virtual Northern data shows that IGFBP2 is overexpressed in brain and ovarian cancers and underexpressed in colon cancer. Our RNAi experiments showed that siRNA "B" and "C" promote TRAIL- and Fas-induced apoptosis (Fig. 12A and B). Interestingly, they also promoted apoptosis in untreated cells, suggesting an induction of the intrinsic apoptosis pathway. The phenotypes correlated with a specific knockdown of the IGFBP2 message as measured by real-time PCR (Fig. 12C). The same siRNA against this transcript also produced a TRAIL- and Fas-induced apoptotic phenotype in DU-145 prostate cancer cells (data not shown). These results demonstrate a role for IGFBP2 in cancer in accordance with previous publications.

2. SURVIVAL SELECTION AFTER TREATMENT BY EXTERNAL APOPTOTIC STIMULI TO IDENTIFY PROAPOPTOTIC GENES

As stated earlier, while selection for apoptotic cells yields potential druggable oncogene targets, the process is more challenging due to the negative selection nature. Survival selection is much more efficient with significantly less background, and the following examples demonstrate cases that were validated using survival selection.

a. Genes Involved in Fas-Induced Apoptosis in Cervical Cancer HeLa Cells

Taira and colleagues examined the Fas apoptosis pathway using their hybrid hammerhead ribozyme library (Kawasaki and Taira, 2002; Kawasaki *et al.*, 2002a), using a survival selection scheme in which HeLa cells that resist apoptosis were selected (Fig. 13). Since HeLa does not express Fas, a Fas transgene expression cassette was introduced into HeLa cells to generate HeLa-Fas for the study. The HeLa-Fas cells were transiently transfected with the ribozyme library and subsequently treated with α-Fas antibody. After five rounds of selection, the ribozymes enriched in the survival cells were rescued and sequenced. As expected, the authors identified a number

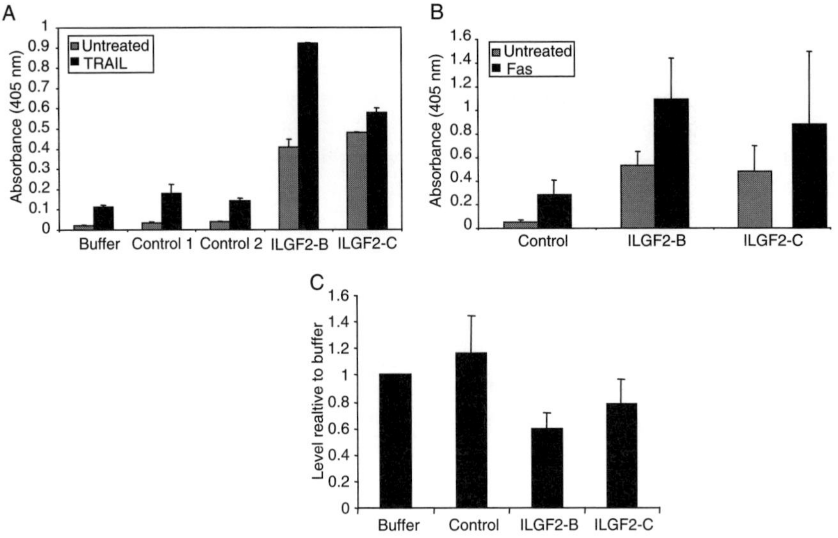

Fig. 12 Genes involved in TRAIL apoptosis pathway. (A) TRAIL-induced apoptosis in A2058 cells transfected with siRNA against IGFBP2. Cells were transfected with 10 nM of the indicated siRNA. Two days later, cells were either left untreated or incubated with 10 μg/ml TRAIL for 16 h. Apoptosis was then measured by DNA-fragmentation ELISA. The results are from one experiment and were reproduced in two other experiments. The IGFBP2 target sequences for each siRNA are as follows: ILGF2-B, AACCTCAAACAGTGCAAGATG, nucleotides 905–925; and ILGF2-C, AACGGAGAGTGCTTGGGTGGT, nucleotides 1142–1162. (B) Fas-induced apoptosis in A2058 cells transfected with siRNA against IGFBP2. The experiment is identical to "A" except agonistic Fas antibody was used instead of TRAIL extrinsic pathway apoptosis induction. The results are from three experiments. IGFBP2 siRNA "B" and "C" produce a significant (p-value < 0.05) phenotype in both untreated and Fas-treated cells. (C) Knockdown of IGFBP2 mRNA levels by siRNA. A2058 cells were seeded into T25 flasks and transfected with 10 nM of each siRNA using Oligofectamine at ~30% confluence. Forty-eight hours after transfection, the cells were harvested and RNA was isolated. Real-time PCR was performed for 40 cycles and the values were normalized to 18S rRNA. A knockdown of IGFBP2 mRNA is observed that correlates with the observed apoptosis phenotype. This graph summarizes the data from two separate transfections. Data are normalized to untransfected "buffer" only cells. "Control" refers to cells transfected with an siRNA with no known target.

of known genes involved in Fas-mediated apoptosis pathway such as Fas-associated death domain protein (FADD), caspases 3, 8, and 9, Bik, Apaf-1, and DNase CAD. They also identified a number of novel genes (Kawasaki and Taira, 2002; Kawasaki *et al.*, 2002a).

b. Genes Involved in TNF-α-Induced Apoptosis

Taira and colleagues also described a survival selection scheme using TNF-α as external stimuli to identify genes involved in the TNF-α-induced

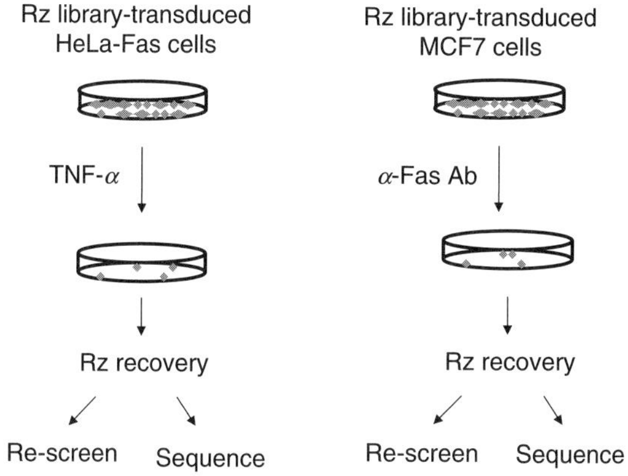

Fig. 13 Selection scheme based on cell survival for genes involved in apoptosis.

apoptosis pathway (Kawasaki *et al.*, 2002a). TNF-α, a pleotropic cytokine that is related to TRAIL and FasL, also induces apoptosis in some cell lines. TNF-α acts by binding to its receptors, TNFR-1 and -2, on the cell surface. This binding triggers signal transduction eventually leading to activation of caspase 8 and subsequent proteolytic cascade which ultimately induces cell death. Breast cancer cell line MCF-7 cells were transduced with the ribozyme gene library (Rz-A60 library), and then treated with TNF-α along with cycloheximide which enhances TNF-α-induced apoptosis. The ribozymes recovered from the surviving clones were sequenced to identify putative ribozyme targets. Among the genes identified include previously known genes involved in the TNF-α apoptosis pathway (TRADD, Caspases 2, 8 and RIP, RAIDD, and so on), and 30 new genes.

C. Identification of Genes Involved in Cell Invasion Using an *In Vitro* Invasion Assay

The fatality of the most cancer malignancies is attributed to cancer metastasis. Little is known regarding mechanisms underlying the metastasis, although comparative expression profiling of metastatic cells and nonmetastatic cells has recently been used to identify metastatic genes (Fidler and Radinsky, 1996). Cell invasion is an important contributing factor to the metastatic potential of cancer. Invasion through the extracellular matrix (EMC) involves cell adhesion to and penetration of the wall of blood vessel (migration), a process that involves the proteolytic activities of matrix metalloproteinases (MMPs) and adhesion molecules. The invasion potential of cancer cells can be assessed through an *in vitro* invasion assay (Chemicon

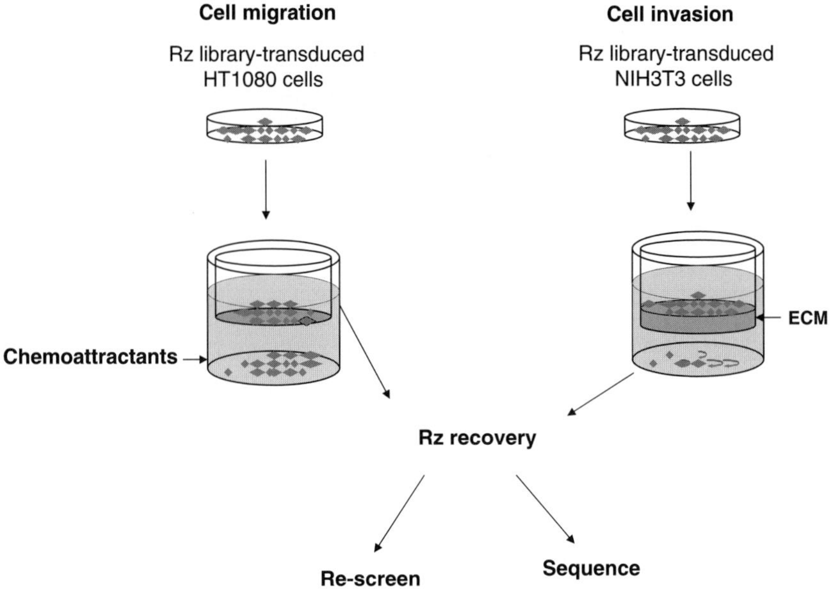

Fig. 14 Selection schemes for cells with distinct migratory or invasion potential.

Cell Invasion Assay Kit), while migration can be assessed by a Boyden chamber chemotaxis and wound migration assay. Both assays can be readily adapted as live cell selection for combinatorial ribozyme library selection (Fig. 14).

Taira and colleagues recently reported two studies using the *in vitro* migration and invasion assays with the phenotype-driven ribozyme library approach to identify genes involved in migration and invasion (Suyama *et al.*, 2003a,b). In their first study, the highly invasive HT1080 human fibrosarcoma cell line was transduced with the A60-hybrid hammerhead ribozyme library. The transduced cells were subjected to the transwell chemotaxis migration assay in Boyden chambers (Suyama *et al.*, 2003a). The cells and the assay conditions were chosen such that the majority of cells (99.9%) migrated toward the chemoattractant (bottom well) and few stayed for minimizing selection background. The ribozymes from the nonmigratory cell populations (top well) (Fig. 14) were recovered and enriched by repeated selections. Two enriched ribozymes were confirmed in both migration and invasion assays and their sequences were determined to identify the target gene. They found that these two ribozymes cleave the ROCK1 gene, a regulator of the actin cytoskeleton. ROCK1 and the related ROCK2 genes were previously implicated in metastasis, but not cell proliferation (Itoh *et al.*, 1999).

In the second study (Suyama *et al.*, 2003b), the authors identified genes involved in cell invasion by *in vivo* invasion selection. Noninversive or weakly invasive murine NIH3T3 fibroblasts were transduced Rz-A60 library, and the transduced cells were subjected to invasion selection in Boyden chambers coated with ECM gel porous filters that prevented migration of noninvasive cells. The invasive 3T3 cells (bottom well) were harvested and the ribozymes recovered (Fig. 14). Repeated selections yielded eight enriched ribozymes that promote NIH3T3 cell invasion. Since cell invasiveness can be attributed to two factors, migration and dissolution of ECM, these ribozymes were also tested in the migration assay. Six of the eight ribozymes enhanced migration of 3T3. A few candidate gene targets have been identified including GEM-GTPase, a gene previously implicated in cell invasion.

D. Other Inverse Genomics® Screenings for Cancer Targets

We have previously described a few other *in vitro* cancer cell model systems for identifying cancer genes, for example, NIH3T3 cell focus formation (Li *et al.*, 2000b) and BRCA1 promoter up-regulation (Beger *et al.*, 2001). Evidence also suggests that ribozyme-mediated gene inactivation could potentially be used to evaluate *in vivo* models such as the tumor xenograft (see preceding section) and transgenic models (L'Huillier *et al.*, 1996; Luo *et al.*, 2003). Therefore, Inverse Genomics® can also be adapted to *in vivo* selection to identify cancer targets. We have preliminarily attempted an *in vivo* selection using pulmonary metastasis/colonization of human osteosarcoma SAOS-2 cells in immunocompromised nude mouse (Jia *et al.*, 1999; Habita *et al.*, unpublished) (Fig. 15). We identified functional ribozymes responsible for SAOS-2 cells metastasis/colonization, suggesting the possibility that genes promoting metastasis and invasion *in vivo* can be identified using Inverse Genomics®. Taira and colleagues recently reported identification of metastasis-related genes in a murine B16F0 melanoma cell pulmonary metastasis mouse model using a library of randomized ribozymes (Suyama *et al.*, 2004) similar to the above SAOS-2 model. Their studies have identified a number of genes potentially involved in metastasis.

V. SUMMARY

Cancers are genetic diseases resulting from mutations in key genes that alter gene expression and/or gene product activity. Antagonizing these aberrations through pharmacological means leads to cancer cell death and

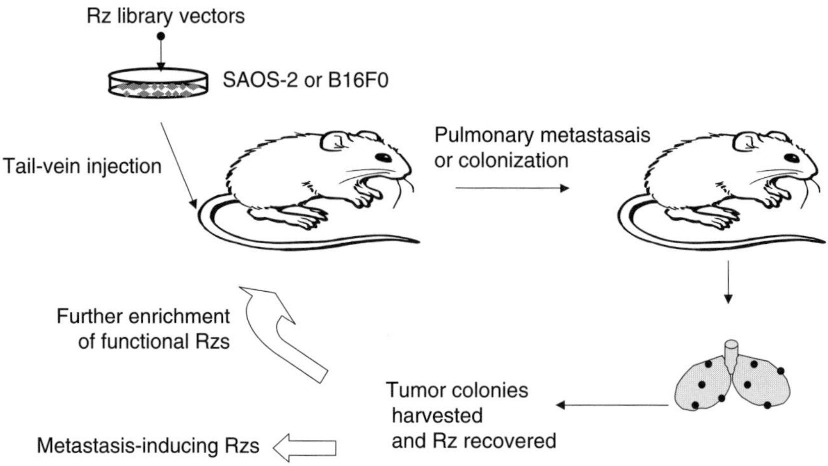

Fig. 15 Selection schemes for cells with metastatic potential. Non- or weakly metastatic human osteosarcoma cell line SAOS-2 or murine melanoma cell line B16F0 cells were transduced with hairpin ribozyme library or hammerhead ribozyme library. The transduced cells were tail vein injected into immunocomprised mouse *nu/nu* for SAOS-2 and synergic C57BL/6NCrj mouse for B16F0 cells. Two weeks later, the tumor colonies in mouse lungs were observed, counted, and harvested for *in vitro* culture and Rz rescue.

effective cancer therapy. This requires identifying the critical and specific genetic changes in cancer cells. This report describes a genomic approach, we dubbed Inverse Genomics®, which utilizes combinatorial ribozyme gene inactivation for concurrent cancer gene target identification and validation. Another key to successful cancer target discovery and validation is the development of the effective cell-based phenotype assays. In light of this, we have described some assays that have proven effective.

The concept of using ribozyme-mediated gene inactivation for cancer gene target identification and validation has contributed to the rapid application of siRNA gene knockdown technologies for this same purpose. We and others are currently using siRNA to identify novel cancer gene targets that may lead to new cancer therapies. It is widely known that siRNA-based technology has many advantages as compared to ribozyme-based technology, including more robust silencing, higher specificity, increased stability, easier site selection, and so on. This is the primary reason siRNA has been broadly accepted as the tool of choice for gene silencing. However, ribozymes still have certain features that siRNA may lack. For example, ribozyme is strand specific, and less likely to cause interferon response. Furthermore, ribozyme library approach may be more specific due to the requirement of GUC or NUH at the restriction site. Additionally, siRNA can enter the miRNA pathway to inhibit protein translation with much less stringent sequence requirements. Ribozyme libraries are also much more easily constructed. Lastly, since RNAi is a cellular mechanism and requires numerous cell factors, its effects under certain situations can potentially be restricted, whereas ribozymes would not have the same issue. Therefore, in spite of the power of siRNA as a gene inactivation tool, ribozymes are still valuable tools for functional genomics studies.

ACKNOWLEDGMENTS

The authors would like to thank Dr. Helmut Zarbl of Fred Hutchinson Cancer Research Center for kindly providing HeLaHF cells; we thank Drs. Richard Tritz, Gisela Claassen, Dehua Yu, Wufang Fan, Mirta Grifman, Xiuyuan Hu, Celia Habita, Kristin DeFife, and Joan Robbins for helpful suggestions; we thank Dr. Roshni Sundaram for critical reading of the chapter; we thank Aaron Albers, Lindsay Bezet, Ben Keily, Gina Cruz-Aranda, Violet Abraham, Jeremy To, Rebecca Lynn, and Erik Pendleton for excellent technical assistance; and we also thank Bernd Mayhack and Arndt Brachat of Novartis Pharma AG for helpful suggestions.

REFERENCES

Andres, A. C., Reid, H. H., Zurcher, G., Blaschke, R. J., Albrecht, D., and Ziemiecki, A. (1994). Expression of two novel eph-related receptor protein tyrosine kinases in mammary gland development and carcinogenesis. *Oncogene* **9**, 1461–1467.

Athanassiou, M., Hu, Y., Jing, L., Houle, B., Zarbl, H., and Mikheev, A. M. (1999). Stabilization and reactivation of the p53 tumor suppressor protein in nontumorigenic revertants of HeLa cervical cancer cells. *Cell Growth Differ.* **10**, 729–737.

Beger, C., Pierce, L. N., Kruger, M., Marcusson, E. G., Robbins, J. M., Welcsh, P., Welch, P. J., Welte, K., King, M. C., Barber, J. R., and Wong-Staal, F. (2001). Identification of Id4 as a regulator of BRCA1 expression by using a ribozyme-library-based inverse genomics approach. *Proc. Natl. Acad. Sci. USA* **98**, 130–135.

Bennett, B. D., Wang, Z., Kuang, W. J., Wang, A., Groopman, J. E., Goeddel, D. V., and Scadden, D. T. (1994). Cloning and characterization of HTK, a novel transmembrane tyrosine kinase of the EPH subfamily. *J. Biol. Chem.* **269**, 14211–14218.

Berclaz, G., Andres, A. C., Albrecht, D., Dreher, E., Ziemiecki, A., Gusterson, B. A., and Crompton, M. R. (1996). Expression of the receptor protein tyrosine kinase myk-1/htk in normal and malignant mammary epithelium. *Biochem. Biophys. Res. Commun.* **226**, 869–875.

Bertrand, E., Castanotto, D., Zhou, C., Carbonnelle, C., Lee, N. S., Good, P., Chatterjee, S., Grange, T., Pictet, R., Kohn, D., Engelke, D., and Rossi, J. J. (1997). The expression cassette determines the functional activity of ribozymes in mammalian cells by controlling their intracellular localization. *RNA* **3**, 75–88.

Boylan, M. O., Athanassiou, M., Houle, B., Wang, Y., and Zarbl, H. (1996). Activation of tumor suppressor genes in nontumorigenic revertants of the HeLa cervical carcinoma cell line. *Cell Growth Differ.* **7**, 725–735.

Bras, A., Albar, J. P., Leonardo, E., de Buitrago, G. G., and Martinez, A. C. (2000). Ceramide-induced cell death is independent of the Fas/Fas ligand pathway and is prevented by Nur77 overexpression in A20 B cells. *Cell Death Differ.* **7,** 262–271.

Cech, T. R., Zaug, A. J., and Grabowski, P. J. (1981). *In vitro* splicing of the ribosomal RNA precursor of Tetrahymena: Involvement of a guanosine nucleotide in the excision of the intervening sequence. *Cell* **27,** 487–496.

Chang, Z., Westaway, S., Li, S., Zaia, J. A., Rossi, J. J., and Scherer, L. J. (2002). Enhanced expression and HIV-1 inhibition of chimeric tRNA(Lys3)-ribozymes under dual U6 snRNA and tRNA promoters. *Mol. Ther.* **6,** 481–489.

Chatterton, J. E., Hu, X., and Wong-Staal, F. (2004). Ribozymes in gene identification, target validation and drug discovery. *DDT:TARGETS* **3,** 10–17.

Cheng, L. E., Chan, F. K., Cado, D., and Winoto, A. (1997). Functional redundancy of the Nur77 and Nor-1 orphan steroid receptors in T-cell apoptosis. *EMBO J.* **16,** 1865–1875.

Claassen, G., Ke, N., Yu, D. H., Chatterton, J. E., Hu, X., Albers, A., Nguy, V., Chionis, J., Truong, T., Meyhack, B., Wong-Staal, F., and Li, Q. (2004). Comprehensive assessment of cell growth properties affected by various agents using an integrated and non-subjective approach. *Preclinica* **2,** 435–439.

Feng, Y., Leavitt, M., Tritz, R., Duarte, E., Kang, D., Mamounas, M., Gilles, P., Wong-Staal, F., Kennedy, S., Merson, J., Yu, M., and Barber, J. R. (2000). Inhibition of CCR5-dependent HIV-1 infection by hairpin ribozyme gene therapy against CC-chemokine receptor 5. *Virology* **276,** 271–278.

Fidler, I. J., and Radinsky, R. (1996). Search for genes that suppress cancer metastasis. *J. Natl. Cancer Inst.* **88,** 1700–1703.

Fleischer, A., Ayllon, V., Dumoutier, L., Renauld, J. C., and Rebollo, A. (2002a). Proapoptotic activity of ITM2B(s), a BH3-only protein induced upon IL-2-deprivation which interacts with Bcl-2. *Oncogene* **21,** 3181–3189.

Fleischer, A., Ayllon, V., and Rebollo, A. (2002b). ITM2BS regulates apoptosis by inducing loss of mitochondrial membrane potential. *Eur. J. Immunol.* **32,** 3498–3505.

Gerety, S. S., Wang, H. U., Chen, Z. F., and Anderson, D. J. (1999). Symmetrical mutant phenotypes of the receptor EphB4 and its specific transmembrane ligand ephrin-B2 in cardiovascular development. *Mol. Cell* **4,** 403–414.

Godi, A., Di Campli, A., Konstantakopoulos, A., Di Tullio, G., Alessi, D. R., Kular, G. S., Daniele, T., Marra, P., Lucocq, J. M., and De Matteis, M. A. (2004). FAPPs control Golgi-to-cell-surface membrane traffic by binding to ARF and PtdIns(4)P. *Nat. Cell Biol.* **6,** 393–404.

Guerrier-Takada, C., Gardiner, K., Marsh, T., Pace, N., and Altman, S. (1983). The RNA moiety of ribonuclease P is the catalytic subunit of the enzyme. *Cell* **35,** 849–857.

Hampel, A., and Tritz, R. (1989). RNA catalytic properties of the minimum (-)sTRSV sequence. *Biochemistry* **28,** 4929–4933.

Hampel, A., Tritz, R., Hicks, M., and Cruz, P. (1990). 'Hairpin' catalytic RNA model: Evidence for helices and sequence requirement for substrate RNA. *Nucleic Acids Res.* **18,** 299–304.

Helbling, P. M., Saulnier, D. M., and Brandli, A. W. (2000). The receptor tyrosine kinase EphB4 and ephrin-B ligands restrict angiogenic growth of embryonic veins in *Xenopus laevis*. *Development* **127,** 269–278.

Hendry, P., McCall, M. J., and Lockett, T. J. (1997). Design of hybridizing arms in hammerhead ribozymes. *Methods Mol. Biol.* **74,** 253–264.

Inada, T., Iwama, A., Sakano, S., Ohno, M., Sawada, K., and Suda, T. (1997). Selective expression of the receptor tyrosine kinase, HTK, on human erythroid progenitor cells. *Blood* **89,** 2757–2765.

Itoh, K., Yoshioka, K., Akedo, H., Uehata, M., Ishizaki, T., and Narumiya, S. (1999). An essential part for Rho-associated kinase in the transcellular invasion of tumor cells. *Nat. Med.* **5**, 221–225.

Jia, S. F., Worth, L. L., and Kleinerman, E. S. (1999). A nude mouse model of human osteosarcoma lung metastases for evaluating new therapeutic strategies. *Clin. Exp. Metastasis* **17**, 501–506.

Katoh, M. (2003). FNBP2 gene on human chromosome 1q32.1 encodes ARHGAP family protein with FCH, FBH, RhoGAP and SH3 domains. *Int. J. Mol. Med.* **11**, 791–797.

Katoh, M. (2004). Characterization of human ARHGAP10 gene in silico. *Int. J. Oncol.* **25**, 1201–1206.

Kawasaki, H., and Taira, K. (2002). Identification of genes by hybrid ribozymes that couple cleavage activity with the unwinding activity of an endogenous RNA helicase. *EMBO Rep.* **3**, 443–450.

Kawasaki, H., Onuki, R., Suyama, E., and Taira, K. (2002a). Identification of genes that function in the TNF-alpha-mediated apoptotic pathway using randomized hybrid ribozyme libraries. *Nat. Biotechnol.* **20**, 376–380.

Kawasaki, H., Tsunemi, M., Iyo, M., Oshima, K., Minoshima, H., Hamada, A., Onuki, R., Suyama, E., and Taira, K. (2002b). A functional gene discovery in cell differentiation by hybrid ribozyme and siRNA libraries. *Nucleic Acids Res. Suppl.* **2**, 275–276.

Ke, N., Albers, A., Claassen, G., Yu, D. H., Chatterton, J. E., Hu, X., Meyhack, B., Wong-Staal, F., and Li, Q. X. (2004a). One-week 96-well soft agar growth assay for cancer target validation. *Biotechniques* **36**, 826–828, 830, 832–833.

Ke, N., Claassen, G., Yu, D. H., Albers, A., Fan, W., Tan, P., Grifman, M., Hu, X., Defife, K., Nguy, V., Brachat, A., Meyhack, B., *et al.* (2004b). Nuclear hormone receptor NR4A2 is involved in cell transformation and apoptosis. *Cancer Res.* **64**, 8208–8212.

Kim, I., Ryu, Y. S., Kwak, H. J., Ahn, S. Y., Oh, J. L., Yancopoulos, G. D., Gale, N. W., and Koh, G. Y. (2002). EphB ligand, ephrinB2, suppresses the VEGF- and angiopoietin 1-induced Ras/mitogen-activated protein kinase pathway in venous endothelial cells. *FASEB J.* **16**, 1126–1128.

Kruger, K., Grabowski, P. J., Zaug, A. J., Sands, J., Gottschling, D. E., and Cech, T. R. (1982). Self-splicing RNA: Autoexcision and autocyclization of the ribosomal RNA intervening sequence of Tetrahymena. *Cell* **31**, 147–157.

Kruger, M., Beger, C., Li, Q. X., Welch, P. J., Tritz, R., Leavitt, M., Barber, J. R., and Wong-Staal, F. (2000). Identification of eIF2Bgamma and eIF2gamma as cofactors of hepatitis C virus internal ribosome entry site-mediated translation using a functional genomics approach. *Proc. Natl. Acad. Sci. USA* **97**, 8566–8571.

Kuwabara, T., Warashina, M., Koseki, S., Sano, M., Ohkawa, J., Nakayama, K., and Taira, K. (2001). Significantly higher activity of a cytoplasmic hammerhead ribozyme than a corresponding nuclear counterpart: Engineered tRNAs with an extended 3′ end can be exported efficiently and specifically to the cytoplasm in mammalian cells. *Nucleic Acids Res.* **29**, 2780–2788.

L'Huillier, P. J., Soulier, S., Stinnakre, M. G., Lepourry, L., Davis, S. R., Mercier, J. C., and Vilotte, J. L. (1996). Efficient and specific ribozyme-mediated reduction of bovine alpha-lactalbumin expression in double transgenic mice. *Proc. Natl. Acad. Sci. USA* **93**, 6698–6703.

Law, S. W., Conneely, O. M., DeMayo, F. J., and O'Malley, B. W. (1992). Identification of a new brain-specific transcription factor, NURR1. *Mol. Endocrinol.* **6**, 2129–2135.

Le, W., Conneely, O. M., He, Y., Jankovic, J., and Appel, S. H. (1999). Reduced Nurr1 expression increases the vulnerability of mesencephalic dopamine neurons to MPTP-induced injury. *J. Neurochem.* **73**, 2218–2221.

Li, H., Kolluri, S. K., Gu, J., Dawson, M. I., Cao, X., Hobbs, P. D., Lin, B., Chen, G., Lu, J., Lin, F., Xie, Z., Fontana, J. A., *et al.* (2000a). Cytochrome c release and apoptosis induced by mitochondrial targeting of nuclear orphan receptor TR3. *Science* **289,** 1159–1164.

Li, Q. X., Robbins, J. M., Welch, P. J., Wong-Staal, F., and Barber, J. R. (2000b). A novel functional genomics approach identifies mTERT as a suppressor of fibroblast transformation. *Nucleic Acids Res.* **28,** 2605–2612.

Liu, W., Ahmad, S. A., Jung, Y. D., Reinmuth, N., Fan, F., Bucana, C. D., and Ellis, L. M. (2002). Coexpression of ephrin-Bs and their receptors in colon carcinoma. *Cancer* **94,** 934–939.

Liu, Z. G., Smith, S. W., McLaughlin, K. A., Schwartz, L. M., and Osborne, B. A. (1994). Apoptotic signals delivered through the T-cell receptor of a T-cell hybrid require the immediate-early gene nur77. *Nature* **367,** 281–284.

Luo, X., Gong, X., and Tang, C. K. (2003). Suppression of EGFRvIII-mediated proliferation and tumorigenesis of breast cancer cells by ribozyme. *Int. J. Cancer* **104,** 716–721.

Masters, J. R. (2002). HeLa cells 50 years on: The good, the bad and the ugly. *Nat. Rev. Cancer* **2,** 315–319.

Mathews, D. H., Sabina, J., Zuker, M., and Turner, D. H. (1999). Expanded sequence dependence of thermodynamic parameters improves prediction of RNA secondary structure. *J. Mol. Biol.* **288,** 911–940.

Mikheev, A. M., Mikheeva, S. A., Liu, B., Cohen, P., and Zarbl, H. (2004). A functional genomics approach for the identification of putative tumor suppressor genes: Dickkopf-1 as suppressor of HeLa cell transformation. *Carcinogenesis* **25,** 47–59.

Munarini, N., Jager, R., Abderhalden, S., Zuercher, G., Rohrbach, V., Loercher, S., Pfanner-Meyer, B., Andres, A. C., and Ziemiecki, A. (2002). Altered mammary epithelial development, pattern formation and involution in transgenic mice expressing the EphB4 receptor tyrosine kinase. *J. Cell Sci.* **115,** 25–37.

Nikolova, Z., Djonov, V., Zuercher, G., Andres, A. C., and Ziemiecki, A. (1998). Cell-type specific and estrogen dependent expression of the receptor tyrosine kinase EphB4 and its ligand ephrin-B2 during mammary gland morphogenesis. *J. Cell Sci.* **111**(Pt. 18), 2741–2751.

Ohkura, N., Hijikuro, M., Yamamoto, A., and Miki, K. (1994). Molecular cloning of a novel thyroid/steroid receptor superfamily gene from cultured rat neuronal cells. *Biochem. Biophys. Res. Commun.* **205,** 1959–1965.

Pan, W. H., Xin, P., Bui, V., and Clawson, G. A. (2003). Rapid identification of efficient target cleavage sites using a hammerhead ribozyme library in an iterative manner. *Mol. Ther.* **7,** 129–139.

Scherr, M., and Rossi, J. J. (1998). Rapid determination and quantitation of the accessibility to native RNAs by antisense oligodeoxynucleotides in murine cell extracts. *Nucleic Acids Res.* **26,** 5079–5085.

Scherr, M., Reed, M., Huang, C. F., Riggs, A. D., and Rossi, J. J. (2000). Oligonucleotide scanning of native mRNAs in extracts predicts intracellular ribozyme efficiency: Ribozyme-mediated reduction of the murine DNA methyltransferase. *Mol. Ther.* **2,** 26–38.

Scherr, M., LeBon, J., Castanotto, D., Cunliffe, H. E., Meltzer, P. S., Ganser, A., Riggs, A. D., and Rossi, J. J. (2001). Detection of antisense and ribozyme accessible sites on native mRNAs: Application to NCOA3 mRNA. *Mol. Ther.* **4,** 454–460.

Shahi, S., and Banerjea, A. C. (2002). Multitarget ribozyme against the S1 genome segment of reovirus possesses novel cleavage activities and is more efficacious than its constituent mono-ribozymes. *Antiviral Res.* **55,** 129–140.

Stephenson, S. A., Slomka, S., Douglas, E. L., Hewett, P. J., and Hardingham, J. E. (2001). Receptor protein tyrosine kinase EphB4 is up-regulated in colon cancer. *BMC Mol. Biol.* **2,** 15.

Steube, K. G., Meyer, C., Habig, S., Uphoff, C. C., and Drexler, H. G. (1999). Expression of receptor tyrosine kinase HTK (hepatoma transmembrane kinase) and HTK ligand by human leukemia-lymphoma cell lines. *Leuk. Lymphoma* **33,** 371–376.

Suenobu, S., Takakura, N., Inada, T., Yamada, Y., Yuasa, H., Zhang, X. Q., Sakano, S., Oike, Y., and Suda, T. (2002). A role of EphB4 receptor and its ligand, ephrin-B2, in erythropoiesis. *Biochem. Biophys. Res. Commun.* **293,** 1124–1131.

Suyama, E., Kawasaki, H., Kasaoka, T., and Taira, K. (2003a). Identification of genes responsible for cell migration by a library of randomized ribozymes. *Cancer Res.* **63,** 119–124.

Suyama, E., Kawasaki, H., Nakajima, M., and Taira, K. (2003b). Identification of genes involved in cell invasion by using a library of randomized hybrid ribozymes. *Proc. Natl. Acad. Sci. USA* **100,** 5616–5621.

Suyama, E., Wadhwa, R., Kaur, K., Miyagishi, M., Kaul, S. C., Kawasaki, H., and Taira, K. (2004). Identification of metastasis-related genes in a mouse model using a library of randomized ribozymes. *J. Biol. Chem.* **279,** 38083–38086.

Suzuki, S., Suzuki, N., Mirtsos, C., Horacek, T., Lye, E., Noh, S. K., Ho, A., Bouchard, D., Mak, T. W., and Yeh, W. C. (2003). Nur77 as a survival factor in tumor necrosis factor signaling. *Proc. Natl. Acad. Sci. USA* **100,** 8276–8280.

Tritz, R., Leavitt, M., and Barber, J. R. (1999). Screening promoters for optimal expression of ribozymes. *In* "Intracellular Ribozyme Applications: Principle and Protocols" (J. J. Rossi and L. Couture, Eds.), pp. 115–123. Horizon Scientific Press, Norfolk, England.

Vogelstein, B., and Kinzler, K. W. (2004). Cancer genes and the pathways they control. *Nat. Med.* **10,** 789–799.

Warashina, M., Kuwabara, T., Kato, Y., Sano, M., and Taira, K. (2001). RNA-protein hybrid ribozymes that efficiently cleave any mRNA independently of the structure of the target RNA. *Proc. Natl. Acad. Sci. USA* **98,** 5572–5577.

Welch, P. J., Marcusson, E. G., Li, Q. X., Beger, C., Kruger, M., Zhou, C., Leavitt, M., Wong-Staal, F., and Barber, J. R. (2000). Identification and validation of a gene involved in anchorage-independent cell growth control using a library of randomized hairpin ribozymes. *Genomics* **66,** 274–283.

Woronicz, J. D., Calnan, B., Ngo, V., and Winoto, A. (1994). Requirement for the orphan steroid receptor Nur77 in apoptosis of T-cell hybridomas. *Nature* **367,** 277–281.

Yu, D. H., Chatterton, J. E., Ke, N., Nguy, V., Bliesath, J. R., Hu, X., Meyhack, B., Wong-Staal, F., and Li, Q. X. (2005). A 96 well surrogate survival assay coupled with a special RNAi vector strategy for cancer gene target identification and validation with enhanced signal/noise ratio. *Assay Drug Dev. Technol.* **3,** 401–411.

Yu, D. H., Fan, W., Liu, G., Nguy, V., Chatterton, J. E., Long, S., Ke, N., Meyhack, B., Bruengger, A., Brachat, A., Wong-Staal, F., and Li, Q. N. (2006). PHTS, a novel putative tumor suppressor, is involved in the transformation reversion of HeLaHF cells independently of the p53 pathway. *Exp. Cell Res.* **312,** 865–876.

Yu, Q., and Burke, J. M. (1997). Design of hairpin ribozymes for *in vitro* and cellular applications. *Methods Mol. Biol.* **74,** 161–169.

Cancer Cell-Based Genomic and Small Molecule Screens

Jeremy S. Caldwell

Genomics Institute of the Novartis Research Foundation, San Diego, California 92121

I. Introduction
II. Drugs, Druggability, and Target Validation
III. Post-genomic Discovery of Novel Targets
  A. Functional cDNA Screening and Oncology
  B. RNA Interference Screening and Oncology
IV. Small Molecule Screens in Oncology
  A. Parallel Cellular Screens
  B. MOA Determination
  C. Data Analysis of Multidimensional Cellular Datasets
  D. Mechanisms for New Drugs
V. Conclusions
References

This chapter focuses on the promising post-genomic technologies being used for discovery of new, safer, and better cancer drugs and drug targets. Since cancer is largely a disease of the cell, usually involving unrestricted cell proliferation as a result of heritable genetic changes such as mutation, this chapter will focus on cell-centric technologies and their utility in addressing major questions in cancer biology. Recent advances in cell-based technology, including phenotypic assays, image-based readouts, primary tumor cell growth and maintenance *in vitro*, gene and small molecule delivery tools, and automated systems for cell manipulation, provide a novel means to understand the etiology and mechanisms of cancer as never before. In addition to the abundant tool sophistication, many aspects of cancer can be emulated and monitored in cell systems, which makes them ideal vehicles for exploitation to discover new targets and drugs. This chapter will first handle nomenclature and provide a context for a "good drug target" within the framework of the human genome, then overview functional genomic gene-based library screening approaches with specific applications to cancer target discovery. Second, small molecule screening applications will be handled, with an emphasis on the new paradigm of massively parallel screening and resultant multidimensional dataset analysis approaches to identify drug candidates, assign mechanism of action, and address problems in deriving selective and safe chemical entities. © 2007 Elsevier Inc.

Advances in CANCER RESEARCH

0065-230X/07 $35.00
DOI: 10.1016/S0065-230X(06)96006-0

I. INTRODUCTION

A fundamental premise in the scientific enterprise is that technology has a powerful impact on research, allowing scientists to ask old questions in new and exciting ways. The recent sequencing of the complete human genome represented a fundamental shift in how biomedical research is performed such that biological phenomena can now be addressed with more global, comprehensive sets of tools. In parallel, advances in automation, high-throughput screening, gene-based libraries, combinatorial chemistry, cell biology, and computation coupled with the genome sequence have spawned the post-genomic era in which biomedical discovery is advancing at an unprecedented rate. In particular, comprehensive genome-wide gene libraries [e.g., cDNA overexpression and small interfering RNA (siRNA) or gene knockdown libraries], collections of diverse chemicals (e.g., natural products, kinase, protease, and other gene family targeted compound collections, known drugs), advances in cell technology, and automated robotic systems to screen these libraries in cell-based assays are leading to discoveries of novel drug targets, new drug candidates, and a means to properly triage these targets and drug candidates to minimize attrition rates and clinical failure.

Although the post-genomic technologies have applications to many biomedical areas, cancer and diseases of cell proliferation are possibly best positioned to benefit in the short term. Properly poised, these tools may be able to address the most fundamental problems in the treatment and cure of cancer. The major problems to be addressed center around the fact that although a significant number of new chemical entities are generated annually to target a variety of cancer types, the majority of treatments are non-specific chemotherapies which wreak havoc on other major organ systems beyond the intended target. The etiology of such toxic effects and narrow therapeutic indices are largely unknown, but current thought implicates the ubiquitous tissue distribution of the drug and drug target throughout the body. Drugs such as cisplatin and taxol target DNA and microtubules, respectively. Neither of these targets is found preferentially in tumor versus normal, healthy cells. Furthermore, these drugs distribute throughout the body, affecting significant systemic toxicity. In contrast, the highly successful Glivec (STI571) is a specific, molecularly targeted drug, directed toward BCR/ABL, found only in lymphoid cells which have undergone a rare translocation event, leading to chronic myeloid leukemia (CML) (Druker *et al.*, 1996). The question then becomes, can more specific drugs and drug targets be discovered? If so, how can tools of the post-genome lead us to them expeditiously?

II. DRUGS, DRUGGABILITY, AND TARGET VALIDATION

For the first time, access to a list of nearly all human genes is available of which many may represent targets for disease intervention. Some estimate 5000–10,000 new drug targets may have come to light based on the full human genome sequence, while more conservative estimates are in the thousands (Bailey *et al.*, 2001; Hopkins and Groom, 2002). The sheer number of new possible targets has led to the impression that more targets exist than can be handled appropriately. The glut of new targets might congest the pipeline, inflate research and development (R&D) costs, and make drug discovery less efficient.

The question then posed by the genome opportunity is how to most quickly turn human genomic information into validated targets without sacrificing efficiency? There are at least two ways of addressing this question: first, by prioritizing which targets are the most druggable (i.e., what is the likelihood of discovering a modulatory small molecule or antibody against the target?) and second by determining which targets are best validated (i.e., what is the likelihood that target modulation will result in a tangible mitigation of disease or symptoms?). The target boon begs for a framework to assess which "targets" represent both "druggable" and "validated" targets, an important question given the associated risk involved in launching drug discovery campaigns on poorly validated targets.

Historically, good drug targets have been defined in two ways: first, as targets with a proven, causative role in a human disease, which when addressed with a small molecule leads to reversal of the disease, or second, as targets for already successful drugs, in which hindsight defines "good." Many drug targets, such as the immunosuppressant drug cyclosporine A (CsA), metformin, and even penicillin, were used as probes themselves to discover their targets (Ghuysen, 1977; Handschumacher *et al.*, 1984; Zhou *et al.*, 2001). To define "good drug targets" as those for which existing drugs provide direct rationale would limit the number to only 120 based on all marketed drugs (Hopkins and Groom, 2002). While the list of known drugs provide a means to a reverse definition of a druggable target, in light of the genome project, how do we determine a "forward" definition of a druggable target? Can we establish criteria for target druggability and relevance to disease?

Lipinski *et al.* (2001) established a set of criteria for chemical druggability based on physicochemical and pharmacokinetic characteristics of known drugs recognized as the "rule-of-five" for oral bioavailability. By analogy then, in the biological world, are there guidelines for druggable targets based on gene family, structure, expression, tissue distribution, function, and biological pathway connectivity? Hopkins and Groom assessed the number of "druggable" targets using the guideline that the target must be physically capable of binding compounds, preferably to those with drug-like features as set forth by the Lipinski "rule-of-five." The analysis resulted in 3051 "druggable" targets; roughly 10% of the predicted ~30,000 or so genes (this analysis does not take into account nonprotein targets). Of the ~3000 "druggable" targets, many fall within protein families (GPCR, kinase, protease) that suggests the potential for nonspecific, off-target effects of compounds on related targets which can often lead to untoward and significant side effects. An understanding of the consequences of such off-target modulation, knowing which targets to avoid, requires functional definition of these targets. Unfortunately, less than half of all proteins encoded by the genome have any documented functional annotation (Su *et al.*, 2004). In fact, without an understanding of the function of a biological target in disease versus normal cells, the point of "druggability" is moot.

III. POST-GENOMIC DISCOVERY OF NOVEL TARGETS

How then does one identify putative targets based on function and relevance to disease? Given the significant number of genes in the genome, what methods exist to functionally annotate all possible genes?

Is there a high-throughput methodology to ascribe functional definition to a gene to the degree necessary to launch a drug discovery effort? Genetics provides a means to determine gene function in the context of a whole organism (gene knockout, transgenesis, and organismal mutagenesis), however, the methods employed are typically low throughput and the evolutionary distance from human (i.e., yeast, worms, flies, fish, and mice) detracts from the goal. Alternatively, human cell systems represent a higher throughput vehicle with superior application to human disease. Based on this inference, several groups have devised methods to comprehensively scan genes at a genome-wide scale for function using cellular assays. By arraying thousands of distinct human genes one-by-one in mammalian expression vectors in microtiter plates, transfecting massively in parallel into cellular reporter assays, each gene can be tested for functional effects of overexpression. The method termed Genome Functionalization through Arrayed cDNA Transduction (GFAcT) or gene-by-gene screening has been used to assign function to hundreds of genes in cellular assays for cell proliferation, cell-cycle arrest, and other phenotypes (Chanda *et al.*, 2003; Iourgenko *et al.*, 2003). Similarly, comprehensive libraries of siRNAs targeting individual genes of the human genome for knockdown or reduction of mRNA levels have been arrayed and screened for function in cell-based assays as well (Huesken *et al.*, 2005) (Fig. 1). These systems have potential to reduce the genome down to smaller more manageable subpopulations based on activity in functional assays relevant to pathways, physiology, and disease.

Given that many signal transduction pathways and cell-intrinsic biological networks are intact in mammalian cells, whole cell-based assays have further utility in elaborating biological specificity of a given gene. Overexpressing or knocking down a given gene in the context of cells affords the ability to scan for effects in multiple signaling pathways within a given cell, or across a panel of cellular assays. In the context of the theory

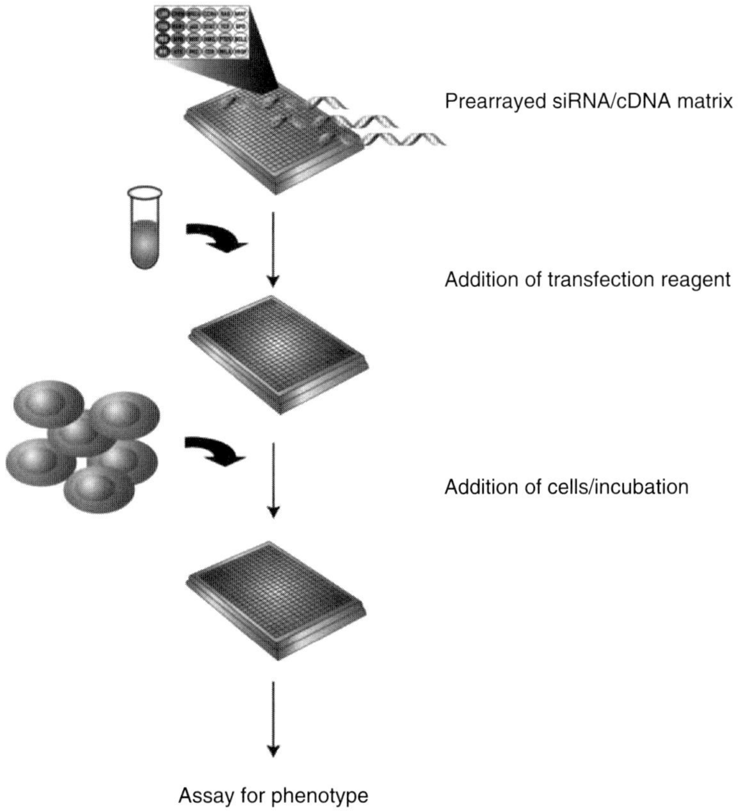

Fig. 1 Screening of arrayed gene libraries for pathway and target identification. For high-throughput functional genomic analysis, distinct cDNAs or siRNAs are arrayed in multiwell format. Transfection reagent is subsequently added to the nucleic acids, and the transfection is completed by the addition of cells (retrotransfection). Effects of gain of function (cDNAs) and loss of function (siRNAs) of genes across the genome on a particular cellular phenotype are then measured.

of biological connectivity and node theory in which signaling networks are composed of at least two distinct types of interactions, a low number of highly linked nodes that interface with a high number of lowly linked nodes and a high number of lowly linked nodes that interface with a small number of highly linked nodes, the importance of phenotypic specificity is manifest (Clemons, 2004) (Fig. 2). An example of a highly linked node is classical AP-1 (a complex of fos and jun proteins), which can be modulated by dozens of stimuli (such as mitogens, cytokines, UV, and so on), since perturbations of highly linked nodes can have pleiotropic effects of which some are desirable (i.e., disease alleviation) and some undesirable (i.e., toxicity). Other examples of highly linked nodes include NF-κB, p53, and ras, which represent meeting points for many signaling pathways involved in cell growth, proliferation, and adhesion, in a variety of systems regulating growth, inflammation, neuronal function, and other important physiology (Ghosh and Karin, 2002; Gomez-Lazaro *et al.*, 2004; Olson and Marais, 2000). In contrast, perturbation of lowly linked nodes would have

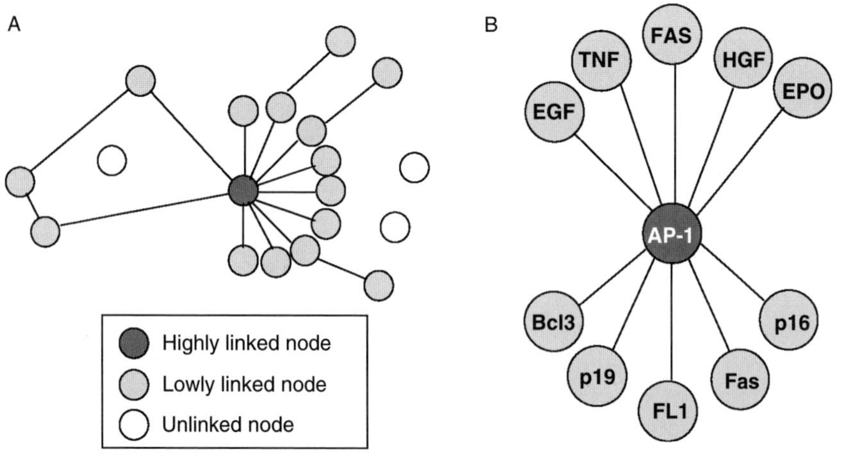

Fig. 2 Node network theory schematic. (A) Biological networks can be represented as an array of nodes including highly linked nodes (dark gray circles), lowly linked nodes (light gray circles), and unlinked nodes (white circles). Theoretically, highly linked nodes would control multiple processes and thus connect with many other nodes. Perturbation of such highly linked nodes is likely to have pleiotropic effects because of their influence on many processes. Perturbation of lowly linked nodes would likely have fewer pleiotropic effects because of their isolation (as would be the case of perturbing the disconnected or unconnected nodes). (B) Examples of a highly connected node would be the transcription factor complex AP-1 which is involved in regulating many processes including proliferation and inflammation. Disruption of a lowly linked node upstream of AP-1 activation, such as hepatoctye growth factor (HGF) inhibition, would not impede AP-1 activation through other effectors such as TNF and FAS and thus have fewer side effects than direct inhibition of AP-1.

less overall impact across signaling systems, and more likely to achieve a specific desired outcome. While examples of both types of nodes exist as drug targets, the ability to characterize the effect of any given gene across diverse pathways can provide information for target prioritization. Parallel screens of libraries of cDNAs or siRNAs, or those subsets which score positive in a phenotypic assay can reveal biological specificity of a hit and perhaps narrow the list of putative targets from a genome-wide screen.

A. Functional cDNA Screening and Oncology

Genetic dissection of signaling pathways in human disease is the current challenge in drug target identification. The ability to generate libraries of cDNAs from human tissue sources and clone into mammalian expression vectors is enabling this field. Harnessing viruses to introduce libraries blanketly into cells, paired with advances in human tissue culture systems and phenotypic readouts have ushered in the age of human cell genetics. In the late 1980s, Brian Seed established the expression screening approach in mammalian cells and laid the groundwork for functional gene identification in cell systems relevant to human physiology (Allen and Seed, 1989). In the technology's infancy, functional cDNA expression screening identified genes such as LFG, an antiapoptotic gene, Toso, a T-cell surface receptor that blocks FAS-induced apoptosis, multiple members of the JAK and STAT interferon signaling cascade, and NF-κB activating genes such as IKK-gamma (Darnell *et al.*, 1994; Hitoshi *et al.*, 1998; Yamaoka *et al.*, 1998). These screens were done by ectopic overexpression of pooled cDNAs, looking for genes which confer "gain of function." Despite obvious utility, this pooled screening strategy is analogous to "finding a needle in a haystack"; first the chance of finding something in the haystack is very slim based on the abundance of the gene of interest, and second, one is not even assured the gene of interest is in the proverbial stack in the first place. More precisely, the limitations include: (1) strong dependence on an assay's signal-to-noise ratio; (2) nonnormalized libraries make rare, low abundance genes more difficult to find; (3) libraries usually tissue specific and lack many genes; (4) are ridden with cDNA fragment; and (5) rescuing the causal cDNA from the cell of interest can be difficult.

As a result of the completed human genome sequence, an effort to clone all human genes, in expression-ready full-length form are becoming available to facilitate genome-wide functional screens in mammalian cells. This new approach circumvents many of the problems associated with pool screening. Full-length cDNAs, encoded by mammalian expression vectors, including both known and predicted genes are arrayed in individual wells of microtiter plates, and systematically introduced in parallel into reporter cells.

These libraries can include the 12,000 plus human genes assembled by the Mammalian Genome Collection (http://mgc.nci.nih.gov/), the 24,000 unique member Origene collection (http://www.origene.com/cdna/), all of which exist in mammalian expression ready vectors for high-level expression via transient transfection or in virus-based vectors for screens in primary cells (Gerhard *et al.*, 2004; Strausberg *et al.*, 2002). The advantages of arrayed-based formats over pooled-library approaches are: (1) the reporter assays do not require a very high signal-to-noise ratio, since each gene is tested individually for function; (2) the libraries are normalized such that rare genes have an equal opportunity to register as hits; and (3) rescuing the causal gene is unnecessary since the identity of each gene in each well is already known. Another advantage soon to be realized from a comprehensive set of all human genes arrayed for functional screening is that one will no longer have to wonder if the gene of interest resides in the library if no hits score as positives—in this case, perhaps a gene with the desired function does not exist.

The arrayed functional screening approach is beginning to offer new targets in oncology. This gain-of-function screening approach has particular utility in oncology screens given that the etiology of many cancers stems from misregulation and overexpression of oncogenes. Chanda *et al.* (2003) reported a strategy to identify positive regulatory effectors of AP-1-dependent mitogenesis using an AP-1-responsive reporter vector upstream of a reporter gene encoding. Michiels *et al.* (2002) used a 13,000 member-arrayed adenoviral expression library to identify novel regulators of osteogenesis, metastasis, and angiogenesis. Iourgenko *et al.* (2003) used a similar methodology to identify genes which activate expression of interleukin-8, a cytokine with etiological bases in asthma, arthritis, and cancer. Other reports utilize various reporter constructs in genome-wide functional analysis to analyze NF-κB signaling pathways, to identify p53 regulators or to identify new components of the Wnt signaling pathway, an important controller of embryonic development and possibly cancer (Huang *et al.*, 2004; Liu *et al.*, 2005; Matsuda *et al.*, 2003). The main advantage of the arrayed approach is that one typically discovers more genes than from the pooled approach, which serves to increase the probability of finding one that satisfies "good target" criteria.

B. RNA Interference Screening and Oncology

Whereas cDNA screens are useful for identifying genes that cause cell transformation and possibly cancer, RNA interference (RNAi) inhibition screens may be used to identify which genes can be inhibited to reverse the disease phenotype. Since RNAi inhibition of gene expression is in some respects analogous to inhibition of the same gene product by a small molecule antagonist, the utility of RNAi for drug target identification and validation has caught the interest of the drug discovery community. The phenomenon of RNAi, first discovered in the nematode *Caenorhabditis elegans*, has become a favored method for studying "loss-of-function" phenotypes in a high throughput and unbiased manner (Fire *et al.*, 1998). The direct use of RNAi in mammalian tissue culture cells has now become a feasible approach for functional genomics studies. The utility of this approach stems largely from efficient design and delivery of siRNA to mammalian cells, and the availability of the human genome sequence to design targeting-oligos for whole genome screens in mammalian cells.

The availability of the human genome sequence enables design of specific inhibitory RNA sequences against all of the predicted human genes, and advances in screening methodologies and automation allow one to interrogate the effects of knocking down every gene, one-by-one, in parallel in cellular assays just as described for cDNAs. In mammalian cells, gene silencing has been achieved by transient transfection of synthetic short double-stranded (ds) siRNA [*in vitro* synthesized siRNA, and plasmid-based short hairpin RNAs (shRNAs) (Aza-Blanc *et al.*, 2003; Berns *et al.*, 2004; Brummelkamp *et al.*, 2002; Hsieh *et al.*, 2004; Paddison *et al.*, 2004; Yang *et al.*, 2002; Zheng *et al.*, 2004]. The short hairpins are processed by ribonuclease III activity of the Dicer enzyme to generate effective 21–22 nt siRNA (Bernstein *et al.*, 2001). Besides Moloney-based retroviral delivery of shRNAs, other viral delivery systems including adenoviral RNAi vectors or lentiviral vectors have additional benefits (Arts *et al.*, 2003; Berns *et al.*, 2004; Paddison *et al.*, 2004; Rubinson *et al.*, 2003). Lentiviruses, in contrast to retroviruses, have the capacity to integrate into the host genome of nondividing cells and enable stable expression of the delivered shRNAs (Naldini *et al.*, 1996a,b). Adenoviral vectors are also used to achieve long-term expression and high gene transfer efficiency. Both types of RNAi reagents, including chemically synthesized siRNAs and vector-encoded shRNAs, have their own unique limitations that impact the success of the screen. For instance, it has been shown that some siRNAs, and more often shRNAs are prone to induce a nonspecific interferon-mediated response, which can complicate readouts (Bridge *et al.*, 2003; Sledz *et al.*, 2003). In addition, excessive concentrations of RNAi reagents can lead to significant off-target effects (Persengiev *et al.*, 2004). The ability to control the concentration of exogenous siRNA in contrast to the noncontrollable effective dose of shRNA transcribed inside the cell minimizes this risk of nonspecific effects.

Toward oncology drug targets, initial proof of concept for an arrayed synthetic siRNA library and the feasibility of large-scale RNAi screens in mammalian cells was first demonstrated by Aza-Blanc *et al.* (2003). siRNA oligos targeting 510 human kinases were tested for their ability to modulate the induction of apoptosis by TNF-related apoptosis inducing-ligand (TRAIL, a "biologic" drug in clinical development for oncology applications) in a cell viability assay. The goal of the screen was to identify genes which when selectively knocked down could enhance sensitivity of the cells to TRAIL-mediated apoptosis, or alternatively block apoptosis. Both known and previously uncharacterized genes, including *do*wnstream of *bi*d (DOBI) and MIRSA, were identified. In addition, a functional linkage between MYC, WNT, JNK, and BMK1/ERK5 genes to the TRAIL-mediated response pathways was revealed. A genome-wide library comprising ~49,000 synthetic siRNAs spotted in an arrayed format (two siRNAs per gene) has also been designed (Huesken *et al.*, 2005). An artificial intelligence algorithm (BIOPRED*si*, www.biopredsi.org) that computationally predicts 21-nt siRNA sequences that have an optimal knockdown effect for a given gene was used. BIOPRED*si* implements a neuronal network, based on the Stuttgart Neural Net Simulator (http://www-ra.informatik.uni-tuebingen.de/SNNS/). This collection was used to interrogate the response to hypoxia by induction/suppression of HIF-1α levels measured by the use of a luciferase reporter gene containing the hypoxia-response element. HIF-1α and aryl hydrocarbon receptor nuclear translocator (ARNT, HIF-1β), which heterodimerizes with HIF-1α on hypoxic shock, scored among the top hits, thus validating this approach. These examples demonstrate the powerful nature of siRNA design when coupled with genome-wide screening. Genes required for cell division in HeLa cells were screened by using a high-throughput cell viability assay followed by a high content video-microscopy assay to quantify the frequency of cells in mitosis for arrest phenotypes. Thirty-seven genes were identified, including several splicing factors whose silencing generated mitotic spindle defects. Another example of a vector library directed against the family of deubiquitinating enzymes revealed CYLD, the familial cylindromatosis tumor suppressor gene, as a suppressor of NF-κB, thus establishing a direct link between a tumor suppressor gene and NF-κB signaling (Brummelkamp *et al.*, 2003). It was also shown that inhibition of CYLD enhanced protection from apoptosis and can be reversed by aspirin derivatives, such as salicylate, which are established NF-κB inhibitory molecules. This result led to the suggestion of a therapeutic strategy for treating cylindromatosis with existing drugs and provides an insight as to how genetic screening approaches can reveal potential drug targets and intervention strategies.

As a result of the power of siRNA-based genetic screens, several groups have combined their efforts to expand this resource. The RNAi consortium (TRC), a collaborative effort among six research institutions and five international life sciences organizations, released a lentiviral shRNA library consisting of ~35,000 shRNA constructs targeting 5300 human and 2200 mouse genes. The first-generation library named MISSION™ TRC-Hs 1.0 (Sigma-Aldrich) or Expression Arrest™ TRC (Open Biosystems) is based on the self-inactivating lentiviral vector pLKO.1, wherein expression of the hairpin sequence is driven by the U6 promoter (Stewart *et al.*, 2003). The effectiveness of this newly released lentiviral library has not been reported, however, the results of several aforementioned screens indicate the potential utility of this resource.

IV. SMALL MOLECULE SCREENS IN ONCOLOGY

While it is appreciated that oncogenesis arises as a result of the accumulation of selectively advantageous somatic mutations, many of which can be mapped and sequenced, therapies that effectively reverse the unchecked tumor growth phenotype are sorely needed. Although cDNA screens can identify proto-oncogenes, siRNA libraries can identify potential targets for therapeutic intervention, small molecule screens in cellular cancer models to selectively impair tumor cell growth may identify both pathways, targets and drug candidates for cancer therapy. An outgrowth of the genomics revolution has also rejuvenated an age-old approach to drug discovery, cell-based screening. The ability to survey small molecule space in disease-relevant cellular pathway screens affords the opportunity to rapidly identify bioactive compounds for target identification (either by affinity approaches or in rare instances informatics) and tumor cell therapy. High-throughput screening of large, diverse chemical libraries has become fairly commonplace in academic and industrial settings (Clemons, 2004; Edwards *et al.*, 2004; Eggert *et al.*, 2004). Due to the significant literature covering this space, this chapter will focus more on the advances and uses of parallel screens of small molecules against cellular assays and analysis of the resulting multidimensional datasets.

A. Parallel Cellular Screens

Over the past 10 years advances in high-throughput screening, robotics, combinatorial chemistry, and assay technology have set new standards in drug discovery lead generation (Cox *et al.*, 2000; Floyd *et al.*, 1999; Sundberg, 2000; Weinstein *et al.*, 1992). Despite the capabilities and the enormous data they create, the failure rate of drug candidates in the clinical and preclinical phase is still high, primarily due to unexpected biological side effects such as toxicity (Schuster *et al.*, 2005). Large-scale screening databases routinely contain records of millions of compounds and their activities across a multitude of biological assays. Recently, it has been realized that interscreen data analysis across large-scale screening databases can be usefully mined for additional information on the activity of drug candidates. Analyzing the results of many screens at once affords the ability to negatively select compounds which have problematic properties such as nonspecific toxicity. Compounds with attractive properties such as cell- or gene-dependent apoptosis can be quickly identified for follow-up. Profiling of lead compounds across batteries of biologically relevant assays could identify untoward side effects early in preclinical discovery but also beneficial properties such as alternative indications, orphan target modulation, and new biology (Butcher *et al.*, 2004; Plavec *et al.*, 2004; Rabow *et al.*, 2002). Comparing compound activities across cellular data is inherently richer than biochemical or direct enzymatic measurements in a cell-free environment since cells encode the genetic networks and pathway architecture necessary to measure most physiology, especially that relevant to cancer such as growth, proliferation, checkpoint control, apoptosis, cell migration, and so on.

At its most basic level, parallel screens comparing as few as two assays side-by-side is significant (Clemons *et al.*, 2004). This approach has been used successfully to define new compounds with potential therapeutic importance in the area of oncology. For instance, Torrance *et al.* performed parallel simultaneous screens against isogenic pairs of colon cancer cell lines; one encoding mutant K-ras, the other a deletion of mutant K-ras, comparing data from about 30,000 unique compounds. By comparing the two screens, the strategy unveiled a new, differentially cytotoxic cytidine-containing compound which selectively eradicated mutant K-ras expressing cells (Torrance *et al.*, 2001). Dolma *et al.* (2003) screened ~24,000 novel compounds against two cell lines: an engineered tumorigenic cell versus its normal, untransformed counterpart, to identify a potent small molecule inhibitor of cell growth, termed erastin. Fantin *et al.* (2002) screened immortalized epithelial cell pairs either lacking or expressing the oncogene neuT to identify compounds which could selectively induce cell type-specific BrDu-incorporation. These examples demonstrate the concept of parallel screens whereby straightforward subtractive analysis derived data to rapidly select compounds with desired qualities.

Comparison across a multitude of cellular assays can derive more incisive information for follow-up work. For instance, compound profiling across a panel of phenotypic assays can portray the global effects of molecules from a given structural series from a lead optimization program, and guide chemical design to achieve desirable biological effects. As proof of concept, Kim *et al.* (2004) tested a library of structurally related compounds synthesized by diversity-oriented combinatorial methods, using a combination of monocyclized molecules and their corresponding bicyclic complement against a series of cellular growth-related readouts including DNA synthesis, overall cellular esterase activity, and mitochondrial membrane potential and intracellular reducing potential. Parallel tests of ~250 molecules and a small set of known drugs against the four cellular readouts in a total of 40 cellular assays resulted in a matrix of ~20,000 data points. In order to simplify the results, the matrix was subdivided into various descriptor sets such as number of rotatable bonds per compound. The overall cellular activity patterns of compounds migrated with their mono or bicyclic counterparts. Further analysis showed that subtle stereochemical alterations of the functional groups decorating the scaffold ultimately determined cell effects. These studies importantly demonstrate the possibility of using parallel cellular assays for quantitative measurement of the effects of discrete chemical modification.

Parallel cellular profiling can be used to prioritize compounds readily. Shi *et al.* (1998a,b) showed correlation between activity patterns and molecular structure of 112 ellipticine analogues tested in the National Cancer Institute (NCI) tumor cell panel. The population subdivided into patterns associated with either normal ellipticines or N2-alkyl substituted ellipticiniums. The outliers were all dependent on specific structural modifications. Interestingly, a correspondence between p53-status and glioblastoma-like tumor types was manifest in the patterns, where the ellipticiniums were generally more active than the ellipticines in cells with mutant p53. The distinction provides the researcher with a rational means to prioritize and choose next steps in a compound progression scheme based on p53 status or on structural liabilities of these compounds. Haggarty *et al.* (2004) analyzed an antihistone deacetylase 1,3 dioxane-based diversity-oriented synthetic library using molecular descriptors to generate a multidimensional matrix relating chemical structure with activity. The results identified structural elements important for SAR relationships distinguished by tubulin- or histone-dependent substrate preference. These studies show again that multidimensional profiling can elucidate relationships between small structural perturbations and specific cellular activity, and define a roadmap for further studies.

Based on the promise of this approach, engineers and biologists at the Genomics Institute of the Novartis Research Foundation (GNF) have built an automated robotic system for simultaneous parallel cellular profiling of hundreds of cellular assays against large, diverse molecular libraries. The industrial scale system, termed an *a*utomated *c*ompound *p*rofiling (ACP) system combines automated tissue culture for propagating cell-based assays with a system for automatically performing miniaturized assays in 384- or 1536-well microplates (Fig. 3). The ACP can rapidly test thousands of compounds, in replicates, simultaneously in dose–response against hundreds of unique cell-based assays in a single experiment. The system

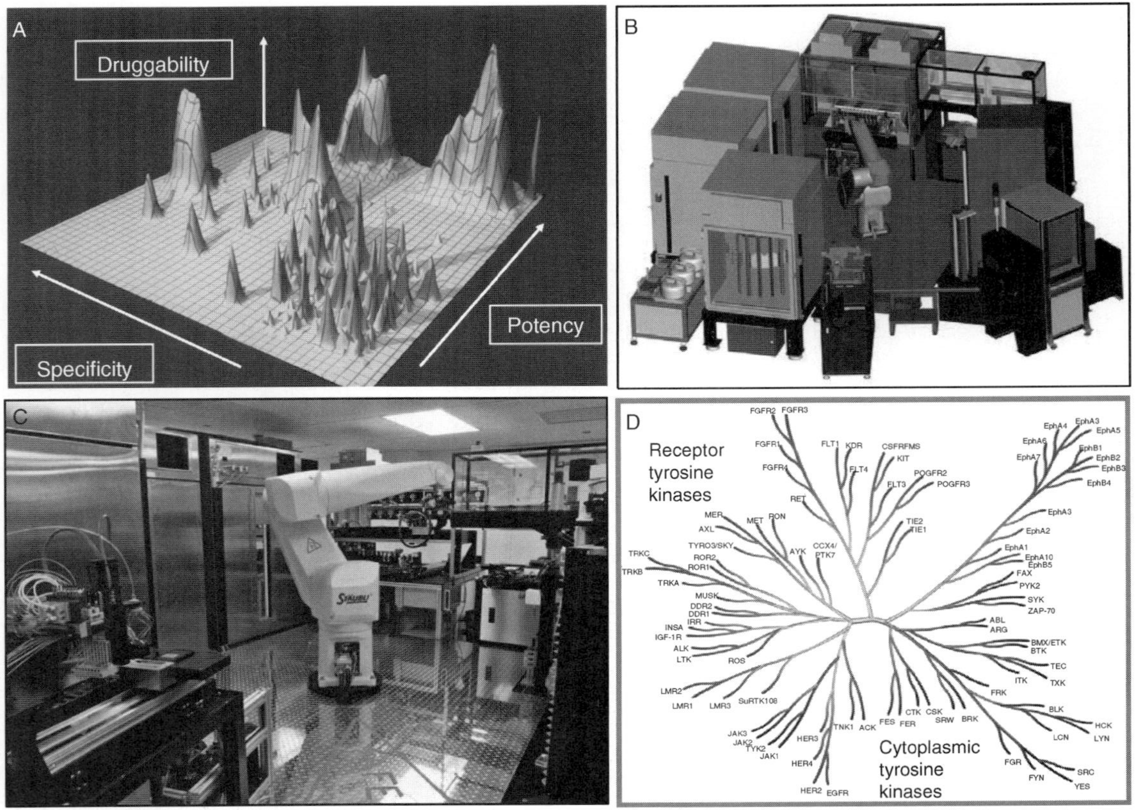
A
Druggability
Specificity
Potency
B
C
D
Receptor tyrosine kinases
Cytoplasmic tyrosine kinases
FGFR2 FGFR3
FGFR1
FGFR4
RET
FLT1 KDR
FLT4
CSFRFMS
KIT
FLT3
POGFR2
POGFR3
TIE2
TIE1
MER
AXL
MET
RON
TYRO3/SKY
AYK
CCX4/
PTK7
TRKC
TRKB
ROR2
ROR1
TRKA
MUSK
DDR2
DDR1
IRR
INSA
IGF-1R
ALK
LTK
ROS
LMR2
LMR1
LMR3
SuRTK108
JAK3
JAK2
TYK2
JAK1
HER3
HER4
HER2
EGFR
TNK1
ACK
FES
FER
CTK
CSK
SRW
BRK
EphA4
EphA3
EphA5
EphA6
EphB1
EphA7
EphB2
EphB3
EphB4
EphA3
EphA2
EphA1
EphA10
EphB5
FAX
PYK2
SYK
ZAP-70
ABL
ARG
BMX/ETK
BTK
TEC
ITK
TXK
FRK
BLK
HCK
LYN
LCN
FGR
FYN
SRC
YES

was first applied to the problem of identifying kinase-selective small molecule inhibitors that target protein tyrosine kinases. Tyrosine kinase in general play key roles in many cell growth processes and their misregulation leads to a number of disorders, most notably cancer (Blume-Jensen and Hunter, 2001; Hunter, 1998). Spontaneous translocations in which tyrosine kinase domains become fused to other genes (such as NPM, Tel, and BCR) have been identified as the basis for oncogenesis and several B-cell lymphomas, including the Philadelphia chromosome BCR/ABL and CML (Baserga and Castoldi, 1973; Tkachuk *et al.*, 1990). Due to the difficulty of identifying selective small molecules that inhibit these tyrosine kinases, a parallel cellular screen may provide a more fruitful approach.

In the first ACP experiment, several thousand small molecule kinase inhibitors including 2,6,9-trisubstituted purines were profiled against an array of 35 novel tyrosine kinase-dependent cell-based assays constructed in the mouse pro-B Ba/F3 cell background (Ding *et al.*, 2001; Melnick *et al.*, 2006). The experiment discovered selective small molecule inhibitors for a variety of tyrosine kinases, but also demonstrated structure activity relationships within subclusters of compounds which could be used to improve potency and selectivity against a given set of related targets. Interestingly, the approach showed that the relationship between tyrosine kinases based on sequence data did not superimpose directly with a dendrogram based on compound data. This type of phylogenographic analysis could reveal which side activities for a given set of compounds might arise through further optimization, either positive or negative for oncology drug development.

The dataset also contained dose–response information for a select set of known drugs and drug candidates developed for a variety of disease indications. As the industry is realizing through abundant examples of

Fig. 3 An automated system for profiling compounds versus panels of cellular assays. (A) Compound profiling provides multidimensional data for more informed selection of lead drug candidates, the depicted landscape represents specificity and potency of a panel of compound scaffolds two dimensions (x and y), with a z-axis representing druggability or the likelihood that the associated chemical scaffold can be optimized to become a drug. The schematic makes the point that choosing compounds based on multidimensional data may lead to more optimal drugs (z-axis). (B) The components of the automated compound profiling (ACP) system. The system comprises three custom environmentally controlled incubators, a 1536-well plate cell dispenser and reagent dispenser, a control station consisting of a computer running custom scheduling software, a CCD-plate reader, an inverted microscope with an automated stage and image analysis software, a compound transfer station, a tissue culture station for splitting cells, and a Staubli RX130 anthropomorphic robot with a custom plate gripper. Together these components integrate to perform high-throughput assays on panels of cells versus arrayed molecular libraries. (C) A photograph of the ACP installed at the Genomics Institute of the Novartis Research Foundation. (D) The tyrosine kinome representing the first interrogated protein family using automated profiling technology.

"polypharmacy," drugs developed for one purpose may have advantages in other diseases against other targets, the data was examined for unexpected side activities and novel properties. Glivec, designed to target BCR/ABL, for instance has described activity against pdgf and c-kit as well, which are the justification for pursuing Glivec in GIST and asthma. In this experiment, interesting cross-activities for other, less well characterized small molecule kinase inhibitors currently under clinical investigation were discovered. The pan-p38 inhibitor BIRB796 was shown to block Tie2, a protein implicated in promoting angiogenesis. Extending the inhibitor's activity to Tie2 could potentially increase the utility of this compound in cancer treatment. In another example, the dual inhibitor of Src-family kinases, BMS-354825, inhibited several Ephrin receptors implicated in both tumor angiogenesis and tumor cell survival (Brantley-Sieders and Chen, 2004; Brantley-Sieders *et al.*, 2004; Kullander and Klein, 2002; Lombardo *et al.*, 2004). The results of this profiling experiment indicate the potential for expansion into other tumor types. The ACP experiment demonstrates the power of multidimensional profiling not only as a strategy for detailed SAR and a framework for optimal chemical modification, but also to identify new therapeutic opportunities for small molecules. In fact, some of the most promising oncology drugs may currently be available as prescription drugs or generics for other indications.

The idea of parallel cellular screens is, however, not entirely novel. The US NCI has been pioneering the approach of parallel screening for the past several decades. The NCI carries out an anticancer drug discovery program in which more than 100,000 different chemical entities have been screened, and another 10,000 novel chemicals are screened annually, in a panel of more than 60 human cancer cell lines from a variety of different cancer-disease tissues. As testimony to the success of this approach, at least five compounds from the NCI screen have entered into phase I trials for oncology including the DNA intercalator, quinocarmycin, a DNA interactor termed spicamycin, KRN-5000 which disrupts glycoprotein processing the kinase inhibitors UCN-1 and flavopiridol, and depsipeptide NSC 630176, a histone deacetylase inhibitor (Bunnell *et al.*, 2001; Dorr *et al.*, 1988; Lee *et al.*, 1995; Sedlacek, 2001; Seynaeve *et al.*, 1994).

The massive dataset achieved by the NCI has provided ample information for another type of analysis which can only be performed using parallel chemical-profiling data. These methods are used to extract another dimension of information which describes more than selectivity, potency, and even alternative indications. Higher-order analysis of parallel profiling experiments can suggest if not deduce the molecular mechanism of action (MOA) of a drug whose target was completely unknown. The following sections describe the growing importance of this powerful analytical technique and the promise it holds for deriving better molecular-defined cancer drugs.

B. MOA Determination

As a brief history, in early drug discovery, experimentalists tested small collections of chemicals first in cellular and animal models then man independent of whether the compound's target or MOA was known. Cyclosporin, aspirin, and penicillin are all examples of this approach (Hoyem, 1961; Kopp and Ghosh, 1994; Schreiber, 1991). Later, the advent of cloning and protein expression allowed one to test many compounds against a single enzyme in a cell-free environment in search of active modulators which could then be tested in cellular and animal models, in this case with target and MOA information at hand. Due to the relatively meager increases in successful drug discovery with the modern approach, given the diminishing number of new chemical entities produced by major pharmaceutical companies each year, a return to the "old" days of drug discovery using cells is being reexamined.

Inherent in the cell-based screening approach is a level of target blindness where the interesting compounds and drug candidates may not have an established MOA. For instance, one may screen for compounds that selectively ablate a cancer cell type by testing thousands of small molecules in a cell-based assay for inhibition of proliferation, as is the case for the NCI screen, and others described before by Kinzler *et al.* and Dolma *et al.* (2003). Unfortunately, despite very interesting molecules being discovered, neither the exact molecular target in the cell, the target type, nor the pathway modulated is known. The only information one has is that said molecule kills said cells, which makes further characterization, especially safety of the compound, a major black box to circumnavigate. Given the growing concern over small molecule drug toxicity and FDA restrictions, an increased importance has been placed on knowing a compound's MOA well before going to market, however, methods to detect the target and MOA have not kept up with the many advances in the post-genomic era (Monks *et al.*, 1997). High-throughput proteomics, affinity methods using cDNA libraries on phage, chromatographic purification, or genetic complementation assays using whole genomic cDNAs to identify the target of a small molecule are interesting approaches, but have been relatively unsuccessful. A means to accurately predict a compound's target and MOA is sorely needed for the cell-based strategy to have an impact.

Multidimensional profiling may provide a means to informatically predict a drug's MOA. An understanding of the definition of a compound's activity profile is necessary to appreciate how compound activity across a battery of cellular assays can be used to predict its MOA. In essence, any compound can be described by its unique set of activities in a variety of biological test systems. This unique set of activities can be represented as a signature or fingerprint, herein termed an activity profile for the compound. Just as a person can be uniquely identified by his/her fingerprint, a compound can be identified or traced by its signature activity profile. The NCI has determined activity profiles for over 100,000 compounds against a panel of patient-derived tumor cells, just as the GNF has determined activity profiles for thousands of kinase inhibitors across a panel of tyrosine kinase assays. Both the NCI and GNF datasets provide ample evidence that small molecules with similar structures have almost superimposable activity profiles (Melnick *et al.*, 2006; Rabow *et al.*, 2002). The differences that do exist are linked to small alterations in chemical structure and are unlikely to create significant incongruence over a large number of screening assays. Likewise, compounds with similar mechanisms of action are likely to have similar activity profiles even if they have completely distinct chemical structures. Compounds with overlapping profiles will generally have the same or similar molecular targets, while entities similar in chemical structure with different mechanisms of action will have different and nonoverlapping activity profiles.

These principles clearly agree with the observation that, although the activity data for a single biological assay contains some useful information, activity profiles derived over a broad panel of assays encode incisive information which equate with a compound's MOA (Weinstein *et al.*, 1997). These conclusions are the general result of the NCI effort. The screening strategy relies on the premise that agents tested will show reproducible patterns of differential response among the 60 cell lines and that these differences underscore differences in chemical structure and thus MOA. In the widely mined NCI60 dataset (*vide infra*), all of the cellular assays are antiproliferative or toxicity assays performed in dose response under nearly identical conditions, differing in little other than the cell line used. Based on this set of rules, it is fairly obvious that by comparing activity profiles of known drugs with known targets, new compounds with identical or highly similar activity patterns are likely to hit the same target. Therefore, comparison of panels of reference compounds to new ones can begin to functional classify and assign putative targets for the latter. The lesson from the NCI profiling database, which represents the records of the screens and describe a comprehensive set of cellular experiments, is that if appropriately curated the data can be mined in clever ways to extract knowledge regarding compound MOA and testable hypotheses in cancer research.

C. Data Analysis of Multidimensional Cellular Datasets

Understanding appropriate statistical analysis to mine multidimensional profiling data is of paramount importance to derive the most useful, predictive hypotheses. Sophisticated methods to analyze multidimensional datasets have come to the fore including techniques of hierarchical clustering analysis (HCA), multidimensional scaling (MDS), self-organizing maps (SOMs), and methods based on neural networks. Each of these methods has been adapted to parallel cell-based screening to extract knowledge, largely applied to if not inspired by the NCI's drug screening program of more than 60 different human tumors (Brown *et al.*, 2000; Kohonen, 1999; Meyer and Cook, 2000; Rabow *et al.*, 2002; Sneath and Sokal, 1962). In general, methods such as principal component analysis (PCA), MDS, and HCA cluster compounds from large multidimensional cellular datasets to assign putative functional relationships between compounds. PCA takes linear combinations of a data matrix such that the first principal component (PC) explains as much of the overall variation as possible, the second PC explains the next most variation subject to being orthogonal to the first, and so on (Shi *et al.*, 2000). One of the powerful features of PCA is its ability to identify outliers, random equipment abnormalities, and data entry errors (Shi *et al.*, 2000). In contrast, using MDS, members of a group (i.e., compounds) are separated to reduce the number of dimensions associated with the data for ease of visual inspection. HCA clusters based on similarity to generate cluster maps, akin to a phylogenogram to quickly identify relationships between groups of compounds.

Appropriate cross-screen normalization and data analysis methods are of paramount importance since the results of these analyses lead to testing of small molecules in animals and man. Toward a global analysis of the NCI screening data, Paull *et al.* (1992) developed COMPARE, a computer program that scans for useful information in each compound's activity pattern. COMPARE calculates linear correlation coefficients between the data over all cell lines for the pattern of interest (called the seed) and all searchable sets of data in the database and then sorts by the correlation coefficient. For a designated "seed" compound, COMPARE searches the database for compounds whose activity pattern most closely resembles the seed pattern. By this methodology, compounds matched by pattern often have similar chemical structures and a correspondingly related *in vitro* biochemical MOA thus lending this strategy to generate testable hypotheses (Weinstein *et al.*, 1997; Zaharevitz *et al.*, 2002). For example, when the activity pattern for anthrax lethal factor (NSC 678519) was used as a "seed" the most outstanding correlation was with NSC 679828, a known MAP kinase kinase (MKK) inhibitor. This correlation suggested that the anthrax lethal factor was eliminating MKK activity via degradation of the enzyme, which was later confirmed experimentally (Duesbery *et al.*, 1998). COMPARE calculations are versatile, allowing one to search for compounds with the strongest positive correlation or the largest negative correlation, and provides the ability to eliminate cell line data either irrelevant to a particular comparison or of suspect quality.

Researchers at GNF developed a set of analysis methods similar to COMPARE, devised to classify compounds by mechanism and to identify putative MOAs and targets for previously undescribed compounds. The GNF dataset is distinct from that of the NCI's in that it comprises single point activity data for over 1 million compounds in over 200 high-throughput screening assays, making it more comparable to the databases most pharmaceutical companies retain (Zhou *et al.*, unpublished result). Unlike the PCA analysis, the variance for each assay is not normalized to one. For each pairing between the seed compounds and queried compounds, two similarity metrics are computed over the assays those compounds have in common: an uncentered Pearson correlation coefficient and a count-normalized Euclidean distance. Using these analysis methods, predictably, comparisons between signatures of known compounds demonstrate a high degree of structural similarity. The similarity of activity profiles between structurally distinct compounds with reportedly distinct therapeutic modalities and treatment populations, including cytotoxic compounds and calcium channel modulators was also highly significant suggesting a mechanistic link between calcium channel inhibitors and cytotoxic compounds, a result which has been corroborated by the literature (Root *et al.*, 2003).

COMPARE and associated methods have been successful for discovery of promising lead candidates but may be limited by treating the compounds and targets one pair at a time (Fang *et al.*, 2004). A more comprehensive approach was devised by Weinstein *et al.* (1997) called DISCOVERY which compares dissimilar types of data relating to the compounds. As a first pass, to generate a human logical pattern describing compound data and interrelationships, an algorithm termed "ClusCor" was employed. This algorithm takes into consideration three datasets (compound, target knowledge, and activity) as a statistical matrix for which each compound's activity and target data are normalized by its mean and standard deviation. The normalized data are multiplied to obtain the "matrix transpose," relating activity of the compound to target patterns. DISCOVERY was used to identify a cluster of structurally related platinum analogues (such as cisplatin and carboplatin) which share a common mechanism, but also a series of structurally related diaminocyclohexyl platinum compounds with a completely unique MOA. The sensitivity implicit in this methodology affords the identification of more subtle associations between structure and mechanism.

For even more subtle associations and treatment of completely disparate cell-based assay datasets, SOMs methods have been employed (Rabow *et al.*, 2002). While PCA, MDS, HCA, and related methods maximize signal-to-noise, the SOM method is best suited for datasets with significant random noise such as cell-based assays. The SOM method both clusters data in high-dimensional space and projects into lower dimensional space (Rabow *et al.*, 2002). SOMs were used to analyze the complete NCI database to identify relationships between chemo-types of tested compounds and their effect on four types of cellular activity: membrane transport/integrity, nucleic acid synthesis, mitosis, and kinase-dependent cell-cycle regulation were identified. The map effectively distinguishes effectors of uridine biosynthesis, cytidine synthesis, purine biosynthesis, ribonucleotide reduction, and nucleic acid intercalation from one another demonstrating the strength of the SOM in dividing nucleotide biosynthetic processes into discrete categories (Rabow *et al.*, 2002).

Computational tools can only generate hypotheses of biological significance as good as the datasets being analyzed. Understanding the influence of different factors composing the dataset is necessary to determine the utility of associated predictions. For instance, the NCI dataset is composed of a battery of cell lines which have inherently different growth rates—how does this factor influence interpretation of the data? To address this problem, Rabow *et al.* (2002) designed a method of data conditioning whereby Z-score normalization (a function of distance from sample mean to standard deviation) is performed to enhance the signal-to-noise ratio. Using Z-score transformation data clustering on the NCI dataset, quality was improved ~15% versus the raw data and bounded any data obscuring resulting from the most sensitive cell lines. Another enhancing method is to limit the maximum and minimum allowed GI_{50} values, which prevents domination by a nominal set of cell lines or compounds with extreme effects. In contrast, filling in missing data with the group average will compromise the clustering. In essence, Z-score normalization of each cell line's response to all tested compounds establishes a common reference point such that a rigorous comparison of the relative contribution of experimental design parameters can be assessed.

These are the mathematical tools used to efficiently mine multidimensional datasets, however, what is their true value in the grand scheme of cancer drug discovery? Can the datasets be used to make discoveries and predictions of importance, such as MOA, for completely novel compounds? Can relationships underlying multidimensional datasets be queried in refined ways to convey permissible hypotheses and concurrently fortify correct ones through learning? This line of inquiry begs for an infrastructure analogous to the architecture of the brain and neural circuitry which in contrast to computer algorithms, which are programmed to get the right answer, actually "learn" from the examples set forth in a dataset to get the right answer. Weinstein and others have explored this possibility which is covered in the following section.

D. Mechanisms for New Drugs

Possibly the strongest examples of systematic MOA determination of small molecules from blinded pathway screens are found in the NCI's effort to identify new drugs for cancer treatment. The NCI strategy is to profile annotated, mechanistically described antiproliferative compounds against panels of cell-based assays to generate a reference panel by which new compounds can be compared (by activity profiles) to infer MOA (i.e., purine biosynthesis, antifolates, apoptosis) or even precise target family [i.e., topoisomerase, cyclin-dependent kinases (CDKs); Rabow *et al.*, 2002].

The NCI anticancer screen database was first determined to be of utility for predicting MOA prediction when the COMPARE algorithm allowed identification of unique structures which when compared to the reference dataset revealed their MOA, including antitubulin-binding activity and antitopoisomerase II activity (Bai *et al.*, 1991; Cleaveland *et al.*, 1995; Jayaram *et al.*, 1992; Leteurtre *et al.*, 1994; Paull *et al.*, 1992). Subsequently, Fang *et al.* designed a web-based tool similar to COMPARE to mine the NCI database for novel targets and MOA. The study observes a correlation between *in vitro* anticancer activity of the drugs in the NCI anticancer database and protein levels and mRNA levels of the molecules they target. Shao *et al.* used a similar strategy to predict novel protein kinase C (PKC) ligands, such as NSC 631939 and NSC 631941, which led to use of the nanomolar affinity PKC ligand mezerein as a "seed," to find new PKC ligands from a series of 7269 compounds based on overlapping GI_{50}s across a panel of tumor cell lines (Fang *et al.*, 2004; Shao *et al.*, 2001). This analysis identified two new PKC ligands including NSC 266186 (huratoxin) and NSC 654239 (cytoblastin), an analogue of teleocidin (Fang *et al.*, 2004). Rabow *et al.* (2002) using SOM analysis were able to identify a number of powerful and previously unappreciated structure–function relationships and MOAs. As an example, areas of the SOM corresponding to CDK inhibition by known CDK inhibitors such as UCN-01, and olomoucine, were several paullone-like compounds with previously unknown CDK inhibition characteristics (Rabow *et al.*, 2002). A phosphatase 2A inhibition subregion reclassified the topoisomerase inhibitors fostriecin and cytostatin as PP2A inhibitors. Using the SOM, advised with nothing more than a known structure and the NCI panel profile, an uncharacterized compound can be assigned a target class. Based on the power of this approach, the cancer community sends compounds of unknown mechanism to the NCI screen to obtain MOA predictions.

The pharmaceutical community and others have emulated the NCI approach by testing their own internal-derived compounds against cell-based assay panels. Yamori *et al.* established a panel of 39 cell lines to include stomach cell lines outside the NCI panel, based on the prevalence of this disease in Japan, and discovered a novel anticancer agent, MS-247. A netropsin-like moiety and an alkylating residue in MS-247 provided a structural link to the profile of camptothecin, suggesting the two drug groups may have overlapping modes of action (Dan *et al.*, 2003; Yamori *et al.*, 1999). Subsequently, MS-247 was shown to have antitumor activity in xenograft models: i.v. injection of MS-247 significantly inhibited the growth of lung, colon, stomach, breast, and ovarian cancer xenografts and was more efficacious than other chemotherapeutics such as cisplatin, adriamycin, 5-fluorouracil, cyclophosphamide, VP-16, and vincristine. Therefore, MOA and precise target definition predicted by multiparameter cell-profiling approaches has been borne out in empirical tests, serving as additional proof of concept for the approach.

Publicly available data represents another source of cell panel versus compound data to be mined for novel cancer drugs. A computer-based screening method was constructed based on literature documented compound MOA associations. Assembling information from the literature Poroikov *et al.* (2003) built a database connecting compound structure, MOA, pharmacological effects, known toxicities, and other descriptors for a significant fraction of the publicly available literature dating back to 1972 (Lee *et al.*, 1995). This database can be queried using the prediction of activity spectra for substances (PASS) application which comprises about 250,000 compounds of the NCI Open Database and over 64 million PASS predictions in the enhanced NCI database (Poroikov *et al.*, 2003). By preassembling the public literature into a cluster map, queries for compounds which are likely to possess user-defined desirable properties (biological, pharmacological, or MOA) can be made. For instance, a search for "angiogenesis" inhibitors yielded multiple unique compound structures with a given probability of blocking angiogenesis empirically. This search identified seven candidate angiogenesis inhibitors of which four remarkably demonstrated the predicted inhibitory activity. Although the number of tested compounds is too low to show statistical relevance, this tool does provide a free method of mining public data for possible novel leads in cancer drug discovery.

V. CONCLUSIONS

The oncology research community is in need of strategies and useful tools to combat cancer. The number of anticancer drugs in the arsenal is small and to date insufficient to address cancer patients effectively. The hope is that the significant technological advances made in the past several years in academia, governmental, and industrial discovery labs will lead to new cures for cancer and repositioning of old drugs for unmet cancer therapeutic needs.

This chapter outlines several of the advances, focused in the cellular technologies, to discover new drug targets, drugs, and strategies to uncover novel medications rapidly. Namely, the post-genomic tools assembled based on the human genome sequence are being used to comprehensively scan and identify new druggable targets implicated in cancer etiology and maintenance. Similarly, advances in small molecule discovery, high-throughput instrumentation for cell-based screening, and the computational methods to mine compound data will hopefully pave the way to new chemical entities and the next cadre of oncology drugs. Perhaps the efficiency of cellular panel profiling will lead to the necessary commoditization and export of the systems outlined in this chapter to practitioners of cancer drug discovery such that lead optimization campaigns throughout the industry can exploit them for guidance in optimal lead selection. Indeed, a combination of the approaches elaborated by the NCI, GNF, and others which robustly interrogate multiple cancer indications simultaneously in high throughput will meet with the target identification and safety requirements demanded by the modern pharmaceutical engine.

REFERENCES

Allen, J. M., and Seed, B. (1989). Isolation and expression of functional high-affinity Fc receptor complementary DNAs. *Science* **243**(4889), 378–381.

Arts, G. J., Langenmeijer, E., Tissingh, R., Ma, L., Pavliska, H., Dokic, K., Dooijes, R., Mesic, E., Clasen, R., Michiels, F., van der Schueren, J., Lambrecht, M., *et al.* (2003). Adenoviral vectors expressing siRNAs for discovery and validation of gene function. *Genome Res.* **13**(10), 2325–2332.

Aza-Blanc, P., Cooper, C. L., Wagner, K., Batalov, S., Deveraux, Q. L., and Cooke, M. P. (2003). Identification of modulators of TRAIL-induced apoptosis via RNAi-based phenotypic screening. *Mol. Cell* **12**(3), 627–637.

Bai, R. L., Paull, K. D., Herald, C. L., Malspeis, L., Pettit, G. R., and Hamel, E. (1991). Halichondrin B and homohalichondrin B, marine natural products binding in the vinca domain of tubulin. *J. Biol. Chem.* **266**(24), 15882–15889.

Bailey, D., Zanders, E., and Dean, P. (2001). The end of the beginning for genomic medicine. *Nat. Biotechnol.* **19**(3), 207–209.

Baserga, A., and Castoldi, G. L. (1973). The Philadelphia chromosome. *Biomedicine* **18**(2), 89–94.

Berns, K., Hijmans, E. M., Mullenders, J., Brummelkamp, T. R., Velds, A., Heimerikx, M., Kerkhoven, R. M., Madiredjo, M., Nijkamp, W., Weigelt, B., Agami, R., Ge, W., *et al.* (2004). A large-scale RNAi screen in human cells identifies new components of the p53 pathway. *Nature* **428**(6981), 431–437.

Bernstein, E., Caudy, A. A., Hammond, S. M., and Hannon, G. J. (2001). Role for a bidentate ribonuclease in the initiation step of RNA interference. *Nature* **409**(6818), 363–366.

Blume-Jensen, P., and Hunter, T. (2001). Oncogenic kinase signalling. *Nature* **411**(6835), 355–365.

Brantley-Sieders, D. M., and Chen, J. (2004). Eph receptor tyrosine kinases in angiogenesis: From development to disease. *Angiogenesis* **7**(1), 17–28.

Brantley-Sieders, D. M., Schmidt, S., Parker, M., and Chen, J. (2004). Eph receptor tyrosine kinases in tumor and tumor microenvironment. *Curr. Pharm. Des.* **10**(27), 3431–3442.

Bridge, A. J., Pebernard, S., Ducraux, A., Nicoulaz, A. L., and Iggo, R. (2003). Induction of an interferon response by RNAi vectors in mammalian cells. *Nat. Genet.* **34**(3), 263–264.

Brown, M. P., Grundy, W. N., Lin, D., Cristianini, N., Sugnet, C. W., Furey, T. S., Ares, M., Jr., and Haussler, D. (2000). Knowledge-based analysis of microarray gene expression data by using support vector machines. *Proc. Natl. Acad. Sci. USA* **97**(1), 262–267.

Brummelkamp, T. R., Bernards, R., and Agami, R. (2002). A system for stable expression of short interfering RNAs in mammalian cells. *Science* **296**(5567), 550–553.

Brummelkamp, T. R., Nijman, S. M., Dirac, A. M., and Bernards, R. (2003). Loss of the cylindromatosis tumour suppressor inhibits apoptosis by activating NF-kappaB. *Nature* **424**(6950), 797–801.

Bunnell, C. A., Supko, J. G., Eder, J. P., Jr., Clark, J. W., Lynch, T. J., Kufe, D. W., and Shulman, L. N. (2001). Phase I clinical trial of 7-cyanoquinocarcinol (DX-52-1) in adult patients with refractory solid malignancies. *Cancer Chemother. Pharmacol.* **48**(5), 347–355.

Butcher, E. C., Berg, E. L., and Kunkel, E. J. (2004). Systems biology in drug discovery. *Nat. Biotechnol.* **22**(10), 1253–1259.

Chanda, S. K., White, S., Orth, A. P., Reisdorph, R., Miraglia, L., Thomas, R. S., DeJesus, P., Mason, D. E., Huang, Q., Vega, R., Yu, D. H., Nelson, C. G., *et al.* (2003). Genome-scale functional profiling of the mammalian AP-1 signaling pathway. *Proc. Natl. Acad. Sci. USA* **100**(21), 12153–12158.

Cleaveland, E. S., Monks, A., Viagro-Wolff, A., Zaharevitz, D. W., Paull, K., Ardalan, K., Cooney, D. A., and Ford, H., Jr. (1995). Site of action of two novel pyrimidine biosynthesis inhibitors accurately predicted by the compare program. *Biochem. Pharmacol.* **49**(7), 947–954.

Clemons, P. A. (2004). Complex phenotypic assays in high-throughput screening. *Curr. Opin. Chem. Biol.* **8**(3), 334–338.

Cox, B., Denyer, J. C., Binnie, A., Donnelly, M. C., Evans, B., Green, D. V., Lewis, J. A., Mander, T. H., Merritt, A. T., Valler, M. J., and Watson, S. P. (2000). Application of high-throughput screening techniques to drug discovery. *Prog. Med. Chem.* **37**, 83–133.

Dan, S., Shirakawa, M., Mukai, Y., Yoshida, Y., Yamazaki, K., Kawaguchi, T., Matsuura, M., Nakamura, Y., and Yamori, T. (2003). Identification of candidate predictive markers of anticancer drug sensitivity using a panel of human cancer cell lines. *Cancer Sci.* **94**(12), 1074–1082.

Darnell, J. E., Jr., Kerr, I. M., and Stark, G. R. (1994). Jak-STAT pathways and transcriptional activation in response to IFNs and other extracellular signaling proteins. *Science* **264**(5164), 1415–1421.

Ding, S., Gray, N. S., Ding, O., and Schultz, P. G. (2001). A concise and traceless linker strategy toward combinatorial libraries of 2,6,9-substituted purines. *J. Org. Chem.* **66**(24), 8273–8276.

Dolma, S., Lessnick, S. L., Hahn, W. C., and Stockwell, B. R. (2003). Identification of genotype-selective antitumor agents using synthetic lethal chemical screening in engineered human tumor cells. *Cancer Cell* **3**(3), 285–296.

Dorr, F. A., Kuhn, J. G., Phillips, J., and von Hoff, D. D. (1988). Phase I clinical and pharmacokinetic investigation of didemnin B, a cyclic depsipeptide. *Eur. J. Cancer Clin. Oncol.* **24**(11), 1699–1706.

Druker, B. J., Tamura, S., Buchdunger, E., Ohno, S., Segal, G. M., Fanning, S., Zimmermann, J., and Lydon, N. B. (1996). Effects of a selective inhibitor of the Abl tyrosine kinase on the growth of Bcr-Abl positive cells. *Nat. Med.* **2**(5), 561–566.

Duesbery, N. S., Webb, C. P., Leppla, S. H., Gordon, V. M., Klimpel, K. R., Copeland, T. D., Ahn, N. G., Oskarsson, M. K., Fukasawa, K., Paull, K. D., and Vande Woude, G. F. (1998). Proteolytic inactivation of MAP-kinase-kinase by anthrax lethal factor. *Science* **280**(5364), 734–737.

Edwards, B. S., Oprea, T., Prossnitz, E. R., and Sklar, L. (2004). Flow cytometry for high-throughput, high-content screening. *Curr. Opin. Chem. Biol.* **8**(4), 392–398.

Eggert, U. S., Kiger, A. A., Richter, C., Perlman, Z. E., Perrimon, N., Mitchison, T. J., and Field, C. M. (2004). Parallel chemical genetic and genome-wide RNAi screens identify cytokinesis inhibitors and targets. *PLoS Biol.* **2**(12), e379.

Fang, X., Shao, L., Zhang, H., and Wang, S. (2004). Web-based tools for mining the NCI databases for anticancer drug discovery. *J. Chem. Inf. Comput. Sci.* **44**(1), 249–257.

Fantin, V. R., Verardi, M. J., Scorrano, L., Korsmeyer, S. J., and Leder, P. (2002). A novel mitochondriotoxic small molecule that selectively inhibits tumor cell growth. *Cancer Cell* **2**(1), 29–42.

Fire, A., Xu, S., Montgomery, M. K., Kostas, S. A., Driver, S. E., and Mello, C. C. (1998). Potent and specific genetic interference by double-stranded RNA in *Caenorhabditis elegans*. *Nature* **391**(6669), 806–811.

Floyd, C. D., Leblanc, C., and Whittaker, M. (1999). Combinatorial chemistry as a tool for drug discovery. *Prog. Med. Chem.* **36**, 91–168.

Gerhard, D. S., Wagner, L., Feingold, E. A., Shenman, C. M., Grouse, L. H., Schuler, G., Klein, S. L., Old, S., Rasooly, R., Good, P., Guyer, M., Peck, A. M., *et al.* (2004). The status, quality, and expansion of the NIH full-length cDNA project: The Mammalian Gene Collection (MGC). *Genome Res.* **14**(10B), 2121–2127.

Ghosh, S., and Karin, M. (2002). Missing pieces in the NF-kappaB puzzle. *Cell* **109**(Suppl.), S81–S96.

Ghuysen, J. M. (1977). The concept of the penicillin target from 1965 until today. The thirteenth marjory stephenson memorial lecture. *J. Gen. Microbiol.* **101**(1), 13–33.

Gomez-Lazaro, M., Fernandez-Gomez, F. J., and Jordan, J. (2004). p53: Twenty five years understanding the mechanism of genome protection. *J. Physiol. Biochem.* **60**(4), 287–307.

Haggarty, S. J., Clemons, P. A., Wong, J. C., and Schreiber, S. L. (2004). Mapping chemical space using molecular descriptors and chemical genetics: Deacetylase inhibitors. *Comb. Chem. High Throughput Screen* **7**(7), 669–676.

Handschumacher, R. E., Harding, M. W., Rice, J., Drugge, R. J., and Speicher, D. W. (1984). Cyclophilin: A specific cytosolic binding protein for cyclosporin A. *Science* **226**(4674), 544–547.

Hitoshi, Y., Lorens, J., Kitada, S. I., Fisher, J., LaBarge, M., Ring, H. Z., Francke, U., Reed, J. C., Kinoshita, S., and Nolan, G. P. (1998). Toso, a cell surface, specific regulator of Fas-induced apoptosis in T cells. *Immunity* **8**(4), 461–471.

Hopkins, A. L., and Groom, C. R. (2002). The druggable genome. *Nat. Rev. Drug Discov.* **1**(9), 727–730.

Hoyem, T. (1961). The biochemical mechanism of action of penicillin. *Nord. Vet. Med.* **13**(Jul–Aug), 433–438.

Hsieh, A. C., Bo, R., Manola, J., Vazquez, F., Bare, O., Khvorova, A., Scaringe, S., and Sellers, W. R. (2004). A library of siRNA duplexes targeting the phosphoinositide 3-kinase pathway: Determinants of gene silencing for use in cell-based screens. *Nucleic Acids Res.* **32**(3), 893–901.

Huang, Q., Raya, A., DeJesus, P., Chao, S. H., Quon, K. C., Caldwell, J. S., Chanda, S. K., Izpisua-Belmonte, J. C., and Schultz, P. G. (2004). Identification of p53 regulators by genome-wide functional analysis. *Proc. Natl. Acad. Sci. USA* **101**(10), 3456–3461.

Huesken, D., Lange, J., Mickanin, C., Weiler, J., Asselbergs, F., Warner, J., Meloon, B., Engel, S., Rosenberg, A., Cohen, D., Labow, M., Reinhardt, M., *et al.* (2005). Design of a genome-wide siRNA library using an artificial neural network. *Nat. Biotechnol.* **23**(8), 995–1001.

Hunter, T. (1998). The role of tyrosine phosphorylation in cell growth and disease. *Harvey Lect.* **94**, 81–119.

Iourgenko, V., Zhang, W., Mickanin, C., Daly, I., Jiang, C., Hexham, J. M., Orth, A. P., Miraglia, L., Meltzer, J., Garza, D., Chirn, G. W., McWhinnie, E., *et al.* (2003). Identification of a family of cAMP response element-binding protein coactivators by genome-scale functional analysis in mammalian cells. *Proc. Natl. Acad. Sci. USA* **100**(21), 12147–12152.

Jayaram, H. N., Gharehbaghi, K., Jayaram, N. H., Rieser, J., Krohn, K., and Paull, K. D. (1992). Cytotoxicity of a new IMP dehydrogenase inhibitor, benzamide riboside, to human myelogenous leukemia K562 cells. *Biochem. Biophys. Res. Commun.* **186**(3), 1600–1606.

Kim, Y. K., Arai, M. A., Arai, T., Lamenzo, J. O., Dean, E. F., III, Patterson, N., Clemons, P. A., and Schreiber, S. L. (2004). Relationship of stereochemical and skeletal diversity of small molecules to cellular measurement space. *J. Am. Chem. Soc.* **126**(45), 14740–14745.

Kohonen, T. (1999). Comparison of SOM point densities based on different criteria. *Neural Comput.* **11**(8), 2081–2095.

Kopp, E., and Ghosh, S. (1994). Inhibition of NF-kappa B by sodium salicylate and aspirin. *Science* **265**(5174), 956–959.

Kullander, K., and Klein, R. (2002). Mechanisms and functions of Eph and ephrin signalling. *Nat. Rev. Mol. Cell Biol.* **3**(7), 475–486.

Lee, Y. S., Nishio, K., Ogasawara, H., Funayama, Y., Ohira, T., and Saijo, N. (1995). *In vitro* cytotoxicity of a novel antitumor antibiotic, spicamycin derivative, in human lung cancer cell lines. *Cancer Res.* **55**(5), 1075–1079.

Leteurtre, F., Kohlhagen, G., Paull, K. D., and Pommier, Y. (1994). Topoisomerase II inhibition and cytotoxicity of the anthrapyrazoles DuP 937 and DuP 941 (Losoxantrone) in the national cancer institute preclinical antitumor drug discovery screen. *J. Natl. Cancer Inst.* **86**(16), 1239–1244.

Lipinski, C. A., Lombardo, F., Dominy, B. W., and Feeney, P. J. (2001). Experimental and computational approaches to estimate solubility and permeability in drug discovery and development settings. *Adv. Drug Deliv. Rev.* **46**(1–3), 3–26.

Liu, J., Bang, A. G., Kintner, C., Orth, A. P., Chanda, S. K., Ding, S., and Schultz, P. G. (2005). Identification of the Wnt signaling activator leucine-rich repeat in Flightless interaction protein 2 by a genome-wide functional analysis. *Proc. Natl. Acad. Sci. USA* **102**(6), 1927–1932.

Lombardo, L. J., Lee, F. Y., Chen, P., Norris, D., Barrish, J. C., Behnia, K., Castaneda, S., Cornelius, L. A., Das, J., Doweyko, A. M., Fairchild, C., Hunt, J. T., *et al.* (2004). Discovery of N-(2-chloro-6-methyl-phenyl)-2-(6-(4-(2-hydroxyethyl)-piperazin-1-yl)-2-methylpyrimidin-4-ylamino)thiazole-5-carboxamide (BMS-354825), a dual Src/Abl kinase inhibitor with potent antitumor activity in preclinical assays. *J. Med. Chem.* **47**(27), 6658–6661.

Matsuda, A., Suzuki, Y., Honda, G., Muramatsu, S., Matsuzaki, O., Nagano, Y., Doi, T., Shimotohno, K., Harada, T., Nishida, E., Hayashi, H., and Sugano, S. (2003). Large-scale identification and characterization of human genes that activate NF-kappaB and MAPK signaling pathways. *Oncogene* **22**(21), 3307–3318.

Melnick, J. S., Janes, J., Kim, S., Chang, J. Y., Sipes, D. G., Gunderson, D., Jarnes, L., Matzen, J. T., Garcia, M. E., Hood, T. L., Beigi, R., Xia, G., *et al.* (2006). An efficient rapid system for profiling the cellular activities of molecular libraries. *Proc. Natl. Acad. Sci. USA* **103**(9), 3153–3158.

Meyer, R. D., and Cook, D. (2000). Visualization of data. *Curr. Opin. Biotechnol.* **11**(1), 89–96.

Michiels, F., van Es, H., van Rompaey, L., Merchiers, P., Francken, B., Pittois, K., van der Schueren, J., Brys, R., Vandersmissen, J., Beirinckx, F., Herman, S., Dokic, K., *et al.* (2002). Arrayed adenoviral expression libraries for functional screening. *Nat. Biotechnol.* **20**(11), 1154–1157.

Monks, A., Scudiero, D. A., Johnson, G. S., Paull, K. D., and Sausville, E. A. (1997). The NCI anti-cancer drug screen: A smart screen to identify effectors of novel targets. *Anticancer Drug Des.* **12**(7), 533–541.

Naldini, L., Blomer, U., Gage, F. H., Trono, D., and Verma, I. M. (1996a). Efficient transfer, integration, and sustained long-term expression of the transgene in adult rat brains injected with a lentiviral vector. *Proc. Natl. Acad. Sci. USA* **93**(21), 11382–11388.

Naldini, L., Blomer, U., Gallay, P., Ory, D., Mulligan, R., Gage, F. H., Verma, I. M., and Trono, D. (1996b). *In vivo* gene delivery and stable transduction of nondividing cells by a lentiviral vector. *Science* **272**(5259), 263–267.

Olson, M. F., and Marais, R. (2000). Ras protein signalling. *Semin. Immunol.* **12**(1), 63–73.

Paddison, P. J., Silva, J. M., Conklin, D. S., Schlabach, M., Li, M., Aruleba, S., Balija, V., O'Shaughnessy, A., Gnoj, L., Scobie, K., Chang, K., Westbrook, T., *et al.* (2004). A resource for large-scale RNA-interference-based screens in mammals. *Nature* **428**(6981), 427–431.

Paull, K. D., Lin, C. M., Malspeis, L., and Hamel, E. (1992). Identification of novel antimitotic agents acting at the tubulin level by computer-assisted evaluation of differential cytotoxicity data. *Cancer Res.* **52**(14), 3892–3900.

Persengiev, S. P., Zhu, X., and Green, M. R. (2004). Nonspecific, concentration-dependent stimulation and repression of mammalian gene expression by small interfering RNAs (siRNAs). *RNA* **10**(1), 12–18.

Plavec, I., Sirenko, O., Privat, S., Wang, Y., Dajee, M., Melrose, J., Nakao, B., Hytopoulos, E., Berg, E. L., and Butcher, E. C. (2004). Method for analyzing signaling networks in complex cellular systems. *Proc. Natl. Acad. Sci. USA* **101**(5), 1223–1228.

Poroikov, V. V., Filimonov, D. A., Ihlenfeldt, W. D., Gloriozova, T. A., Lagunin, A. A., Borodina, Y. V., Stepanchikova, A. V., and Nicklaus, M. C. (2003). PASS biological activity spectrum predictions in the enhanced open NCI database browser. *J. Chem. Inf. Comput. Sci.* **43**(1), 228–236.

Rabow, A. A., Shoemaker, R. H., Sausville, E. A., and Covell, D. G. (2002). Mining the national cancer institute's tumor-screening database: Identification of compounds with similar cellular activities. *J. Med. Chem.* **45**(4), 818–840.

Root, D. E., Flaherty, S. P., Kelley, B. P., and Stockwell, B. R. (2003). Biological mechanism profiling using an annotated compound library. *Chem. Biol.* **10**(9), 881–892.

Rubinson, D. A., Dillon, C. P., Kwiatkowski, A. V., Sievers, C., Yang, L., Kopinja, J., Rooney, D. L., Ihrig, M. M., McManus, M. T., Gertler, F. B., Scott, M. L., and Van Parijs, L. (2003). A lentivirus-based system to functionally silence genes in primary mammalian cells, stem cells and transgenic mice by RNA interference. *Nat. Genet.* **33**(3), 401–406.

Schreiber, S. L. (1991). Chemistry and biology of the immunophilins and their immunosuppressive ligands. *Science* **251**(4991), 283–287.

Schuster, D., Laggner, C., and Langer, T. (2005). Why drugs fail—a study on side effects in new chemical entities. *Curr. Pharm. Des.* **11**(27), 3545–3559.

Sedlacek, H. H. (2001). Mechanisms of action of flavopiridol. *Crit. Rev. Oncol. Hematol.* **38**(2), 139–170.

Seynaeve, C. M., Kazanietz, M. G., Blumberg, P. M., Sausville, E. A., and Worland, P. J. (1994). Differential inhibition of protein kinase C isozymes by UCN-01, a staurosporine analague. *Mol. Pharmacol.* **45**(6), 1207–1214.

Shao, L., Lewin, N. E., Lorenzo, P. S., Hu, Z., Enyedy, I. J., Garfield, S. H., Stone, J. C., Marner, F. J., Blumberg, P. M., and Wang, S. (2001). Iridals are a novel class of ligands for phorbol eter receptors with modest selectivity for the Ras GRP receptor subfamily. *J. Med. Chem.* **44**(23), 3872–3880.

Shi, L. M., Fan, Y., Myers, T. G., O'Connor, P. M., Paull, K. D., Friend, S. H., and Weinstein, J. N. (1998a). Mining the NCI anticancer drug discovery databases: Genetic function approximation for the QSAR study of anticancer ellipticine analogues. *J. Chem. Inf. Comput. Sci.* **38**(2), 189–199.

Shi, L. M., Myers, T. G., Fan, Y., O'Connor, P. M., Paull, K. D., Friend, S. H., and Weinstein, J. N. (1998b). Mining the national cancer institute anticancer drug discovery database: Cluster analysis of ellipticine analogs with p53-inverse and central nervous system-selective patterns of activity. *Mol. Pharmacol.* **53**(2), 241–251.

Shi, L. M., Fan, Y., Lee, J. K., Waltham, M., Andrews, D. T., Scherf, U., Paull, K. D., and Weinstein, J. N. (2000). Mining and visualizing large anticancer drug discovery databases. *J. Chem. Inf. Comput. Sci.* **40**(2), 367–379.

Sledz, C. A., Holko, M., de Veer, M. J., Silverman, R. H., and Williams, B. R. (2003). Activation of the interferon system by short-interfering RNAs. *Nat. Cell Biol.* **5**(9), 834–839.

Sneath, P. H., and Sokal, R. R. (1962). Numerical taxonomy. *Nature* **193**, 855–860.

Stewart, S. A., Dykxhoorn, D. M., Palliser, D., Mizuno, H., Yu, E. Y., An, D. S., Sabatini, D. M., Chen, I. S., Hahn, W. C., Sharp, P. A., Weinberg, R. A., and Novina, C. D. (2003). Lentivirus-delivered stable gene silencing by RNAi in primary cells. *RNA* **9**(4), 493–501.

Strausberg, R. L., Feingold, E. A., Grouse, L. H., Derge, J. G., Klausner, R. D., Collins, F. S., Wagner, L., Shenmen, C. M., Schuler, G. D., Altschul, S. F., Zeeberg, B., Buetow, K. H., *et al.* (2002). Generation and initial analysis of more than 15,000 full-length human and mouse cDNA sequences. *Proc. Natl. Acad. Sci. USA* **99**(26), 16899–16903.

Su, A. I., Wiltshire, T., Batalov, S., Lapp, H., Ching, K. A., Block, D., Zhang, J., Soden, R., Hayakawa, M., Kreiman, G., Cooke, M. P., Walker, J. R., *et al.* (2004). A gene atlas of the mouse and human protein-encoding transcriptomes. *Proc. Natl. Acad. Sci. USA* **101**(16), 6062–6067.

Sundberg, S. A. (2000). High-throughput and ultra-high-throughput screening: Solution- and cell-based approaches. *Curr. Opin. Biotechnol.* **11**(1), 47–53.

Tkachuk, D. C., Westbrook, C. A., Andreeff, M., Donlon, T. A., Cleary, M. L., Suryanarayan, K., Homge, M., Redner, A., Gray, J., and Pinkel, D. (1990). Detection of bcr-abl fusion in chronic myelogeneous leukemia by *in situ* hybridization. *Science* **250**(4980), 559–562.

Torrance, C. J., Agrawal, V., Vogelstein, B., and Kinzler, K. W. (2001). Use of isogenic human cancer cells for high-throughput screening and drug discovery. *Nat. Biotechnol.* **19**(10), 940–945.

Weinstein, J. N., Kohn, K. W., Grever, M. R., Viswanadhan, V. N., Rubinstein, L. V., Monks, A. P., Scudiero, D. A., Welch, L., Koutsoukos, A. D., Chiausa, A. J., *et al.* (1992). Neural computing in cancer drug development: Predicting mechanism of action. *Science* **258**(5081), 447–451.

Weinstein, J. N., Myers, T. G., O'Connor, P. M., Friend, S. H., Fornace, A. J., Jr., Kohn, K. W., Fojo, T., Bates, S. E., Rubinstein, L. V., Anderson, N. L., Buolamwini, J. K., van Osdol, W. W., *et al.* (1997). An information-intensive approach to the molecular pharmacology of cancer. *Science* **275**(5298), 343–349.

Yamaoka, S., Courtois, G., Bessia, C., Whiteside, S. T., Weil, R., Agou, F., Kirk, H. E., Kay, R. J., and Israel, A. (1998). Complementation cloning of NEMO, a component of the IkappaB kinase complex essential for NF-kappaB activation. *Cell* **93**(7), 1231–1240.

Yamori, T., Matsunaga, A., Sato, S., Yamazaki, K., Komi, A., Ishizu, K., Mita, I., Edatsugi, H., Matsuba, Y., Takezawa, K., Nakanishi, O., Kohno, H., *et al.* (1999). Potent antitumor activity of MS-247, a novel DNA minor groove binder, evaluated by an *in vitro* and *in vivo* human cancer cell line panel. *Cancer Res.* **59**(16), 4042–4049.

Yang, D., Buchholz, F., Huang, Z., Goga, A., Chen, C. Y., Brodsky, F. M., and Bishop, J. M. (2002). Short RNA duplexes produced by hydrolysis with *Escherichia coli* RNase III mediate effective RNA interference in mammalian cells. *Proc. Natl. Acad. Sci. USA* **99**(15), 9942–9947.

Zaharevitz, D. W., Holbeck, S. L., Bowerman, C., and Svetlik, P. A. (2002). COMPARE: A web accessible tool for investigating mechanisms of cell growth inhibition. *J. Mol. Graph. Model.* **20**(4), 297–303.

Zheng, L., Liu, J., Batalov, S., Zhou, D., Orth, A., Ding, S., and Schultz, P. G. (2004). An approach to genomewide screens of expressed small interfering RNAs in mammalian cells. *Proc. Natl. Acad. Sci. USA* **101**(1), 135–140.

Zhou, G., Myers, R., Li, Y., Chen, Y., Shen, X., Fenyk-Melody, J., Wu, M., Ventre, J., Doebber, T., Fujii, N., Musi, N., Hirshman, M. F., *et al.* (2001). Role of AMP-activated protein kinase in mechanism of metformin action. *J. Clin. Invest.* **108**(8), 1167–1174.

Tumor Antigens as Surrogate Markers and Targets for Therapy and Vaccines

Angus Dalgleish and Hardev Pandha

St. George's University of London, Cranmer, London SW17 0RE, United Kingdom

I. Introduction: The Immune System and the Concept of Tumor Antigens
II. How Tumor Antigens Are Recognized
III. Cancer Antigens
IV. Novel TAA Identification Techniques
 A. hTERT
 B. NY-ESO as a Target
V. Effective Targeting of TAA in the Clinic
VI. TAAs as Therapeutic Targets
VII. Cancer Vaccines
VIII. Tumor Antigens and Cancer Vaccines
IX. Tumor Antigens as Surrogate Markers and Targets for Therapy and Vaccines
X. Summary and Future Directions
References

There are a large number of tumor antigens, which may either be specific to the tumor or inappropriately expressed or processed (tumor-associated antigen, TAA). Over the last few years, hundreds of new TAAs have been identified. Some of these represent good targets for both passive (antibody based) and active (vaccine based) therapies. Antibody treatments targeted on tumor-specific antigens, such as Herceptin and Cetuximab, have been effective in clinical trials and are now licensed. In addition, TAAs act as good surrogate markers for use in both the diagnosis and assessment of treatment in cancer patients. © 2007 Elsevier Inc.

I. INTRODUCTION: THE IMMUNE SYSTEM AND THE CONCEPT OF TUMOR ANTIGENS

The immune system evolved to protect the host against infection. However, its role in surveillance of cancer is more controversial. The identification of tumor-specific antigens (TSAs) or tumor-associated antigens (TAAs) and the control of tumors in preclinical models have led to renewed interest in the use of tumor antigens as targets for both passive (antibodies) and active (vaccine) immunotherapies.

Advances in CANCER RESEARCH

0065-230X/07 $35.00
DOI: 10.1016/S0065-230X(06)96009-6

There are basically two overlapping arms to the immune response in man: the innate and the adaptive (or acquired) immune response, which clearly evolved much later than the innate response. Adaptive immunity has two major advantages compared to the innate immune response being antigen specificity and immunological memory. In addition, adaptive immunity comprises two main functions, namely humoral immunity and cell-mediated immunity. The humoral response is directed at native, usually extracellular antigens, whereas the cellular immune response can eliminate intracellular pathogens (such as mycobacteria and chronic viral infections) and transformed cells. Advances in basic research exploring the interactions of the innate and adaptive immune responses have facilitated potential new approaches for immunotherapy for cancer. This evidence base includes model systems confirming immune surveillance implicating B-cells, T-cells, natural killer (NK) cells, and NKT cells with key roles for the interferon-γ response pathway. Second, there has been an increase in rational identification of TAAs using specific technologies. These TAAs may be targeted for cellular- and antibody-based immune treatment to decrease patient morbidity and improve their survival.

TAAs comprise a large number of different categories. In addition to oncogenic viral antigens and TSAs, for example fusion genes (BCR/ABL) in chronic myeloid leukemia (CML), there are a number of TAAs which exhibit limited normal expression in the testes and placenta, for example, MAGE, and so on. Others are oncofetal in nature, for example, CEA and αFP (i.e., expressed in development but not in mature tissues). In addition, there are a number of molecules which are widespread on many cell types but whose phenotype is altered in malignant cells, thus giving rise to antigens only expressed on tumors such as the ganglioside and mucin molecules. With regards specific "TAA," there are now hundreds of reported examples and only those which may pose a suitable therapeutic target will be mentioned here. (For a table of different subtypes and examples see Table I.) Before going into detail, it is necessary to review how a tumor antigen is recognized and presented to the immune system to induce both humoral and cytotoxic T-lymphocyte (CTL) cell function.

II. HOW TUMOR ANTIGENS ARE RECOGNIZED

In order to manipulate the immune system to attack tumor cells, it is first necessary to understand the mechanism of CTL activation. Antigen naive T-cells must be primed in an antigen-specific manner to become functional effector cells. In the current model of CTL priming, two signals are required. The first is antigen-specific triggering of the T-cell receptor

Table I Tumor Antigens—Major Classifications

Viral-associated tumors (nonself)	EBV—(Burkitt's and nasopharangeal) HPV—cervical and perineal HBV—hepatoma (prophylactic vaccine available) HCV—hepatoma HTLV-1—T-cell leukemia lymphoma
"Altered self" TAA	Normal expression on normal tissues increased in tumors, for example: HER-2/neu Tyrosinase MART
Abnormal levels of expression usually present only in ontogeny and in limited mature tissues such as the testes	MAGE family (family is associated with many other tumor types, e.g., PAGE for prostate cancer) Oncofetal antigens, CEA and FP
Tumor-specific antigens	Mutated versions of self, for example, ras, p53, and so on Altered self epitopes, for example, gangliosides and mucins

(TCR) by binding to the peptide–MHC complex, and the second involves so-called costimulatory signaling, such as the interaction between CD80/CD86 (B7.1/B7.2) on antigen-presenting cells (APCs) with CD28 on T-cells. For the generation of signal 1 there must be interaction between MHC–peptide and a specific TCR. The functional outcome of TCR–peptide–MHC interaction depends on costimulation (Fig. 1). Dendritic cells (DCs) are the key players in providing costimulation, and thereby provide the link between innate and adaptive immunity. DCs sample their local antigenic environment and present their findings as short peptides loaded onto MHC molecules. DCs respond to "danger" signals as opposed to just detecting "nonself" antigens. The problem with cancer antigens (CAs) is that they fail to trigger a danger signal and this has to be provided when presenting CAs as a vaccine—usually in the form of an adjuvant which stimulates special receptors that makes the DC think it is dealing with an exogenous infectious agent. The T-cells scan these peptide–MHC complexes and respond when recognizing a specific peptide sequence. Once a CTL is primed and activated, it is driven through multiple rounds of division and the induction of specific effector function even in the subsequent absence of peptide–MHC complex or costimulation (Kaech and Ahmed, 2001; van Stipdonk *et al.*, 2001).

The current model therefore has a central role for APC in the priming of T-cells. Naive T-cells circulate between secondary lymphoid organs, such as

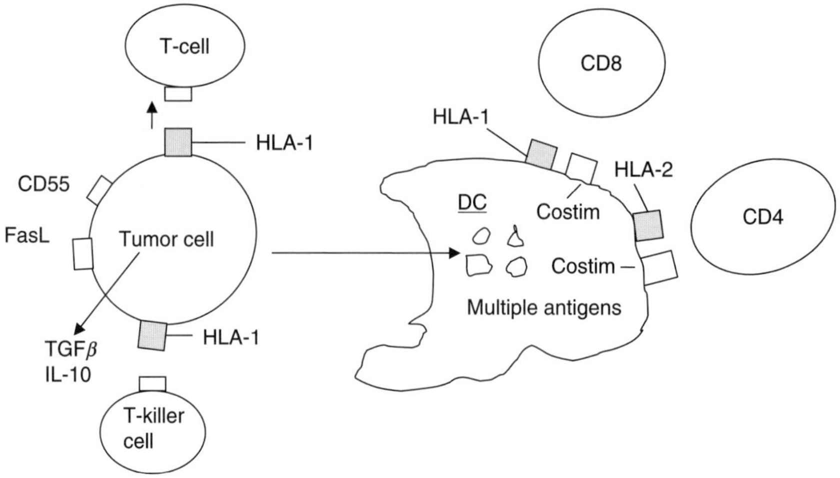

Fig. 1 Tumor cells presenting antigen in the absence of costimulatory molecules can induce anergy. However, if ingested by dendritic cells (DCs) the antigens can be presented to both CD4 and CD8 cells with costimulatory help which can break tolerance. An effective T-cell response may still be defended against by the expression of FasL and the secretion of immunosuppressive cytokines. CD55 expressed in most colorectal and prostate cancers defends against complement dependent cytotoxity.

lymph nodes and the spleen, via the blood or lymphatics and do not enter nonlymphoid tissues effectively (Butcher and Picker, 1996). As CTL priming generally occurs in lymphoid organs, the APC must migrate from the site of antigen uptake to the draining lymph nodes. DCs are the most likely bone marrow-derived APC to be responsible for the priming of naive T-cells. DCs have a distinct morphology characterized by the presence of numerous membrane processes. Additional morphological features include high concentrations of intracellular structures associated with antigen processing such as endosomes and lysosomes. DCs are also characterized by the presence, on their surface, of large amounts of class II MHC antigens and the absence of T-cell, B-cell, and NK cell markers. Activated DCs are motile and express a variety of different chemokine receptors, for example CCR7, that result in homing to lymph nodes; they have been demonstrated to generate both $CD4^+$ and $CD8^+$ T-cell responses and are essential to the development of immunological T-cell memory (Banchereau *et al.*, 2000).

The phenotype of the activated T-cells is profoundly affected by cytokines in the local environment. Conventionally, these are divided into two groups based on the subset of T-helper (Th) lymphocytes that produce them. Th-1 cells broadly promote CTL responses, whereas Th-2 cells promote humoral responses. It is generally considered that shifting the balance from Th-2 to Th-1 is beneficial in the treatment of solid tumors (Hrouda *et al.*, 1998).

III. CANCER ANTIGENS

The aim of targeted immunotherapy for cancer is to generate an immune-mediated specific anti-TAA response resulting in the abrogation or elimination of tumor growth. The choice of antigen may be TAA derived from whole protein, truncated long peptide, short peptide, or DNA sequences encoding specific epitopes, delivered to APCs by viral or nonviral vectors. DCs can be grown and modified *ex vivo* and exposed to antigens in the form of peptides mRNA, proteins, and so on, and then injected back into the host in order to activate a killer CTL response. Numerous variations on this approach on virtually all tumor types are under intense evaluation as targeted therapies. However, regardless of the therapeutic approach, the T-cell repertoire induced by immunotherapy will be determined by the nature of the TAA with unique, mutated or viral antigens potentially differing to, and more immunogenic than, overexpressed or inappropriately expressed normal proteins (differentiation or oncofetal antigens, as shown in Table I).

An effective immune response will result in immune evasion strategies by the tumor. The tumor may be subject to immune selection which can result in TAA-negative or MHC-low/negative tumor variants which can escape the attention of tumor-specific T-cells. Down-regulation may be epitope and HLA allele specific demonstrating highly focused (and effective) CTL targeting. Other mechanisms include induction of T-cell anergy and the prevention of T-cell activation by the secretion of immunosuppressive factors by the tumor (e.g., IL-10, TGF-2) or the up-regulation of ligands which inhibit T-cell- or complement-based killing, for example FasL or CD55 ligands (up-regulation offers a further TAA target!).

Definitive proof that human cancers express antigens which can be specifically targeted by cellular immunity has been a crucial stage in the development of numerous strategies for lymphocyte-based anticancer immunotherapy over the last decade. Despite the fact that most tumor antigens described are restricted to a few tumor types, and indeed to a subset of patients who have these tumors, the demonstration that TAAs do exist in human cancer also provides part of the rationale behind the long-standing pursuit of tumor cell-based therapies in which the target antigens are not completely defined (e.g., whole cell vaccination). Whole cell vaccines have been shown to induce clinically significant results in randomized studies of both renal and colorectal cancers. Unfortunately, these vaccines are autologous tumor cell based. The more practical approach using allogeneic vaccines are being pursued in melanoma and prostate cancer.

Since the early 1990s, it has been clear that TAAs exist outside viral oncogenic proteins, oncogenes, or the immunoglobulin idiotype in B-cell tumors.

It is important to note the phrase the "tumor-associated antigen" cannot be used interchangeably with "tumor-rejection antigen" and "tumor-regression antigen" as clearly the expression of a TAA does not necessarily guarantee an antigen-specific response in the host which may lead to the rejection of that tumor (Gilboa, 1999). "Tumor rejection antigen" is a term which can be used to describe how well an immune response elicited against a tumor antigen will impact on the tumor growth. This will clearly depend on the nature of antigen but also on the nature of the tumor response to this antigen. The potential tumor rejection antigen would ideally elicit high-avidity T-cell responses, recruit a high frequency of T-cells with a high diversity in TCR usage, and establish immunology memory. The identification and molecular characterization of tumor antigens that elicit specific immune responses in the tumor-bearing host has been a key step in the development of anticancer immunotherapy. The analysis of humoral and cellular immune responses against such antigens in cancer patients has indicated that cancer-specific antigens do exist and are recognized by the host immune system (Renkvist *et al.*, 2001).

IV. NOVEL TAA IDENTIFICATION TECHNIQUES

To define the molecular nature of these antigens, cloning techniques were developed that used established CTL clones or high titred circulating antibodies as probes for screening tumor-derived expression libraries (Sahin *et al.*, 1995; van der Bruggen *et al.*, 1991). This rationale led to the development of SEREX (serological analysis of tumor antigens by recombinant cDNA expression cloning), and allows a systematic search for antibody responses against proteins and the direct molecular definition of respective tumor antigens based on their reactivity with autologous patient serum. For SEREX, cDNA expression libraries are constructed from fresh tumor specimens, cloned into λ phage expression vectors which are used to transfect *Escherichia coli*. Recombinant proteins expressed by the *E. coli* are transferred onto nitrocellulose membranes and incubated with the patients serum. Clones reactive with high titred antibodies are identified (using a second antibody specific for human IgG) and sequenced. SEREX allows sequence data to be quickly compared with databases to reveal identity, homology with known genes, and to identify domains and motifs. SEREX has identified shared tumor antigens, differentiation antigens, mutated genes, splice variants, viral antigens, overexpression, gene amplification, and even underexpressed genes (Minenkova *et al.*, 2003).

The biggest practical problem of currently described TAAs is the lack of expression across malignancies of different origin. Clinical studies have been limited to strategies testing one malignancy at a time and in some cases, as in the immunoglobulin idiotype antigen in B-cell malignances, patient by patient. There are a number of algorithms publicly available for predicting the MHC class I binding affinities of peptides. Using these, T-cell peptide epitopes can be chosen based on predicted binding affinities of peptide to MHC and then scrutinized for immunogenicity based on the capacity of experimentally generated peptide-specific T-lymphocytes to kill tumors *in vitro* and *in vivo*. Two broadly expressed examples will be mentioned here, human telomerase reverse transcriptase (hTERT) and NY-ESO.

A. hTERT

Using this approach, a prototype of a universal tumor antigen hTERT, the most widely expressed tumor antigen yet described, has been evaluated (Kim *et al.*, 1994). This enzyme mediates the RNA-dependent synthesis of telomeric DNA. Telomeres at the distal ends of eukaryotic chromosomes stabilize the chromosomes during cell division and prevent end-to-end fusion. The telomerase catalytic subunit, hTERT, is the rate-limiting component in the telomerase complex and is most closely correlated with telomerase activity. More than 85% of human cancers have telomerase activity and express hTERT, whereas most normal adult human cells do not maintain the lengths of their telomeres. Therefore, telomerase was considered to be an attractive candidate target antigen for the development of immunotherapies for the treatment of patients with a variety of human cancers. In addition, telomerase expression has been directly linked to the ability of tumor cells to replicate indefinitely (Hahn *et al.*, 1999a,b; Herbert *et al.*, 1999). Therefore, if a T-cell response could be directed against peptide epitopes processed from telomerase, it seems likely that any immune escape variants that did not express this protein would not be able to survive. hTERT is expressed in more than 85% of all human tumors but rarely in normal cells. In tumors that expressed telomerase activity, hTERT appears not only to be found in nearly all tumor cells within a human lesion (Shay and Bacchetti, 1997), but hTERT expression appears to increase as tumor progresses from carcinoma *in situ* to primary tumors to metastatic tumors (Kolquist *et al.*, 1998). Peptides derived from hTERT were shown to be naturally processed by tumor cells presented in the MHC class I restricted fashion and function as a target for antigen-specific CTL (Minev *et al.*, 2000; Vonderheide *et al.*, 1999). CTLs generated *in vitro* using an hTERT peptide killed a wide range of hTERT positive tumor cells and primary tumors in a peptide-specific MHC-restricted fashion (Vonderheide *et al.*, 1999). There are currently clinical trials targeting patients with prostate and breast cancer using hTERT sequences pulsed onto autologous DCs (Su *et al.*, 2005; Vonderheide *et al.*, 2004).

B. NY-ESO as a Target

NY-ESO-1 is a tumor antigen discovered through SEREX and belongs to the cancer testis class of antigens. It is expressed in a wide range of tumors and in normal germ cell tissue. The discovery of naturally occurring strong humoral NY-ESO-1-specific immunity was followed by evidence of T-cell immunity (both HLA class I and class II restricted) to a number of peptide epitopes within the NY-ESO-1 protein (Romero *et al.*, 2001). Spontaneous immune responses to the cancer testis antigen NY-ESO-1 are frequently found in cancer patients bearing antigen-expressing tumors (Shang *et al.*, 2004; van Rhee *et al.*, 2005). In HLA-A2-expressing patients, naturally elicited NY-ESO-1-specific, tumor-reactive CTLs are mostly directed against an immunodominant epitope corresponding to peptide NY-ESO-1 157–165 (Gnjatic *et al.*, 2002; Held *et al.*, 2004; Webb *et al.*, 2004). NY-ESO-1-specific CTLs can also be induced by synthetic peptide vaccines, but they are heterogeneous in terms of functional avidity and tumor reactivity. There is currently an international collaboration evaluating a number of approaches in a range of malignancies targeting class I and class II epitopes using peptide (p157–167, p157–165, and p155–163) and DNA vaccination.

V. EFFECTIVE TARGETING OF TAA IN THE CLINIC

The idea of a magic bullet specifically aimed at a TSA was first suggested by Paul Erlich. This was first considered a realistic possibility with the development of monoclonal antibodies (MABs) by Milstein and Kohler in the mid-seventies. Unfortunately, early optimism was dashed by a strong immune response to the murine sequences. It has taken over two decades of molecular engineering (together with the discovery and defining of TAA as targets) to get to the position we are in today, whereby there are now several effective MABs available as therapies licensed for the treatment of cancer (Table II). The first two candidates were developed to a mutation of the EGFR noted in hormonally resistant breast cancer called HER-2/neu. An MAB developed against this has been shown to give a survival benefit in patients with advanced breast cancer, an effect that is enhanced if given with chemotherapy (Emens, 2005; Rastetter *et al.*, 2004). Moreover, the clinical benefit is thought to be associated with an additional vaccination effect induced by the antibody.

Table II Approved Monoclonal Antibodies for Cancer Therapy[a]

Generic name	Proprietary name	Target antigen	Isotype	Indication	Date of US FDA approval
Rituximab	Rituxan/ Mabthera	CD20	Chimeric IgG1	B-cell lymphoma	November 1997
Alemtuzumab	Campath	CD52	Humanized IgG1	B-CLL	May 2001
Trastuzumab	Herceptin	HER-2/neu	Humanized 1gG1	Breast cancer	September 1998
Cetuximab	Erbitux	EGFR	Chimeric IgG1	Colorectal	February 2004
Bevacizumab	Avastin	VEGF	Humanized IgG1	Colorectal	February 2004

[a]FDA, food and drug administration; Ig, immunoglobulin; CLL, chronic lymphocytic leukemia; EGFR, epidermal growth factor receptor; VEGF, vascular endothelial growth factor.

Early antibodies were all murine with short plasma half-lives and poor interaction with FcRn with poor recruitment of effector mechanisms and which were often immunogenic. New molecular biology technology has allowed the creation of a chimeric antibody whereby the Fc backbone is human and the light chains are murine. Humanized antibodies have very little murine sequences and the most recent are completely human antibodies.

The second major TAA being targeted by MABs is the CD20 ligand, which is expressed in B-cell lymphomas. Again, response and survival benefit has been reported in patients with non-Hodgkin's lymphomas (NHLs) which, like Herceptin, is more marked when given with chemotherapy.

Other effective MABs with clinical approval target mutations seen in a number of different tumor types. Several tumor types express mutations in the EGF pathway and an MAB Cetuximab has been approved for metastatic colorectal cancer and is also reported to have marked activity in other tumor types such as head and neck cancers.

The ideal universal tumor antigen does not exist, although there are several candidates present in several tumor types. One possible shared weakness in most tumor types is their dependence on generating new vessels (angiogenesis) in order to grow and metastasize. A number of drugs with partial activity against cancer, such as Interferon and Thalidomide, have weak antiangiogenic activity. A major feature of these new vessels is the expression of receptors for vascular endothelial growth factor (VEGF). A monoclonal to this ligand known as Avastin is active in a number of tumors with marked vascular components, although it is only registered for colorectal cancer at the time of writing (although several other indications, such as renal cell cancer, are likely in the future).

The rapid rise in MABs as treatments for cancer has underscored the existence of TAA and the practicality of targeting them. They also underscore the rationale for targeting these same antigens with cancer-based vaccines.

Indeed, combinations of MABs (given with chemotherapy) followed by vaccination against the same epitope could lead to better clinical outcomes.

VI. TAAs AS THERAPEUTIC TARGETS

Ever since the effectiveness of chemotherapy on some tumor types was accepted, the proclivity of responding tumors to become resistant to the treatment has been recognized as a major limitation. Cancer is a stochastic process, having escaped all normal checks and controls and becomes both anarchic and the "criminal on the run." It is therefore unlikely that targeting one pathway, however dominant, will eliminate the tumor even if 99.9% of the tumor is reduced. This has been recognized even when highly specific treatments, such as Gleevec (Glivec), are used as the main oncogenic pathway mutation. This means that tumors will default (escape) to use other oncogenic pathways. In spite of the potential for using almost any growth-promoting pathway, there are a number of shared ones active in many different tumor types. It therefore makes reasonable sense to target the most dominant of these in combination or sequentially with other similar candidates.

A major pathway which becomes permanently activated in several cancers is the EGFR pathway which is associated with several steps in cellular activation. There are now several drugs which inhibit this pathway (Iressa, Tarceva) in addition to MABs (Cetuximab), which attach to extracellular components of the receptor and kill the cell via antibody-directed cell-mediated toxicity. Unfortunately, in spite of dramatic clinical responses in the case of Iressa and non-small-cell lung cancer, the effectiveness of this was very disappointing in randomized studies using the small molecule Iressa. One explanation is that default pathways may need to be targeted at the same time or sequentially and there are already protocols targeting two major shared pathways such as EGFR and VEGFR treatments.

VII. CANCER VACCINES

The use of "reverse immunology" to establish CTL recognized epitopes, such as MAGE and MART, together with SEREX to identify many others, such as NY-ESO as a target, has led to the identification of hundreds of epitopes which could be used to induce an effective immune response. These epitopes are all HLA restricted and most genes now have many different epitopes for different HLA backgrounds, for example, MAGE 1, 2, 3, gp100, MART-1, PSA, tyrosinase, HER-2/neu, MUC, and so on (for a detailed list of all peptide sequences see Novellino *et al.*, 2005).

VIII. TUMOR ANTIGENS AND CANCER VACCINES

Prior to the identification of specific TSA and TAA sequences, tumor antigens were provided from tumor cells, either as lysates or as irradiated cell lines derived from either autologous or allogeneic tumors. The perception is that good anecdotal and phase I/II studies have never translated into positive randomized trials. However, there are two positive randomized vaccine trials, one for colorectal cancer and the other for renal, both published in the *Lancet* (Jocham *et al.*, 2004; Vermorken *et al.*, 1999).

Vaccines made from autologous tissue are "procedure" vaccines as opposed to "product" vaccines. An alternative to autologous vaccines is the use of established allogeneic cell lines which express shared antigens. Concerns about HLA matching do not appear to be a negative concern as preclinical models show that allogeneic cell lines are often better than autologous—presumably because "allo" represents a danger signal and is more likely to break tolerance.

There are several allogeneic cell line-based vaccines. The most advanced is the triple melanoma cell line vaccine of Donald Morton of the John Wayne Cancer Institute, Santa Monica (Faries and Morton, 2005; Morton *et al.*, 2002). Single institution results in phase II show a clear survival benefit but in a multicenter randomized study the Data Monitoring Committee recommended halting the stage IV trial due to similarity of both treatment and the control arm. The stage III results are not known at the time of writing.

Prostate cancer has provided an ideal candidate for cancer vaccines with a superb surrogate marker in prostate specific antigen (PSA) levels and an absence of effective nontoxic treatment after hormone failure. At least three candidates use cells as the basis of the vaccine, Dendreon uses the patients own DCs pulsed with a prostate alkaline phosphatase-based antigen. In a randomized study, patients on the vaccine arm had a 20% improval in mean survival. Cellgenesys are using two allogeneic cell lines transfected with GM-CSF as a vaccine, with no published results at time of writing. Onyvax have used three allogeneic cell lines which have been enhanced for antigenicity and immunogenicity without using gene transfer technologies. A 42% response of prolonged production in the rate of rise of PSA has been reported in a hormonally refractory study where mean time to progression was 58 weeks (Michael *et al.*, 2005). Therion, using a viral vector-based delivery system using PSA as the antigen, has also claimed efficacy, and a recent randomized study suggests that the best results with regards TTP are when the vaccine is given first and anti-androgen (AA) treatment added in after 6 months if the PSA is still climbing. The TTP difference in this arm was 26 months when compared to 16 months in the arm that got AA therapy first (Arlen *et al.*, 2005).

In addition to the above there are DNA vaccines being developed, at least two using PSMA as the antigen. Early studies often used cell lysates which do not appear as effective as irradiated whole cells. Today the commonest cell-based vaccine is the DC technique, as used by Dendreon. However, it is a very labor and material expensive way of expanding and priming autologous APCs. The technique is excellent as a research tool but still requires an antigen for it to present. Trials have been conducted using peptides, proteins, mRNA, whole cells, lysates, and so on. There are over 300 DC trials in the literature which all use different details in the protocols. However, a number of generalizations can be made:

1. There are occasional spectacular clinical responses in virtually all tumor types. Unfortunately, these appear to be uncommon.
2. The most effective outcome would appear to be stable disease in young and generally fit patients.
3. Immunotherapy may be synergistic with other modalities of treatment.

(For further reviews see Grunebach *et al.*, 2005; Nestle *et al.*, 2005.)

IX. TUMOR ANTIGENS AS SURROGATE MARKERS AND TARGETS FOR THERAPY AND VACCINES

A surrogate marker is a biomarker intended to substitute for a clinical endpoint. It is expected to predict clinical benefit or harm or lack thereof. In addition to providing targets for therapy, tumor antigens have provided ideal surrogate markers for several tumor types. PSA is used to define the effectiveness of treatment for prostate cancer, with a baseline low level, which starts to steadily increase over time representing relapse in previously diagnosed and treated patients. A gradually increasing PSA in a patient in complete androgen blockade is an indication that the disease has become hormonally resistant, even if there is no other evidence of disease activity. A sustained reduction following treatment is likely to indicate that the treatment is effective. With regards screening, a rising PSA may reflect an enlarging organ, hence the need for a biopsy to confirm diagnosis. There are now several examples of tumor markers which are present in the serum and can hence be easily measured and which are accepted as reliable surrogates for determining therapy (Table III).

TAAs may also be useful surrogate markers in non-serum sources, such as in sputum and fecal samples, as well as being targeted in imaging techniques.

As previously mentioned, with regards to therapeutic targets there are now so many TAAs discovered through a variety of techniques, for example, SEREX, proteomics, metabolomics, etc. Thus it is impossible to list all

Table III Examples of Tumor Antigen Surrogate Markers[a]

TAA	Disease
βHCG	Choriocarcinoma and testicular cancer
αFP	Hepatic and testicular cancer
CA125	Ovarian cancer
CEA	Colorectal and pancreatic cancer
CA19.9	Pancreatic cancer
PSA	Prostate cancer

[a]These are some of the markers currently available in widespread clinical use. Due to the heterogenicity of tumors, some tumors do not secrete these proteins and that targeting them will not be effective. Moreover, in mixed clonality, tumor targeting may leave negative clones to survive/be selected for.

possibilities. The examples mentioned here are in current clinical use or are potential candidates for the future.

The "common" TAA assays are very useful in clinical practice providing the limitations are acknowledged (e.g., TAA-negative tumors or part thereof) but many other TAAs are being used in development of new drugs. For instance, the HER-2/neu marker, which if it was not used to select patients suitable to treat with Herceptin, would not have given a positive endpoint in breast cancer patients! Many other different "marker" classes are being explored in determining activity and end points in new clinical entities in phase I/II clinical trials.

X. SUMMARY AND FUTURE DIRECTIONS

Tumor antigens have already become established as not only surrogate markers for determining the effects of therapy (e.g., PSA) but also as targets for both active and passive therapies. The dramatic success of Herceptin and Rituximab will no doubt herald a new generation of dozens of such finely engineered antibodies targeted to a number of critical tumor ligands. The field of cancer vaccines has not yet developed a candidate blockbuster and is at the same stage as MABs were 5–10 years ago with a few good phase II studies but lacking success at the phase III level for numerous reasons, some of which may mirror the problems solved by the antibody development strategy, for example correct patient selection and increased sophistication of development and manufacturing technology. The discovery of new TAAs and TSAs by "reverse" immunology and SEREX has revealed hundreds of new TAAs, which may be useful targets for therapy in future strategies. Two of the most promising have been specifically mentioned here as examples, that is NY-ESO and hTERT, both of which could be useful for a broad range of cancers.

Genomics and Proteomics are capable of determining more targets than can possibly be tried clinically and the role of Bioinformatics as well as Bayesian clinical judgment will be crucial in developing the next generation of targets and new treatment approaches.

REFERENCES

Arlen, P. M., Gulley, J. L., Todd, N., Lieberman, R., Steinberg, S. M., Morin, S., Bastian, A., Marte, J., Tsang, K. Y., Beetham, P., Grosenbach, D. W., Schlom, J., *et al.* (2005). Antiandrogen, vaccine and combination therapy in patients with nonmetastatic hormone refractory prostate cancer. *J. Urol.* **174,** 539–546.

Banchereau, J., Briere, F., Caux, C., Davoust, J., Lebecque, S., Liu, Y. J., Pulendran, B., and Palucka, K. (2000). Immunobiology of dendritic cells. *Annu. Rev. Immunol.* **18,** 767–811.

Butcher, E. C., and Picker, L. J. (1996). Lymphocyte homing and homeostasis. *Science* **272,** 60–66.

Emens, L. A. (2005). Trastuzumab: Targeted therapy for the management of HER-2/neu-overexpressing metastatic breast cancer. *Am. J. Ther.* **12,** 243–253.

Faries, M. B., and Morton, D. L. (2005). Therapeutic vaccines for melanoma: Current status. *BioDrugs* **19,** 247–260.

Gilboa, E. (1999). The makings of a tumor rejection antigen. *Immunity* **11,** 263–270.

Gnjatic, S., Jager, E., Chen, W., Altorki, N. K., Matsuo, M., Lee, S. Y., Chen, Q., Nagata, Y., Atanackovic, D., Chen, Y. T., Ritter, G., Cebon, J., *et al.* (2002). CD8(+) T cell responses against a dominant cryptic HLA-A2 epitope after NY-ESO-1 peptide immunization of cancer patients. *Proc. Natl. Acad. Sci. USA* **99,** 11813–11818.

Grunebach, F., Muller, M. R., and Brossart, P. (2005). New developments in dendritic cell-based vaccinations: RNA translated into clinics. *Cancer Immunol. Immunother.* **54,** 517–525.

Hahn, W. C., Counter, C. M., Lundberg, A. S., Beijersbergen, R. L., Brooks, M. W., and Weinberg, R. A. (1999a). Creation of human tumour cells with defined genetic elements. *Nature* **400,** 464–468.

Hahn, W. C., Stewart, S. A., Brooks, M. W., York, S. G., Eaton, E., Kurachi, A., Beijersbergen, R. L., Knoll, J. H., Meyerson, M., and Weinberg, R. A. (1999b). Inhibition of telomerase limits the growth of human cancer cells. *Nat. Med.* **5,** 1164–1170.

Held, G., Matsuo, M., Epel, M., Gnjatic, S., Ritter, G., Lee, S. Y., Tai, T. Y., Cohen, C. J., Old, L. J., Pfreundschuh, M., Reiter, Y., and Hoogenboom, H. R., *et al.* (2004). Dissecting cytotoxic T cell responses towards the NY-ESO-1 protein by peptide/MHC-specific antibody fragments. *Eur. J. Immunol.* **34,** 2919–2929.

Herbert, B., Pitts, A. E., Baker, S. I., Hamilton, S. E., Wright, W. E., Shay, J. W., and Corey, D. R. (1999). Inhibition of human telomerase in immortal human cells leads to progressive telomere shortening and cell death. *Proc. Natl. Acad. Sci. USA* **96,** 14276–14281.

Hrouda, D., Souberbielle, B. E., Kayaga, J., Corbishley, C. M., Kirby, R. S., and Dalgleish, A. G. (1998). Mycobacterium vaccae (SRL172): A potential immunological adjuvant evaluated in rat prostate cancer. *Br. J. Urol.* **82,** 870–876.

Jocham, D., Richter, A., Hoffmann, L., Iwig, K., Fahlenkamp, D., Zakrzewski, G., Schmitt, E., Dannenberg, T., Lehmacher, W., von Wietersheim, J., and Doehn, C. (2004). Adjuvant autologous renal tumour cell vaccine and risk of tumour progression in patients with renal-cell carcinoma after radical nephrectomy: Phase III, randomised controlled trial. *Lancet* **363,** 594–599.

Kaech, S. M., and Ahmed, R. (2001). Memory CD8+ T cell differentiation: Initial antigen encounter triggers a developmental program in naive cells. *Nat. Immunol.* **2,** 415–422.

Kim, N. W., Piatyszek, M. A., Prowse, K. R., Harley, C. B., West, M. D., Ho, P. L., Coviello, G. M., Wright, W. E., Weinrich, S. L., and Shay, J. W. (1994). Specific association of human telomerase activity with immortal cells and cancer. *Science* **266,** 2011–2015.

Kolquist, K. A., Ellisen, L. W., Counter, C. M., Meyerson, M., Tan, L. K., Weinberg, R. A., Haber, D. A., and Gerald, W. L. (1998). Expression of TERT in early premalignant lesions and a subset of cells in normal tissues. *Nat. Genet.* **19,** 182–186.

Michael, A., Ball, G., Quatan, N., Wushishi, F., Russell, N., Whelan, J., Chakraborty, P., Leader, D., Whelan, M., and Pandha, H. (2005). Delayed disease progression after allogeneic cell vaccination in hormone-resistant prostate cancer and correlation with immunologic variables. *Clin. Cancer Res.* **11,** 4469–4478.

Minenkova, O., Pucci, A., Pavoni, E., De Tomassi, A., Fortugno, P., Gargano, N., Cianfriglia, M., Barca, S., De Placido, S., Martignetti, A., Felici, F., Cortese, R., *et al.* (2003). Identification of tumor-associated antigens by screening phage-displayed human cDNA libraries with sera from tumor patients. *Int. J. Cancer* **106,** 534–544.

Minev, B., Hipp, J., Firat, H., Schmidt, J. D., Langlade-Demoyen, P., and Zanetti, M. (2000). Cytotoxic T cell immunity against telomerase reverse transcriptase in humans. *Proc. Natl. Acad. Sci. USA* **97,** 4796–4801.

Morton, D. L., Hsueh, E. C., Essner, R., Foshag, L. J., O'Day, S. J., Bilchik, A., Gupta, R. K., Hoon, D. S., Ravindranath, M., Nizze, J. A., Gammon, G., Wanek, L. A., *et al.* (2002). Prolonged survival of patients receiving active immunotherapy with Canvaxin therapeutic polyvalent vaccine after complete resection of melanoma metastatic to regional lymph nodes. *Ann. Surg.* **236,** 438–448; discussion 448–449.

Nestle, F. O., Farkas, A., and Conrad, C. (2005). Dendritic-cell-based therapeutic vaccination against cancer. *Curr. Opin. Immunol.* **17,** 163–169.

Novellino, L., Castelli, C., and Parmiani, G. (2005). A listing of human tumor antigens recognized by T cells: March 2004 update. *Cancer Immunol. Immunother.* **54,** 187–207.

Rastetter, W., Molina, A., and White, C. A. (2004). Rituximab: Expanding role in therapy for lymphomas and autoimmune diseases. *Annu. Rev. Med.* **55,** 477–503.

Renkvist, N., Castelli, C., Robbins, P. F., and Parmiani, G. (2001). A listing of human tumor antigens recognized by T cells. *Cancer Immunol. Immunother.* **50,** 3–15.

Romero, P., Dutoit, V., Rubio-Godoy, V., Lienard, D., Speiser, D., Guillaume, P., Servis, K., Rimoldi, D., Cerottini, J. C., and Valmori, D. (2001). CD8+ T-cell response to NY-ESO-1: Relative antigenicity and *in vitro* immunogenicity of natural and analogue sequences. *Clin. Cancer Res.* **7,** 766s–772s.

Sahin, U., Tureci, O., Schmitt, H., Cochlovius, B., Johannes, T., Schmits, R., Stenner, F., Luo, G., Schobert, I., and Pfreundschuh, M. (1995). Human neoplasms elicit multiple specific immune responses in the autologous host. *Proc. Natl. Acad. Sci. USA* **92,** 11810–11813.

Shang, X. Y., Chen, H. S., Zhang, H. G., Pang, X. W., Qiao, H., Peng, J. R., Qin, L. L., Fei, R., Mei, M. H., Leng, X. S., Gnjatic, S., Ritter, G., *et al.* (2004). The spontaneous CD8+ T-cell response to HLA-A2-restricted NY-ESO-1b peptide in hepatocellular carcinoma patients. *Clin. Cancer Res.* **10,** 6946–6955.

Shay, J. W., and Bacchetti, S. (1997). A survey of telomerase activity in human cancer. *Eur. J. Cancer* **33,** 787–791.

Su, Z., Dannull, J., Yang, B. K., Dahm, P., Coleman, D., Yancey, D., Sichi, S., Niedzwiecki, D., Boczkowski, D., Gilboa, E., and Vieweg, J. (2005). Telomerase mRNA-transfected dendritic cells stimulate antigen-specific CD8+ and CD4+ T cell responses in patients with metastatic prostate cancer. *J. Immunol.* **174,** 3798–3807.

van der Bruggen, P., Traversari, C., Chomez, P., Lurquin, C., De Plaen, E., Van den Eynde, B., Knuth, A., and Boon, T. (1991). A gene encoding an antigen recognized by cytolytic T lymphocytes on a human melanoma. *Science* **254**, 1643–1647.

van Rhee, F., Szmania, S. M., Zhan, F., Gupta, S. K., Pomtree, M., Lin, P., Batchu, R. B., Moreno, A., Spagnoli, G., Shaughnessy, J., and Tricot, G. (2005). NY-ESO-1 is highly expressed in poor-prognosis multiple myeloma and induces spontaneous humoral and cellular immune responses. *Blood* **105**, 3939–3944.

van Stipdonk, M. J., Lemmens, E. E., and Schoenberger, S. P. (2001). Naive CTLs require a single brief period of antigenic stimulation for clonal expansion and differentiation. *Nat. Immunol.* **2**, 423–429.

Vermorken, J. B., Claessen, A. M., van Tinteren, H., Gall, H. E., Ezinga, R., Meijer, S., Scheper, R. J., Meijer, C. J., Bloemena, E., Ransom, J. H., Hanna, M. G., Jr., and Pinedo, H. M. (1999). Active specific immunotherapy for stage II and stage III human colon cancer: A randomised trial. *Lancet* **353**, 345–350.

Vonderheide, R. H., Domchek, S. M., Schultze, J. L., George, D. J., Hoar, K. M., Chen, D. Y., Stephans, K. F., Masutomi, K., Loda, M., Xia, Z., Anderson, K. S., Hahn, W. C., *et al.* (2004). Vaccination of cancer patients against telomerase induces functional antitumor CD8+ T lymphocytes. *Clin. Cancer Res.* **10**, 828–839.

Vonderheide, R. H., Hahn, W. C., Schultze, J. L., and Nadler, L. M. (1999). The telomerase catalytic subunit is a widely expressed tumor-associated antigen recognized by cytotoxic T lymphocytes. *Immunity* **10**, 673–679.

Webb, A. I., Dunstone, M. A., Chen, W., Aguilar, M. I., Chen, Q., Jackson, H., Chang, L., Kjer-Nielsen, L., Beddoe, T., McCluskey, J., Rossjohn, J., and Purcell, A. W. (2004). Functional and structural characteristics of NY-ESO-1-related HLA A2-restricted epitopes and the design of a novel immunogenic analogue. *J. Biol. Chem.* **279**, 23438–23446.

Practices and Pitfalls of Mouse Cancer Models in Drug Discovery

Andrew L. Kung

Department of Pediatric Oncology, Dana-Farber Cancer Institute and Children's Hospital, Harvard Medical School, Boston, Massachusetts 02115

I. Why Are Animal Models Needed?
II. Model Types: Tumor Locations
  A. Subcutaneous Xenograft Models
  B. Hollow Fiber Models
  C. Orthotopic Models
  D. Genetically Defined Tumor Models
III. Tumor Models: Cell Types
IV. Study Endpoints
  A. Efficacy Endpoints: Is Tumor Growth Effected?
  B. Functional and Molecular Endpoints: Is the Target Modulated?
V. Animal Modeling in the Post-genomics Age
VI. Conclusions
References

Mouse models of cancer are critical tools for elucidating mechanisms of cancer development, as well as for assessment of putative cancer therapies. However, there are ongoing concerns about the value of mouse cancer models for predicting therapeutic efficacy in humans. This chapter reviews the most commonly used transplanted tumor models, including subcutaneous and orthotopic tumors in mice. It also reviews commonly utilized *in vivo* study endpoints. Even small improvements in predictive value achieved through careful selection of models and endpoints have the potential to have large impacts on productivity and overall drug development costs. © 2007 Elsevier Inc.

I. WHY ARE ANIMAL MODELS NEEDED?

Drug development is a costly and risky endeavor. It is estimated that each new approved drug requires an expenditure of US$800 million in resources over 10–12 years in time between conception and approval (DiMasi *et al.*, 2003). Of all new chemical entities (NCEs) that enter into clinical testing, the rate of Food and Drug Administration (FDA) approval is 21.5% overall (DiMasi *et al.*, 2003), but only 10% for antineoplastic therapies (Von Hoff, 1998). The approval rate in oncology is among the lowest for all disease areas. While there are many reasons for attrition, lack of efficacy in clinical

Advances in CANCER RESEARCH

0065-230X/07 $35.00
DOI: 10.1016/S0065-230X(06)96007-2

trials is the single most common cause for discontinuation of NCE development (Frank and Hargreaves, 2003). Lack of efficacy in late-phase clinical trials and rejection by the FDA are by far the most costly points of failure. Shifting the decision to terminate development forward to phase I clinical trials, for example through the use of clinical biomarkers, would produce significant savings (DiMasi, 2002). Improving the ability of preclinical models to predict clinical efficacy would have even greater impact since the majority of drug development costs are incurred in clinical testing, with out-of-pocket costs of the preclinical period being less than half of that for clinical testing (DiMasi *et al.*, 2003). Since most NCEs for oncology enter into clinical testing with the backing of at least some evidence of efficacy in animal models, the prevalence of failure due to lack of efficacy in humans and the low rate of eventual FDA approval is an indictment of the predictive value of traditional animal cancer models.

Modern genomic technologies hold the promise of greatly accelerating target discovery and target validation. Concurrently, combinatorial chemistries, high information content screening, and robotics hold the promise of improving the throughput of compound development. These technologies will undoubtedly increase productivity in the discovery phases. However, they will almost certainly also increase research and development costs (DiMasi *et al.*, 2003). As modern drug development technologies increase the numbers of candidate therapeutics, and as the costs for preclinical and clinical testing escalate, the use of model systems to help prioritize compounds for clinical investigation becomes increasingly important.

The use of mouse models for the advancement of cancer drug discovery began in the 1950s at the National Cancer Institute (NCI), where a large-scale screen for compounds with antitumor activity was implemented using three mouse-transplanted tumor models (Goldin *et al.*, 1961a,b; Stock *et al.*, 1960a,b), with a particular emphasis on the L1210 leukemia model. Discovery of the ability to establish human tumors in the athymic nude mouse (Rygaard and Povlsen, 1969) led to the incorporation of human tumor xenografts in the NCI screening program in the mid 1970s (Venditti *et al.*, 1984). While the vast contribution of mouse models to advancing our understanding of cancer biology is inarguable, there have been ongoing concerns about the value of mouse models for predicting drug efficacy in humans (Takimoto, 2001). Furthermore, there has been a major shift in the focus of modern drug discovery away from cytotoxic therapies to therapies directed at cancer-associated molecular targets, further calling into question the use of conventional xenograft models where the primary endpoint is whether tumor growth is impacted by the therapeutic agent. There is an obvious need for better tumor models, and perhaps equally importantly, better-defined questions and endpoints with which animal models are used to interrogate.

II. MODEL TYPES: TUMOR LOCATIONS

A. Subcutaneous Xenograft Models

The ability to grow human tumors cells in immunodeficient mice was established over 30 years ago (Giovanella *et al.*, 1972; Rygaard and Povlsen, 1969; Shimosato *et al.*, 1976). Since that time, the use of xenograft tumors has become an integral part of the drug discovery process, both in academia and in industry, in large part because of the technical ease of such models. Typically, the "standard" model uses established human cell lines that can be easily propagated in tissue culture, for example, from the NCI 60 panel of cell lines (Shoemaker *et al.*, 1988). Cells are injected into the subcutaneous (SQ) tissue of immunodeficient mice, commonly athymic nude mice, which cannot reject the species-mismatched cells. Such models offer good throughput by utilizing cell lines that are readily replenished in tissue culture and by allowing assessment of tumor burden by simple caliper measurements of the superficial tumors. In an effort to more closely approximate the treatment of established tumors, most studies allow tumors to grow for a defined period of time, or to a specific volume (commonly 100–200 mm^3), prior to commencing treatments. Serial caliper measurements over the treatment course allow for comparison of tumor growth in the drug-treated cohort of mice to controls (vehicle treated), with the ratio of treated to control tumor volume (% T/C) as a common metric for antitumor efficacy.

Over the years, serious concerns have been raised as to whether such SQ xenograft models are adequate predictors of drug efficacy in human patients. Since 90% of antineoplastic NCEs eventually fail in clinical testing (Von Hoff, 1998) despite evidence of antitumor efficacy in preclinical models, it is clear that the predictive value of these models is low. The Developmental Therapeutics Program at the NCI has utilized a number of *in vitro* and *in vivo* models to screen for potential antitumor therapeutics. To assess the predictive value of SQ xenograft models, a retrospective analysis of 39 clinical therapeutics was undertaken, comparing animal antitumor efficacy with clinical efficacy in phase II human clinical trials (Johnson *et al.*, 2001). Of the 10 broad cancer types (breast, non-small cell lung cancer, melanoma, ovary, brain, colon, gastric, head and neck, pancreas, and renal), a statistically significant correlation between SQ xenograft efficacy and human clinical trial results was only found in non-small cell lung cancer (NSCLC). When animal efficacy across all models was considered in aggregate, 45% of agents with activity in greater than 1/3 of all models tested also had activity in human trials. By comparison, no agents with efficacy in less than 1/3 of all models tested had any clinical efficacy

($p = 0.04$). These data suggest that SQ xenograft model may have some predictive value, but it should be noted that the agents tested were largely cytotoxic therapies that might be expected to have activity across many tumor types. It is doubtful that broad antitumor activity across a spectrum of SQ xenografts will be a useful metric for targeted therapies where the relevant molecular targets are likely highly restricted in their expression pattern. For example, targeted therapies such as imatinib, trastizumab, and rituximab would all fail to meet the criterion of broad antitumor activity across greater than 1/3 of all tumor models.

Other retrospective studies have arrived at similar conclusions about the overall poor predictive value of SQ xenograft models. Of interest, however, when considered as disease-specific models, xenograft models of NSCLC do appear to predict for clinical efficacy in a number of studies (Johnson *et al.*, 2001; Mattern *et al.*, 1988; Voskoglou-Nomikos *et al.*, 2003). Xenograft models of breast (Bailey *et al.*, 1980; Inoue *et al.*, 1983; Johnson *et al.*, 2001; Mattern *et al.*, 1988; Voskoglou-Nomikos *et al.*, 2003) and ovarian (Johnson *et al.*, 2001; Mattern *et al.*, 1988; Taetle *et al.*, 1987; Voskoglou-Nomikos *et al.*, 2003) cancer have resulted in conflicting results. It is not clear why the predictive value of NSCLC models appear to be superior to other disease models.

Together these studies paint an unfavorable picture in terms of the use of xenograft SQ tumor models to predict clinical antitumor efficacy. By virtue of their study design, these results largely assess the positive predictive value of SQ xenograft models (i.e., if a compound has efficacy in mouse models, how likely is it that it will have efficacy in humans). These data are not as clear about the negative predictive value of these models (i.e., if there is no efficacy in mouse models, what is the likelihood that it will likewise lack efficacy in humans). Furthermore, these studies were largely assessments of cytotoxic therapies, utilizing antitumor efficacy as the only endpoints. It is not clear whether these results extend to the testing of targeted therapies, and/or whether alternative endpoints, such as pharmacodynamic markers, improve the predictive value of SQ xenograft models.

B. Hollow Fiber Models

Typical tumor implant studies require several weeks to months to perform: inject tumor cells, let tumors establish 1–2 weeks, then follow response to treatment over 3–6 weeks. Given the time and resource requirements, a shorter-term *in vivo* assay may be useful for prioritizing compounds for further testing. The hollow fiber assay was developed by the NCI for this purpose (Hollingshead *et al.*, 1995). In the hollow fiber assay, tumor cells are injected into 1 mm × 2 cm PVDF hollow tubes which are sealed to prevent cell leakage, but remain permeable to nutrients and small molecules. Multiple hollow fibers are implanted into the SQ or intraperitoneal (IP) space of nude mice and allowed to establish for 2–3 days. Treatment then commences, and after 4 days of treatment, hollow fibers are removed from the mouse and cell numbers quantified using a tetrazolium-based colorimetric assay. Total assay time is thus roughly 1 week.

The predictive value of hollow fiber activity was evaluated by analyzing the antitumor activity of 564 compounds tested using both a hollow fiber assay as well as an SQ xenograft model (Johnson *et al.*, 2001). SQ-implanted hollow fibers had no predictive value, whereas antitumor activity in multiple IP-implanted hollow fibers had a statistically significant predictive value for antitumor activity in SQ xenograft models. Overall, 8% of the 564 compounds tested had antitumor efficacy in SQ xenograft models, compared to 20% in the subgroup of compounds that had activity in the hollow fiber assay ($p < 0.0001$). Given the poor value of hollow fiber assays for predicting SQ xenograft efficacy (20% at best), and given the concerns about SQ xenograft models for predicting human clinical efficacy, it is certain that the hollow fiber model is not an improvement in terms of overall clinical prognostication. However, while the hollow fiber model is even more artificial than SQ xenografts, the short assay time and potential for moderate throughput has led some to adopt it as an intermediary method to screen multiple compounds, to help prioritize compounds for further *in vivo* testing.

C. Orthotopic Models

One of the primary rationales for the use of *in vivo* models is to replicate aspects of tumorigenesis that are not reflected in tissue culture. For example, nonmalignant stromal cells play an important role in promoting tumorigenesis, including providing critical growth factors, nutrients, and angiogenesis (Condon, 2005; De Wever and Mareel, 2003; Kim *et al.*, 2005). Although SQ xenografts may recapitulate certain aspects of the tumor–host microenvironment, it is clear that there are other aspects of the tumor microenvironment that are more closely approximated by implanting tumor cells into the organs and anatomical sites from which they originally arose. By the simplest measure, there are certain tumor cell lines that will only form tumors when implanted in an orthotopic site, but not in the SQ compartment, including for example certain prostate cell lines (Stephenson *et al.*, 1992).

Brain tumor models also illustrate potential increased fidelity of orthotopic models in comparison to SQ (ectopic) models. Although the normal blood–brain barrier is not fully intact in patients with brain tumors (Vajkoczy and Menger, 2004), there is a partial blood–tumor barrier (BTB) that excludes certain therapeutics such as large biologics (Neuwelt *et al.*, 1986). In animal models, implantation of brain tumor cells orthotopically in the brain results in formation of a BTB, in contrast to implantation of the same cells ectopically in the SQ space, where no BTB is formed (Hobbs *et al.*, 1998; Yuan *et al.*, 1994). Formation of a BTB is not a nonspecific effect of intracranial implantation, since other tumor cell types implanted in the brain do not induce formation of a BTB. These results have obvious implications in terms of the creation of clinically relevant models for studying brain tumor therapeutics.

Site-specific differences in vascular biology may not be limited to the brain. There is a growing interest in therapeutic strategies that target tumor angiogenesis (Kerbel and Folkman, 2002), especially with completion of clinical trials demonstrating proof of concept in humans (Hurwitz *et al.*, 2004). To this end, it is important to note that the vasculature in SQ tumors may be very different by comparison to the same tumor cells implanted in an orthotopic location. For example, a study of the vasodilator hydralazine revealed vast differences in modulating blood flow to SQ tumor by comparison to orthotopic tumors (or metastases) established with the same cell lines (Cowen *et al.*, 1995).

The propensity for metastasis is also influenced by tumor implant sites. Experimental modeling of metastasis can be accomplished by direct inoculation of tumor cells into the target organs or by intravenous or intracardiac inoculation (Hoffman, 1999; Manzotti *et al.*, 1993). However, these approaches bypass the initial steps of metastasis, including local-regional invasion. A more complete modeling of metastasis is achieved by implantating tumor cells in a primary site and allowing for “spontaneous” metastasis. SQ xenograft tumors seldom produce metastases, and the primary tumor often reaches maximum limits before metastases occur. In many cases, orthotopic implantation of tumor cells has been found to greatly enhance metastasis, by comparison to the same cells implanted subcutaneously (Fidler, 1986; Fidler *et al.*, 1990; Manzotti *et al.*, 1993; Naito *et al.*, 1987a,b; Stephenson *et al.*, 1992; Waters *et al.*, 1995).

The sensitivity of xenografted tumors to therapeutics may also be modulated by their location (Killion *et al.*, 1998). For example, the response of tumor cells to doxorubicin differs depending on site of implantation, likely due to tissue-specific modulation of *mdr1* expression (Dong *et al.*, 1994; Fidler *et al.*, 1994). In another example, sensitivity of small cell lung cancer (SCLC) cells to clinically relevant chemotherapeutics was found to predict clinical utility in a lung orthotopic model, but not in an SQ model (Kuo *et al.*, 1993). In this example, an orthotopic SCLC model revealed *in vivo* sensitivity to cisplatin but not to mytomycin C, similar to the clinical utility of these agents. In contrast, an SQ model established using the same cells found the inverse—sensitivity to mytomycin C, but not cisplatin.

Together, these results suggest that orthotopic transplant models may be superior to SQ transplant models, at least in recapitulating certain aspects of the tumor–host interactions. The major drawback of orthotopic models, however, is the loss of caliper measurements as a means of quantifying tumor burden for most disease sites. Tumor burden can be assessed by serial sacrifice of cohorts of mice if tumor formation is highly penetrant and synchronous. However, this necessitates large numbers of animals. Noninvasive methods for assessing tumor burden, including small animal imaging, overcome these limitations, as discussed below.

D. Genetically Defined Tumor Models

In recent years, one major approach to improving tumor models has been to develop genetically engineered mouse (GEM) models in which tumor formation is driven by clinically relevant oncogenes, or loss of tumor suppressors. The use of GEM cancer models for elucidating cancer biology and drug discovery has been previously reviewed (Van Dyke and Jacks, 2002). The major advantages of GEM models include the ability to drive tumorigenesis with defined genetic changes and syngeneic tumor–host compartments. However, at the current time, there are still intellectual property issues complicating the use of GEM in drug discovery (Marshall, 2002). Another major disadvantage of GEM models for drug discovery is the asynchronous development of tumors. In a setting where one wishes to test multiple doses of multiple compounds in appropriately powered studies, the synchronous development of tumors in sufficient numbers of mice to conduct such studies is difficult. The use of models in which tumor formation can be induced in a conditional manner (Jackson *et al.*, 2001) improves the likelihood of having sufficient cohorts of mice. However, the CRE-lox system, which is the most commonly used approach for conditional induction of tumors, also carries additional intellectual property issues (Wadman, 1998).

To overcome the problems associated with asynchronous tumor development in GEM, it is possible to use GEM as a source for tumor cells to be transplanted into coisogenic wild-type mice. In this case, a tumor arising in a GEM is collected, disaggregated, or fragmented, and then transplanted into large numbers of naïve wild-type recipients. Tumors arise in a synchronous manner in the recipients, and the advantages of syngeneic tumor–host compartments are maintained. Orthotopic implantation into the originating anatomical site can further restore tissue-specific microenvironment. In many cases, cell lines can be established from tumors arising in GEM, and this may provide an alternative source of material for syngeneic transplant studies. Alternatively, there are a number of existing mouse tumor cell lines (e.g., Lewis lung carcinoma, B16 melanoma) derived from spontaneous or chemically induced tumors that can be transplanted into naive mice.

The major disadvantage of the latter is a lack of genetic definition as to the underlying tumor-inducing changes. As with any cell line, prolonged propagation in cell culture may select for cells with characteristics that are different than *in situ* tumors as described below.

An alternative method for generating genetically defined mouse tumors is to begin with normal mouse cells and engineer them *ex vivo*, prior to transplantation back into syngeneic naive mice. This approach has been extensively used for studying the mechanisms of cell transformation, for example, in ovarian cancer (Orsulic *et al.*, 2002) and hematologic malignancies (Lavau *et al.*, 1997, 2000; Schwaller *et al.*, 1998). More recently, such approaches have been used to test novel therapeutic strategies, for example, small molecule kinase inhibitors in activated FLT3-induced myeloproliferative disease (Weisberg *et al.*, 2002).

III. TUMOR MODELS: CELL TYPES

One potential source for the lack of predictability in xenograft models is the use of established cell lines that have been selected for growth in cell culture. While providing a steady source of cells, selection over years of propagation results in outgrowth of cells with characteristics often quite different than *in situ* human tumors (Engelholm *et al.*, 1985; Hausser and Brenner, 2005; Nielsen *et al.*, 1994; Smith *et al.*, 1989). Patient tumor material can be implanted directly into immunodeficient mice, although subsequent tumor growth will occur only in a minority of cases (Scholz *et al.*, 1990; Steel *et al.*, 1983; Winograd *et al.*, 1987). Testing of chemotherapeutics in such models suggests good correlation with patient outcome, for example, with up to a 90% predictive value for clinical response and a 97% predictive value for resistance (Fiebig *et al.*, 2004; Scholz *et al.*, 1990). As a practical matter, however, primary tumor explants are extremely time- and resource-intensive, unpredictable, and there are multiple barriers to incorporating these methods into a high-throughput drug discovery flow.

For the purposes of testing targeted therapies, it is critical that the tumor model be driven by, or at least expresses, the target of interest. When using existing cell lines, it is thus critical that they be characterized for the relevant mutations or expression of the target protein. Alternatively, engineering tumors to be driven by the genetic changes of interest can be accomplished in either mouse or mouse cells as described above. While there are advantages to maintaining a syngeneic relationship between tumor and host compartments, for certain therapeutics it may be critical that the tumor targets are human in origin. On the most basic level, if the therapeutic target is not sufficiently conserved, or the therapeutic is highly specific for the human target, testing in a fully mouse system may not demonstrate efficacy due to a lack of effect on the orthologous target (e.g., an antibody therapeutic that only recognizes the human target). In fact, retrospective analysis has shown that 30% of agents with activity in human xenografts have no activity in mouse syngeneic transplant models (Venditti *et al.*, 1984). While the poor predictive value of xenograft models leads one to wonder if the lack of efficacy in the mouse models actually reflects better predictive value, it is notable that taxol was one of the agents that was "missed" by the mouse syngeneic model.

In recent years, several groups have identified finite numbers of genetic changes that can be introduced *ex vivo* into normal human cells resulting in full oncogenic transformation (Boehm and Hahn, 2005). This raises the possibility of creating xenograft tumor models in which the transformed human cells have been generated *de novo* by introduction of relevant oncogenic changes into essentially normal (or immortalized) human cells. It is currently not clear if xenograft tumors of such engineered cancer cells more closely resemble *in situ* human tumors by comparison to cell line xenografts. However, in certain cases, one can ensure that the relevant tumor targets are present either because the target is one of the transforming changes *per se* or if expression of the target is tightly linked to the engineered genetic change.

Regardless of the source, injection of human cancer cells results in a species mismatch between the tumor cells and their microenvironment (e.g., nontransformed stromal cells). Since stromal cells contribute to tumorigenesis, fidelity of xenografts might be improved by establishing a fully human microenvironment. Recent studies have shown that when nonmalignant stromal cells are coinjected with transformed human cancer cells, the human stromal cells do contribute to eventual tumor formation (Berger *et al.*, 2004; Chudnovsky *et al.*, 2005). The histology of these nonmalignant cells resembles that found in *in situ* human tumors. These studies have provided important insights into mechanisms responsible for tumor formation. However, further studies are needed to determine whether these models are an improvement over conventional xenograft or orthograft models for assessing therapeutic agents.

IV. STUDY ENDPOINTS

A. Efficacy Endpoints: Is Tumor Growth Effected?

The majority of agents currently utilized in the clinic were discovered on the basis of their ability to kill tumor cells in tissue culture, that is, cytotoxics. Since tumor cell killing was the primary *in vitro* endpoint, extrapolation to animal models necessitated tumor cell killing as the primary endpoint as well *in vivo*. While there is a general perception that mouse-transplanted tumor models are highly susceptible to cytotoxic therapies, in fact most cytotoxic therapies have little efficacy in mouse solid tumor models (Corbett *et al.*, 1987). Nonetheless, as described above, there are numerous concerns about the predictive value of antitumor efficacy in mouse models.

The predictive value may depend on the criterion used to define efficacy. When comparing the growth of tumors in treated to control animals, a ratio of 20–30% (% T/C) is a common threshold utilized for establishing efficacy. However, the predictive value may be improved in mouse models by adopting tumor regression as the efficacy endpoint, as opposed to decreased tumor growth. In one comparison of six clinically relevant therapies, the three agents that induced tumor regression in a mouse breast cancer model (i.e., melphalan, cyclophosphamide, and 5-fluorouracil) had clinical efficacy in patients, by comparison to 6-thioguanine, Ara-C, and *N*-phosphonacetylaspartate which did not induce regression in the mouse model and likewise did not demonstrate clinical efficacy (Stolfi *et al.*, 1988).

Antitumor efficacy is simple to measure in SQ tumors, thus accounting for their popularity. Caliper measurements of length and width can be used to calculate tumor volume. In the case of orthotopically placed tumors, breast tumors are easily measured due to their superficial location. Tumors situated within internal organs (e.g., prostate, kidney, brain, lungs, or hematologic malignancies) are not accessible to direct physical measurement. In some models, if the tumor establishment rate is high and the growth characteristics are predictable, tumors can be implanted and allowed to establish for a predetermined period of time. Animals are divided into treatment groups, and drug treatment administered either until animals become "sick" or for a predetermined period of time. The problems with this approach include the fact that for many tumor models, a baseline "nontake" rate necessitates large group sizes to overcome the statistical variance resulting from nontakers. Furthermore, using death (or even moribund state) as an endpoint is generally viewed as unethical. Finally, survival studies can take a very long time as tumor burden may have to be quite significant in some locations before resulting in constitutional symptoms.

Noninvasive methods of tumor quantification can obviate many of the problems associated with orthotopic tumor models. One alternative is to engineer tumor cells to produce an ectopic secreted protein that can be detected in the serum or urine, such as urinary β-hCG (Shih *et al.*, 2000). For solid tumors, anatomical imaging modalities can provide localization and volumetric information. In general, these are clinically utilized imaging modalities that have been adapted for small animal use (Rudin and Weissleder, 2003; Weissleder, 2002), including magnetic resonance imaging (MRI) (Evelhoch *et al.*, 2000; Gillies *et al.*, 2000; Nelson *et al.*, 2003), X-ray computed tomography (CT) (Paulus *et al.*, 2000), ultrasound (US), and plain radiographs. These technologies are available commercially, in generally user-friendly interfaces. However, space and resource requirements may be prohibitive, and throughput may be inadequate for drug discovery purposes, as these modalities generally allow imaging of only one animal at a time. For tumors not localized to a single location, for example hematologic malignancies, anatomical imaging has limited applicability.

Fluorescence (Hoffman, 2002a,b) and bioluminescence (Edinger *et al.*, 2002; McCaffrey *et al.*, 2003) optical imaging have been increasingly adopted by many laboratories for tumor burden quantification in mice. Resource and space requirements are modest by comparison to MRI. Anatomical definition is poor, but there is tight correlation with volumetric measurements, validating their use for tumor burden quantification. Reagents necessary to render cells fluorescent or luminescent are widely available in most laboratories. Fluorescence-based imaging offers the possibility of increased information content (e.g., simultaneous imaging of differently fluorescent cells), however, the emission spectrum is greatly attenuated by tissue, thus limiting usefulness for certain orthotopic models (e.g., brain tumors). Luciferase-mediated bioluminescence has significant emission at wavelengths greater than 650 nm, which has superior tissue penetration in comparison to green fluorescent protein (GFP) that has peak emission at $\sim$509 nm, and is thus preferable for deep orthotopic models.

Another factor that may lead to a lack of predictability is species-specific differences in pharmacokinetics. For many agents, maximally tolerated levels in mice may exceed achievable or tolerable levels in humans and may thus produce greater efficacy. In a series of studies in which multiple drugs were administered at "clinically equivalent dose" (i.e., reaching clinically achievable concentrations), antitumor efficacy was highly predictive of clinical efficacy in gastric (Inaba *et al.*, 1988), lung (Tashiro *et al.*, 1989), breast (Inaba *et al.*, 1989), and glioma (Maruo *et al.*, 1990) SQ xenograft tumor models. Such strategies may be very useful when testing existing or approved therapeutics for a potential new indication. In the development of an NCE, achievable human levels will not be known until phase I clinical trials. It might be reasonable to return to animal testing when such information is known, especially if there are multiple potential disease indications.

B. Functional and Molecular Endpoints: Is the Target Modulated?

In the era of targeted therapies, inhibition of tumor growth may no longer be the most relevant endpoint. If a cancer target is well validated (e.g., strong epidemiology), an equally, if not more, relevant question may be whether the therapeutic target is modulated *in vivo*. In the absence of knowing whether a drug modulates its intended target, antitumor efficacy is impossible to interpret. For example, if antitumor efficacy is observed, yet the intended target is not modulated, the implication is that antitumor efficacy may be the result of an off-target effect. To the extent that the intended target is known, assessment of target modulation can be accomplished by multiple methods.

Conventional molecular analysis can be accomplished by sacrificing cohorts of mice over time, and analyzing the target, or a biomarker of a given pathway, using conventional biochemical and microscopic techniques (e.g., Western blot, immunohistochemistry). For example, the effects of a kinase inhibitor might be assessed using activation-state-specific antibodies to assess either the kinase itself or downstream signaling mediators. The major disadvantage of these approaches is that they provide static data points, and a dynamic picture must be constructed by looking at the averages of cohorts of mice over time. Thus, large numbers of mice and significant time and resources need to be expended to establish a dynamic assessment of drug effect.

An extension of the hollow fiber assay can be utilized for assessing molecular endpoints. After establishing fibers, treating animals, and then recovering fibers in the usual manner, cells can be recovered from hollow fibers and used for assessing molecular endpoints (Suggitt *et al.*, 2004). This is a facile method for rapidly isolating tumor cells, but clearly suffers from a rather artificial tumor–host microenvironment.

Noninvasive imaging can also be used to assess molecular effects of drugs (i.e., molecular imaging). By virtue of being noninvasive and nonlethal, molecular imaging can provide serial measurements over time in animals, establishing a true pharmacodynamic readout. As with anatomical imaging, clinically useful functional imaging modalities have been adapted for small animal imaging. These include micro-positron emission tomography (micro-PET) (Mankoff *et al.*, 2000, 2005; Ray *et al.*, 2003; Yang *et al.*, 2003), and MRI (Nelson *et al.*, 2003). PET can provide metabolic information, such as glucose consumption using standard ^{18}F-fluorodeoxyglucose (FDG), as is commonly utilized in humans. Furthermore, since many drugs can be radiolabeled with ^{11}C or ^{18}F, PET offers the possibility of directly monitoring drug distribution and clearance (Weissleder, 2002). It is also possible to design PET probes that bind to specific drug targets, for example to monitor levels of HER2 expression (Smith-Jones *et al.*, 2004), or to probe integrin expression in tumors (Chen *et al.*, 2004; Haubner *et al.*, 2001). Obviously, these approaches require significant radiological, biological, and chemical expertise.

Bioluminescence and fluorescence imaging can also be used for molecular imaging. For example, reporter cells can be engineered whereby luciferase expression is stimulus dependent, and luminescence is thus a readout of the activity of certain molecular pathways (Bhaumik and Gambhir, 2002; Contag *et al.*, 1997; Kung *et al.*, 2004). The reagents necessary to create reporter cell lines are commonly available in many laboratories (i.e., luciferase- or GFP-based reporter constructs). These modalities are strictly research tools, as clinical translation will not be possible using existing technologies. In the realm of optical imaging, near-infrared imaging is the only modality that has the potential for clinical utility in the near future. Examples of molecular imaging using near-infrared probes include caged probes that become fluorescent after enzymatic cleavage *in vivo*, allowing for noninvasive assessment of enzyme activities (Mahmood and Weissleder, 2003; Petrovsky *et al.*, 2003).

V. ANIMAL MODELING IN THE POST-GENOMICS AGE

The dichotomization of therapies as either cytotoxic or targeted therapies is misnomered. In fact, the molecular targets for most cytotoxics are known (e.g., microtubules, nucleotide pools, topoisomerases, and so on). Conversely, many therapies developed to inhibit specific molecular targets may have widespread effects in both malignant and nonmalignant cells. In either case, animal models serve the same purposes in the drug discovery flow. Fundamentally, we want animal models to tell us: (1) is the target important for tumor homeostasis or growth, (2) does the drug modulate its intended target *in vivo*, and (3) does a drug have an effect on tumor growth. While it may not be possible to create models that are perfectly predictive of clinical efficacy, if we can alter our use of animal models to be more informative, even small incremental increases in the predictive value in a preclinical setting may translate to marked improvements in productivity and substantial savings in drug development costs (DiMasi, 2002).

The drug discovery flow generally commences with the identification of a target, or pathway, that is felt to be important for maintaining the malignant phenotype, or for growth of human tumors. Targets may be suggested by clinical epidemiology [e.g., recurrent activating mutations of epidermal growth factor receptor (EGFR)]. Targets may also be suggested by extrapolation from basic research (e.g., the role of telomerase in transformation). Before commencing on a search for chemical or biologic inhibitors, targets are generally validated *in vitro*, and preferably *in vivo*. For *in vitro* validation, the target of interest may be manipulated by a variety of methods, including overexpression of dominant-negative isoforms, small interfering RNA (siRNA) knockdown, neutralizing antibodies, or the use of tool compound inhibitors. Many of these approaches can also be used to assess effects on *in vivo* tumor growth. For example, the growth of tumors in which a putative target is stably knocked down with a lentiviral short hairpin RNA (shRNA) can be compared to control tumors expressing an irrelevant shRNA (e.g., to GFP). There are other mechanisms of action that can only be evaluated *in vivo*, such as mediators of metastasis or modulators of angiogenesis. For both *in vitro* and *in vivo* experiments, it is obviously critical to establish that the target of interest is expressed in the cell line used. Alternatively, it may be possible to transform normal mouse or human cells with a set of oncogenes that include the target of interest, or is inextricably linked to activation or expression of the target. Thus, animal models play an important role even at the very earliest stages of the drug discovery flow for validating candidate targets.

Once a putative therapeutic has been identified, careful consideration is again required for the disease model and the endpoints to be measured. In most cases, targets will arise out of the clinical epidemiology of a specific disease, but for *in vivo* testing, it is not sufficient to just seek out models of the underlying disease without regard for the specific molecular target of interest. For example, testing of EGFR inhibitors should utilize not just any lung cancer model, but specifically lung cancer models where the tumor is driven by mutated EGFR. In the case of xenografts or syngeneic tumor transplants, it will be critical to fully characterize the cell lines and verify that they at least express the target of interest. In the case of GEM, tumor cells derived from GEM, or engineered human cells, it is likewise important to establish that the model is driven by the target of interest.

Next, the location of tumor implantation or transplant should be considered. Although the exact impact on predictive value is not known, there are ample examples where the biology of orthotopic tumors seems to have higher fidelity by comparison to SQ models. Orthotopic models in which tumor location is superficial (e.g., breast, skin, muscle) can be monitored using conventional caliper measurements. Orthotopic models with deep visceral placement (e.g., brain, lung, liver, prostate, colon) require alternative methods of tumor burden quantification. Anatomical imaging modalities (e.g., US, CT, MRI) can provide volumetric measurements of solid tumors. Cells can also be engineered to express ectopic biomarkers, including secreted proteins (e.g., β-HCG), GFP for noninvasive fluorescence imaging, or luciferase for bioluminescence imaging. For hematologic malignancies, total body tumor burden can best be assessed by optical imaging, since tumor distribution is disseminated (e.g., throughout the marrow space).

Efficacy studies are generally meant to evaluate the impact of a putative therapeutic on established tumors. While the specific starting criteria will vary according to the model type, at the very least the tumor must be detectable, and ideally should be clearly increasing in size. For SQ or superficial tumors, or solid tumors followed by anatomical imaging, this might be on reaching a predetermined tumor volume (e.g., 100 mm^3). For tumors followed by fluorescence or bioluminescence imaging, serial imaging might be used to identify animals with increasing tumor burden. Drugs may have very different effects on bulky tumors by comparison to small tumors, and there is some concern about starting treatments too early when using highly sensitive imaging modalities.

While attenuation of tumor growth is a standard endpoint, it is by itself not sufficient. It is imperative to determine whether the therapeutic of interest actually modulates its molecular target *in vivo*. That is, if a drug is found to have no effect on tumor growth, interpretation of the results is highly dependent on whether the drug "worked," that is, did it hit its target. Conversely, even if there is antitumor efficacy, if the molecular target is not modulated, enthusiasm may be greatly dampened for an entity that is "working" through off-target effects. Evaluation of informative pharmacodynamic markers is thus critical. This can be accomplished through conventional means by collecting tumor samples from animals in treated and control groups, and then using conventional biochemical and microscopic methods to interrogate the target of interest [e.g., immunostaining, Western blot, enzyme-linked immunosorbent assay (ELISA)]. Modification-specific antibodies (e.g., phosphorylation, acetylation) may be used to evaluate enzyme targets (e.g., histone acetylation, phosphorylation state of kinase substrates), or as biomarkers of activity (e.g., phosphorylation as a marker of the activation state of kinases). To evaluate signaling pathways, the expression of downstream targets of the pathway can be assessed. Protein levels can be assessed by bulk means such as Western blot or ELISA of tumor lysates. Immunohistochemistry can be used to evaluate protein levels *in situ* within fixed tumors. Flow cytometry can be used to evaluate cell surface as well as intracellular protein levels on a single cell basis.

As with *in vitro* studies, evaluating a number of pharmacodynamic markers may increase confidence over a single marker. To expand the analysis to a more global level, tumor cells can be isolated from tumors and subjected to gene expression or proteomic analysis. In most solid tumors, gross dissection of the tumor results in fairly good enrichment for tumor cells. For greater purity, tumors can be disaggregated and tumor cells specifically isolated by fluorescence-activated cell sorting or immunomagnetic separation using antibodies to tumor-specific (or species-specific) surface antigens. For fixed material, laser-capture microdissection can be used to isolate tumor cells, although subsequent analysis will be complicated by small cell numbers. With these methods, one could examine the effect of drug treatment not only on handfuls of markers, but more globally at whole expression signatures associated with specific pathways.

At the very least, evaluation should include cohorts of drug-treated versus control animals to establish whether the drug target is modulated at a specific time point, for example, around the time of peak drug levels. A dynamic assessment of drug effects can be reconstructed using conventional means by analysis of groups of treated and untreated mice sacrificed in a timed manner after drug administration. Alternatively, functional or molecular imaging can be used to evaluate drug effects in a dynamic manner. With these approaches, the goal is to not just ask whether the target is modulated, but to establish the time course of drug effects.

These studies should provide two types of information: does the drug modulate its target and does it shrink the tumor? Even if the cell line or model used is absolutely dependent on the target (e.g., the cells undergo apoptosis or arrest with shRNA knockdown of the target *in vitro*), it may still be possible to modulate the target *in vivo* without an overt antitumor effect. Targets where episodic inhibition is insufficient or where alternative pathways are activated *in vivo* are examples where there may be a disconnect between *in vivo* and *in vitro* antitumor efficacy.

There is considerable controversy as to what should happen if a candidate therapeutic modulates its target, but does not produce antitumor efficacy in mouse models. Some argue that if the target is robustly validated (e.g., with strong clinical epidemiology), then lack of efficacy in preclinical models should not preclude further clinical development. Others argue that mouse models are generally overly sensitive, and thus lack of efficacy would dampen enthusiasm for further development. Careful validation of the models used for such studies may make such discordance less likely. That is, if a cell line or GEM is fully validated to be dependent on the target of interest (e.g., loss of function results in apoptosis or arrest), then tumor viability and growth may be very closely linked to biological modulation of the target.

The other perplexing situation is how to interpret antitumor efficacy in the setting of not being able to show an effect on a pharmacodynamic marker. Here, it is important to make sure the pharmacodynamic marker is as closely linked to the target as possible. Phosphorylation-state-specific antibodies will be better indicators of the inhibition of a receptor tyrosine kinase, for example, in comparison to the expression level of some downstream gene (which might be activated by redundant signaling pathways). Furthermore, one must consider effects on cellular compartments other than the tumor cells, for example effects on endothelial cells resulting in an antiangiogenic effect.

In addition to pharmacodynamic readouts, experimental manipulation of the tumor model may help provide evidence for specificity of action. For example, a lack of activity in cells that are not dependent on the target (but of the same tumor type) provides support for specificity of action. Even more convincing, introduction of genetic elements that are redundant to or bypasses the targeted protein, followed by demonstration of resulting insensitivity to the drug in a model that was previously sensitive, is a powerful control for specificity.

In late stages of preclinical testing, animal models may also be important testing grounds for biomarkers to be used in the initial clinical trials. For example, the development of informative immunodetection reagents, functional imaging modalities, and identifying molecular signatures of response may all be amenable to testing in preclinical models. While species-specific differences may preclude direct translation of certain reagents (e.g., detection antibodies), animal models can be used to establish proof of concept for a biomarker strategy, much as tool compounds may be used for proof of concept experiments prior to a search for a pharmacologic inhibitor.

VI. CONCLUSIONS

In the age of cytotoxic therapies, where drugs were discovered by virtue of their ability to kill tumor cells in cell culture, the analogous endpoint in animal models was whether the growth of tumors was inhibited. The paradigm of modern drug discovery is one of "targeted therapies," where therapeutics are directed against well-characterized tumor targets. It is thus antithetical to maintain the endpoints of tumor killing *in vivo* as the only, or perhaps even primary, endpoint. Equally, if not more important is to use animal models to determine if the target of interest is modulated *in vivo*. By crafting models that more closely recapitulate human tumors, and by asking more specific questions with such models (not just whether a tumor shrinks), animal models will continue to play an important role in validating cancer targets and evaluating putative cancer therapies. That is, we must use the right models to ask the right questions about the right therapies. While it is unrealistic to expect that any animal model will perfectly predict human outcome, even small improvements will translate into increased productivity and significant cost savings.

REFERENCES

Bailey, M. J., Gazet, J. C., Smith, I. E., and Steel, G. G. (1980). Chemotherapy of human breast-carcinoma xenografts. *Br. J. Cancer* **42**(4), 530–536.

Berger, R., Febbo, P. G., Majumder, P. K., Zhao, J. J., Mukherjee, S., Signoretti, S., Campbell, K. T., Sellers, W. R., Roberts, T. M., Loda, M., Golub, T. R., and Hahn, W. C. (2004). Androgen-induced differentiation and tumorigenicity of human prostate epithelial cells. *Cancer Res.* **64**(24), 8867–8875.

Bhaumik, S., and Gambhir, S. S. (2002). Optical imaging of Renilla luciferase reporter gene expression in living mice. *Proc. Natl. Acad. Sci. USA* **99**(1), 377–382.

Boehm, J. S., and Hahn, W. C. (2005). Understanding transformation: Progress and gaps. *Curr. Opin. Genet. Dev.* **15**(1), 13–17.

Chen, X., Tohme, M., Park, R., Hou, Y., Bading, J. R., and Conti, P. S. (2004). Micro-PET imaging of alphavbeta3-integrin expression with ^{18}F-labeled dimeric RGD peptide. *Mol. Imaging* **3**(2), 96–104.

Chudnovsky, Y., Adams, A. E., Robbins, P. B., Lin, Q., and Khavari, P. A. (2005). Use of human tissue to assess the oncogenic activity of melanoma-associated mutations. *Nat. Genet.* **37**(7), 745–749.

Condon, M. S. (2005). The role of the stromal microenvironment in prostate cancer. *Semin. Cancer Biol.* **15**(2), 132–137.

Contag, C. H., Spilman, S. D., Contag, P. R., Oshiro, M., Eames, B., Dennery, P., Stevenson, D. K., and Benaron, D. A. (1997). Visualizing gene expression in living mammals using a bioluminescent reporter. *Photochem. Photobiol.* **66**(4), 523–531.

Corbett, T. H., Valeriote, F. A., and Baker, L. H. (1987). Is the P388 murine tumor no longer adequate as a drug discovery model? *Invest. New Drugs* **5**(1), 3–20.

Cowen, S. E., Bibby, M. C., and Double, J. A. (1995). Characterisation of the vasculature within a murine adenocarcinoma growing in different sites to evaluate the potential of vascular therapies. *Acta Oncol.* **34**(3), 357–360.

De Wever, O., and Mareel, M. (2003). Role of tissue stroma in cancer cell invasion. *J. Pathol.* **200**(4), 429–447.

DiMasi, J. A. (2002). The value of improving the productivity of the drug development process: Faster times and better decisions. *Pharmacoeconomics* **20**(Suppl. 3), 1–10.

DiMasi, J. A., Hansen, R. W., and Grabowski, H. G. (2003). The price of innovation: New estimates of drug development costs. *J. Health Econ.* **22**(2), 151–185.

Dong, Z., Radinsky, R., Fan, D., Tsan, R., Bucana, C. D., Wilmanns, C., and Fidler, I. J. (1994). Organ-specific modulation of steady-state mdr gene expression and drug resistance in murine colon cancer cells. *J. Natl. Cancer Inst.* **86**(12), 913–920.

Edinger, M., Cao, Y. A., Hornig, Y. S., Jenkins, D. E., Verneris, M. R., Bachmann, M. H., Negrin, R. S., and Contag, C. H. (2002). Advancing animal models of neoplasia through *in vivo* bioluminescence imaging. *Eur. J. Cancer* **38**(16), 2128–2136.

Engelholm, S. A., Vindelov, L. L., Spang-Thomsen, M., Brunner, N., Tommerup, N., Nielsen, M. H., and Hansen, H. H. (1985). Genetic instability of cell lines derived from a single human small cell carcinoma of the lung. *Eur. J. Cancer Clin. Oncol.* **21**(7), 815–824.

Evelhoch, J. L., Gillies, R. J., Karczmar, G. S., Koutcher, J. A., Maxwell, R. J., Nalcioglu, O., Raghunand, N., Ronen, S. M., Ross, B. D., and Swartz, H. M. (2000). Applications of magnetic resonance in model systems: Cancer therapeutics. *Neoplasia* **2**(1–2), 152–165.

Fidler, I. J. (1986). Rationale and methods for the use of nude mice to study the biology and therapy of human cancer metastasis. *Cancer Metastasis Rev.* **5**(1), 29–49.

Fidler, I. J., Naito, S., and Pathak, S. (1990). Orthotopic implantation is essential for the selection, growth and metastasis of human renal cell cancer in nude mice [corrected]. *Cancer Metastasis Rev.* **9**(2), 149–165.

Fidler, I. J., Wilmanns, C., Staroselsky, A., Radinsky, R., Dong, Z., and Fan, D. (1994). Modulation of tumor cell response to chemotherapy by the organ environment. *Cancer Metastasis Rev.* **13**(2), 209–222.

Fiebig, H. H., Maier, A., and Burger, A. M. (2004). Clonogenic assay with established human tumour xenografts: Correlation of *in vitro* to *in vivo* activity as a basis for anticancer drug discovery. *Eur. J. Cancer* **40**(6), 802–820.

Frank, R., and Hargreaves, R. (2003). Clinical biomarkers in drug discovery and development. *Nat. Rev. Drug Discov.* **2**(7), 566–580.

Gillies, R. J., Bhujwalla, Z. M., Evelhoch, J., Garwood, M., Neeman, M., Robinson, S. P., Sotak, C. H., and Van Der Sanden, B. (2000). Applications of magnetic resonance in model systems: Tumor biology and physiology. *Neoplasia* **2**(1–2), 139–151.

Giovanella, B. C., Yim, S. O., Stehlin, J. S., and Williams, L. J., Jr. (1972). Development of invasive tumors in the "nude" mouse after injection of cultured human melanoma cells. *J. Natl. Cancer Inst.* **48**(5), 1531–1533.

Goldin, A., Venditti, J. M., Kline, I., and Mantel, N. (1961a). Evaluation of antileukemic agents employing advanced leukemia L1210 in mice. IV. *Cancer Res.* **21**(3, Pt. 2), 27–92.

Goldin, A., Venditti, J. M., Kline, I., and Mantel, N. (1961b). Evaluation of chemical agents against carcinoma CA-755 in mice. *Cancer Res.* **21**, 617–691.

Haubner, R., Wester, H. J., Burkhart, F., Senekowitsch-Schmidtke, R., Weber, W., Goodman, S. L., Kessler, H., and Schwaiger, M. (2001). Glycosylated RGD-containing peptides: Tracer for tumor targeting and angiogenesis imaging with improved biokinetics. *J. Nucl. Med.* **42**(2), 326–336.

Hausser, H. J., and Brenner, R. E. (2005). Phenotypic instability of Saos-2 cells in long-term culture. *Biochem. Biophys. Res. Commun.* **333**(1), 216–222.

Hobbs, S. K., Monsky, W. L., Yuan, F., Roberts, W. G., Griffith, L., Torchilin, V. P., and Jain, R. K. (1998). Regulation of transport pathways in tumor vessels: Role of tumor type and microenvironment. *Proc. Natl. Acad. Sci. USA* **95**(8), 4607–4612.

Hoffman, R. M. (1999). Orthotopic metastatic mouse models for anticancer drug discovery and evaluation: A bridge to the clinic. *Invest. New Drugs* **17**(4), 343–359.

Hoffman, R. M. (2002a). Green fluorescent protein imaging of tumour growth, metastasis, and angiogenesis in mouse models. *Lancet Oncol.* **3**(9), 546–556.

Hoffman, R. M. (2002b). Green fluorescent protein imaging of tumor cells in mice. *Lab. Anim. (NY)* **31**(4), 34–41.

Hollingshead, M. G., Alley, M. C., Camalier, R. F., Abbott, B. J., Mayo, J. G., Malspeis, L., and Grever, M. R. (1995). *In vivo* cultivation of tumor cells in hollow fibers. *Life Sci.* **57**(2), 131–141.

Hurwitz, H., Fehrenbacher, L., Novotny, W., Cartwright, T., Hainsworth, J., Heim, W., Berlin, J., Baron, A., Griffing, S., Holmgren, E., Ferrara, N., Fyfe, G., *et al.* (2004). Bevacizumab plus irinotecan, fluorouracil, and leucovorin for metastatic colorectal cancer. *N. Engl. J. Med.* **350**(23), 2335–2342.

Inaba, M., Kobayashi, T., Tashiro, T., Sakurai, Y., Maruo, K., Ohnishi, Y., Ueyama, Y., and Nomura, T. (1989). Evaluation of antitumor activity in a human breast tumor/nude mouse model with a special emphasis on treatment dose. *Cancer* **64**(8), 1577–1582.

Inaba, M., Tashiro, T., Kobayashi, T., Sakurai, Y., Maruo, K., Ohnishi, Y., Ueyama, Y., and Nomura, T. (1988). Responsiveness of human gastric tumors implanted in nude mice to clinically equivalent doses of various antitumor agents. *Jpn. J. Cancer Res.* **79**(4), 517–522.

Inoue, K., Fujimoto, S., and Ogawa, M. (1983). Antitumor efficacy of seventeen anticancer drugs in human breast cancer xenograft (MX-1) transplanted in nude mice. *Cancer Chemother. Pharmacol.* **10**(3), 182–186.

Jackson, E. L., Willis, N., Mercer, K., Bronson, R. T., Crowley, D., Montoya, R., Jacks, T., and Tuveson, D. A. (2001). Analysis of lung tumor initiation and progression using conditional expression of oncogenic K-ras. *Genes Dev.* **15**(24), 3243–3248.

Johnson, J. I., Decker, S., Zaharevitz, D., Rubinstein, L. V., Venditti, J. M., Schepartz, S., Kalyandrug, S., Christian, M., Arbuck, S., Hollingshead, M., and Sausville, E. A. (2001). Relationships between drug activity in NCI preclinical *in vitro* and *in vivo* models and early clinical trials. *Br. J. Cancer* **84**(10), 1424–1431.

Kerbel, R., and Folkman, J. (2002). Clinical translation of angiogenesis inhibitors. *Nat. Rev. Cancer* **2**(10), 727–739.

Killion, J. J., Radinsky, R., and Fidler, I. J. (1998). Orthotopic models are necessary to predict therapy of transplantable tumors in mice. *Cancer Metastasis Rev.* **17**(3), 279–284.

Kim, J. B., Stein, R., and O'Hare, M. J. (2005). Tumour-stromal interactions in breast cancer: The role of stroma in tumourigenesis. *Tumour Biol.* **26**(4), 173–185.

Kung, A. L., Zabludoff, S. D., France, D. S., Freedman, S. J., Tanner, E. A., Vieira, A., Cornell-Kennon, S., Lee, J., Wang, B., Wang, J., Memmert, K., Naegeli, H. U., *et al.* (2004). Small molecule blockade of transcriptional coactivation of the hypoxia-inducible factor pathway. *Cancer Cell* **6**(1), 33–43.

Kuo, T. H., Kubota, T., Watanabe, M., Furukawa, T., Kase, S., Tanino, H., Saikawa, Y., Ishibiki, K., Kitajima, M., and Hoffman, R. M. (1993). Site-specific chemosensitivity of human small-cell lung carcinoma growing orthotopically compared to subcutaneously in SCID mice: The importance of orthotopic models to obtain relevant drug evaluation data. *Anticancer Res.* **13**(3), 627–630.

Lavau, C., Du, C., Thirman, M., and Zeleznik-Le, N. (2000). Chromatin-related properties of CBP fused to MLL generate a myelodysplastic-like syndrome that evolves into myeloid leukemia. *EMBO J.* **19**(17), 4655–4664.

Lavau, C., Szilvassy, S. J., Slany, R., and Cleary, M. L. (1997). Immortalization and leukemic transformation of a myelomonocytic precursor by retrovirally transduced HRX-ENL. *EMBO J.* **16**(14), 4226–4237.

Mahmood, U., and Weissleder, R. (2003). Near-infrared optical imaging of proteases in cancer. *Mol. Cancer Ther.* **2**(5), 489–496.

Mankoff, D. A., Dehdashti, F., and Shields, A. F. (2000). Characterizing tumors using metabolic imaging: PET imaging of cellular proliferation and steroid receptors. *Neoplasia* **2**(1–2), 71–88.

Mankoff, D. A., Shields, A. F., and Krohn, K. A. (2005). PET imaging of cellular proliferation. *Radiol. Clin. North Am.* **43**(1), 153–167.

Manzotti, C., Audisio, R. A., and Pratesi, G. (1993). Importance of orthotopic implantation for human tumors as model systems: Relevance to metastasis and invasion. *Clin. Exp. Metastasis* **11**(1), 5–14.

Marshall, E. (2002). Intellectual property. DuPont ups ante on use of Harvard's OncoMouse. *Science* **296**(5571), 1212.

Maruo, K., Ueyama, Y., Inaba, M., Emura, R., Ohnishi, Y., Nakamura, O., Sato, O., and Nomura, T. (1990). Responsiveness of subcutaneous human glioma xenografts to various antitumor agents. *Anticancer Res.* **10**(1), 209–212.

Mattern, J., Bak, M., Hahn, E. W., and Volm, M. (1988). Human tumor xenografts as model for drug testing. *Cancer Metastasis Rev.* **7**(3), 263–284.

McCaffrey, A., Kay, M. A., and Contag, C. H. (2003). Advancing molecular therapies through *in vivo* bioluminescent imaging. *Mol. Imaging* **2**(2), 75–86.

Naito, S., Giavazzi, R., Walker, S. M., Itoh, K., Mayo, J., and Fidler, I. J. (1987a). Growth and metastatic behavior of human tumor cells implanted into nude and beige nude mice. *Clin. Exp. Metastasis* **5**(2), 135–146.

Naito, S., von Eschenbach, A. C., and Fidler, I. J. (1987b). Different growth pattern and biologic behavior of human renal cell carcinoma implanted into different organs of nude mice. *J. Natl. Cancer Inst.* **78**(2), 377–385.

Nelson, A. L., Algon, S. A., Munasinghe, J., Graves, O., Goumnerova, L., Burstein, D., Pomeroy, S. L., and Kim, J. Y. (2003). Magnetic resonance imaging of patched heterozygous and xenografted mouse brain tumors. *J. Neurooncol.* **62**(3), 259–267.

Neuwelt, E. A., Specht, H. D., and Hill, S. A. (1986). Permeability of human brain tumor to 99mTc-gluco-heptonate and 99mTc-albumin. Implications for monoclonal antibody therapy. *J. Neurosurg.* **65**(2), 194–198.

Nielsen, K. V., Madsen, M. W., and Briand, P. (1994). *In vitro* karyotype evolution and cytogenetic instability in the non-tumorigenic human breast epithelial cell line HMT-3522. *Cancer Genet. Cytogenet.* **78**(2), 189–199.

Orsulic, S., Li, Y., Soslow, R. A., Vitale-Cross, L. A., Gutkind, J. S., and Varmus, H. E. (2002). Induction of ovarian cancer by defined multiple genetic changes in a mouse model system. *Cancer Cell* **1**(1), 53–62.

Paulus, M. J., Gleason, S. S., Kennel, S. J., Hunsicker, P. R., and Johnson, D. K. (2000). High resolution X-ray computed tomography: An emerging tool for small animal cancer research. *Neoplasia* **2**(1–2), 62–70.

Petrovsky, A., Schellenberger, E., Josephson, L., Weissleder, R., and Bogdanov, A., Jr. (2003). Near-infrared fluorescent imaging of tumor apoptosis. *Cancer Res.* **63**(8), 1936–1942.

Ray, P., Wu, A. M., and Gambhir, S. S. (2003). Optical bioluminescence and positron emission tomography imaging of a novel fusion reporter gene in tumor xenografts of living mice. *Cancer Res.* **63**(6), 1160–1165.

Rudin, M., and Weissleder, R. (2003). Molecular imaging in drug discovery and development. *Nat. Rev. Drug Discov.* **2**(2), 123–131.

Rygaard, J., and Povlsen, C. O. (1969). Heterotransplantation of a human malignant tumour to "Nude" mice. *Acta Pathol. Microbiol. Scand.* **77**(4), 758–760.

Scholz, C. C., Berger, D. P., Winterhalter, B. R., Henss, H., and Fiebig, H. H. (1990). Correlation of drug response in patients and in the clonogenic assay with solid human tumour xenografts. *Eur. J. Cancer* **26**(8), 901–905.

Schwaller, J., Frantsve, J., Aster, J., Williams, I. R., Tomasson, M. H., Ross, T. S., Peeters, P., Van Rompaey, L., Van Etten, R. A., Ilaria, R., Jr., Marynen, P., and Gilliland, D. G. (1998). Transformation of hematopoietic cell lines to growth-factor independence and induction of a fatal myelo- and lymphoproliferative disease in mice by retrovirally transduced TEL/JAK2 fusion genes. *EMBO J.* **17**(18), 5321–5333.

Shih, I. M., Torrance, C., Sokoll, L. J., Chan, D. W., Kinzler, K. W., and Vogelstein, B. (2000). Assessing tumors in living animals through measurement of urinary beta-human chorionic gonadotropin. *Nat. Med.* **6**(6), 711–714.

Shimosato, Y., Kameya, T., Nagai, K., Hirohashi, S., Koide, T., Hayashi, H., and Nomura, T. (1976). Transplantation of human tumors in nude mice. *J. Natl. Cancer Inst.* **56**(6), 1251–1260.

Shoemaker, R. H., Monks, A., Alley, M. C., Scudiero, D. A., Fine, D. L., McLemore, T. L., Abbott, B. J., Paull, K. D., Mayo, J. G., and Boyd, M. R. (1988). Development of human tumor cell line panels for use in disease-oriented drug screening. *Prog. Clin. Biol. Res.* **276**, 265–286.

Smith, A., van Haaften-Day, C., and Russell, P. (1989). Sequential cytogenetic studies in an ovarian cancer cell line. *Cancer Genet. Cytogenet.* **38**(1), 13–24.

Smith-Jones, P. M., Solit, D. B., Akhurst, T., Afroze, F., Rosen, N., and Larson, S. M. (2004). Imaging the pharmacodynamics of HER2 degradation in response to Hsp90 inhibitors. *Nat. Biotechnol.* **22**(6), 701–706.

Steel, G. G., Courtenay, V. D., and Peckham, M. J. (1983). The response to chemotherapy of a variety of human tumour xenografts. *Br. J. Cancer* **47**(1), 1–13.

Stephenson, R. A., Dinney, C. P., Gohji, K., Ordonez, N. G., Killion, J. J., and Fidler, I. J. (1992). Metastatic model for human prostate cancer using orthotopic implantation in nude mice. *J. Natl. Cancer Inst.* **84**(12), 951–957.

Stock, C. C., Clarke, D. A., Philips, F. S., and Barclay, R. K. (1960a). Cancer chemotherapy screening data. V. Sarcoma 180 screening data. *Cancer Res.* **20**(3, Pt. 2), 1–192.

Stock, C. C., Clarke, D. A., Philips, F. S., Barclay, R. K., and Myron, S. A. (1960b). Sarcoma 180 screening data. *Cancer Res.* **20**(5, Pt. 2), 193–381.

Stolfi, R. L., Stolfi, L. M., Sawyer, R. C., and Martin, D. S. (1988). Chemotherapeutic evaluation using clinical criteria in spontaneous, autochthonous murine breast tumors. *J. Natl. Cancer Inst.* **80**(1), 52–55.

Suggitt, M., Swaine, D. J., Pettit, G. R., and Bibby, M. C. (2004). Characterization of the hollow fiber assay for the determination of microtubule disruption *in vivo*. *Clin. Cancer Res.* **10**(19), 6677–6685.

Taetle, R., Rosen, F., Abramson, I., Venditti, J., and Howell, S. (1987). Use of nude mouse xenografts as preclinical drug screens: *In vivo* activity of established chemotherapeutic agents against melanoma and ovarian carcinoma xenografts. *Cancer Treat. Rep.* **71**(3), 297–304.

Takimoto, C. H. (2001). Why drugs fail: Of mice and men revisited. *Clin. Cancer Res.* 7(2), 229–230.

Tashiro, T., Inaba, M., Kobayashi, T., Sakurai, Y., Maruo, K., Ohnishi, Y., Ueyama, Y., and Nomura, T. (1989). Responsiveness of human lung cancer/nude mouse to antitumor agents in a model using clinically equivalent doses. *Cancer Chemother. Pharmacol.* **24**(3), 187–192.

Vajkoczy, P., and Menger, M. D. (2004). Vascular microenvironment in gliomas. *Cancer Treat. Res.* **117**, 249–262.

Van Dyke, T., and Jacks, T. (2002). Cancer modeling in the modern era: Progress and challenges. *Cell* **108**(2), 135–144.

Venditti, J. M., Wesley, R. A., and Plowman, J. (1984). Current NCI preclinical antitumor screening *in vivo*: Results of tumor panel screening, 1976–1982, and future directions. *Adv. Pharmacol. Chemother.* **20**, 1–20.

Von Hoff, D. D. (1998). There are no bad anticancer agents, only bad clinical trial designs—twenty-first Richard and Hinda Rosenthal Foundation Award Lecture. *Clin. Cancer Res.* **4**(5), 1079–1086.

Voskoglou-Nomikos, T., Pater, J. L., and Seymour, L. (2003). Clinical predictive value of the *in vitro* cell line, human xenograft, and mouse allograft preclinical cancer models. *Clin. Cancer Res.* **9**(11), 4227–4239.

Wadman, M. (1998). DuPont opens up access to genetics tool. *Nature* **394**(6696), 819.

Waters, D. J., Janovitz, E. B., and Chan, T. C. (1995). Spontaneous metastasis of PC-3 cells in athymic mice after implantation in orthotopic or ectopic microenvironments. *Prostate* **26**(5), 227–234.

Weisberg, E., Boulton, C., Kelly, L. M., Manley, P., Fabbro, D., Meyer, T., Gilliland, D. G., and Griffin, J. D. (2002). Inhibition of mutant FLT3 receptors in leukemia cells by the small molecule tyrosine kinase inhibitor PKC412. *Cancer Cell* **1**(5), 433–443.

Weissleder, R. (2002). Scaling down imaging: Molecular mapping of cancer in mice. *Nat. Rev. Cancer* **2**(1), 11–18.

Winograd, B., Boven, E., Lobbezoo, M. W., and Pinedo, H. M. (1987). Human tumor xenografts in the nude mouse and their value as test models in anticancer drug development (review). *In Vivo* **1**(1), 1–13.

Yang, H., Berger, F., Tran, C., Gambhir, S. S., and Sawyers, C. L. (2003). MicroPET imaging of prostate cancer in LNCAP-SR39TK-GFP mouse xenografts. *Prostate* **55**(1), 39–47.

Yuan, F., Salehi, H. A., Boucher, Y., Vasthare, U. S., Tuma, R. F., and Jain, R. K. (1994). Vascular permeability and microcirculation of gliomas and mammary carcinomas transplanted in rat and mouse cranial windows. *Cancer Res.* **54**(17), 4564–4568.

Pharmacodynamic Biomarkers for Molecular Cancer Therapeutics

Debashis Sarker and Paul Workman

Signal Transduction and Molecular Pharmacology Team, Cancer Research UK Centre for Cancer Therapeutics, The Institute of Cancer Research, Haddow Laboratories, Sutton, Surrey SM2 5NG, United Kingdom

I. Introduction
II. Types of Biomarkers
III. The Pharmacological Audit Trail
IV. Methodological Issues
V. Rationale for Use of PD Markers to Facilitate Drug Development
A. PD Markers in Preclinical Drug Development
B. PD in Clinical Trials
VI. The Need for Multidisciplinary and Broad-Based Collaborative Research and Development
VII. Biomarker Methodology
A. Minimally Invasive Imaging
B. Invasive Molecular Endpoints
C. Immunohistochemistry
D. Gene Expression Microarrays and Proteomics
VIII. Examples of PD Biomarkers for Specific New Drug Classes
A. Imatinib Mesylate as a Paradigm for the Development of Molecular Therapeutics
B. EGFR Inhibitors
C. Inhibitors of the HSP90 Molecular Chaperone
D. Inhibitors of the PI3K-AKT-mTOR Pathway
IX. Combining Chemotherapy with Molecularly Targeted Agents
X. Conclusions and Future Perspective
References

Rational and efficient development of new molecular cancer therapeutics requires discovery, validation, and implementation of informative biomarkers. Measurement of molecular target status, pharmacokinetic (PK) parameters of drug exposure, and pharmacodynamic (PD) endpoints of drug effects on target, pathway, and downstream biological processes are extremely important. These can be linked to therapeutic effects in what we term a "pharmacological audit trail." Using biomarkers in preclinical drug discovery and development facilitates optimization of PK, PD, and therapeutic properties so that the best agent is selected for clinical evaluation. Applying biomarkers in early clinical trials helps identify the most appropriate patients; provides proof of concept for target modulation; helps test the underlying hypothesis; informs the rational selection of dose and schedule; aids decision making, including key go/no go questions; and may explain or predict clinical outcomes. Despite many successes such as trastuzumab and imatinib, exemplifing the value of targeting specific cancer defects, only 5% of oncology drugs that enter the clinic make it to marketing approval. Use of biomarkers should

Advances in CANCER RESEARCH

0065-230X/07 $35.00
DOI: 10.1016/S0065-230X(06)96008-4

reduce this high level of attrition and bring forward key decisions (e.g., "fail fast"), thereby reducing the spiraling costs of drug development and increasing the likelihood of getting innovative and active drugs to cancer patients. In this chapter, we focus primarily on PD endpoints that demonstrate target modulation, including both invasive molecular assays and functional imaging technology. We also discuss related clinical trial design issues. Implementation of biomarkers in trials remains disappointingly low and we emphasize the need for greater cooperation between various stakeholders to improve this.

I. INTRODUCTION

We are now entering a new era of anticancer drug development led primarily by an increased understanding of the genetics and cancer and the molecular basis of malignant progression (Hanahan and Weinberg, 2000). Of particular recent significance has been the sequencing of the human genome and its oncological successor, the Cancer Genome Project (Futreal *et al.*, 2001; Lander *et al.*, 2001; Venter *et al.*, 2001). Together with the information from decades of basic research into molecular biology and genetics of cancer, these important technical achievements have the potential to transform the management of cancer from empirically based treatment to a predictive, individualized model based on the molecular classification of a tumor and the selection of appropriate targeted therapy (Ramaswamy, 2004; Sawyers, 2003a; Workman, 2005a). This approach is outlined in Fig. 1.

The challenge ahead for those involved in cancer research is to discover in precise detail the series of molecular abnormalities that arise in the genomes of all types of cancer, to use that information to understand the process of malignancy, and then to develop more rational and effective strategies for diagnosis and treatment (Workman, 2005c; Workman and Kaye, 2002). The development of "targeted" agents in oncology has exploited our growing knowledge of cancer genes and oncogenic pathways to provide innovative cancer therapies which offer the potential not only for improved therapeutic efficacy, but also for less severe toxicity compared with the previous generation of cytotoxic agents (Workman, 2003b, 2005c). The regulatory approval of drugs like imatinib (Gleevec), trastuzumab (Herceptin), gefitinib (Iressa), erlotinib (Tarceva), bevacizumab (Avastin), cetuximab (Erbitux), sorafenib (Nexavar), and sunitinib (Sutent) have provided clinical validation for this molecularly targeted approach.

Despite technological developments and the high-profile successes listed above, drug discovery remains an expensive, slow, and high-risk enterprise (Kelloff and Sigman, 2005; Kola and Landis, 2004; Reichert, 2003; Workman, 2003a). The average cost of completing a successful drug development project is in the range of US$700–1700 million, with a typical timescale of at least 8–10 years from preclinical discovery research to regulatory approval. Data available for the period from 1990 to 2000 show that only

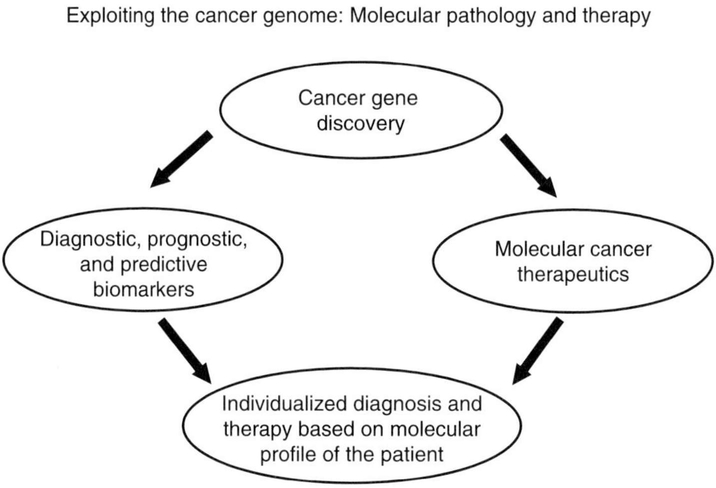

Fig. 1 How to exploit the cancer genome in the twenty-first century. The discovery and exploitation of cancer genes has the potential to usher in a new era of individualized diagnosis and therapy. The two critical steps in this process are: (1) the successful development of diagnostic, prognostic predictive and pharmacodynamic biomarkers; and (2) effective molecularly targeted therapeutics. The close integration of the discovery, development, and application of the molecular biomarkers and molecular therapeutics is key to future success. Adapted from P. Workman, *Eur. J. Cancer* 2002.

1 in 20 cancer drugs entering clinical trial gained regulatory approval: 70% failed at the phase II stage, 59% at phase III, and 70% at the registration phase (Kola and Landis, 2004). Reasons for the poor success for drugs across all disease areas have been documented (Kola and Landis, 2004). In 1991, poor pharmacokinetic (PK) and bioavailability properties were the main reasons for attrition. This problem has to a large extent been addressed by paying greater attention to optimizing these properties in preclinical development. Major causes for failure are now inadequate therapeutic activity (30%) and toxicity (30%) and hence these particular areas to focus attention on, while not neglecting properties such as PK and drug potency.

One of the most critical elements in improving success rates is the development of pharmacodynamic (PD) biomarkers of drug effects. Whereas PK endpoints are concerned with "what the body does to the drug," PD biomarkers focus on "what the drug does to the body" (Workman, 2002, 2003a). PD endpoints can be used during preclinical development through early clinical trials to answer critical questions. Incorporation of PD and PK endpoints allows drug development to proceed in a rational, hypothesis-testing fashion and enables sensible "go/no go" decisions to facilitate modern drug development. Drugs with the properties consistent with clinical effectiveness sufficient for regulatory approval can be prioritized. Those with liabilities can be identified early ("fail fast") and resources transferred to more deserving candidates.

Since the new generation of molecular therapeutics target the precise mechanisms that are causally responsible for the disease, early clinical trials with appropriately optimized drugs can be seen as providing a test of the hypothesis that a particular oncogenic mechanism is indeed driving the particular cancer. PD markers are critical to allow conclusions to be drawn about the reason for success or failure in the clinic. They also provide a basis for informed decisions on how and when to administer the drug.

In this chapter, we describe the rationale for using PD biomarkers for molecularly targeted anticancer agents, the methods involved, and the challenges ahead. In addition, examples of the use of important biomarkers for molecular cancer therapeutics will be described. Although PD endpoints are markers of drug effect, it is important to stress from the very outset that, as we will discuss, they do not necessarily correlate with or predict a therapeutic impact on the disease. Since, as we will show, PD endpoints should not be viewed in isolation, we will also, where appropriate, discuss biomarkers for patient selection and prognosis.

II. TYPES OF BIOMARKERS

With the growth of biomarkers in clinical trials, a lack of agreement has been apparent in the definition of the varying types of markers. Consensus in the use of terminology is important to help clarity of thinking. An expert working group was created by the US National Institutes of Health to propose terms, definitions, and a conceptual model (Biomarkers Definitions Working Group, 2001; Frank and Hargreaves, 2003). A summary of the recommended terminology is given in the following paragraph:

Biological marker or biomarker: A characteristic that is objectively measured and evaluated as an indicator of normal biological processes, pathogenic processes, or pharmacological responses to a therapeutic intervention.

Clinical endpoint: A characteristic or variable that reflects how a patient feels or functions, or how long a patient survives.

Surrogate endpoint: A biomarker intended to substitute for a clinical endpoint. A surrogate endpoint is expected to predict clinical benefit or harm, or lack of benefit or harm, based on epidemiologic, therapeutic, pathophysiologic, or other scientific evidence.

These descriptions have been expanded further to reflect recent thinking by scientists, regulators, and clinical trialists. The additional terms below encompass the use of biomarkers in early drug development, cohort selection and patient management, as well as in determining clinical benefit (Kelloff and Sigman, 2005; Ludwig and Weinstein, 2005).

Clinical correlates: These are endpoints that can be used for obtaining regulatory drug approvals. They are commonly used in clinical practice, and historically have demonstrated their relevance in that setting. Among the most widely used are tumor response rate, disease-free survival, and time to progression. These require measurement of treated cancer, commonly with anatomical imaging [e.g., computerized tomography (CT)]. Tumor regression is frequently not ideal for assessing the efficacy of molecularly targeted agents, since these often do not cause tumor shrinkage and are more likely to be "cytostatic" in nature.

Prognostic biomarkers: These are correlated with clinical outcome and can be divided into *biological progression markers* and *risk biomarkers*.

Biological progression markers are measures of tumor burden and are commonly circulating cellular proteins that are associated with tumor progression. Among the most commonly used of these "tumor markers" are CA-125 for ovarian cancer and prostate-specific antigen (PSA) for prostate cancer (Bubley *et al.*, 1999; Rustin, 2003; Rustin *et al.*, 2004).

Risk biomarkers are usually implicated in the mechanisms of disease causality or neoplastic progression, and are increasingly used in drug development to identify populations likely to be responsive to a given drug treatment. The foremost example of this is *ERBB2/HER2* gene amplification in 25% of patients with invasive breast cancer, which correlates with inferior patient survival (Slamon *et al.*, 1987). Trastuzumab is a humanized monoclonal antibody that binds to ERBB2 and inhibits the growth of ERB2-overexpressing cells, and has demonstrated significant clinical benefit in both the adjuvant and advanced disease settings (Piccart-Gebhart *et al.*, 2005; Slamon *et al.*, 2001). It is critical to emphasize that the therapeutic activity of trastuzumab would likely have been obscured if the agent was given to patients unselected for their ERBB2 status [as measured by either a standardized, semiquantitative immunohistochemistry (IHC) scoring system, or by detection of *HER2* gene amplification by fluorescence *in situ* hybridization] (Park *et al.*, 2004).

PD biomarkers: These are biomarkers which measure the effects of a drug or other intervention, and include molecular, cellular, histopathological, and imaging parameters. PD biomarkers are used to characterize and ideally to quantitate molecular and functional effects produced by a drug that may or may not correlate with biological and clinical effects.

The biological effects that are used as PD endpoints are often measures of altered activity or expression of a molecular target in response to a mechanism-based therapy. These PD biomarkers are frequently "proximal" to the effect of the agent, as in for example decreased phosphorylation of a protein substrate immediately downstream from a target kinase. Other endpoints, often cellular, histopathological, and imaging biomarkers that measure events occurring during neoplastic progression, can be used to measure the more "distal" effects of the drug that occur further downstream of its immediate molecular target. Examples include changes in proliferation using Ki67 expression, apoptosis using the TUNNEL assay, alterations in gene expression profiles, and functional or molecular imaging changes (Kelloff and Sigman, 2005). Both proximal and distal PD biomarker are predominately used as endpoints in preclinical and early phase clinical trials.

PD biomarkers have many potential uses in all phases of the drug development process, from demonstrating action on the target in the laboratory through preclinical studies to pivotal clinical trials (Hidalgo and Messersmith, 2004; Workman, 2003a). Potential applications of PD studies include:

–Providing proof of mechanism of action of a drug
–Selection of optimal dose and schedule of administration of the drug, in conjunction with factors such as PK and drug toxicity
–Gaining an understanding of response/resistance mechanisms
–Designing rational combination therapies
–Predicting outcome

These applications will be discussed further in the following sections.

III. THE PHARMACOLOGICAL AUDIT TRAIL

To allow proper evaluation of molecularly targeted agents, it is essential to determine not only critical PD endpoints, but also key PK measurements of drug exposure or metabolism. Used together, PK and PD endpoints facilitate the construction of what we have described as a "pharmacological audit trail" such that all of the key stages in drug development—from target status and drug administration through the biological effect to the clinical outcome—can be monitored, linked, and interpreted (Workman, 2002, 2003a). The audit trail also provides a basis for selecting appropriate patients, demonstrating proof of concept for the agent under investigation, together with the ability to help rational decision making (including go/no go questions) during preclinical and clinical drug development. In addition, an understanding of PK–PD relationships also underpins the selection of optimal drug dose and schedule. The critical questions are outlined in Fig. 2 and summarized in the following section:

Is the molecular target expressed or mutated and what is the activity of the pathway? Determination of the presence or status of the molecular target allows the most relevant models to be used in preclinical studies as well as the selection of the most appropriate patients to treat in clinical trials. In addition, the relationship between say target expression/mutation or pathway activity and response to the agent can be explored further.

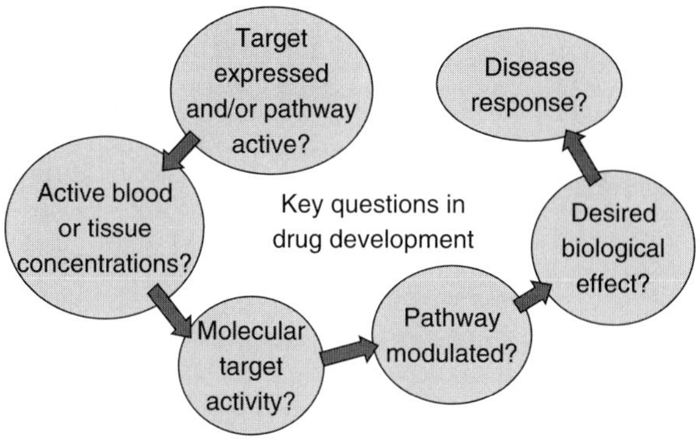

Fig. 2 The pharmacological audit trail for molecular cancer therapeutics. Use of the pharmacological audit trail provides a conceptual and practical framework that allows key questions in drug discovery and development to be addressed by using appropriate diagnostic, PK, PD, and response biomarkers. The audit trail provides a rational basis for assessing the risk of failure at any particular stage, with the likelihood of failure decreasing as the hierarchy of sequential questions are successfully answered. It also provides the basis for making informed decisions such as which patients to treat, what is the optimum dose and schedule, and whether to progress or terminate a drug development program. Adapted from P. Workman, *Eur. J. Cancer* 2002.

Is the drug achieving sufficient concentrations in plasma, blood, and tumor tissue? The measurement of absorption, distribution, metabolism, and excretion is crucial for the drug development process (van de Waterbeemd and Gifford, 2003). If active drug concentrations are not achieved then modifications in scheduling or formulation are required before proceeding further. Alternatively, it may be necessary to modify the chemical structure to produce an analogue to overcome problems in PK and metabolism. Active concentrations should be defined in preclinical models so that target levels or exposures can be defined for clinical studies.

Is there activity on the desired molecular target? This may involve, for example, demonstration of inhibition of a particular kinase or protease. As with PK requirements, it is important to use preclinical models to define a level of activity that is required to achieve the desired cellular or functional outcome. For example, it may be necessary to inhibit a kinase by say 80% for at least 24 h in order to achieve a measurable biological outcome. To emphasize, it is essential to define PK–PD relationships in preclinical models so that we know what to aim for in subsequent clinical studies.

Is there modulation of the biochemical pathway in which the target functions? It is important to determine whether the pathway in which the target operates is being altered and by how much. Modulation of some targets will have more impact than others, depending on multiple factors, including rate constants and feedback mechanisms. Perhaps just as important as "on-target" effects from both a therapeutic and toxicological viewpoint is the assessment of "off-target" modulation of alternative pathways. Techniques such as multiplex immunoassays, gene expression microarrays, and proteomics are potentially able to provide such information on a very broad scale (see later).

Is there achievement of the desired biological effect? The key aim here is to provide a readout of the expected biological outcome of the drug on downstream cellular properties such as cell cycle control, apoptosis, angiogenesis, invasion, or metastasis (Hanahan and Weinberg, 2000). It is possible that there is more than one cellular outcome, either through the intended mechanism or via a molecular side effect.

Do the above effects translate into a relevant clinical response? The audit trail is not complete until the drug-induced events described above can be linked to disease response and patient outcome. The most important outcome is usually patient survival, but this can only be determined in large phase III trials. In early phase trials, surrogates of clinical benefit should be used. Commonly used surrogates of benefit include radiological disease response, duration of stable disease, and assays of circulating tumor markers (e.g., as mentioned earlier, CA-125 for ovarian cancer and PSA for prostate cancer).

IV. METHODOLOGICAL ISSUES

It will be clear from the above discussion that PD markers have the potential to play a key role in the rational and mechanistically informed contemporary drug development process. However, there are a number of important methodological issues that should be discussed at this stage, relating in particular to the need for these assays to be rigorously validated and carefully implemented (Frank and Hargreaves, 2003). In their review of this key area, Hidalgo and Messersmith (2004) outlined three factors of particular importance.

Assay-related factors. Before a PD marker can be considered for clinical use, the assay should be appropriately validated. In particular, the workup should focus on both scientific validation, that is, the linkage of the PD endpoint to the particular pathway and mechanism of action, and how the endpoint relates to clinical outcome. In addition, technical validation is essential. Guidelines of Good Laboratory Practice (GLP) and Good Clinical Laboratory Practice (GCLP) should be followed. This is essential where PD biomarkers are used as primary, decision-making endpoints. Important parameters that reflect the performance of the method include sensitivity, specificity, precision, accuracy, and linearity (Stiles *et al.*, 2005; see also http://www.fda.gov/oc/gcp/guidance.html).

Marker- and sample-related factors. These relate to what is being measured and where. The intrinsic variability of the parameter to be measured in clinical samples needs to be known, and the assay to be employed should have minimal intra- and interobserver patient variability. In this respect, it is desirable to measure modulation of a marker quantitatively to determine the difference in the marker during treatment compared to baseline. Thus baseline sampling is critical. However, pre- and posttreatment tumor biopsies can be difficult to perform, as these are often uncomfortable for the patient and carry extra risks of hemorrhage and infection, raising ethical and logistical issues (Agrawal and Emanuel, 2003). As mentioned earlier, surrogate normal tissues can be used, for example peripheral blood mononuclear cells (PBMCs), skin, and buccal mucosa, but these may not give the same response as cancer cells. Monitoring responses in circulating tumor cells is of considerable interest, but this technology not yet fully validated on a broad scale (Cristofanilli *et al.*, 2004). Investigator training, data calibration, and adjudication procedures may be required to minimize variability (Kelloff and Sigman, 2005). Variability is particularly important for measurement of targets in tumor tissue given that this can be highly heterogeneous. For example, varying quantities of tumor cells, stroma, and endothelial or inflammatory cells may be present. Necrotic or fibrotic regions may also be found. Microdissection may be used. Other key issues relate to the optimal specimen required for the assay, quantity of sample, and the collection, processing, and preservation of samples. It is important to establish that the stability of the marker in the sample can be maintained during the preparation, storage, and performance of the test.

Statistics and clinical trial issues. An important issue that is often overlooked with early PD biomarker studies is the consideration of quantitation, statistics, and trial size. To date, the majority of studies using PD assays have paid little regard to performing statistically rigorous analyses. This may be due to lack of critical information in the development stages of the assay with regard to variability of the assay and the target, making statistical estimations difficult. PD studies are often inadequately powered in clinical trials, most likely because biomarkers measurements are usually performed as secondary objectives. Lack of statistics is almost inevitably the case in phase I studies where the number of patients treated at a particular dose level will often be quite small. If possible it is important to obtain data to indicate the nature of the dose–effect relationship. This generally requires obtaining PD data during the dose escalation.

A further complication is the practical difficulty of performing repeated PD samples, in particular tumor biopsies, as discussed earlier. A balance must be struck between the ideal scientific dataset required for the complete pharmacological audit trail, the minimum information needed for intelligent decision making, and the logistic and practical aspects that allow timely completion of the main clinical study and the achievement of its primary objectives. New clinical designs are now being implemented with molecular cancer therapeutics (see later). With an agent that is expected to offer clinical benefit without significant tumor shrinkage, an expanded phase II trial of 100–300 patients may be appropriate, depending on the exact question being asked by the trial and the required statistical power. This would allow further opportunity for refinement, validation, and assessment of PD assays based on increasing clinical experience, as well increasing the statistical power of the biomarker data.

V. RATIONALE FOR USE OF PD MARKERS TO FACILITATE DRUG DEVELOPMENT

Early clinical trials are increasingly seen to represent a continuation of the preclinical and clinical discovery and development process. It is important to consider the role of PD markers in both preclinical drug development and also as part of the "experimental medicine" component of early clinical trials.

A. PD Markers in Preclinical Drug Development

In considering the role of PD endpoints in the preclinical phase, it is useful to consider the stages involved in the contemporary development of molecular cancer therapeutics. Preclinical drug discovery is now heavily focused on the identification of a molecular target involved in the pathophysiology of a cancer, followed by the development of an antibody or small molecule inhibitor. Starting points for the latter are often obtained by robotic high-throughput screening of diverse compound collections. A hit is a compound that shows activity against the target in such a screen, conducted using a particular assay format under a defined set of conditions (Garrett *et al.*, 2003). Assays are often referred to as either "biochemical" or "cell-based." Biochemical screens usually measure the ability of recombinant proteins, or proteins purified from cells or tissues, to perform a desired biochemical activity. Cell-based screens use assays to measure a particular cellular activity or phenotype. Alongside high-throughput screening (HTS), virtual "*in silico*" screening methods, together with rapid screening (e.g., of fragments) by nuclear magnetic resonance and high-throughput crystallography can be used together to increase the chance of finding promising hits (Blundell *et al.*, 2002).

The initial hit compounds are usually of low potency and regarded as chemical starting points for optimization that will eventually produce a development candidate for clinical testing. This process involves close collaboration between medicinal chemists and bioscientists with multiple iterative rounds of chemical synthesis and progressive refinement of the chemical structure based on biological feedback. This measurement of the biological properties of compounds is achieved by a series of hierarchically arranged tests, information from which is then used in guiding improvements to the structure of the compound so that a molecule is identified that, subject to an appropriate dosing formulation being achieved and an acceptable toxicological profile demonstrated, can be put forward to enter phase I clinical trials.

The series of hierarchical tests is often referred to as a test cascade (Garrett *et al.*, 2003). Along with drug metabolism and PK studies, critical among these tests are cellular assays to show modulation of the intended molecular target and achievement of the desired biological effect (e.g., inhibition of proliferation, induction of cell-cycle arrest, apoptosis, and so on), as described earlier in the discussion of the pharmacological audit trail. Cellular assays are usually performed in molecularly characterized human tumor cell line panels (Lu *et al.*, 2000; Scherf *et al.*, 2000) or in isogenic pairs of cell lines designed to be with or without a particular molecular feature (Sharp *et al.*, 2000; Torrance *et al.*, 2001). Methods such as Western blotting or enzyme-linked immunosorbant assay (ELISA) provide useful readouts for cellular inhibition of a molecular target. For example, measurement of ERK 1/2 phosphorylation by Western blotting using appropriate phosphospecific antibodies was used with a new MEK 1/2 inhibitor (Herrera and Sebolt-Leopold, 2002). RB phosphorylation can be used to measure the effects of cyclin-dependent kinase (CDK) inhibitors (Whittaker *et al.*, 2004). Importantly, total substrate levels as well as the level of phosphorylation of a protein on a particular amino acid site should always be measured to monitor specific signal loss. Inhibition of the HSP90 molecular chaperone can be monitored by measuring depletion of client proteins such as C-RAF and CDK4, together with mechanism-based up-regulation of HSP70 (Hostein *et al.*, 2001).

The use of microarray technology to profile the expression thousands of genes at the level of messenger RNA (mRNA) is having a major impact, not only on understanding the process of oncogenesis and in target identification and validation, but also on discovering novel or additional biological readouts of drug activity (Clarke *et al.*, 2000, 2001, 2004). Thus both on- and off-target effects can be defined.

It is important to mention that whatever molecular assay is employed at this *in vitro* stage of the drug discovery process, this will often also be used during *in vivo* testing in animal models as well as in clinical trials to confirm biological effects of the compound at these later stages of the drug development process. Therefore, the development and use of a well-validated, easily reproducible, ideally quantitative assay is crucial. For example, the use of RB phosphorylation and other endpoints of CDK inhibition was validated in a human colon cancer xenograft model (Raynaud *et al.*, 2005). Similarly, the molecular signature of HSP90 inhibition was validated in a human ovarian tumor xenograft model as part of the establishment of PK–PD relationships prior to their use in phase I clinical trials (Banerji *et al.*, 2005a,b).

It is particularly important to demonstrate target-related biological effects in animal models *in vivo*. Human tumor xenografts continue to be the mainstay for the evaluation of antitumor effects and provide an important measure of tumor versus normal tissue selectivity in a metabolically intact organism (Peterson and Houghton, 2004; Sausville and Burger, 2006). However, human xenografts have their limitations, most notably due to their somewhat artificial nature and their limited predictiveness for clinical activity. Furthermore, there is the potential problem of human tumor cells growing in the context of the mouse host stroma, and also the lack of a complete immune response. In addition, human tumor xenografts are frequently grown subcutaneously, although orthotopic sites can also be used.

Transgenic or knockout mouse models have some advantages (Van Dyke and Jacks, 2002) and offer the possibility of evaluating molecularly targeted therapies against spontaneous tumors arising from specific gene defects and arising within the host animal itself rather than being transplanted (Becher and Holland, 2006; Jonkers and Berns, 2002). A lively debate continues to be carried out regarding the suitability of human tumor xenograft versus transgenic models and particularly their relevance to, and predictiveness for, the clinic. Human tumor xenografts are often criticized because it is felt that they have not predicted for activity of cytotoxic agents in a given tumor type. However, the ability of both xenograft and transgenic models to predict for the activity of targeted molecular cancer therapeutics in the clinic is now being evaluated prospectively. With particular respect to the development of biomarkers and PK–PD relationships, it is likely that both types of model will be useful. The important point is that each model should be fully characterized with respect to its molecular pathology, especially the key drivers of its malignant properties, and these should be related directly to the molecular therapeutics in question. Models used in drug discovery need to be robust and to operate over a sensible timescale. Animal welfare issues should be considered as part of model selection, and guidelines on this have been published (Workman *et al.*, 1998). The application of such guidelines is consistent with good scientific practice. Use of PK–PD endpoints can decrease the number that need to be used for therapy experiments.

It is important to mention that the development of PD endpoints in this preclinical stage should consider not only tumor samples, but also surrogate tissues such as PBMCs, skin, and buccal mucosa. For example, with the HSP90 inhibitor 17-allylamino-17-demethoxygeldanamycin (17-AAG), mouse PBMCs were validated alongside human tumor xenograft material as appropriate surrogate normal tissue for measuring depletion of HSP90 client proteins C-RAF and CDK4 and the induction of HSP70; these assays were then utilized in the subsequent clinical trials (Banerji *et al.*, 2005a,b). Preclinically in both surrogate and tumor tissue it is crucial to determine whether it is possible to observe the desired activity on the molecular target, subsequent modulation of the desired biochemical pathway and the biological effect (see Section III). Measurement of biological effects can be done using assays of proliferation, apoptosis, cell cycle, angiogenesis, invasion, and metastasis. The model systems discussed above are especially important to relate the eventual clinical experience to preclinical data, recognizing that there may be potential differences. Provision of target PK–PD parameters to aim for in the clinic is very important.

B. PD in Clinical Trials

The development of new molecularly targeted agents has created a challenge for the oncology community to provide an appropriate trial framework by which these drugs can be properly assessed. The use of appropriate PD endpoints is critical to this process. It is useful to review briefly the traditional approach and then to consider the modifications that are needed.

1. TRADITIONAL CLINICAL TRIAL DESIGN FOR CYTOTOXICS

The clinical development of traditional cytotoxic agents has followed a time-honored approach of conventional phase I, II, and III trials. The phase I trial of a cytotoxic agent has traditionally been a dose-finding study to determine maximum tolerated dose (MTD) by identification of dose-limiting toxicities (DLTs), and subsequently establishing a recommended dose for further trials. PK behavior is usually determined in these trials.

Phase II studies were typically used as a screening test for subsequent phase III studies, and these focus on a particular tumor type. The main endpoint of traditional phase II trials is usually defined in terms of response rate. Evaluation of response to therapy is commonly determined by radiological assessment of tumor size, most commonly using the Response Evaluation

Criteria in Solid Tumors (RECIST) guidelines (Therasse *et al.*, 2000). Other methods include the use of validated circulating tumor markers, for example CA-125 and PSA (Park *et al.*, 2004). Classical phase III studies were designed to determine evidence of superiority of the new therapy (usually as improvement in overall survival or quality of life) compared with the current gold standard treatment. It has been uncommon for molecular biomarkers to be measured as part of the development of cytotoxic agents in these conventional trials. However, other conventional surrogate endpoints were used, for example bone marrow suppression was often a satisfactory endpoint for decision making about recommended dose in phase I trials.

2. TRIAL DESIGN FOR MOLECULAR CANCER THERAPEUTICS

When designing trials for molecular therapeutics, there are a number of key differences compared to the traditional design described above.

For phase I studies, the paradigm described above is challenged in a number of ways. In phase I studies of cytotoxics, toxicities against rapidly dividing normal tissues serve as biological surrogates and often define the MTD (Eckhardt *et al.*, 2003). For molecularly targeted therapeutics, these clinical effects on normal tissues cannot be reliably used as surrogates, thus making the assessment of toxicological endpoints more complex. In addition, the standard design of increasing drug dose to toxicity may be unnecessary for optimal drug effect, and the use of MTD as a surrogate of effective dose may be inappropriate in the phase I setting (Parulekar and Eisenhauer, 2004). This has led to use of the term "optimum biological dose" to define a dose, often significantly below the MTD, which can be assessed by measurement of PD markers of biological activity of drug in tumor and surrogate healthy tissues. Another key difference between trials with molecularly targeted agents versus cytotoxics relates to optimal scheduling of drug. As indicated earlier, molecularly targeted agents are often cytostatic rather than cytotoxic in nature. Hence they are preferably given as a continuous (probably oral) administration rather than as intermittent pulsed cycles. This also takes into consideration the fact that there is likely to be less need for recovery of normal proliferating tissues (especially bone marrow) with targeted therapies compared to cytotoxics.

However, although the determination of optimum biological dose provides a logical framework within which to conduct phase I clinical studies of targeted agents, the experience to date of conducting these trials has been chastening. In a survey by Parulekar and Eisenhauer (2004), 57 phase I publications concerning 31 novel targeted agents were reviewed with regard to patient population, starting dose, methods of dose escalation, determination of recommended phase II dose (RPIID), and inclusion of correlative studies. Of the 57 studies, 36 used toxicity and 7 used PK data to halt dose escalation. Nontraditional endpoints, such as molecular effects on surrogate tissues, were rarely incorporated into trial design, and in only two trials was a targeted endpoint or surrogate tissue biomarker used as to determine RPIID. This study and others demonstrate the challenges in using nontraditional endpoints in the design of phase I trial (Gelmon *et al.*, 1999; Korn, 2004). One of the most critical issues is the lack of appropriately validated PD biomarkers. This may relate to difficulties defining the desired target effect, and to practical issues in measuring these effects once they have been defined, for example, because of a lack of reliable assays or the problems in obtaining the required tumor specimens (Korn *et al.*, 2001). Quite often there is simply a failure to plan ahead and implement validation of biomarkers so that they are available for use in the clinical development phase.

In addition, the fact remains that the new generation of targeted therapies can still cause significant toxicity. In certain circumstances, this is only apparent outside the time window of phase I DLT definitions. An example is the hypersensitivity pneumonitis seen with gefitinib (Konishi *et al.*, 2005). For this reason, and the difficulty of obtaining validated PD biomarkers due to lack of appropriate preclinical data, some investigators still feel that it is not prudent to base definition of MTD on a biomarker (Adjei and Hidalgo, 2005). This does not mean, however, that obtaining biomarker data is not of scientific or practical value.

So what is the optimal current way to incorporate PD endpoints into phase I studies? Critical to the process is the development of hypothesis-based trials (Adjei and Hidalgo, 2005). One approach, which we favor, is to conduct a two-stage approach of "dose estimation" based on effects on normal tissues, followed by "dose confirmation" based on effects on tumor tissues (Hidalgo, 2004). PD assays that have been validated preclinically and are readily feasible should be used in conjunction with relevant PK analyses to construct meaningful PD–PK relationships. In practice, therefore, once surrogate tissue data demonstrate inhibition of target, it is important to determine whether the surrogate tissue results correlate with findings in tumor tissue. This is probably best conducted at or near the RPIID based on traditional toxicity endpoints, often in an expanded cohort of patients (e.g., 10–12 patients are often used). Functional imaging can be incorporated if indicated. In this way, multiple endpoints can be included and an RPIID can be incorporated based on both MTD and optimum biological dose. However, tolerability should probably still be used as the primary determinant (Eckhardt *et al.*, 2003).

In response to the shift toward earlier use of correlative laboratory studies, and the need to reduce the time and resources expended during early drug development on candidates that are unlikely to succeed, regulatory authorities have streamlined requirements for exploratory trials (Collins, 2005). Recent draft guidance from the US Food and Drug Administration

(FDA) has considered exploratory Investigational New Drug (IND) studies as an appropriate tool to distinguish promising drug candidates from those less likely to succeed (http://www.fda.gov/cder/guidance/6384dft.htm#_Toc100638018). These are "early phase I exploratory approaches that are consistent with regulatory requirements, but that will enable sponsors to move ahead more efficiently with the development of promising candidate products while maintaining needed human subject protections." For example, exploratory IND studies can help sponsors gain an understanding of the relationship between a specific mechanism of action and the treatment of a disease, or select the most promising lead product from a group of candidates designed to interact with a particular therapeutic target in humans, or explore a product's biodistribution characteristics using various imaging technologies.

For phase II studies, one of the key issues is the question of tumor response. We have already discussed the fact that novel molecular cancer therapeutics are more likely to act in a cytostatic manner as a result of mechanism-based cell-cycle arrest or the induction of generally modest increases in apoptosis. They will not produce several logs of cell kill as was seen with alkylating agents or radiation in responsive cancers. As a result, these agents may be active on prolonged administration without causing significant tumor shrinkage, creating difficulty in assessment of these agents by the traditional phase II endpoint of radiological response. The challenge is therefore to produce innovative designs and endpoints for phase II studies, as argued by Ratain and Eckhardt (2004). One possible solution for such agents is to use progression-free survival as an alternative endpoint and to carry out the trials with crossover or randomized discontinuation designs and involving a range of doses including placebo. The randomized discontinuation design is increasingly being used, and was pivotal in the development of the multitargeted kinase inhibitor sorafenib. Using this design, patients with renal cell carcinoma showed significant disease-stabilizing activity compared to placebo and led to the subsequent approval of this agent (Ratain *et al.*, 2006). However, despite the innovative trial design, PD biomarker data were not reported, which has led to uncertainly as to which kinase or kinases are the key targets for the drug. When biomarker studies are included in randomized phase II studies, the presence of a control group which is also sampled for the biomarkers could be helpful in further defining whether effects observed in the treated patients are truly related to the drug, and also in determining whether the presence of a biological change is related to a beneficial effect in the patient (Eckhardt *et al.*, 2003).

The use of appropriate PD endpoints is one of the main ways that negative phase III studies can be avoided. These expensive, usually multinational studies have blighted the oncology literature in recent years, with a range of targeted therapies producing a series of high-profile failures, including matrix metalloproteinase inhibitors, epidermal growth factor receptor (EGFR) inhibitors, and farnesyltransferase inhibitors. The rapid movement from phase I to phase III trials without any knowledge of whether the drug was hitting the target, and in some cases (e.g., gefitinib) uncertainty as to optimal drug dosage, has proved a chastening experience. Therefore, to minimize the risk of negative phase III trials of novel agents, the use of properly conducted phase II studies with informative biomarkers is crucial.

As mentioned earlier in this chapter, with an agent that is expected to be clinically effective without causing tumor regression, an expanded phase II trial of 100–300 patients can be very valuable. In addition, to showing clinical benefit with improved statistical power, it may be possible to use the data from the larger number of patients to identify a subgroup of patients who are especially sensitive to the agent based on the molecular characteristics of the tumors, for example patients with mutations in the *EGFR* gene predicting response to gefitinib (Lynch *et al.*, 2004).

VI. THE NEED FOR MULTIDISCIPLINARY AND BROAD-BASED COLLABORATIVE RESEARCH AND DEVELOPMENT

The successful development of PD biomarkers for molecularly targeted agents requires a true multidisciplinary approach, with effective collaborations between industry, academia, and regulatory authorities. The reason for this is set out in Section I of this chapter: the staggering cost of and length of time involved bringing a new drug to market and the very high failure rates for oncology drugs (Kola and Landis, 2004). Both FDA and industry have cited lack of clearly defined and measurable endpoints as major factors in the rising costs and late stage failures in drug development (Lesko and Woodcock, 2004). In the US this has prompted the fostering of collaborative efforts between the FDA, the National Cancer Institute (NCI), academic and pharmaceutical industry scientists, together with representatives from the American Society of Clinical Oncology (ASCO) and the American Association of Cancer Research (AACR), to develop strategies and guidance on endpoints in specific clinical settings (Kelloff *et al.*, 2004). One of the key ways that collaborative efforts can achieve success is in the facilitation of access to both novel technologies (e.g., genomic, proteomic, and functional imaging) and clinical samples (e.g., serum and tumor biopsies). Current examples of collaborative projects of note include:

–The NCI's Specialized Programs of Research Excellence (SPORE), which encourage multidisciplinary teams to address biomarker identification and validation (http://spores.nci.nih.gov/)

–The NCI's Cancer Bioinformatics Information Grid (caBIG) project, which has created a bioinformatics platform to aid the collection and sharing of key data elements across cancer research institutes (https://cabig.nci.nih.gov/)
–The UK National Translational Cancer Research Network (NTRAC; http://www.ntrac.org.uk/). This translational research initiative has been extended recently with increased funding from the government Department of Health and the charity Cancer Research UK to support a network of Experimental Cancer Medicine Centres and operates under the auspices of the UK National Cancer Research Institute (www.ncri.org.uk)
–The International Genomics Consortium focuses on standardizing sample collection and genomic array analyses for cancer (http://www.intgen.org/)

VII. BIOMARKER METHODOLOGY

A. Minimally Invasive Imaging

A range of invasive and relatively noninvasive techniques are available to provide PD endpoints. However, because of the logistical and ethical issues associated with invasive measurements such as tumor biopsies, there is increasingly a need for minimally invasive techniques that genuinely add value and aid decision making rather than being purely "decorative" in nature (Collins, 2003). Minimally invasive imaging is being used increasingly as a tool for PD assessment. These methods will therefore be discussed first.

For an in-depth analysis of the status and potential of minimally invasive technologies in hypothesis-testing early clinical trials of molecular cancer therapeutics, the reader is directed to a recent review (Workman *et al.*, 2006a). This outlines the findings of the Pharmacodynamic/Pharmacokinetic Technologies Advisory Committee (PTAC) of Cancer Research UK, which was established to review from a PK–PD perspective all applications to Cancer Research UK for phase I and II trials of new cancer drugs. They noted that submissions for phase I trials of new cancer drugs in the United Kingdom often lack detailed information about PK and/or PD endpoints, which can lead to suboptimal information being obtained in those trials or to delays in starting the trials while PK–PD methods are developed and validated. The development of minimally invasive PK–PD technologies provides potential logistic and ethical advantages over more invasive technologies. PTAC recommended that particular priority should be given to assays that would provide relatively generic readouts for important oncogenic pathways and biological effects. They argued that this would be more cost-effective than developing methods for individual molecular targets.

The ideal, of course, would be to have both so that the audit trail can be more complete.

1. POSITRON EMISSION TOMOGRAPHY

Positron emission tomography (PET) measures the three-dimensional distribution of a positron-labeled compound within the living body, dynamically and in a surgically noninvasive manner (Workman *et al.*, 2006a). The radioisotope decays after injection and emits a positron which combines with an electron resulting in the two particles being annihilated. This produces two 511-keV photons emitted 180° apart, and can be detected within a short time. Mathematical reconstruction methods, corrected for photon attenuation and scatter, can estimate the location and quantity of positron-emitting radionuclides within an object (Kelloff *et al.*, 2005). The use of PET in oncology, in particular with 2-[^{18}F]-fluoro-2-deoxyglucose (FDG), has increased dramatically in the last few years, primarily as a staging tool and for assessing response to therapy. One of the key demonstrations of the importance of PET in oncology was the use of FDG-PET in monitoring treatment of gastrointestinal stromal tumor (GIST) with the c-KIT inhibitor imatinib (see in a later section). These advances have provided the basis for suggesting that PET has a key role in accelerating the drug development process, for example, by providing the potential to support go/no go decisions in phase II trials, or allowing dosage adjustments or early identification of responders.

PD studies that can be performed by PET may be grouped into those employing generic or specific biological endpoints (Aboagye and Price, 2003; Kelloff *et al.*, 2005; Workman *et al.*, 2006a). Generic endpoints can measure the following changes:

- Cellular proliferation with ^{11}C-thymidine and [^{18}F]-fluorothymidine (FLT) (Eary *et al.*, 1999; Shields *et al.*, 1998)
- Glucose utilization with [^{18}F]-FDG (Findlay *et al.*, 1996)
- Tissue perfusion with [^{15}O]-H_2O (Wilson *et al.*, 1992)
- Blood volume with [^{15}O]-CO (Beaney *et al.*, 1984)
- Assessment of apoptosis with [^{124}I]-annexin V (Collingridge *et al.*, 2003; Keen *et al.*, 2005)

These methods will be used increasingly to assess the effects of novel molecularly targeted therapies in the future. Specific PET endpoints are also being validated to provide proof of principle for the proposed mechanism of action of novel therapeutics. These include:

- Vascular endothelial growth factor (VEGF)/VEGF receptor expression with [^{124}I]-labeled antibodies and peptides (Collingridge *et al.*, 2002)

- Measuring the degradation of ERBB2 receptors with a [^{68}Ga]-labeled anti-ERBB2 antibody fragment after treatment with HSP90 inhibitors (Smith-Jones *et al.*, 2004)

A number of methodological issues are worthy of discussion with regard to PET (Workman *et al.*, 2006a).

- PET images have low anatomical resolution and so coregistration of PET data with computed tomography (CT) and magnetic resonance imaging (MRI) is often performed; the development of integrated PET-CT systems may reduce the extent of this problem.
- There can be lack of chemical resolution with PET, making interpretation difficult especially for extensively metabolized drugs.
- PET isotopes may have short physical half-lifes, for example for ^{15}O this value is 2 min and this can limit the type of biological information available. Usually for patient comfort imaging times should be limited to 1–3 h.
- The spatial resolution attainable in PET is limited to 2–3 mm on current clinical scanners.
- The cost of PET scanning is high.
- Guidelines for standardizing the methodology for PET data collection and analysis are required, and standardization of disparate scanner technologies is needed. To date, the EORTC Functional Imaging Group has produced a set of guidelines for FDG response assessment, but there have been few other validated guidelines available (Young *et al.*, 1999).

In summary, PET is a minimally invasive technique that, although possessing some limitations, has enormous potential in drug development to measure molecular target expression and as a PD tool to assess and modulation of a particular molecular target, pathway, or biological process.

2. MAGNETIC RESONANCE SPECTROSCOPY

Magnetic resonance spectroscopy (MRS) has the ability to determine concentrations of endogenous compounds, tracers, or drugs and their metabolites in a way that is minimally invasive. Data are produced and visualized in the form of a spectrum in which peaks correspond to different chemicals. Endogenous metabolites that can be measured as PD bioamarkers include phosphomonoesters (PMEs), phosphocholine, or inorganic phosphate using [^{31}P]-MRS, or lactate, choline, or inositol compounds using [^{1}H]-MRS (Workman *et al.*, 2006a). Clinical [^{31}P]-MRS studies have been carried out in a number of tumor types. For example, breast cancer patients have shown a reduction in the PME peak that correlate with response to treatment and this may be used as a marker for changes in proliferation (Leach *et al.*, 1998). [^{1}H]-MRS has also been used recently to investigate response to therapy; for example, a fall in total choline (tCho) has been shown to correlate with changes in vascular permeability and response to temozolomide in glioma (Murphy *et al.*, 2004). In addition, MRS can provide measures of specific molecular processes *in vivo*. For example, an increase in 1,6-bisphosphate is a marker of apoptosis in cell systems, rising early in the apoptotic process and probably reflecting activation of poly (ADP-ribose) polymerase (Ronen *et al.*, 1999). Treatment of human breast cancer cells with the protoype PI3 kinase (PI3K) inhibitors LY294002

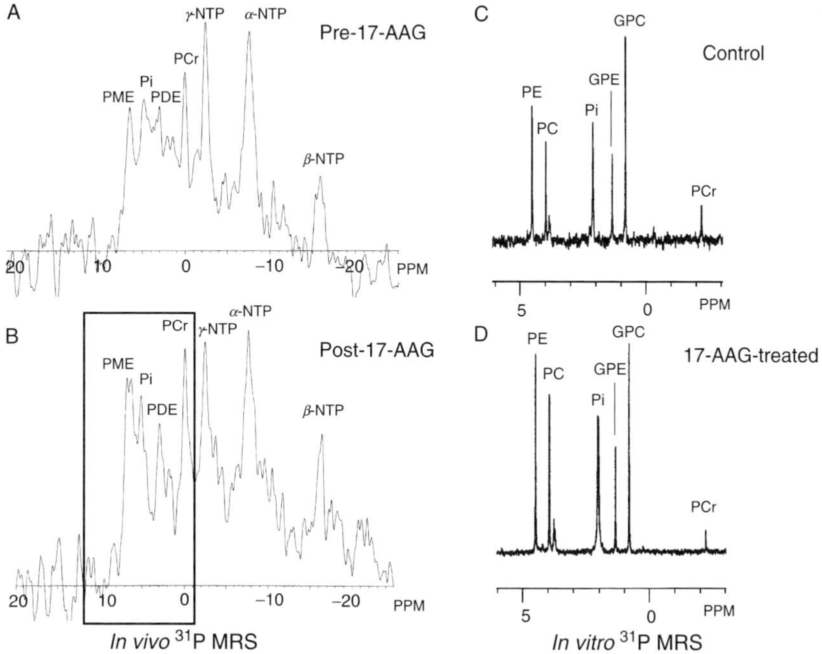

Fig. 3 Magnetic resonance spectroscopy changes in response to treatment with the HSP90 inhibitor 17-AAG. Panels (A) and (B): *In vivo* ^{31}P MR spectra of HT29 human colon tumor xenografts (A) before and (B) after 17-AAG treatment. Panels (C) and (D): Expanded *in vitro* ^{31}P MR spectra of extracts from (C) a control and (D) a 17-AAG-treated HT29 tumor. The box in panel B indicates the region of the *in vivo* spectrum corresponding to the *in vitro* spectra shown in (C) and (D). Peak assigned as: phosphomonoester (PME), comprising mainly of phosphocholine (PC) and phosphoethanolamine (PE); inorganic phosphate (Pi); phosphodiester (PDE), comprising glycerophosphocholine (GPC) and glycerophosphoethanolamine (GPE); phosphocreatine (PCr); α-, β-, γ-nucleoside triphosphates (α-, β-, γ-NTP). Figure courtesy of Dr. Yuen-Li Chung and colleagues, St. George's Hospital Medical School, London. For more details see Chung *et al.*, *J. Ntl. Cancer Inst.* 2003.

and wortmannin is associated with a significant fall in phosphocholine (Beloueche-Babari *et al.*, 2006). Inhibition of MEK 1/2 leads to a fall in phosphocholine that correlates with inhibition of phospho-ERK 1/2 (Beloueche-Babari *et al.*, 2005). Treatment with the HSP90 molecular chaperone inhibitor 17-AAG was shown to increase PME in human colon cell lines and xenografts (see Fig. 3) (Chung *et al.*, 2003). This method is currently being evaluated in ongoing phase I and II studies of 17-AAG.

Potential problems with MRS are associated with poor resolution of PME signal and assessment of PME changes being nonquantitative. In addition, the use of changes in PME and phosphodiesters (PDEs) as markers of specific pathway inhibition may be complicated by the possibility that effects on different signal transduction pathways may cause similar PME changes.

Recommendations for standardization of clinical MRS measurements have been published (Leach *et al.*, 1994).

3. DYNAMIC CONTRAST-ENHANCED MAGNETIC RESONANCE IMAGING

The use of MRI contrast agents which have magnetic properties that change image signal intensity provides a further way of assessing the functional characteristics of tumors. In particular, low-molecular-weight contrasts, such as gadolinium diethyltriaminepentaacetic acid (Gd-DTPA), are increasingly being used in the development of antivascular agents. They allow key changes in tumor microvasculature to be measured, such as transfer constant (K^{trans}), initial area under the gadolinium curve (IAUC), and leakage space (v_e) (Galbraith, 2003). Recent recommendations for the use of MRI in the assessment of novel antivascular and antiangiogenic agents have been published (Leach *et al.*, 2005).

Assessment of permeability and blood flow are developing favor as important PD endpoints, and has been incorporated into phase I trials of the antivascular agents combretastatin-A4 phosphate (Galbraith *et al.*, 2003), 5,6-dimethylxanthenone-4-acetic acid (Galbraith *et al.*, 2002), and ZD6126 (Evelhoch *et al.*, 2004). In these studies, a decrease in the kinetics of contrast agent uptake was interpreted in terms of reduction of tumor perfusion (Workman *et al.*, 2006a). In the phase I trials of combretastatin-A4 phosphate, the use of dynamic contrast-enhanced magnetic resonance imaging (DCE-MRI) contributed to decisions regarding phase II dose and schedule selection, in addition to the usual toxicity profile (Collins, 2003; Galbraith *et al.*, 2003). This was in part because extensive preclinical work had been done comparing the dose–response with DCE-MRI to that with a technique for measuring absolute blood flow in the same animal model (Maxwell *et al.*, 2002). In addition, DCE-MRI has been used as a PD marker of inhibition of VEGF-targeted therapies (Liu *et al.*, 2005; Morgan *et al.*, 2003). In the phase I trial of the human anti-VEGF antibody HuMV833, a median decrease of 44% in K_{fp} (first pass permeability) was seen. However, in this trial with a range of tumor types, the authors concluded that in those cohorts with three or fewer patients the heterogeneity of response is such that discrimination of one dose level from another becomes difficult (Jayson *et al.*, 2002). This important point perhaps suggests that in trials with an imaging endpoint, the use of single tumor types may have an advantage; in addition, larger cohorts, perhaps at two or three dose levels that are well tolerated, would provide robust and more statistically significant data (Galbraith, 2003). The practical implications of conducting such a study mean that multiple institutions would need to be involved; datasets from different machines would require standardization and a common imaging protocol, together with appropriate quality assurance (Workman *et al.*, 2006a).

B. Invasive Molecular Endpoints

A large range of methods are available for making molecular measurements in, for example, tissue biopsies. A full review of these is beyond the scope of the present review. For examples of potential methods, including their application to histone deacetylase inhibitors, please see the following references: Chung *et al.*, 2006; Tuomela *et al.*, 2005.

Western blotting is widely used but is only semiquantitative and difficult to validate to GCLP requirements (Cummings *et al.*, 2006). Considerable care should be taken in the selection of specific antibodies. ELISA methodology is more quantitative for specific protein estimation. Commercial equipment is now available for multiplex ELISA assays. One format that we have evaluated successfully is the Mesoscale Discovery (MSD) platform. This is based on sandwich immunoassays consisting of a capture antibody, labeled with an electrochemilumiscent compound. For mRNA analysis, RT-PCR provides reproducible and semiquantitative data.

C. Immunohistochemistry

IHC has become one of the most popular techniques for assessing biomarker expression because of its versatility and the fact that it could be used on paraffin-embedded as well as frozen tissue (Baselga, 2003). Many thousands of papers have been published which show an association between protein expression by IHC and clinical outcome or response to therapy. However, different laboratories optimize IHC conditions using different methods, and there is now increasing concern regarding the disparate nature of some of the results obtained (Henson, 2005). For example,

IHC analysis of p53 in breast cancer and association with clinical outcome appears to vary from study to study (Elledge and Allred, 1998). The association between expression of specified biomarkers (ERBB2, p53, and estrogen receptor) and clinical outcome was shown to change very significantly depending on the concentration of the biomarker antibody used for IHC (McCabe *et al.*, 2005). This appears to be because the concentration range of the antibody-based IHC assay is insufficient to span the range of expression of biomarker proteins in tissues. In addition, it is well known that IHC staining quality declines with duration of storage and antibodies against the same marker but acting at a different epitope can produce conflicting results (Henson, 2005). Particular care needs to be taken when using antibodies against phosphorylated proteins, as dephosphorylation can occur rapidly by phosphatases if the samples are not processed quickly, including addition of phosphatase inhibitors. Great care must also be taken in the selection of the antibody with regard to its specificity for the antigen of interest. Positive and negative controls need to be included.

An advantage of IHC is that it provides direct visualization of staining in tumor cells with the ability to give spatial information. In a leading example, Baselga *et al.* (2002, 2005) have provided proof of concept studies using IHC in tumor and normal skin for gefitinib and the mammalian target of rapamycin (mTOR) inhibitor RAD001 (see specific later sections). A recent paper used IHC to implicate EGFRvIII and PTEN mutations as molecular determinants of the sensitivity of glioblastomas to EGFR kinase inhibitors (Mellinghoff *et al.*, 2005). Paraffin-embedded tissue sections underwent IHC analysis for PTEN and EGFRvIII, and the results were verified independently by two pathologists using a predetermined set of scoring criteria. Quantitative image analysis to confirm the pathologists' scoring was also performed with the use of Soft Imaging System software.

Despite its usefulness, IHC has some documented limitations in terms of accuracy, quality control, and quality assurance required for a properly validated biomarker. Techniques that can measure the exact amounts of multiple proteins in specific subcellular compartments, together with more stringent validation procedures for a particular assay, will be required in the future. Although they lack spatial information, multiplex ELISA-based methods or flow cytometry techniques are of interest, although the latter requires disaggregation for solid tissues (Irish *et al.*, 2006). Use of these other techniques alongside IHC can provide reassurance.

D. Gene Expression Microarrays and Proteomics

The use of gene expression microarrays to generate molecular signatures and explore molecular mechanisms of drug action is having a significant impact in the drug discovery and development process. The reader is directed to comprehensive reviews by Clarke *et al.* (2001, 2004) for a full commentary on the current status of gene expression microarray technology. The main technology platforms are Affymetrix and spotted cDNA arrays. Gene expression profiling of cells in response to novel molecular therapeutics can play an important role in identifying PD markers, which can be subsequently validated and analyzed on other technology platforms. In addition, the gene expression signature itself can be used.

A major advantage is that array techniques provide unbiased genome-wide analysis. Not only does gene expression array profiling allow the detailed investigation of cellular mechanism of action of molecularly targeted agents, but crucially both on- and off-target effects can be determined. This contributes to the construction of the pharmacological audit trail described earlier. In addition, analysis of gene expression profiles can play an important role in the clinical setting with respect to more accurate determination of prognosis. In young women with early stage breast cancer, the use of a 70-gene expression signature to classify patients into either a good or poor prognosis group was a more powerful predictor of the outcome of disease than standard systems based on clinical and histologic criteria (van de Vijver *et al.*, 2002).

cDNA gene expression profiling has been used successfully to identify transcript changes in rectal cancer patients treated with cytotoxic therapy (Clarke *et al.*, 2003).

The use of proteomic methods for the large-scale analysis of proteins is also receiving considerable attention. The reader is directed to recent reviews on this topic (Carr *et al.*, 2004; Petricoin *et al.*, 2005). Particular attention is drawn to the controversy surrounding the use of proteomic patterns for detection of ovarian cancer (Petricoin *et al.*, 2002).

The hardware for gene expression and proteomic profiling is now becoming much more robust. Increasingly, a major focus of attention is the development of bioinformatic methodology to facilitate the handling and interpretation of the vast quantities of data that are produced (Petricoin *et al.*, 2005).

VIII. EXAMPLES OF PD BIOMARKERS FOR SPECIFIC NEW DRUG CLASSES

In this section, we will provide selected examples of the use of PD endpoints with targeted molecularly targeted agents. We also describe the mechanism of action these agents, since as will be seen, a mechanistic understanding is closely interrelated to PD biomarkers.

A. Imatinib Mesylate as a Paradigm for the Development of Molecular Therapeutics

The case of imatinib (Gleevec; STI571) has shown how the use of a powerful, highly validated preclinical biomarker provided a convincing rationale for a novel targeted therapeutic; furthermore, its subsequent clinical development was also based around this key biomarker (Park *et al.*, 2004). Imatinib was discovered in the late 1980s, and emerged as the lead compound from a series initially optimized against the platelet-derived growth factor (PDGFR) tyrosine kinase. Imatinib also selectively inhibits the ABL tyrosine kinases *in vitro* and blocks cellular proliferation and tumor growth of cells expressing BCR-ABL or v-ABL (Buchdunger *et al.*, 2000).

Chronic myeloid leukemia (CML) is characterized by the BCR-ABL fusion protein that is formed by a reciprocal translocation between the long arms of chromosomes 9 and 22 (the Philadelphia chromosome). The BCR-ABL fusion protein functions as a constitutively activated tyrosine kinase. In the first phase I trials of imatinib in CML patients who were resistant to interferon-α, treatment with >300 mg/day had complete hematological responses (defined as a decrease in marrow blasts to 5% or less of total cellularity, a disappearance of blasts from the peripheral blood, an absolute neutrophil count of more than 1000 cm^{-3}, and a platelet count of more than 100,000 cm^{-3}) in 98% of patients with chronic phase and 55% of those in blast phase (Druker *et al.*, 2001). These striking results have been confirmed in further clinical testing, including a phase III trial showing statistically superior rates of cytogenetic (determined by percentage of Philadelphia chromosome–positive cells in metaphase in bone marrow) and hematological response with imatinib over standard therapy in patients with chronic-phase disease, as well as prolonging time to progression of accelerated-phase or blast-crisis disease (Kantarjian *et al.*, 2002; O'Brien *et al.*, 2003; Sawyers *et al.*, 2002).

In the initial trials of imatinib, dose selection was guided by the rational selection of an appropriate PD endpoint: this involved assessment of BCR-ABL tyrosine kinase inhibition in circulating peripheral blood leukemic cells (Druker, 2002; Druker *et al.*, 2001). Previous studies had demonstrated that in BCR-ABL expressing cells, the SH2, SH3 domain adaptor protein CRKL is the major tyrosine phosphorylated protein (Oda *et al.*, 1994). The form of CRKL that is phosphorylated by BCR-ABL migrates more slowly on gel electrophoresis than the unphosphorylated form, allowing for the development of an assay that examined the relative proportion of phosphorylated CRKL before and during treatment (Senechal *et al.*, 1998). Immunoblot assays demonstrating the degree of phosphorylation of CRKL showed that there is a plateau in inhibition of BCR-ABL above 250 mg. Together with the PK data, a dose of at least 400 mg was therefore recommended

(Druker *et al.*, 2001). An MTD was not achieved with the agent, which has a high therapeutic index. In a subsequent phase III trial, quantitative real-time RT-PCR was used to measure levels of *BCR-ABL* transcripts in patients achieving complete cytogenetic response (Hughes *et al.*, 2003). Risk of disease progression was found to inversely correlate with reduction of *BCR-ABL* mRNA compared with pretherapeutic levels. The proportion of patients with CML who had a reduction in *BCR-ABL* transcript levels of at least 3 log by 12 months of therapy had a negligible risk of disease progression during the subsequent 12 months (Deininger *et al.*, 2005; Hughes *et al.*, 2003). Use of RT-PCR has since been increasingly accepted for monitoring and assessment of response to imatinib, and the use of this molecular endpoint as a basis for clinical decision making has proved to be feasible and safe both in multicenter clinical trials (Hess *et al.*, 2005; Martinelli *et al.*, 2006).

Despite the stunning trial results for imatinib, which led to it being the first kinase inhibitor approved by the US FDA, the emergence of resistance to imatinib has shown that the new post-genomic designer drugs will not escape this age-old problem of cancer therapy (Workman, 2005c; Workman and Kaye, 2002). Resistance was first identified in patients who relapsed while receiving imatinib and was associated with point mutations that rendered the ABL-kinase resistant to the drug, or to a lesser extent with *BCR-ABL* gene amplification (Gorre *et al.*, 2001; Krause and Van Etten, 2005). Crystallographic studies have revealed that imatinib binds to the ATP site of ABL only when the activation loop of the kinase is closed and thus stabilizes the protein in this inactive confirmation (Schindler *et al.*, 2000). Mutations at 17 different amino acid positions within the BCR-ABL kinase domain have so far been identified (Branford *et al.*, 2003; Shah *et al.*, 2002). The majority of amino acid substitutions prevent BCR-ABL from achieving the inactive conformational state required for imatinib binding (Shah *et al.*, 2002). Many of the mutations found in experimental systems were also found in treated patients. This has led to the development of several second-generation ABL kinase inhibitors with increased potency and activity against most imatinib-resistant mutants. Foremost among these is dasatinib (BMS-354825, Sprycel), a dual ABL/SRC kinase inhibitor (Shah *et al.*, 2004) which has recently been approved by the FDA. This drug binds to ABL but in an open, active conformation, and has demonstrated clinical activity in patients with a wide range of imatinib-resistant BCR-ABL kinase domain mutations (Shah *et al.*, 2004; Talpaz *et al.*, 2006).

The ability of imatinib to inhibit the mutated c-KIT tyrosine kinase, which is associated with the rare malignancy GIST, has resulted in further success for imatinib. GISTs are highly refractory to standard chemotherapies, but response rates close to 60% in clinical trials were achieved with imatinib, transforming the management of this disease (Demetri *et al.*, 2002;

van Oosterom *et al.*, 2001). The highest activity appears to correlate with activating mutations at exon 11 of the *KIT* gene, with less activity in GISTs expressing exon 9 mutations or wild-type KIT (Heinrich *et al.*, 2003a). Interestingly, 35% of GISTs with wild-type KIT have constitutively active PDGFRα, which is also inhibited by imatinib, most likely explaining the excellent clinical responses to the drug in this cohort (Arteaga and Baselga, 2004; Heinrich *et al.*, 2003b). Tumor genotype is also of major prognostic significance for progression-free survival and overall survival in patients treated with imatinib for advanced GISTs. The presence of exon 9-activating mutations are the strongest adverse prognostic factor for response to imatinib, increasing the relative risk of death by 190% ($P < 0.0001$) when compared with KIT exon 11 mutants. Patients with exon 9 mutations hence benefit the most from the higher (800 mg daily) dose of the drug (Debiec-Rychter *et al.*, 2006).

Confirmation of imatinib's activity in GISTs has been possible through the use of FDG-PET. Initial studies showed high FDG uptake in untreated patients on PET imaging (Antoch *et al.*, 2004; van Oosterom *et al.*, 2001). PET was shown to be superior to CT in detection of the earliest functional parameters indicative of tumor response induced by imatinib therapy (Stroobants *et al.*, 2003; van Oosterom *et al.*, 2001). Overall GIST disease status as evaluated by changes in size, density, and number of tumor nodules and vessels within the lesion correlated best with the reduction of maximum standardized uptake values (SUV) on FDG-PET (Choi *et al.*, 2004). For responding patients, FDG uptake in the tumor decreases markedly from base line as early as 24 h after a single dose of imatinib (Demetri *et al.*, 2002). Other c-KIT and PDGFR tyrosine kinase inhibitors are currently in clinical trials, and have shown activity in imatinib-refractory GIST, with FDG-PET used to confirm tumor response (Sarker *et al.*, 2005, see Fig. 4).

B. EGFR Inhibitors

The family of ERBB family of membrane tyrosine kinase receptors comprises four members: EGFR/ERBB1, ERBB2, ERBB3, and ERBB4. EGFR and ERBB2 are involved in the development of numerous types of human cancer and have been pursued as novel anticancer agents (Hynes and Lane, 2005). In preclinical models, treatment of tumor cells with ERBBB inhibitors leads to down-regulation of AKT, ERK 1/2, SRC and STAT signaling, and inhibits cell proliferation. Small molecule inhibitors or antibodies have shown definite evidence of clinical activity, but their development has also been marked by some high-profile failures. In this section, we review the use of these inhibitors with regard to their clinical use and development of PD endpoints.

EGFR overexpression occurs in a number of malignancies, including non-small cell lung cancer (NSCLC), glioma, prostate, pancreatic, colorectal,

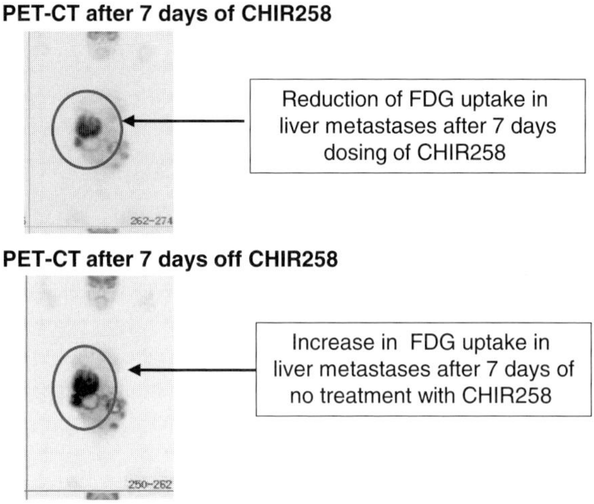
PET-CT after 7 days of CHIR258
Reduction of FDG uptake in liver metastases after 7 days dosing of CHIR258
262-274
PET-CT after 7 days off CHIR258
Increase in FDG uptake in liver metastases after 7 days of no treatment with CHIR258
250-262

Fig. 4 PET-CT of a patient with gastrointestinal stromal (GIST) tumor patient showing a response to a novel c-KIT inhibitor CHIR258. This patient with imatinib-refractory GIST was treated on a phase I trial of a multitargeted tyrosinse kinase inhibitor, with activity against both c-KIT and PDGFR (Sarker *et al.*, 2005). The drug was given on a 7 days on, 7 days off schedule, and PET-CT showed significant decrease in uptake in her liver metastases during treatment compared to 7 days off treatment. The patient experienced prolonged stable disease for 9 months on the drug. Figure courtesy of Dr. Gary Cook, Nuclear Medicine Department, Royal Marsden Hospital, Sutton, UK and Chiron Corporation, Emeryville, CA.

and head and neck tumors (Hynes and Lane, 2005). Indeed, EGFR was the first tyrosine kinase receptor to be linked to human cancer, and linkage of EGFR expression to outcome in NSCLC other tumor types led to the subsequent development of drugs against this target (Gschwind *et al.*, 2004). Among the most prominent have been the small molecule inhibitors, gefitinib (Iressa) and erlotinib (Tarceva), and the monoclonal antibody cetuximab (Erbitux). The development of gefitinib and erlotinib will be discussed here.

Gefitinib and erlotinib are small molecule anilinoquinazolines that are selective, competitive inhibitors of ATP binding by EGFR; both drugs have been approved for use in advanced NSCLC. Gefitinib monotherapy showed partial response rates of 9–19% in phase II trials of patients with refractory advanced NSCLC (Kris *et al.*, 2003). Erlotinib also improved disease-related symptoms, and disease-free and overall survival compared with placebo (Shepherd *et al.*, 2005). However, the addition of either gefitinib or erlotinib to chemotherapy in the initial treatment of NSCLC showed no evidence of survival benefit (Giaccone *et al.*, 2004; Herbst *et al.*, 2004).

It was apparent from the very first studies that a subgroup of NSCLC patients with adenocarcinomas, and specifically those that were Asian, female, and never smoked, often had dramatic, durable responses to gefitinib. Subsequent sequencing of the *EGFR* gene in tumor tissues from these patients showed heterozygous somatic mutations within the tyrosine kinase domains of EGFR; these enhanced the responsiveness of the receptor to EGF ligand and increased its sensitivity to gefitinib and erlotinib, as well as preferentially activating the antiapoptotic AKT and STAT signaling pathways (Lynch *et al.*, 2004; Pao *et al.*, 2004; Sordella *et al.*, 2004). However, in the National Cancer Institute of Canada (NCIC) randomized trial of erlotinib in NSCLC, the accompanying molecular correlates study showed that expression of EGFR and amplification of the *EGFR* gene were significantly associated with response to erlotinib, but not increased survival; this suggests that erlotinib may depend for its activity on other signaling pathways that were not assessed in this study, such as AKT phosphorylation or ERBB3 status (Doroshow, 2005; Engelman *et al.*, 2005; Tsao *et al.*, 2005). The NCIC study contrasts with studies from Asia where EGFR mutant tumors are more common, and are associated with both significantly increased response rate and overall survival (Han *et al.*, 2005; Mitsudomi *et al.*, 2005). These studies exemplify the need for more attention to be placed on the appropriate selection of patients for molecular cancer therapeutics.

A series of highly elegant PD studies were performed in the original phase I trials of gefitinib. In these, skin was proposed as a surrogate tissue because of ease of access and high EGFR expression. In these studies by Baselga, Albanell, and colleagues, paired skin biopsies were taken pretherapy and on-therapy (day 28); immunohistochemical analysis of EGFR status, EGFR phosphorylation, ERK 1/2 phosphorylation, proliferation marker Ki67, $p27^{KIP1}$, keratin 1, and phospho-STAT3 expression were performed in paraffin-embedded sections (Albanell *et al.*, 2002; Baselga *et al.*, 2002). Statistically significant inhibition of EGFR activation in the basal layer of interfollicular epidermis and in hair follicle keratinocytes was achieved in all paired skin samples during gefitinib treatment. ERK 1/2 phosphorylation was assayed as a downstream marker of EGFR signaling and showed a significant reduction in expression at day 28, providing proof of concept for target inhibition. In addition, cell proliferation index (using Ki67) was reduced, and there was increased expression of the CDK2 inhibitor $p27^{KIP1}$ reflecting induction of G1 cell-cycle arrest, as previously demonstrated in preclinical models (Busse *et al.*, 2000). Hematoxylin and eosin staining of skin during therapy showed that the stratum corneum of the epidermis was thinner and contained an increased number of apoptotic cells. All of these effects on the target and EGFR-dependent molecular endpoints were observed at every dose level, indicating a lack of dose–response effect. It is therefore likely that potentially biologically active concentrations were achieved at all dose levels (see Fig. 5).

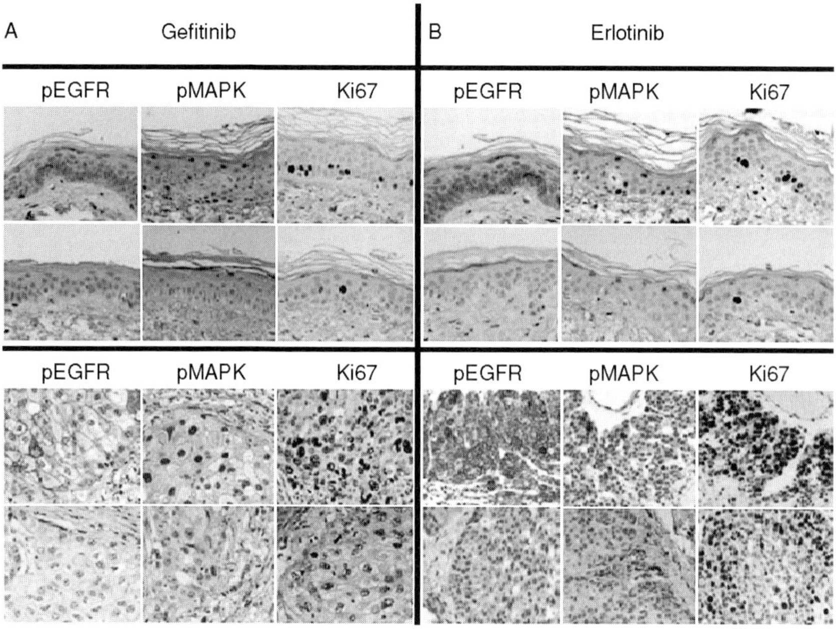

Fig. 5 Immunohistochemical demonstration of pharmacodynamic changes in tumor and skin in response to treatment with EGFR inhibitors. Immunohistochemical sections of skin and tumor biopsies pretherapy versus on-therapy for (A) a patient with breast carcinoma treated with gefitinib and (B) a patient with NSCLC treated with erlotinib. The upper panels show sections of skin and the lower panels show sections of tumor. In each case the pretreatment samples are shown above the posttreatment samples. Both patients show inhibition of the phosphorylation of EGFR and MAPK (ERK 1/2) both in tumor and skin. In addition, a clear decrease of proliferation, as measured by the Ki67 marker, is observed. Figure courtesy of Dr. Federico Rojo and Dr. Jose Baselga, Val d'Hebron University Hospital, Barcelona, Spain. (See Color Insert.)

Further studies are required with respect to the ongoing or completed phase II/III trials in an attempt to correlate the above biomarkers with response, survival, and EGFR mutational status. A number of limitations are apparent. There has been a lack of PD data from tumor tissue. PD effects measured in well-vascularized surrogate tissues such as skin may not predict for similar effects in poorly vascularized tumors. In addition, the activation of various signal transduction pathways in tumors might lead to different PD responses compared to healthy cells. Higher doses might be needed to inhibit EGFR activation in tumors than would be predicted by the effects measured in skin samples (Dancey and Freidlin, 2003).

In a recent phase II study of gefitinib at a dose of 500 mg/day in patients with advanced breast cancer, paired skin and tumor biopsies were obtained pretherapy and after 28 days of treatment, and PD assays performed as in the original phase I studies (Baselga *et al.*, 2005). Gefitinib caused complete inhibition of EGFR phosphorylation in both skin and tumor (see Fig. 5). However, downstream consequences of receptor blockade were distinct in skin and tumor: ERK 1/2 phosphorylation was inhibited in both tissues, while gefitinib caused induction of $p27^{KIP1}$ and a decrease in Ki67 in skin only. In addition, gefitinib did not result in a decrease in phosphorylated AKT in tumors (not done on skin). No complete or partial responses were observed in this study, indicating that in this setting inhibition of EGFR phosphorylation may be an indicator of target inhibition, but not of sensitivity to anti-EGFR agents. While inhibition of target may be required for antitumor effect, sensitivity to EGFR inhibitors is likely to be related to level of receptor dependence or "oncogene addiction" in the individual tumor.

A great deal has been learned from the assessment of biomarkers in the development of gefitinib. Unfortunately much of this was learned too late. The development of EGFR inhibitors illustrates the need for greater efforts to develop standardized assay procedures for assessing and predicting the effects of novel molecularly targeted agents, to standardize methods for the prospective collection of tumor samples, and to incorporate these assays into phase II/III clinical trials to maximize the likelihood of definitive clinical results (Doroshow, 2005).

C. Inhibitors of the HSP90 Molecular Chaperone

HSP90 is a member of the family of heat shock proteins that acts as an ATPase-driven molecular chaperone. HSP90 is required for the stability, conformation, function, and regulation of a number of key oncogenic "client proteins," including C-RAF, CDK4, ERBB2, and AK, as well as several transcription factors, notably steroid receptors, and retinoid receptors, and also mutant and chimeric proteins such as mutated p53 or BCR-ABL (Maloney and Workman, 2002; Whitesell and Lindquist, 2005). HSP90 inhibition leads to the simultaneous combinatorial degradation of the various oncogenic client proteins by the ubiquitin-proteasome pathway. Thus HSP90 inhibitors have the potential to block the essential hallmark traits of cancer and provide a one-step combinatorial therapy against a range of malignancies (Workman, 2003b, 2004a).

The first-in-class HSP90 inhibitor 17-AAG caused cell-cycle arrest and apoptosis in human cancer cells and cytostatic growth arrest in human tumor xenograft models (Hostein *et al.*, 2001; Kelland *et al.*, 1999). Phase I trials have now been completed at our institution and other centers in the USA (Banerji *et al.*, 2005a; Goetz *et al.*, 2005; Grem *et al.*, 2005). These studies provided proof of concept for HSP90 inhibition, as measured by the well-defined molecular signature of the mechanism-based depletion in client protein and induction of heat shock protein expression in tumor tissue of treated patients, in addition to similar effects on PBMCs. The molecular biomarkers that we selected and validated for use in our own trial were based on both our cDNA microarray gene expression and proteomic profiling studies (Clarke *et al.*, 2000; Panaretou *et al.*, 2002), and PK–PD relationships were obtained in a human tumor xenograft model (Banerji *et al.*, 2005b). In the trial of once-weekly 17-AAG performed at our institution, after the demonstration of plasma concentrations above those required for activity in human tumor xenograft models and PD changes in PBMCs, pre- and 24 h posttreatment, tumor biopsies were performed and Western blotting used to demonstrate depletion of C-RAF in four of six patients, CDK4 depletion in eight of nine patients, and HSP70 induction in eight of nine patients at the dose levels 320 and 450 mg/m^2. In addition, IHC was used to demonstrate that the increased expression of HSP70 was present in viable tumor cells within melanoma biopsies. It was not possible to reproducibly demonstrate PD biomarker changes in biopsies taken 5 days after treatment, and it is likely that target inhibition lasted for only 2–3 days on a weekly schedule. Thus the drug would ideally be given on a twice-weekly schedule, but this was not possible because of the cumbersome formulation. In association with our demonstration of the molecular signature of HSP90 in melanoma biopsies, we observed prolonged stable disease in two patients with advanced melanoma (Banerji *et al.*, 2005a). This may relate to the importance of the RAS-RAF-MEK-ERK 1/2 signaling pathway in melanoma and which is inhibited by 17-AAG. Figure 6 illustrates the elements of the pharmacological audit trail that could be measured for 17-AAG in our trial.

It is useful to consider the gene expression profiling studies in a little more detail. cDNA gene expression profiling of human colon cancer cell lines following exposure to 17-AAG showed induction of the HSP90 drug target in cell lines that had reduced sensitivity to 17-AAG, contrasting with low HSP90 expression in cell lines particularly sensitive to 17-AAG (Clarke *et al.*, 2000). Other changes included induction of HSP70 expression in all cell lines and differential effects on the expression of cytoskeletal and signaling genes. The induction of HSP70 family genes was confirmed at the protein level and was also observed in human PBMCs treated with 17-AAG *ex vivo* (Clarke *et al.*, 2000). Thus using a combination of microarray and Western blotting analysis, a molecular signature of HSP90 inhibition, consisting of HSP70 induction and client protein depletion, was then explored further in the clinical studies.

A further detailed study of changes in expression profile in response to 17-AAG was carried out in A2780 human ovarian cancer cells (Maloney *et al.*, 2003). We used both cDNA microarrays and proteomic analysis, and a key observation was the identification of AHA1 as a stress-regulated

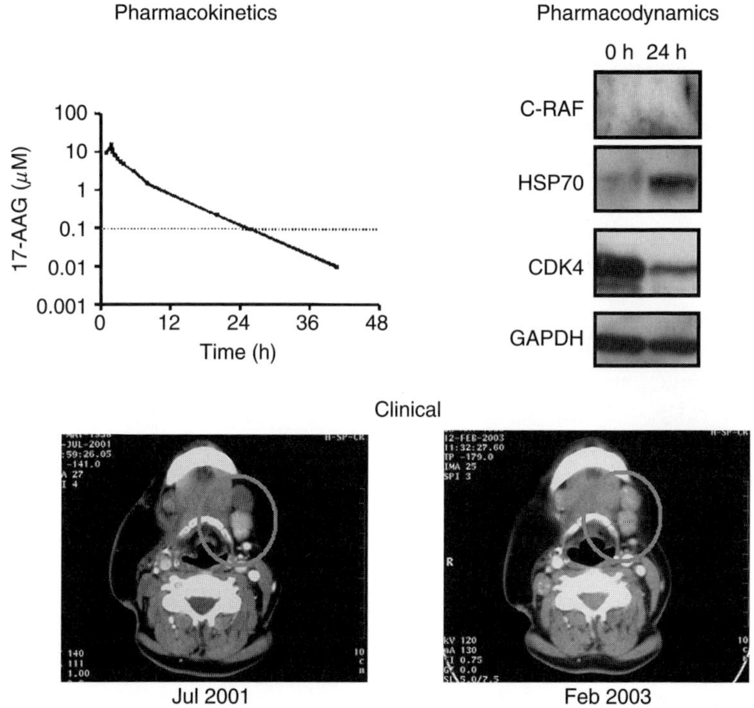

Fig. 6 PK–PD clinical relationship for 17-AAG in a malignant melanoma patient with prolonged stable disease. This figure demonstrates the PK–PD clinical relationships in this patient. *Pharmacokinetics*: The presence of active concentrations of 17-AAG, with peak plasma concentrations of >10 μmol/liter and concentrations of approximately 100 nmol/liter achieved at 24 h. *Pharmacodynamics*: Demonstration of the pharmacodynamic signature of HSP90 inhibition at 24-h postdose in tumor tissue with up-regulation of HSP70, down-regulation of the client protein CDK4 (this patient did not express C-RAF). *Clinical*: prolonged stable disease as shown in the CT images of the patient's left submandibular lymphadenopathy. This patient experienced stable disease with 17-AAG for 3 years. Modified from Banerji *et al.*, *J. Clin. Oncol.* 2005.

cochaperone that activates the essential ATPase of HSP90 (Panaretou *et al.*, 2002). We showed that *AHA1* expression was up-regulated at both the mRNA and protein level by 17-AAG in human cancer cells, potentially affecting sensitivity in patients.

Recently, two novel potential biomarkers of HSP90 inhibition have been identified: insulin-like growth factor binding protein-2 (IGFBP2) and ERBB2 extracellular domain (Zhang *et al.*, 2006). Both are secreted proteins and are derived from or regulated by HSP90 client proteins. In studies with HSP90 inhibitors in sera from BT474 tumor-bearing mice, both IGFBP2 and ERBB2 extracellular domain were down-regulated in a time- and dose-dependent manner. Both IGFBP2 and ERBB2 extracellular domain can be detected in patient sera by ELISA, raising the possibility that they might be sensitive serum markers of HSP90 inhibitor activity. We are also investigating the possibility of measuring circulating HSP70.

As part of a broad cross-platform search for biomarkers of HSP90 inhibition, we have also explored approaches based on MRS and PET. In collaboration with our MRS and MRI colleagues, we identified MRS PD biomarkers, comprising an unusual combination of increases in phosphocholine and phosphomonoester levels after treatment with 17-AAG in human colon cancer xenograft models (see Fig. 3) (Chung *et al.*, 2003). In collaboration with our PET collaborators, we have identified radiolabeled choline as a potential marker for HSP90 inhibition potentially alongside FLT as a marker of cytostasis and radiolabeled annexin V to image apoptosis (Collingridge *et al.*, 2003; Liu *et al.*, 2002). Such assays could be used to image PD effects of HSP90 inhibitors in animal models and patients alongside molecular assays for client protein degradation (Smith-Jones *et al.*, 2004).

Unfortunately 17-AAG has a number of limitations, particularly its low aqueous solubility, necessitating the use of less than ideal formulations. This has led to the search for more soluble analogues of 17-AAG, such as 17-DMAG which is now in phase I clinical trials. In addition, we have developed a novel series of small molecule HSP90 inhibitors using high-throughput screening and structure-based design (McDonald *et al.*, 2006). The use of the PK–PD relationships described above continues to play an important role in the development of new HSP90 inhibitors.

Although good progress has been made on the development of PD endpoints, assays that may be predictive of response to treatment will need to be developed. Since ERBB2 is a very sensitive client protein, it is possible that tumors with high levels of ERBB2 may be particularly sensitive to HSP90 inhibitors and activity has been reported in trastuzumab-resistant disease.

D. Inhibitors of the PI3K-AKT-mTOR Pathway

This pathway is one of the most intensively studied as a potential source of molecular therapeutics for clinical development.

1. PI3K INHIBITORS

The PI3K pathway is frequently deregulated by various means in a wide range of malignancies, and is a key effector of several important cellular processes such as proliferation, growth, apoptosis, and cytoskeletal rearrangement (Bader *et al.*, 2005; Vivanco and Sawyers, 2002). PI3K catalyzes the conversion of phosphatidylinositol 4,5-biphosphate (PIP_2) to phosphatidylinositol 3,4,5-triphosphate (PIP_3). *PIK3CA*, which encodes the p110α catalytic subunit of PI3K, has recently been found to exhibit hot spot mutations in a number of malignancies, including colorectal, breast, gastric, and brain tumors (Broderick *et al.*, 2004; Campbell *et al.*, 2004; Samuels *et al.*, 2004). These mutations result in gain of enzymatic function, and activation of AKT signaling, as well as induction of oncogenic transformation (Kang *et al.*, 2005). In addition to mutations, *PIK3CA* is amplified in some ovarian and cervical cancers (Shayesteh *et al.*, 1999). Activation in the PI3K signaling pathway can also occur from loss of the PI3K negative regulator *PTEN* (Cantley and Neel, 1999). PTEN is a phosphatase which dephosphorylates PIP_3 to PIP_2, and hence loss of PTEN increases PIP_3 levels, thereby activating the pathway. PTEN is the second most commonly mutated tumor-suppressor gene and has been particularly implicated in glioma, endometrial, and prostate cancers (Cully *et al.*, 2006; Vivanco and Sawyers, 2002). This cumulative evidence suggests PI3K is an attractive target for the development of molecular therapeutics (Hennessy *et al.*, 2005; Workman, 2004b).

To date, relatively few PI3K inhibitors have been identified, and these possess considerable limitations, such as lack of selectivity for PI3K isoforms, limited stability, and poor PK and metabolic/stability properties (Hennessy *et al.*, 2005; Workman, 2004b, 2005b). In a collaborative project, we have identified potent and selective inhibitors of class 1A PI3K (Workman, 2005b; Workman *et al.*, 2004). These inhibitors, including the prototype compound PI103 (a representative of the pyridofuropyrimidine series) cause a rapid and concentration-dependent inhibition phosphorylation of the substrate AKT (Fan *et al.*, 2006; Workman, 2005a,b; Workman *et al.*, 2004, 2006b). In addition, decreased phosphorylation of proteins that lie downstream in the PI3K pathway is seen, and down-regulation of cyclin D1 and an increase in p21 expression are observed. A dual strategy of candidate proteins and gene expression profiling has been used for biomarker discovery. Gene expression profiling of PI103 shows relatively few changes; an increase in the expression of integrin β4 by PI103 has been demonstrated, and could possibly be used as a PD marker of PI3K inhibition, as well as indicating effects on cell adhesion. Induction of forkhead (FOXO1) translocation may also provide a readout of PI3K inhibition.

PX-866 is a wortmannin derivative which inhibits PI3K and has shown efficacy in a range of human tumor xenografts. Human hair and skin have been studied as potential surrogate tissues for PI3K inhibition with PX-866 (Williams *et al.*, 2006). Using direct IHC staining of plucked human hair, phosphorylation of AKT was shown to be inhibited by PX-866 in culture. Inhibition of phospho-AKT by PX-866 in mouse hair keratinocytes was greater than inhibition of phospho-AKT in HT29 and A549 xenografts in the same mice. These results suggest that human hair could be a sensitive surrogate tissue for measuring the effect of PI3K inhibition.

Given the importance of the target it seems likely that a range of PI3K inhibitors will enter the clinic. It will be very important to develop markers that will identify potentially sensitive patients (obvious candidates may include *PIK3CA* mutational status, phospho-AKT levels, and PTEN status), as well as PD markers for proof of concept.

2. mTOR INHIBITORS

mTOR is a serine/threonine kinase that lies downstream of PI3K and AKT. mTOR is involved in the regulation of a range of cellular functions, including transcription, translation initiation, membrane trafficking protein degradation, and reorganization of the actin cytoskeleton (Adjei and Hidalgo, 2005; Schmelzle and Hall, 2000). mTOR controls cap-dependent translation initiation through phosphorylating and inactivating eukaryotic initiation factor 4E-binding proteins (4E-BPs) (Bjornsti and Houghton, 2004). In addition, mTOR functions as a sensor of mitogen, energy and nutrient levels, acting as a gatekeeper for cell-cycle progression from G1 to S phase. Rapamycin, a naturally occurring inhibitor of mTOR is an immunosuppressive agent that has significant antiproliferative action against a range of malignancies in preclinical models. A number of rapamycin derivatives have been developed as anticancer agents. In preclinical models they exert tumor growth inhibition, induce apoptosis in certain models, and inhibit angiogenesis (Huang *et al.*, 2003). Preferential activity is generally seen in tumor cells that are addicted to the PI3K-AKT-mTOR pathway, for example by loss of PTEN.

Temsirolimus (CCI-779), an ester of rapamycin, was the first mTOR inhibitor used in oncology, and its development highlights some of the recent problems associated with the development of PD biomarkers in the last few years. The development of temsirolimus started in the late 1990s; phase I and II clinical trials have been completed in a range of tumor types, including glioma, breast, and renal cancers, with promising enough results to warrant randomized phase III trials which are underway (Atkins *et al.*, 2004; Chan *et al.*, 2005; Faivre *et al.*, 2006; Galanis *et al.*, 2005). Preliminary results from a phase III trial in patients with advanced renal cancer, who have poor prognostic features, demonstrated significant improvement in overall survival compared to interferon-α (or the combination) with temsirolimus (Hudes *et al.*, 2006).

The phase I trials of temsirolimus were not targeted to particular patients in which mTOR inhibitors might be expected to have activity. Rather these were empirical studies designed to define MTD and to define toxicities and PK properties. Results showed that it was generally well tolerated with the main toxicities comprising stomatitis, rash, thrombocytopenia, and fatigue, and that it displayed nonlinear kinetics and demonstrated some antitumor effect (Adjei and Hidalgo, 2005; Raymond *et al.*, 2004). However, in the two phase I studies using intravenous administration, the MTD was not determined due to a non-dose-related toxicity of the drug (asthenia, stomatitis, and manic-depressive syndrome) (Hidalgo, 2004; Raymond *et al.*, 2004). Based on responses in breast and renal cancer, randomized phase II studies using multiple doses were carried out in these tumor types to further clarify optimal dose, as well as to gain standard response data (Atkins *et al.*, 2004; Chan *et al.*, 2005). Modest responses were demonstrated, but the optimal dose remained to be defined. Of note here is the lack of PD studies in these trials to help define dose and schedule (Hidalgo, 2004; Sawyers, 2003b). Only by the time the drug was in phase II studies did preclinical PD studies of temosirolimus appear (Dudkin *et al.*, 2001; Peralba *et al.*, 2003). Exposure of Raji lymphoblastoid cells to increasing concentrations of rapamycin resulted in a linear concentration-dependent inhibition of ribosomal p70S6 kinase activity (hereafter referred to as S6K1). This was performed using a quantitative assay for S6K1. Subsequent studies in breast xenograft models using this same assay showed temsirolimus resulted in $a > 80\%$ inhibition of S6K1 activity in PBMCs 72 h after treatment, and importantly, the degree of S6K1 inhibition was identical in PBMCs and simultaneously collected tumor tissue. This suggested that the PBMCs are an adequate surrogate tissue for S6K1 activity *in vivo*.

These PD assays have been now incorporated into clinical studies using a different rapamycin ester analogue, RAD001. Results in preclinical models demonstrated significant inactivation of S6K1 in tumor biopsies, skin biopsies, and PBMCs, providing a validated PD marker for clinical study (Boulay *et al.*, 2004). Pancreatic tumor xengrafts were used to profile mTOR signaling in tumors, skin, and PBMCs after a single dose of RAD001. RAD001 treatment blocked phosphorylation of the downstream translational repressor eukaryotic initiation factor 4E-BP1, as measured by immunoblotting (Boulay *et al.*, 2004). S6K1 protein and activity levels were also analyzed; while total S6K1 protein levels were unaffected by RAD001 treatment, an *in vitro* kinase assay using 40S ribosomal subunits as a substrate revealed a statistically significant reduction in S6K1 activity in all extracts (tumor, skin, and PBMCs). This reduction in S6K1 activity was associated with the dramatic dephosphorylation of its physiological substrate, 40S ribosomal protein S6, in tumor extracts. A similar reduction could not be observed in skin and PBMC extracts because these tissues exhibited no detectable S6 phosphorylation in control animals. There was a correlation between the antitumor efficacy of intermittent RAD001 treatment schedules and prolonged S6K1 inactivation in PBMC, suggesting that long-term monitoring of PBMC-derived S6K1 activity levels could be used for assessing RAD001 treatment schedules in cancer patients.

A phase I dose-escalation study of RAD001 administered once-weekly was performed at our institution to determine the optimum biologically effective dose based on toxicity, PK, and target inhibition as measured by using the biomarker S6K1 in PBMCs (O'Donnell *et al.*, 2003). The drug was well tolerated with mild-moderate toxicities comprising fatigue, anorexia, rash, and mucositis. S6K1 activity in PBMCs was inhibited for 3–5 days at 5- and 10-mg dose levels; at doses >20 mg, 7/8 patients exhibited inhibition for at least 7 days. For the 20-mg dose level, these results, together with plasma concentrations, were equivalent to PK levels and PD changes correlating with antitumor effects in rodents.

A follow-up phase I study with evaluation of PD changes by IHC in serial skin and tumor biopsies at differing doses and schedules was subsequently undertaken. This study investigated weekly cohorts of 20, 50, or 70 mg, or daily dosing of 5 or 10 mg (Tabernero *et al.*, 2005). The drug was well tolerated, with signs of antitumor activity, and showed dose- and schedule-dependent inhibition of phosphorylated S6 and 4E-BP1, with similar effects seen in tumor and skin. This was associated with concomittant decreases in proliferative index as measured by Ki67. However, an initially paradoxical activation of phospho-AKT was shown in tumor at 10 mg daily and >50 mg weekly. The recommended dose of 10 mg daily was based on both toxicity and PD assessment. Use of these PD markers was therefore a useful tool in addressing the downstream effects of RAD001, in the preliminary validation of the mechanism of action in these compartments, and in optimal dose determination.

The unexpected finding of activation of AKT in these tumor biopsies can be explained by reversal of the feedback loop which occurs in tumor cells with constitutive mTOR activation, leading to down-regulation of receptor tyrosine kinase signaling. Reversal of this feedback loop by mTOR inhibitors may attenuate its therapeutic effects by undesirable activation of PI3K-AKT (O'Reilly *et al.*, 2006). Interestingly, this leads to the hypothesis that combination therapy that ablates mTOR function and prevents AKT activation may have improved antitumor activity. The PI3K inhibitor PI103, mentioned above, has recently been reported to be a dual inhibitor of p110α and mTOR, and demonstrates significant antitumor activity in human glioma xenografts (Fan *et al.*, 2006; Knight *et al.*, 2006; Workman, 2005a; Workman *et al.*, 2006b). Thus, this two-hit inhibition of the pathway can abrogate the feedback inhibition of the PI3K pathway and leads to antitumor activity in pathway-addicted cancers (Workman *et al.*, 2006b).

A further PD assay of mTOR inhibition that is being explored is the use of FDG-PET. This is based on the rationale that signaling through the insulin receptor activates PI3K and AKT (Thompson and Thompson, 2004).

This stimulates glucose uptake and glycolysis through activation of mTOR and HIF-1α. These same glycolytic enzymes regulated by PI3K-AKT-mTOR-HIF-1α signaling are responsible for uptake and retention of the PET tracer FDG, hence providing the possibility of a noninvasive readout of PI3K pathway inhibition (Mellinghoff and Sawyers, 2004).

IX. COMBINING CHEMOTHERAPY WITH MOLECULARLY TARGETED AGENTS

Cytotoxic chemotherapy is likely to remain a key component of the oncologist's arsenal for some time. Nevertheless, the future of the medical treatment for cancer is likely to lie with the continued introduction of targeted molecular therapeutics that acts on the genetic and molecular abnormalities that drive the malignant phenotype. In the immediate future, treatment is likely to involve the combination of the new molecular therapeutics with conventional agents. There are considerable challenges with such combinations, particularly with respect to the predictive value of preclinical models and the appropriate design of clinical trials for combinations with cytotoxic agents (Jackman *et al.*, 2004). To date, the clinical trial experience has been sobering, with several high-profile combination failures. Probably foremost among these were the negative IMPACT 1 and 2 trials of gefitinib added to chemotherapy in NSCLC and two similar studies of erlotinib with chemotherapy (Giaccone *et al.*, 2004; Herbst *et al.*, 2004, 2005). A number of hypotheses have been proposed to explain these failures, but two are particularly relevant (Gandara and Gumerlock, 2005). Firstly, lack of patient selection due to lack of suitable biomarker, meaning that trials were not "enriched" with the appropriate population. Secondly, the lack of appropriate preclinical models to determine accurately the effects of scheduling and dose, so that the concurrent administration of EGFR inhibitor and chemotherapy may be antagonistic. For example, EGFR inhibition may induce a G1 arrest that could reduce cell-cycle phase-dependent activity of chemotherapy.

Since the finding of activating mutations of EGFR predicting response to gefitinib and erlotinib, a retrospective subset analysis from one of the erlotinib plus chemotherapy trials has been performed, showing that patients with EGFR mutant tumors have a higher response rate to the combination than those with wild-type tumors (Eberhard *et al.*, 2005).

This is likely to be followed soon by prospective studies using erlotinib with chemotherapy in NSCLC populations enriched for the likelihood of EGFR mutations (never-smokers, adenocarcinomas, Asian ancestry).

In addition, studies are ongoing to investigate intermittent EGFR inhibition with chemotherapy scheduling to achieve PD separation.

Despite the problems described above with EGFR inhibitors and chemotherapy, other molecular therapeutic agents have been more successful in combination. For example, the recombinant monoclonal VEGF-A antibody bevacizumab (Avastin), when combined with irinotecan/5FU chemotherapy as first-line treatment of advanced colorectal cancer, resulted in improved overall survival when compared with chemotherapy alone (Hurwitz *et al.*, 2004). However, it could be argued that even this study, with its traditional empirical design, was somewhat fortuitous in its success, rather than being based on a sound preclinical model with appropriately designed PD biomarkers.

For an excellent example of how chemotherapy should logically be used with targeted agents, preclinical studies combining trastuzumab with a range of chemotherapy agents are an ideal model (Pegram *et al.*, 2004a,b; Sledge, 2004). *In vitro* studies of multiple cell lines with the relevant biology were used, and the agents examined were relevant to the disease state (in this case, metastatic breast cancer). The drug concentrations for the agents examined were clinically relevant, and the *in vitro* data followed up with validated *in vivo* preclinical animal models. Observed phenomena (in this case, the synergistic activity of platinating agents when combined with trastuzumab) were pursued with mechanistic studies (i.e., DNA repair studies).

The clinical trials developed from such *in vitro* and *in vivo* preclinical research should then build on prior trial data, involving the prospective collection of tissues for subsequent biological analysis, and should be sufficiently well powered to provide meaningful results.

The optimal combination of chemotherapy with molecularly targeted agents presents a very significant challenge, but is best approached by a logical combination of molecular characterization of tumors, suitable preclinical models, and PD markers to maximize the likelihood of success.

X. CONCLUSIONS AND FUTURE PERSPECTIVE

PD studies have the potential to have a dramatic impact on the drug development process. When used in conjunction with biomarkers for patient selection and with the concept of the pharmacological audit trail, these provide an intellectual and practical framework for the design of clinical studies and interpretation of data, as well as for critical decision making.

To date, the uptake and use of PD biomarkers in early clinical trials has been disappointing, and their overall impact has been fairly minimal. To fulfill their promise they must be developed early in the drug development process and be appropriately validated. Clinical trials should be designed with PD assays in mind to provide proof of concept for the molecular mode of action of new agents. The range of scientifically novel and powerful PD tools available is ever increasing and it is appropriate that the technical methodology is pushed as hard as possible. Even quite technologically challenging methods can be made much more user friendly and relevant within a very short space of time. At this stage, it is unclear whether complex and data-intensive methodologies like high-throughput gene sequencing and gene expression profiling will become routine in clinical management and, if so, when (Workman and Johnston, 2005). A balance needs to be struck between the value of the detailed information gained and the practical feasibility. For clinical trials where the methods will be used for decision making, compliance with regulatory requirements is important. Although invasive molecular assays are valuable, particular focus should be placed on the use and development of noninvasive techniques, which overcome the logistical and ethical problems associated with obtaining tumor tissue.

Success in advancing the use of biomarkers for rational cancer drug development can only be achieved through collaborative efforts involving funding authorities, academia, regulatory bodies, patient advocacy groups, and the pharmaceutical and biotechnology industry. Collaborations need to address key issues such as facilitating information and technology flow, providing consensus statements for standards, processes, and validation, and the development and validation of suitably robust and informative assays for clinical trials. Improvements in the design and conduct of clinical trials based on the use of biomarkers will benefit all stakeholders. Academic researchers will benefit because the development and use of drugs will become more scientifically based. Pharmaceutical companies will benefit because the decision-making tools should save time and money and reduce attrition in the clinic. Patients will benefit from the rapid development and approval of innovative drugs.

Biomarkers are very high on the agenda of all researchers involved in drug discovery and development. In addition to biomarkers that provide information on the molecular and therapeutic effects of new drugs, there is also considerable interest now in identifying biomarkers that will provide information on potential toxicity. Pharmacogenomic stratification of patients for both efficacy and toxicity is a long-term goal of this approach.

The use of molecular signatures of drug action and drug sensitivity has the potential to transform the development of molecularly targeted agents for cancer treatment. The next few years will be increasingly exciting as the goal of providing rational individualized therapeutics for cancer draws closer.

ACKNOWLEDGMENTS

Work in the authors' laboratory is funded by Cancer Research UK [CUK] Program Grant Number C309/A2187. P.W. is a Cancer Research UK Life Fellow and D.S. is a Cancer Research UK Clinical Fellow. We thank many colleagues and collaborators for helpful discussions.

REFERENCES

Aboagye, E. O., and Price, P. M. (2003). Use of positron emission tomography in anticancer drug development. *Invest. New Drugs* **21,** 169–181.

Adjei, A. A., and Hidalgo, M. (2005). Intracellular signal transduction pathway proteins as targets for cancer therapy. *J. Clin. Oncol.* **23,** 5386–5403.

Agrawal, M., and Emanuel, E. J. (2003). Ethics of phase 1 oncology studies: Reexamining the arguments and data. *JAMA* **290,** 1075–1082.

Albanell, J., Rojo, F., Averbuch, S., Feyereislova, A., Mascaro, J. M., Herbst, R., LoRusso, P., Rischin, D., Sauleda, S., Gee, J., Nicholson, R. I., and Baselga, J. (2002). Pharmacodynamic studies of the epidermal growth factor receptor inhibitor ZD1839 in skin from cancer patients: Histopathologic and molecular consequences of receptor inhibition. *J. Clin. Oncol.* **20,** 110–124.

Antoch, G., Kanja, J., Bauer, S., Kuehl, H., Renzing-Koehler, K., Schuette, J., Bockisch, A., Debatin, J. F., and Freudenberg, L. S. (2004). Comparison of PET, CT, and dual-modality PET/CT imaging for monitoring of imatinib (STI571) therapy in patients with gastrointestinal stromal tumors. *J. Nucl. Med.* **45,** 357–365.

Arteaga, C. L., and Baselga, J. (2004). Tyrosine kinase inhibitors: Why does the current process of clinical development not apply to them? *Cancer Cell* **5,** 525–531.

Atkins, M. B., Hidalgo, M., Stadler, W. M., Logan, T. F., Dutcher, J. P., Hudes, G. R., Park, Y., Liou, S. H., Marshall, B., Boni, J. P., Dukart, G., and Sherman, M. L. (2004). Randomized phase II study of multiple dose levels of CCI-779, a novel mammalian target of rapamycin kinase inhibitor, in patients with advanced refractory renal cell carcinoma. *J. Clin. Oncol.* **22,** 909–918.

Bader, A. G., Kang, S., Zhao, L., and Vogt, P. K. (2005). Oncogenic PI3K deregulates transcription and translation. *Nat. Rev. Cancer* **5,** 921–929.

Banerji, U., O'Donnell, A., Scurr, M., Pacey, S., Stapleton, S., Asad, Y., Simmons, L., Maloney, A., Raynaud, F., Campbell, M., Walton, M., Lakhani, S., *et al.* (2005a). Phase I pharmacokinetic and pharmacodynamic study of 17-allylamino, 17-demethoxygeldanamycin in patients with advanced malignancies. *J. Clin. Oncol.* **23,** 4152–4161.

Banerji, U., Walton, M., Raynaud, F., Grimshaw, R., Kelland, L., Valenti, M., Judson, I., and Workman, P. (2005b). Pharmacokinetic-pharmacodynamic relationships for the heat shock protein 90 molecular chaperone inhibitor 17-allylamino, 17-demethoxygeldanamycin in human ovarian cancer xenograft models. *Clin. Cancer Res.* **11,** 7023–7032.

Baselga, J. (2003). Skin as a surrogate tissue for pharmacodynamic end points: Is it deep enough? *Clin. Cancer Res.* **9,** 2389–2390.

Baselga, J., Rischin, D., Ranson, M., Calvert, H., Raymond, E., Kieback, D. G., Kaye, S. B., Gianni, L., Harris, A., Bjork, T., Averbuch, S. D., Feyereislova, A., *et al.* (2002). Phase I safety, pharmacokinetic, and pharmacodynamic trial of ZD1839, a selective oral epidermal growth factor receptor tyrosine kinase inhibitor, in patients with five selected solid tumor types. *J. Clin. Oncol.* **20,** 4292–4302.

Baselga, J., Albanell, J., Ruiz, A., Lluch, A., Gascon, P., Guillem, V., Gonzalez, S., Sauleda, S., Marimon, I., Tabernero, J. M., Koehler, M. T., and Rojo, F. (2005). Phase II and tumor pharmacodynamic study of gefitinib in patients with advanced breast cancer. *J. Clin. Oncol.* **23**, 5323–5333.

Beaney, R. P., Lammertsma, A. A., Jones, T., McKenzie, C. G., and Halnan, K. E. (1984). Positron emission tomography for *in-vivo* measurement of regional blood flow, oxygen utilisation, and blood volume in patients with breast carcinoma. *Lancet* **1**, 131–134.

Becher, O. J., and Holland, E. C. (2006). Genetically engineered models have advantages over xenografts for preclinical studies. *Cancer Res.* **66**, 3355–3358.

Beloueche-Babari, M., Jackson, L. E., Al-Saffar, N. M., Workman, P., Leach, M. O., and Ronen, S. M. (2005). Magnetic resonance spectroscopy monitoring of mitogen-activated protein kinase signaling inhibition. *Cancer Res.* **65**, 3356–3363.

Beloueche-Babari, M., Jackson, L. E., Al-Saffar, N. M., Eccles, S. A., Raynaud, F. I., Workman, P., Leach, M. O., and Ronen, S. M. (2006). Identification of magnetic resonance detectable metabolic changes associated with inhibition of phosphoinositide 3-kinase signaling in human breast cancer cells. *Mol. Cancer Ther.* **5**, 187–196.

Biomarkers Definitions Working Group (2001). Biomarkers and surrogate endpoints: Preferred definitions and conceptual framework. *Clin. Pharmacol. Ther.* **69**, 89–95.

Bjornsti, M. A., and Houghton, P. J. (2004). The TOR pathway: A target for cancer therapy. *Nat. Rev. Cancer* **4**, 335–348.

Blundell, T. L., Jhoti, H., and Abell, C. (2002). High-throughput crystallography for lead discovery in drug design. *Nat. Rev. Drug Discov.* **1**, 45–54.

Boulay, A., Zumstein-Mecker, S., Stephan, C., Beuvink, I., Zilbermann, F., Haller, R., Tobler, S., Heusser, C., O'Reilly, T., Stolz, B., Marti, A., Thomas, G., *et al.* (2004). Antitumor efficacy of intermittent treatment schedules with the rapamycin derivative RAD001 correlates with prolonged inactivation of ribosomal protein S6 kinase 1 in peripheral blood mononuclear cells. *Cancer Res.* **64**, 252–261.

Branford, S., Rudzki, Z., Walsh, S., Parkinson, I., Grigg, A., Szer, J., Taylor, K., Herrmann, R., Seymour, J. F., Arthur, C., Joske, D., Lynch, K., *et al.* (2003). Detection of BCR-ABL mutations in patients with CML treated with imatinib is virtually always accompanied by clinical resistance, and mutations in the ATP phosphate-binding loop (P-loop) are associated with a poor prognosis. *Blood* **102**, 276–283.

Broderick, D. K., Di, C., Parrett, T. J., Samuels, Y. R., Cummins, J. M., McLendon, R. E., Fults, D. W., Velculescu, V. E., Bigner, D. D., and Yan, H. (2004). Mutations of PIK3CA in anaplastic oligodendrogliomas, high-grade astrocytomas, and medulloblastomas. *Cancer Res.* **64**, 5048–5050.

Bubley, G. J., Carducci, M., Dahut, W., Dawson, N., Daliani, D., Eisenberger, M., Figg, W. D., Freidlin, B., Halabi, S., Hudes, G., Hussain, M., Kaplan, R., *et al.* (1999). Eligibility and response guidelines for phase II clinical trials in androgen-independent prostate cancer: Recommendations from the prostate-specific antigen working group. *J. Clin. Oncol.* **17**, 3461–3467.

Buchdunger, E., Cioffi, C. L., Law, N., Stover, D., Ohno-Jones, S., Druker, B. J., and Lydon, N. B. (2000). Abl protein-tyrosine kinase inhibitor STI571 inhibits *in vitro* signal transduction mediated by c-kit and platelet-derived growth factor receptors. *J. Pharmacol. Exp. Ther.* **295**, 139–145.

Busse, D., Doughty, R. S., Ramsey, T. T., Russell, W. E., Price, J. O., Flanagan, W. M., Shawver, L. K., and Arteaga, C. L. (2000). Reversible G(1) arrest induced by inhibition of the epidermal growth factor receptor tyrosine kinase requires up-regulation of p27(KIP1) independent of MAPK activity. *J. Biol. Chem.* **275**, 6987–6995.

Campbell, I. G., Russell, S. E., Choong, D. Y., Montgomery, K. G., Ciavarella, M. L., Hooi, C. S., Cristiano, B. E., Pearson, R. B., and Phillips, W. A. (2004). Mutation of the PIK3CA gene in ovarian and breast cancer. *Cancer Res.* **64**, 7678–7681.

Cantley, L. C., and Neel, B. G. (1999). New insights into tumor suppression: PTEN suppresses tumor formation by restraining the phosphoinositide 3-kinase/AKT pathway. *Proc. Natl. Acad. Sci. USA* **96,** 4240–4245.

Carr, K. M., Rosenblatt, K., Petricoin, E. F., and Liotta, L. A. (2004). Genomic and proteomic approaches for studying human cancer: Prospects for true patient-tailored therapy. *Hum. Genomics* **1,** 134–140.

Chan, S., Scheulen, M. E., Johnston, S., Mross, K., Cardoso, F., Dittrich, C., Eiermann, W., Hess, D., Morant, R., Semiglazov, V., Borner, M., Salzberg, M., *et al.* (2005). Phase II study of temsirolimus (CCI-779), a novel inhibitor of mTOR, in heavily pretreated patients with locally advanced or metastatic breast cancer. *J. Clin. Oncol.* **23,** 5314–5322.

Choi, H., Charnsangavej, C., de Castro, F. S., Tamm, E. P., Benjamin, R. S., Johnson, M. M., Macapinlac, H. A., and Podoloff, D. A. (2004). CT evaluation of the response of gastrointestinal stromal tumors after imatinib mesylate treatment: A quantitative analysis correlated with FDG PET findings. *AJR Am. J. Roentgenol.* **183,** 1619–1628.

Chung, E. J., Lee, M. J., Lee, S., and Trepel, J. B. (2006). Assays for pharmacodynamic analysis of histone deacetylase inhibitors. *Expert Opin. Drug Metab. Toxicol.* **2,** 213–230.

Chung, Y. L., Troy, H., Banerji, U., Jackson, L. E., Walton, M. I., Stubbs, M., Griffiths, J. R., Judson, I. R., Leach, M. O., Workman, P., and Ronen, S. M. (2003). Magnetic resonance spectroscopic pharmacodynamic markers of the heat shock protein 90 inhibitor 17-allylamino,17-demethoxygeldanamycin (17-AAG) in human colon cancer models. *J. Natl. Cancer Inst.* **95,** 1624–1633.

Clarke, P. A., Hostein, I., Banerji, U., Stefano, F. D., Maloney, A., Walton, M., Judson, I., and Workman, P. (2000). Gene expression profiling of human colon cancer cells following inhibition of signal transduction by 17-allylamino-17-demethoxygeldanamycin, an inhibitor of the hsp90 molecular chaperone. *Oncogene* **19,** 4125–4133.

Clarke, P. A., te Poele, R., Wooster, R., and Workman, P. (2001). Gene expression microarray analysis in cancer biology, pharmacology, and drug development: Progress and potential. *Biochem. Pharmacol.* **62,** 1311–1336.

Clarke, P. A., George, M. L., Easdale, S., Cunningham, D., Swift, R. I., Hill, M. E., Tait, D. M., and Workman, P. (2003). Molecular pharmacology of cancer therapy in human colorectal cancer by gene expression profiling. *Cancer Res.* **63,** 6855–6863.

Clarke, P. A., te Poele, R., and Workman, P. (2004). Gene expression microarray technologies in the development of new therapeutic agents. *Eur. J. Cancer* **40,** 2560–2591.

Collingridge, D. R., Carroll, V. A., Glaser, M., Aboagye, E. O., Osman, S., Hutchinson, O. C., Barthel, H., Luthra, S. K., Brady, F., Bicknell, R., Price, P., and Harris, A. L. (2002). The development of [(124)I]iodinated-VG76e: A novel tracer for imaging vascular endothelial growth factor *in vivo* using positron emission tomography. *Cancer Res.* **62,** 5912–5919.

Collingridge, D. R., Glaser, M., Osman, S., Barthel, H., Hutchinson, O. C., Luthra, S. K., Brady, F., Bouchier-Hayes, L., Martin, S. J., Workman, P., Price, P., and Aboagye, E. O. (2003). *In vitro* selectivity, *in vivo* biodistribution and tumour uptake of annexin V radiolabelled with a positron emitting radioisotope. *Br. J. Cancer* **89,** 1327–1333.

Collins, J. M. (2003). Functional imaging in phase I studies: Decorations or decision making? *J. Clin. Oncol.* **21,** 2807–2809.

Collins, J. M. (2005). Imaging and other biomarkers in early clinical studies: One step at a time or re-engineering drug development? *J. Clin. Oncol.* **23,** 5417–5419.

Cristofanilli, M., Budd, G. T., Ellis, M. J., Stopeck, A., Matera, J., Miller, M. C., Reuben, J. M., Doyle, G. V., Allard, W. J., Terstappen, L. W., and Hayes, D. F. (2004). Circulating tumor cells, disease progression, and survival in metastatic breast cancer. *N. Engl. J. Med.* **351,** 781–791.

Cully, M., You, H., Levine, A. J., and Mak, T. W. (2006). Beyond PTEN mutations: The PI3K pathway as an integrator of multiple inputs during tumorigenesis. *Nat. Rev. Cancer* **6,** 184–192.

Cummings, J., Ranson, M., Lacasse, E., Ganganagari, J. R., St-Jean, M., Jayson, G., Durkin, J., and Dive, C. (2006). Method validation and preliminary qualification of pharmacodynamic biomarkers employed to evaluate the clinical efficacy of an antisense compound (AEG35156) targeted to the X-linked inhibitor of apoptosis protein XIAP. *Br. J. Cancer* **95,** 42–48.

Dancey, J. E., and Freidlin, B. (2003). Targeting epidermal growth factor receptor—are we missing the mark? *Lancet* **362,** 62–64.

Debiec-Rychter, M., Sciot, R., Le Cesne, A., Schlemmer, M., Hohenberger, P., van Oosterom, A. T., Blay, J. Y., Leyvraz, S., Stul, M., Casali, P. G., Zalcberg, J., Verweij, J., *et al.* (2006). KIT mutations and dose selection for imatinib in patients with advanced gastrointestinal stromal tumours. *Eur. J. Cancer* **42,** 1093–1103.

Deininger, M., Buchdunger, E., and Druker, B. J. (2005). The development of imatinib as a therapeutic agent for chronic myeloid leukemia. *Blood* **105,** 2640–2653.

Demetri, G. D., von Mehren, M., Blanke, C. D., Van den Abbeele, A. D., Eisenberg, B., Roberts, P. J., Heinrich, M. C., Tuveson, D. A., Singer, S., Janicek, M., Fletcher, J. A., Silverman, S. G., *et al.* (2002). Efficacy and safety of imatinib mesylate in advanced gastrointestinal stromal tumors. *N. Engl. J. Med.* **347,** 472–480.

Doroshow, J. H. (2005). Targeting EGFR in non-small-cell lung cancer. *N. Engl. J. Med.* **353,** 200–202.

Druker, B. J. (2002). Inhibition of the Bcr-Abl tyrosine kinase as a therapeutic strategy for CML. *Oncogene* **21,** 8541–8546.

Druker, B. J., Sawyers, C. L., Kantarjian, H., Resta, D. J., Reese, S. F., Ford, J. M., Capdeville, R., and Talpaz, M. (2001). Activity of a specific inhibitor of the BCR-ABL tyrosine kinase in the blast crisis of chronic myeloid leukemia and acute lymphoblastic leukemia with the Philadelphia chromosome. *N. Engl. J. Med.* **344,** 1038–1042.

Dudkin, L., Dilling, M. B., Cheshire, P. J., Harwood, F. C., Hollingshead, M., Arbuck, S. G., Travis, R., Sausville, E. A., and Houghton, P. J. (2001). Biochemical correlates of mTOR inhibition by the rapamycin ester CCI-779 and tumor growth inhibition. *Clin. Cancer Res.* **7,** 1758–1764.

Eary, J. F., Mankoff, D. A., Spence, A. M., Berger, M. S., Olshen, A., Link, J. M., O'Sullivan, F., and Krohn, K. A. (1999). 2-[C-11]thymidine imaging of malignant brain tumors. *Cancer Res.* **59,** 615–621.

Eberhard, D. A., Johnson, B. E., Amler, L. C., Goddard, A. D., Heldens, S. L., Herbst, R. S., Ince, W. L., Janne, P. A., Januario, T., Johnson, D. H., Klein, P., Miller, V. A., *et al.* (2005). Mutations in the epidermal growth factor receptor and in KRAS are predictive and prognostic indicators in patients with non-small-cell lung cancer treated with chemotherapy alone and in combination with erlotinib. *J. Clin. Oncol.* **23,** 5900–5909.

Eckhardt, S. G., Eisenhauer, E. A., Parulekar, W. R., Pazdur, R., and Hirschfield, S. (2003). Developmental Therapeutics: Success and Failures of Clinical Trial Designs of Targeted Compounds. Am. Soc. Clin. Oncol. Ed. Book, pp. 209–219.

Elledge, R. M., and Allred, D. C. (1998). Prognostic and predictive value of p53 and p21 in breast cancer. *Breast Cancer Res. Treat.* **52,** 79–98.

Engelman, J. A., Janne, P. A., Mermel, C., Pearlberg, J., Mukohara, T., Fleet, C., Cichowski, K., Johnson, B. E., and Cantley, L. C. (2005). ErbB-3 mediates phosphoinositide 3-kinase activity in gefitinib-sensitive non-small cell lung cancer cell lines. *Proc. Natl. Acad. Sci. USA* **102,** 3788–3793.

Evelhoch, J. L., LoRusso, P. M., He, Z., DelProposto, Z., Polin, L., Corbett, T. H., Langmuir, P., Wheeler, C., Stone, A., Leadbetter, J., Ryan, A. J., Blakey, D. C., *et al.* (2004). Magnetic

resonance imaging measurements of the response of murine and human tumors to the vascular-targeting agent ZD6126. *Clin. Cancer Res.* **10,** 3650–3657.
Faivre, S., Kroemer, G., and Raymond, E. (2006). Current development of mTOR inhibitors as anticancer agents. *Nat. Rev. Drug Discov.* **5,** 671–688.
Fan, Q. W., Knight, Z. A., Goldenberg, D. D., Yu, W., Mostov, K. E., Stokoe, D., Shokat, K. M., and Weiss, W. A. (2006). A dual PI3 kinase/mTOR inhibitor reveals emergent efficacy in glioma. *Cancer Cell* **9,** 341–349.
Findlay, M., Young, H., Cunningham, D., Iveson, A., Cronin, B., Hickish, T., Pratt, B., Husband, J., Flower, M., and Ott, R. (1996). Noninvasive monitoring of tumor metabolism using fluorodeoxyglucose and positron emission tomography in colorectal cancer liver metastases: Correlation with tumor response to fluorouracil. *J. Clin. Oncol.* **14,** 700–708.
Frank, R., and Hargreaves, R. (2003). Clinical biomarkers in drug discovery and development. *Nat. Rev. Drug Discov.* **2,** 566–580.
Futreal, P. A., Kasprzyk, A., Birney, E., Mullikin, J. C., Wooster, R., and Stratton, M. R. (2001). Cancer and genomics. *Nature* **409,** 850–852.
Galanis, E., Buckner, J. C., Maurer, M. J., Kreisberg, J. I., Ballman, K., Boni, J., Peralba, J. M., Jenkins, R. B., Dakhil, S. R., Morton, R. F., Jaeckle, K. A., Scheithauer, B. W., *et al.* (2005). Phase II trial of temsirolimus (CCI-779) in recurrent glioblastoma multiforme: A north central cancer treatment group study. *J. Clin. Oncol.* **23,** 5294–5304.
Galbraith, S. M. (2003). Antivascular cancer treatments: Imaging biomarkers in pharmaceutical drug development. *Br. J. Radiol.* **76**(Spec. No. 1), S83–S86.
Galbraith, S. M., Rustin, G. J., Lodge, M. A., Taylor, N. J., Stirling, J. J., Jameson, M., Thompson, P., Hough, D., Gumbrell, L., and Padhani, A. R. (2002). Effects of 5, 6-dimethylxanthenone-4-acetic acid on human tumor microcirculation assessed by dynamic contrast-enhanced magnetic resonance imaging. *J. Clin. Oncol.* **20,** 3826–3840.
Galbraith, S. M., Maxwell, R. J., Lodge, M. A., Tozer, G. M., Wilson, J., Taylor, N. J., Stirling, J. J., Sena, L., Padhani, A. R., and Rustin, G. J. (2003). Combretastatin A4 phosphate has tumor antivascular activity in rat and man as demonstrated by dynamic magnetic resonance imaging. *J. Clin. Oncol.* **21,** 2831–2842.
Gandara, D. R., and Gumerlock, P. H. (2005). Epidermal growth factor receptor tyrosine kinase inhibitors plus chemotherapy: Case closed or is the jury still out? *J. Clin. Oncol.* **23,** 5856–5858.
Garrett, M. D., Walton, M. I., McDonald, E., Judson, I., and Workman, P. (2003). The contemporary drug development process: Advances and challenges in preclinical and clinical development. *Prog. Cell Cycle Res.* **5,** 145–158.
Gelmon, K. A., Eisenhauer, E. A., Harris, A. L., Ratain, M. J., and Workman, P. (1999). Anticancer agents targeting signaling molecules and cancer cell environment: Challenges for drug development? *J. Natl. Cancer Inst.* **91,** 1281–1287.
Giaccone, G., Herbst, R. S., Manegold, C., Scagliotti, G., Rosell, R., Miller, V., Natale, R. B., Schiller, J. H., Von Pawel, J., Pluzanska, A., Gatzemeier, U., Grous, J., *et al.* (2004). Gefitinib in combination with gemcitabine and cisplatin in advanced non-small-cell lung cancer: A phase III trial—INTACT 1. *J. Clin. Oncol.* **22,** 777–784.
Goetz, M. P., Toft, D., Reid, J., Ames, M., Stensgard, B., Safgren, S., Adjei, A. A., Sloan, J., Atherton, P., Vasile, V., Salazaar, S., Adjei, A., *et al.* (2005). Phase I trial of 17-allylamino-17-demethoxygeldanamycin in patients with advanced cancer. *J. Clin. Oncol.* **23,** 1078–1087.
Gorre, M. E., Mohammed, M., Ellwood, K., Hsu, N., Paquette, R., Rao, P. N., and Sawyers, C. L. (2001). Clinical resistance to STI-571 cancer therapy caused by BCR-ABL gene mutation or amplification. *Science* **293,** 876–880.
Grem, J. L., Morrison, G., Guo, X. D., Agnew, E., Takimoto, C. H., Thomas, R., Szabo, E., Grochow, L., Grollman, F., Hamilton, J. M., Neckers, L., and Wilson, R. H. (2005). Phase I

and pharmacologic study of 17-(allylamino)-17-demethoxygeldanamycin in adult patients with solid tumors. *J. Clin. Oncol.* **23,** 1885–1893.

Gschwind, A., Fischer, O. M., and Ullrich, A. (2004). The discovery of receptor tyrosine kinases: Targets for cancer therapy. *Nat. Rev. Cancer* **4,** 361–370.

Han, S. W., Kim, T. Y., Hwang, P. G., Jeong, S., Kim, J., Choi, I. S., Oh, D. Y., Kim, J. H., Kim, D. W., Chung, D. H., Im, S. A., Kim, Y. T., *et al.* (2005). Predictive and prognostic impact of epidermal growth factor receptor mutation in non-small-cell lung cancer patients treated with gefitinib. *J. Clin. Oncol.* **23,** 2493–2501.

Hanahan, D., and Weinberg, R. A. (2000). The hallmarks of cancer. *Cell* **100,** 57–70.

Heinrich, M. C., Corless, C. L., Demetri, G. D., Blanke, C. D., von Mehren, M., Joensuu, H., McGreevey, L. S., Chen, C. J., Van den Abbeele, A. D., Druker, B. J., Kiese, B., Eisenberg, B., *et al.* (2003a). Kinase mutations and imatinib response in patients with metastatic gastrointestinal stromal tumor. *J. Clin. Oncol.* **21,** 4342–4349.

Heinrich, M. C., Corless, C. L., Duensing, A., McGreevey, L., Chen, C. J., Joseph, N., Singer, S., Griffith, D. J., Haley, A., Town, A., Demetri, G. D., Fletcher, C. D., *et al.* (2003b). PDGFRA activating mutations in gastrointestinal stromal tumors. *Science* **299,** 708–710.

Hennessy, B. T., Smith, D. L., Ram, P. T., Lu, Y., and Mills, G. B. (2005). Exploiting the PI3K/AKT pathway for cancer drug discovery. *Nat. Rev. Drug Discov.* **4,** 988–1004.

Henson, D. E. (2005). Back to the drawing board on immunohistochemistry and predictive factors. *J. Natl. Cancer Inst.* **97,** 1796–1797.

Herbst, R. S., Giaccone, G., Schiller, J. H., Natale, R. B., Miller, V., Manegold, C., Scagliotti, G., Rosell, R., Oliff, I., Reeves, J. A., Wolf, M. K., Krebs, A. D., *et al.* (2004). Gefitinib in combination with paclitaxel and carboplatin in advanced non-small-cell lung cancer: A phase III trial—INTACT 2. *J. Clin. Oncol.* **22,** 785–794.

Herbst, R. S., Prager, D., Hermann, R., Fehrenbacher, L., Johnson, B. E., Sandler, A., Kris, M. G., Tran, H. T., Klein, P., Li, X., Ramies, D., Johnson, D. H., *et al.* (2005). TRIBUTE: A phase III trial of erlotinib hydrochloride (OSI-774) combined with carboplatin and paclitaxel chemotherapy in advanced non-small-cell lung cancer. *J. Clin. Oncol.* **23,** 5892–5899.

Herrera, R., and Sebolt-Leopold, J. S. (2002). Unraveling the complexities of the Raf/MAP kinase pathway for pharmacological intervention. *Trends Mol. Med.* **8,** S27–S31.

Hess, G., Bunjes, D., Siegert, W., Schwerdtfeger, R., Ledderose, G., Wassmann, B., Kobbe, G., Bornhauser, M., Hochhaus, A., Ullmann, A. J., Kindler, T., Haus, U., *et al.* (2005). Sustained complete molecular remissions after treatment with imatinib-mesylate in patients with failure after allogeneic stem cell transplantation for chronic myelogenous leukemia: Results of a prospective phase II open-label multicenter study. *J. Clin. Oncol.* **23,** 7583–7593.

Hidalgo, M. (2004). New target, new drug, old paradigm. *J. Clin. Oncol.* **22,** 2270–2272.

Hidalgo, M., and Messersmith, W. (2004). Pharmacodynamic Studies in Drug Development. Am. Soc. Clin. Oncol. Ed. Book, pp. 106–163.

Hostein, I., Robertson, D., DiStefano, F., Workman, P., and Clarke, P. A. (2001). Inhibition of signal transduction by the Hsp90 inhibitor 17-allylamino-17-demethoxygeldanamycin results in cytostasis and apoptosis. *Cancer Res.* **61,** 4003–4009.

Huang, S., Bjornsti, M. A., and Houghton, P. J. (2003). Rapamycins: Mechanism of action and cellular resistance. *Cancer Biol. Ther.* **2,** 222–232.

Hudes, G., Carducci, M., Tomczak, P., Dutcher, J., Figlin, R., Kapoor, A., Staroslawska, E., O'Toole, T., Kong, S., and Moore, L. (2006). A phase 3, randomized, 3-arm study of temsirolimus (TEMSR) or interferon-alpha (IFN) or the combination of TEMSR + IFN in the treatment of first-line, poor-risk patients with advanced renal cell carcinoma (adv RCC). ASCO Annual Meeting Proceedings Part I. *J. Clin. Oncol.* **24**(18S), LBA4.

Hughes, T. P., Kaeda, J., Branford, S., Rudzki, Z., Hochhaus, A., Hensley, M. L., Gathmann, I., Bolton, A. E., van Hoomissen, I. C., Goldman, J. M., and Radich, J. P. (2003). Frequency of major molecular responses to imatinib or interferon alfa plus cytarabine in newly diagnosed chronic myeloid leukemia. *N. Engl. J. Med.* **349**, 1423–1432.

Hurwitz, H., Fehrenbacher, L., Novotny, W., Cartwright, T., Hainsworth, J., Heim, W., Berlin, J., Baron, A., Griffing, S., Holmgren, E., Ferrara, N., Fyfe, G., *et al.* (2004). Bevacizumab plus irinotecan, fluorouracil, and leucovorin for metastatic colorectal cancer. *N. Engl. J. Med.* **350**, 2335–2342.

Hynes, N. E., and Lane, H. A. (2005). ERBB receptors and cancer: The complexity of targeted inhibitors. *Nat. Rev. Cancer* **5**, 341–354.

Irish, J. M., Kotecha, N., and Nolan, G. P. (2006). Mapping normal and cancer cell signalling networks: Towards single-cell proteomics. *Nat. Rev. Cancer* **6**, 146–155.

Jackman, A., Workman, P., and Kaye, S. (2004). The combination of cytotoxic and molecularly targeted therapies- can it be done? *Drug Discov. Today:Ther. Strateg.* **1**, 445–454.

Jayson, G. C., Zweit, J., Jackson, A., Mulatero, C., Julyan, P., Ranson, M., Broughton, L., Wagstaff, J., Hakannson, L., Groenewegen, G., Bailey, J., Smith, N., *et al.* (2002). Molecular imaging and biological evaluation of HuMV833 anti-VEGF antibody: Implications for trial design of antiangiogenic antibodies. *J. Natl. Cancer Inst.* **94**, 1484–1493.

Jonkers, J., and Berns, A. (2002). Conditional mouse models of sporadic cancer. *Nat. Rev. Cancer* **2**, 251–265.

Kang, S., Bader, A. G., and Vogt, P. K. (2005). Phosphatidylinositol 3-kinase mutations identified in human cancer are oncogenic. *Proc. Natl. Acad. Sci. USA* **102**, 802–807.

Kantarjian, H., Sawyers, C., Hochhaus, A., Guilhot, F., Schiffer, C., Gambacorti-Passerini, C., Niederwieser, D., Resta, D., Capdeville, R., Zoellner, U., Talpaz, M., Druker, B., *et al.* (2002). Hematologic and cytogenetic responses to imatinib mesylate in chronic myelogenous leukemia. *N. Engl. J. Med.* **346**, 645–652.

Keen, H. G., Dekker, B. A., Disley, L., Hastings, D., Lyons, S., Reader, A. J., Ottewell, P., Watson, A., and Zweit, J. (2005). Imaging apoptosis *in vivo* using 124I-annexin V and PET. *Nucl. Med. Biol.* **32**, 395–402.

Kelland, L. R., Sharp, S. Y., Rogers, P. M., Myers, T. G., and Workman, P. (1999). DT-Diaphorase expression and tumor cell sensitivity to 17-allylamino, 17-demethoxygeldanamycin, an inhibitor of heat shock protein 90. *J. Natl. Cancer Inst.* **91**, 1940–1949.

Kelloff, G. J., and Sigman, C. C. (2005). New science-based endpoints to accelerate oncology drug development. *Eur. J. Cancer* **41**, 491–501.

Kelloff, G. J., Bast, R. C., Jr., Coffey, D. S., D'Amico, A. V., Kerbel, R. S., Park, J. W., Ruddon, R. W., Rustin, G. J., Schilsky, R. L., Sigman, C. C., and Woude, G. F. (2004). Biomarkers, surrogate end points, and the acceleration of drug development for cancer prevention and treatment: An update prologue. *Clin. Cancer Res.* **10**, 3881–3884.

Kelloff, G. J., Hoffman, J. M., Johnson, B., Scher, H. I., Siegel, B. A., Cheng, E. Y., Cheson, B. D., O'Shaughnessy, J., Guyton, K. Z., Mankoff, D. A., Shankar, L., Larson, S. M., *et al.* (2005). Progress and promise of FDG-PET imaging for cancer patient management and oncologic drug development. *Clin. Cancer Res.* **11**, 2785–2808.

Knight, Z. A., Gonzalez, B., Feldman, M. E., Zunder, E. R., Goldenberg, D. D., Williams, O., Loewith, R., Stokoe, D., Balla, A., Toth, B., Balla, T., Weiss, W. A., *et al.* (2006). A pharmacological map of the PI3-K family defines a role for p110alpha in insulin signaling. *Cell* **125**, 733–747.

Kola, I., and Landis, J. (2004). Can the pharmaceutical industry reduce attrition rates? *Nat. Rev. Drug Discov.* **3**, 711–715.

Konishi, J., Yamazaki, K., Kinoshita, I., Isobe, H., Ogura, S., Sekine, S., Ishida, T., Takashima, R., Nakadate, M., Nishikawa, S., Hattori, T., Asahina, H., *et al.* (2005). Analysis of the response and toxicity to gefitinib of non-small cell lung cancer. *Anticancer Res.* **25**, 435–441.

Korn, E. L. (2004). Nontoxicity endpoints in phase I trial designs for targeted, non-cytotoxic agents. *J. Natl. Cancer Inst.* **96**, 977–978.

Korn, E. L., Arbuck, S. G., Pluda, J. M., Simon, R., Kaplan, R. S., and Christian, M. C. (2001). Clinical trial designs for cytostatic agents: Are new approaches needed? *J. Clin. Oncol.* **19,** 265–272.

Krause, D. S., and Van Etten, R. A. (2005). Tyrosine kinases as targets for cancer therapy. *N. Engl. J. Med.* **353,** 172–187.

Kris, M. G., Natale, R. B., Herbst, R. S., Lynch, T. J., Jr., Prager, D., Belani, C. P., Schiller, J. H., Kelly, K., Spiridonidis, H., Sandler, A., Albain, K. S., Cella, D., *et al.* (2003). Efficacy of gefitinib, an inhibitor of the epidermal growth factor receptor tyrosine kinase, in symptomatic patients with non-small cell lung cancer: A randomized trial. *JAMA* **290,** 2149–2158.

Lander, E. S., Linton, L. M., Birren, B., Nusbaum, C., Zody, M. C., Baldwin, J., Devon, K., Dewar, K., Doyle, M., FitzHugh, W., Funke, R., Gage, D., *et al.* (2001). Initial sequencing and analysis of the human genome. *Nature* **409,** 860–921.

Leach, M. O., Arnold, D., Brown, T. R., Charles, H. C., de Certaines, J. D., Evelhoch, J. L., Margulis, A. R., Negendank, W. G., Nelson, S. J., and Podo, F. (1994). International workshop on standardization in clinical magnetic resonance spectroscopy measurements: Proceedings and recommendations. *Acad. Radiol.* **1,** 171–186.

Leach, M. O., Verrill, M., Glaholm, J., Smith, T. A., Collins, D. J., Payne, G. S., Sharp, J. C., Ronen, S. M., McCready, V. R., Powles, T. J., and Smith, I. E. (1998). Measurements of human breast cancer using magnetic resonance spectroscopy: A review of clinical measurements and a report of localized 31P measurements of response to treatment. *NMR Biomed.* **11,** 314–340.

Leach, M. O., Brindle, K. M., Evelhoch, J. L., Griffiths, J. R., Horsman, M. R., Jackson, A., Jayson, G. C., Judson, I. R., Knopp, M. V., Maxwell, R. J., McIntyre, D., Padhani, A. R., *et al.* (2005). The assessment of antiangiogenic and antivascular therapies in early-stage clinical trials using magnetic resonance imaging: Issues and recommendations. *Br. J. Cancer* **92,** 1599–1610.

Lesko, L. J., and Woodcock, J. (2004). Translation of pharmacogenomics and pharmacogenetics: A regulatory perspective. *Nat. Rev. Drug Discov.* **3,** 763–769.

Liu, D., Hutchinson, O. C., Osman, S., Price, P., Workman, P., and Aboagye, E. O. (2002). Use of radiolabelled choline as a pharmacodynamic marker for the signal transduction inhibitor geldanamycin. *Br. J. Cancer* **87,** 783–789.

Liu, G., Rugo, H. S., Wilding, G., McShane, T. M., Evelhoch, J. L., Ng, C., Jackson, E., Kelcz, F., Yeh, B. M., Lee, F. T., Jr., Charnsangavej, C., Park, J. W., *et al.* (2005). Dynamic contrast-enhanced magnetic resonance imaging as a pharmacodynamic measure of response after acute dosing of AG-013736, an oral angiogenesis inhibitor, in patients with advanced solid tumors: Results from a phase I study. *J. Clin. Oncol.* **23,** 5464–5473.

Lu, K., Shih, C., and Teicher, B. A. (2000). Expression of pRB, cyclin/cyclin-dependent kinases and E2F1/DP-1 in human tumor lines in cell culture and in xenograft tissues and response to cell cycle agents. *Cancer Chemother. Pharmacol.* **46,** 293–304.

Ludwig, J. A., and Weinstein, J. N. (2005). Biomarkers in cancer staging, prognosis and treatment selection. *Nat. Rev. Cancer* **5,** 845–856.

Lynch, T. J., Bell, D. W., Sordella, R., Gurubhagavatula, S., Okimoto, R. A., Brannigan, B. W., Harris, P. L., Haserlat, S. M., Supko, J. G., Haluska, F. G., Louis, D. N., Christiani, D. C., *et al.* (2004). Activating mutations in the epidermal growth factor receptor underlying responsiveness of non-small-cell lung cancer to gefitinib. *N. Engl. J. Med.* **350,** 2129–2139.

Maloney, A., and Workman, P. (2002). HSP90 as a new therapeutic target for cancer therapy: The story unfolds. *Expert Opin. Biol. Ther.* **2,** 3–24.

Maloney, A., Clarke, P. A., and Workman, P. (2003). Genes and proteins governing the cellular sensitivity to HSP90 inhibitors: A mechanistic perspective. *Curr. Cancer Drug Targets* **3,** 331–341.

Martinelli, G., Iacobucci, I., Rosti, G., Pane, F., Amabile, M., Castagnetti, F., Cilloni, D., Soverini, S., Testoni, N., Specchia, G., Merante, S., Zaccaria, A., *et al.* (2006). Prediction of response to imatinib by prospective quantitation of BCR-ABL transcript in late chronic phase chronic myeloid leukemia patients. *Ann. Oncol.* **17,** 495–502.

Maxwell, R. J., Wilson, J., Prise, V. E., Vojnovic, B., Rustin, G. J., Lodge, M. A., and Tozer, G. M. (2002). Evaluation of the anti-vascular effects of combretastatin in rodent tumours by dynamic contrast enhanced MRI. *NMR Biomed.* **15,** 89–98.

McCabe, A., Dolled-Filhart, M., Camp, R. L., and Rimm, D. L. (2005). Automated quantitative analysis (AQUA) of in situ protein expression, antibody concentration, and prognosis. *J. Natl. Cancer Inst.* **97,** 1808–1815.

McDonald, E., Jones, K., Brough, P. A., Drysdale, M. J., and Workman, P. (2006). Discovery and development of pyrazole-scaffold hsp90 inhibitors. *Curr. Top. Med. Chem.* **6,** 1193–1203.

Mellinghoff, I. K., and Sawyers, C. L. (2004). TORward AKTually useful mouse models. *Nat. Med.* **10,** 579–580.

Mellinghoff, I. K., Wang, M. Y., Vivanco, I., Haas-Kogan, D. A., Zhu, S., Dia, E. Q., Lu, K. V., Yoshimoto, K., Huang, J. H., Chute, D. J., Riggs, B. L., Horvath, S., *et al.* (2005). Molecular determinants of the response of glioblastomas to EGFR kinase inhibitors. *N. Engl. J. Med.* **353,** 2012–2024.

Mitsudomi, T., Kosaka, T., Endoh, H., Horio, Y., Hida, T., Mori, S., Hatooka, S., Shinoda, M., Takahashi, T., and Yatabe, Y. (2005). Mutations of the epidermal growth factor receptor gene predict prolonged survival after gefitinib treatment in patients with non-small-cell lung cancer with postoperative recurrence. *J. Clin. Oncol.* **23,** 2513–2520.

Morgan, B., Thomas, A. L., Drevs, J., Hennig, J., Buchert, M., Jivan, A., Horsfield, M. A., Mross, K., Ball, H. A., Lee, L., Mietlowski, W., Fuxuis, S., *et al.* (2003). Dynamic contrast-enhanced magnetic resonance imaging as a biomarker for the pharmacological response of PTK787/ZK 222584, an inhibitor of the vascular endothelial growth factor receptor tyrosine kinases, in patients with advanced colorectal cancer and liver metastases: Results from two phase I studies. *J. Clin. Oncol.* **21,** 3955–3964.

Murphy, P. S., Viviers, L., Abson, C., Rowland, I. J., Brada, M., Leach, M. O., and Dzik-Jurasz, A. S. (2004). Monitoring temozolomide treatment of low-grade glioma with proton magnetic resonance spectroscopy. *Br. J. Cancer* **90,** 781–786.

O'Brien, S. G., Guilhot, F., Larson, R. A., Gathmann, I., Baccarani, M., Cervantes, F., Cornelissen, J. J., Fischer, T., Hochhaus, A., Hughes, T., Lechner, K., Nielsen, J. L., *et al.* (2003). Imatinib compared with interferon and low-dose cytarabine for newly diagnosed chronic-phase chronic myeloid leukemia. *N. Engl. J. Med.* **348,** 994–1004.

O'Donnell, A., Faivre, S., Judson, I., Delbado, C., Brock, C., Lane, H., Shand, N., Hazell, K., Armand, J.-P., and Raymond, E. (2003). A phase I study of the oral mTOR inhibitor RAD001 as monotherapy to identify the optimal biologically effective dose using toxicity, pharmacokinetic (PK) and pharmacodynamic (PD) endpoints in patients with solid tumours. *Proc. Am. Soc. Clin. Oncol.* **22** (abstr 803).

O'Reilly, K. E., Rojo, F., She, Q. B., Solit, D., Mills, G. B., Smith, D., Lane, H., Hofmann, F., Hicklin, D. J., Ludwig, D. L., Baselga, J., and Rosen, N. (2006). mTOR inhibition induces upstream receptor tyrosine kinase signaling and activates Akt. *Cancer Res.* **66,** 1500–1508.

Oda, T., Heaney, C., Hagopian, J. R., Okuda, K., Griffin, J. D., and Druker, B. J. (1994). Crkl is the major tyrosine-phosphorylated protein in neutrophils from patients with chronic myelogenous leukemia. *J. Biol. Chem.* **269,** 22925–22928.

Panaretou, B., Siligardi, G., Meyer, P., Maloney, A., Sullivan, J. K., Singh, S., Millson, S. H., Clarke, P. A., Naaby-Hansen, S., Stein, R., Cramer, R., Mollapour, M., *et al.* (2002). Activation of the ATPase activity of hsp90 by the stress-regulated cochaperone aha1. *Mol. Cell* **10,** 1307–1318.

Pao, W., Miller, V., Zakowski, M., Doherty, J., Politi, K., Sarkaria, I., Singh, B., Heelan, R., Rusch, V., Fulton, L., Mardis, E., Kupfer, D., *et al.* (2004). EGF receptor gene mutations are common in lung cancers from "never smokers" and are associated with sensitivity of tumors to gefitinib and erlotinib. *Proc. Natl. Acad. Sci. USA* **101,** 13306–13311.

Park, J. W., Kerbel, R. S., Kelloff, G. J., Barrett, J. C., Chabner, B. A., Parkinson, D. R., Peck, J., Ruddon, R. W., Sigman, C. C., and Slamon, D. J. (2004). Rationale for biomarkers and surrogate end points in mechanism-driven oncology drug development. *Clin. Cancer Res.* **10,** 3885–3896.

Parulekar, W. R., and Eisenhauer, E. A. (2004). Phase I trial design for solid tumor studies of targeted, non-cytotoxic agents: Theory and practice. *J. Natl. Cancer Inst.* **96,** 990–997.

Pegram, M. D., Konecny, G. E., O'Callaghan, C., Beryt, M., Pietras, R., and Slamon, D. J. (2004a). Rational combinations of trastuzumab with chemotherapeutic drugs used in the treatment of breast cancer. *J. Natl. Cancer Inst.* **96,** 739–749.

Pegram, M. D., Pienkowski, T., Northfelt, D. W., Eiermann, W., Patel, R., Fumoleau, P., Quan, E., Crown, J., Toppmeyer, D., Smylie, M., Riva, A., Blitz, S., *et al.* (2004b). Results of two open-label, multicenter phase II studies of docetaxel, platinum salts, and trastuzumab in HER2-positive advanced breast cancer. *J. Natl. Cancer Inst.* **96,** 759–769.

Peralba, J. M., DeGraffenried, L., Friedrichs, W., Fulcher, L., Grunwald, V., Weiss, G., and Hidalgo, M. (2003). Pharmacodynamic evaluation of CCI-779, an inhibitor of mTOR, in cancer patients. *Clin. Cancer Res.* **9,** 2887–2892.

Peterson, J. K., and Houghton, P. J. (2004). Integrating pharmacology and *in vivo* cancer models in preclinical and clinical drug development. *Eur. J. Cancer* **40,** 837–844.

Petricoin, E. F., Ardekani, A. M., Hitt, B. A., Levine, P. J., Fusaro, V. A., Steinberg, S. M., Mills, G. B., Simone, C., Fishman, D. A., Kohn, E. C., and Liotta, L. A. (2002). Use of proteomic patterns in serum to identify ovarian cancer. *Lancet* **359,** 572–577.

Petricoin, E. F., III, Bichsel, V. E., Calvert, V. S., Espina, V., Winters, M., Young, L., Belluco, C., Trock, B. J., Lippman, M., Fishman, D. A., Sgroi, D. C., Munson, P. J., *et al.* (2005). Mapping molecular networks using proteomics: A vision for patient-tailored combination therapy. *J. Clin. Oncol.* **23,** 3614–3621.

Piccart-Gebhart, M. J., Procter, M., Leyland-Jones, B., Goldhirsch, A., Untch, M., Smith, I., Gianni, L., Baselga, J., Bell, R., Jackisch, C., Cameron, D., Dowsett, M., *et al.* (2005). Trastuzumab after adjuvant chemotherapy in HER2-positive breast cancer. *N. Engl. J. Med.* **353,** 1659–1672.

Ramaswamy, S. (2004). Translating cancer genomics into clinical oncology. *N. Engl. J. Med.* **350,** 1814–1816.

Ratain, M. J., and Eckhardt, S. G. (2004). Phase II studies of modern drugs directed against new targets: If you are fazed, too, then resist RECIST. *J. Clin. Oncol.* **22,** 4442–4445.

Ratain, M. J., Eisen, T., Stadler, W. M., Flaherty, K. T., Kaye, S. B., Rosner, G. L., Gore, M., Desai, A. A., Patnaik, A., Xiong, H. Q., Rowinsky, E., Abbruzzese, J. L., *et al.* (2006). Phase II placebo-controlled randomized discontinuation trial of sorafenib in patients with metastatic renal cell carcinoma. *J. Clin. Oncol.* **24,** 2505–2512.

Raymond, E., Alexandre, J., Faivre, S., Vera, K., Materman, E., Boni, J., Leister, C., Korth-Bradley, J., Hanauske, A., and Armand, J. P. (2004). Safety and pharmacokinetics of escalated doses of weekly intravenous infusion of CCI-779, a novel mTOR inhibitor, in patients with cancer. *J. Clin. Oncol.* **22,** 2336–2347.

Raynaud, F. I., Whittaker, S. R., Fischer, P. M., McClue, S., Walton, M. I., Barrie, S. E., Garrett, M. D., Rogers, P., Clarke, S. J., Kelland, L. R., Valenti, M., Brunton, L., *et al.* (2005). *In vitro* and *in vivo* pharmacokinetic-pharmacodynamic relationships for the trisubstituted aminopurine cyclin-dependent kinase inhibitors olomoucine, bohemine and CYC202. *Clin. Cancer Res.* **11,** 4875–4887.

Reichert, J. M. (2003). Trends in development and approval times for new therapeutics in the United States. *Nat. Rev. Drug Discov.* **2,** 695–702.

Ronen, S. M., DiStefano, F., McCoy, C. L., Robertson, D., Smith, T. A., Al-Saffar, N. M., Titley, J., Cunningham, D. C., Griffiths, J. R., Leach, M. O., and Clarke, P. A. (1999). Magnetic resonance detects metabolic changes associated with chemotherapy-induced apoptosis. *Br. J. Cancer* **80,** 1035–1041.

Rustin, G. J. (2003). Use of CA-125 to assess response to new agents in ovarian cancer trials. *J. Clin. Oncol.* **21,** 187–193.

Rustin, G. J., Bast, R. C., Jr., Kelloff, G. J., Barrett, J. C., Carter, S. K., Nisen, P. D., Sigman, C. C., Parkinson, D. R., and Ruddon, R. W. (2004). Use of CA-125 in clinical trial evaluation of new therapeutic drugs for ovarian cancer. *Clin. Cancer Res.* **10,** 3919–3926.

Samuels, Y., Wang, Z., Bardelli, A., Silliman, N., Ptak, J., Szabo, S., Yan, H., Gazdar, A., Powell, S. M., Riggins, G. J., Willson, J. K., Markowitz, S., *et al.* (2004). High frequency of mutations of the PIK3CA gene in human cancers. *Science* **304,** 554.

Sarker, D., Evans, T. R. J., Judson, I., Butzberger, P., Marriott, C., Morrison, R., Vora, J., Heise, C., Hannah, A., and de Bono, J. (2005). CHIR-258: First-in-human Phase 1 dose escalating trial of an oral, selectively targeted tyrosine kinase inhibitor in patients with solid tumours. *J. Clin. Oncol.* **23,** 16s (suppl., abstr 3044).

Sausville, E. A., and Burger, A. M. (2006). Contributions of human tumor xenografts to anticancer drug development. *Cancer Res.* **66**(7), 3351–3354.

Sawyers, C. L. (2003a). Opportunities and challenges in the development of kinase inhibitor therapy for cancer. *Genes Dev.* **17,** 2998–3010.

Sawyers, C. L. (2003b). Will mTOR inhibitors make it as cancer drugs? *Cancer Cell* **4,** 343–348.

Sawyers, C. L., Hochhaus, A., Feldman, E., Goldman, J. M., Miller, C. B., Ottmann, O. G., Schiffer, C. A., Talpaz, M., Guilhot, F., Deininger, M. W., Fischer, T., O'Brien, S. G., *et al.* (2002). Imatinib induces hematologic and cytogenetic responses in patients with chronic myelogenous leukemia in myeloid blast crisis: Results of a phase II study. *Blood* **99,** 3530–3539.

Scherf, U., Ross, D. T., Waltham, M., Smith, L. H., Lee, J. K., Tanabe, L., Kohn, K. W., Reinhold, W. C., Myers, T. G., Andrews, D. T., Scudiero, D. A., Eisen, M. B., *et al.* (2000). A gene expression database for the molecular pharmacology of cancer. *Nat. Genet.* **24,** 236–244.

Schindler, T., Bornmann, W., Pellicena, P., Miller, W. T., Clarkson, B., and Kuriyan, J. (2000). Structural mechanism for STI-571 inhibition of abelson tyrosine kinase. *Science* **289,** 1938–1942.

Schmelzle, T., and Hall, M. N. (2000). TOR, a central controller of cell growth. *Cell* **103,** 253–262.

Senechal, K., Heaney, C., Druker, B., and Sawyers, C. L. (1998). Structural requirements for function of the Crkl adapter protein in fibroblasts and hematopoietic cells. *Mol. Cell. Biol.* **18,** 5082–5090.

Shah, N. P., Nicoll, J. M., Nagar, B., Gorre, M. E., Paquette, R. L., Kuriyan, J., and Sawyers, C. L. (2002). Multiple BCR-ABL kinase domain mutations confer polyclonal resistance to the tyrosine kinase inhibitor imatinib (STI571) in chronic phase and blast crisis chronic myeloid leukemia. *Cancer Cell* **2,** 117–125.

Shah, N. P., Tran, C., Lee, F. Y., Chen, P., Norris, D., and Sawyers, C. L. (2004). Overriding imatinib resistance with a novel ABL kinase inhibitor. *Science* **305,** 399–401.

Sharp, S. Y., Kelland, L. R., Valenti, M. R., Brunton, L. A., Hobbs, S., and Workman, P. (2000). Establishment of an isogenic human colon tumor model for NQO1 gene expression: Application to investigate the role of DT-diaphorase in bioreductive drug activation *in vitro* and *in vivo*. *Mol. Pharmacol.* **58,** 1146–1155.

Shayesteh, L., Lu, Y., Kuo, W. L., Baldocchi, R., Godfrey, T., Collins, C., Pinkel, D., Powell, B., Mills, G. B., and Gray, J. W. (1999). PIK3CA is implicated as an oncogene in ovarian cancer. *Nat. Genet.* **21,** 99–102.

Shepherd, F. A., Rodrigues, P. J., Ciuleanu, T., Tan, E. H., Hirsh, V., Thongprasert, S., Campos, D., Maoleekoonpiroj, S., Smylie, M., Martins, R., van Kooten, M., Dediu, M., *et al.* (2005). Erlotinib in previously treated non-small-cell lung cancer. *N. Engl. J. Med.* **353,** 123–132.

Shields, A. F., Grierson, J. R., Dohmen, B. M., Machulla, H. J., Stayanoff, J. C., Lawhorn-Crews, J. M., Obradovich, J. E., Muzik, O., and Mangner, T. J. (1998). Imaging proliferation *in vivo* with [F-18]FLT and positron emission tomography. *Nat. Med.* **4,** 1334–1336.

Slamon, D. J., Clark, G. M., Wong, S. G., Levin, W. J., Ullrich, A., and McGuire, W. L. (1987). Human breast cancer: Correlation of relapse and survival with amplification of the HER-2/neu oncogene. *Science* **235,** 177–182.

Slamon, D. J., Leyland-Jones, B., Shak, S., Fuchs, H., Paton, V., Bajamonde, A., Fleming, T., Eiermann, W., Wolter, J., Pegram, M., Baselga, J., and Norton, L. (2001). Use of chemotherapy plus a monoclonal antibody against HER2 for metastatic breast cancer that overexpresses HER2. *N. Engl. J. Med.* **344,** 783–792.

Sledge, G. W., Jr. (2004). HERe-2 stay: The continuing importance of translational research in breast cancer. *J. Natl. Cancer Inst.* **96,** 725–727.

Smith-Jones, P. M., Solit, D. B., Akhurst, T., Afroze, F., Rosen, N., and Larson, S. M. (2004). Imaging the pharmacodynamics of HER2 degradation in response to Hsp90 inhibitors. *Nat. Biotechnol.* **22,** 701–706.

Sordella, R., Bell, D. W., Haber, D. A., and Settleman, J. (2004). Gefitinib-sensitizing EGFR mutations in lung cancer activate anti-apoptotic pathways. *Science* **305,** 1163–1167.

Stiles, T., Grant, V., and Mawbey, N. British Association of Research Quality Assurance (2005). “Good Clinical Laboratory Practice (GCLP).” Ipswich, UK: BARQA.

Stroobants, S., Goeminne, J., Seegers, M., Dimitrijevic, S., Dupont, P., Nuyts, J., Martens, M., van den Borne, B., Cole, P., Sciot, R., Dumez, H., Silberman, S., *et al.* (2003). 18FDG-Positron emission tomography for the early prediction of response in advanced soft tissue sarcoma treated with imatinib mesylate (Glivec). *Eur. J. Cancer* **39,** 2012–2020.

Tabernero, J., Rojo, F., Burris, H., Casado, E., Macarulla, T., Jones, S., Dimitrijevic, S., Hazell, K., Shand, N., and Baselga, J. (2005). A Phase I study with tumour molecular pharmacodynamic(MPD) evaluation of dose and schedule of the oral mTOR inhibitor Everolimus (RAD001) in patients with adavanced solid tumors. *J. Clin. Oncol.* **23** (suppl, abstr 3007), 16s.

Talpaz, M., Shah, N. P., Kantarjian, H., Donato, N., Nicoll, J., Paquette, R., Cortes, J., O’Brien, S., Nicaise, C., Bleickardt, E., Blackwood-Chirchir, M. A., Iyer, V., *et al.* (2006). Dasatinib in imatinib-resistant Philadelphia chromosome-positive leukemias. *N. Engl. J. Med.* **354,** 2531–2541.

Therasse, P., Arbuck, S. G., Eisenhauer, E. A., Wanders, J., Kaplan, R. S., Rubinstein, L., Verweij, J., Van Glabbeke, M., van Oosterom, A. T., Christian, M. C., and Gwyther, S. G. (2000). New guidelines to evaluate the response to treatment in solid tumors. European Organization for Research and Treatment of Cancer, National Cancer Institute of the United States, National Cancer Institute of Canada. *J. Natl. Cancer Inst.* **92,** 205–216.

Thompson, J. E., and Thompson, C. B. (2004). Putting the rap on Akt. *J. Clin. Oncol.* **22,** 4217–4226.

Torrance, C. J., Agrawal, V., Vogelstein, B., and Kinzler, K. W. (2001). Use of isogenic human cancer cells for high-throughput screening and drug discovery. *Nat. Biotechnol.* **19,** 940–945.

Tsao, M. S., Sakurada, A., Cutz, J. C., Zhu, C. Q., Kamel-Reid, S., Squire, J., Lorimer, I., Zhang, T., Liu, N., Daneshmand, M., Marrano, P., da Cunha, S. G., *et al.* (2005). Erlotinib in lung cancer—molecular and clinical predictors of outcome. *N. Engl. J. Med.* **353,** 133–144.

Tuomela, M., Stanescu, I., and Krohn, K. (2005). Validation overview of bio-analytical methods. *Gene Ther.* **12**(Suppl. 1), S131–S138.

van de Waterbeemd, H., and Gifford, E. (2003). ADMET *in silico* modelling: Towards prediction paradise? *Nat. Rev. Drug Discov.* **2**, 192–204.

van Oosterom, A. T., Judson, I., Verweij, J., Stroobants, S., Donato di Paola, E., Dimitrijevic, S., Martens, M., Webb, A., Sciot, R., Van Glabbeke, M., Silberman, S., and Nielsen, O. S. (2001). Safety and efficacy of imatinib (STI571) in metastatic gastrointestinal stromal tumours: A phase I study. *Lancet* **358**, 1421–1423.

van de Vijver, M. J., He, Y. D., van't Veer, L. J., Dai, H., Hart, A. A., Voskuil, D. W., Schreiber, G. J., Peterse, J. L., Roberts, C., Marton, M. J., Parrish, M., Atsma, D., *et al.* (2002). A gene-expression signature as a predictor of survival in breast cancer. *N. Engl. J. Med.* **347**, 1999–2009.

Van Dyke, T., and Jacks, T. (2002). Cancer modeling in the modern era: Progress and challenges. *Cell* **108**, 135–144.

Venter, J. C., Adams, M. D., Myers, E. W., Li, P. W., Mural, R. J., Sutton, G. G., Smith, H. O., Yandell, M., Evans, C. A., Holt, R. A., Gocayne, J. D., Amanatides, P., *et al.* (2001). The sequence of the human genome. *Science* **291**, 1304–1351.

Vivanco, I., and Sawyers, C. L. (2002). The phosphatidylinositol 3-kinase AKT pathway in human cancer. *Nat. Rev. Cancer* **2**, 489–501.

Whitesell, L., and Lindquist, S. L. (2005). HSP90 and the chaperoning of cancer. *Nat. Rev. Cancer* **5**, 761–772.

Whittaker, S. R., Walton, M. I., Garrett, M. D., and Workman, P. (2004). The cyclin-dependent kinase inhibitor CYC202 (R-roscovitine) inhibits retinoblastoma protein phosphorylation, causes loss of cyclin D1, and activates the mitogen-activated protein kinase pathway. *Cancer Res.* **64**, 262–272.

Williams, R., Baker, A. F., Ihle, N. T., Winkler, A. R., Kirkpatrick, L., and Powis, G. (2006). The skin and hair as surrogate tissues for measuring the target effect of inhibitors of phosphoinositide-3-kinase signaling. *Cancer Chemother. Pharmacol.* **58**, 444–450.

Wilson, C. B., Lammertsma, A. A., McKenzie, C. G., Sikora, K., and Jones, T. (1992). Measurements of blood flow and exchanging water space in breast tumors using positron emission tomography: A rapid and noninvasive dynamic method. *Cancer Res.* **52**, 1592–1597.

Workman, P. (2002). Challenges of PK/PD measurements in modern drug development. *Eur. J. Cancer* **38**, 2189–2193.

Workman, P. (2003a). How much gets there and what does it do?: The need for better pharmacokinetic and pharmacodynamic endpoints in contemporary drug discovery and development. *Curr. Pharm. Des.* **9**, 891–902.

Workman, P. (2003b). The opportunities and challenges of personalized genome-based molecular therapies for cancer: Targets, technologies, and molecular chaperones. *Cancer Chemother. Pharmacol.* **52**(Suppl. 1), S45–S56.

Workman, P. (2004a). Combinatorial attack on multistep oncogenesis by inhibiting the Hsp90 molecular chaperone. *Cancer Lett.* **206**, 149–157.

Workman, P. (2004b). Inhibiting the phosphoinositide 3-kinase pathway for cancer treatment. *Biochem. Soc. Trans.* **32**, 393–396.

Workman, P. (2005a). Drugging the cancer kinome: Progress and challenges in developing personalized molecular cancer therapeutics. *Cold Spring Harb. Symp. Quant. Biol.* **70**, 499–515.

Workman, P. (2005b). Drugging the Cancer Kinome: Successes, Problems, and Emerging Solutions. Am. Soc. Clin. Oncol. Ed. Book, pp. 950–960.

Workman, P. (2005c). Genomics and the second golden era of cancer drug development. *Mol. BioSyst.* **1**(1), 17–26.

Workman, P., and Johnston, P. G. (2005). Genomic profiling of cancer: What next? *J. Clin. Oncol.* **23,** 7253–7256.

Workman, P., and Kaye, S. B. (2002). Translating basic cancer research into new cancer therapeutics. *Trends Mol. Med.* **8,** S1–S9.

Workman, P., Twentyman, P., Balkwill, F., Balmain, A., Chaplin, D., Double, J., Embleton, J., Newell, D., Raymond, R., Stables, J., Stephens, T., and Wallace, J. (1998). United Kingdom co-ordinating committee in cancer research (UKCCR) guidelines for the welfare of animals in experimental neoplasia. *Br. J. Cancer* **77,** 1–10.

Workman, P., Raynaud, F., Clarke, P. A., te Poele, R., Eccles, S., Kelland, L., DiStefano, F., Ahmadi, K., Parker, P., and Waterfield, M. (2004). Pharmacological properties and *in vitro* and *in vivo* antitumour activity of the potent and selective PI3 kinase inhibitor PI-103. *Eur. J. Cancer* **97**(Suppl. 2).

Workman, P., Aboagye, E. O., Chung, Y. L., Griffiths, J. R., Hart, R., Leach, M. O., Maxwell, R. J., McSheehy, P. M., Price, P. M., and Zweit, J. (2006a). Minimally invasive pharmacokinetic and pharmacodynamic technologies in hypothesis-testing clinical trials of innovative therapies. *J. Natl. Cancer Inst.* **98,** 580–598.

Workman, P., Clarke, P. A., Guillard, S., and Raynaud, F. I. (2006b). Drugging the PI3 kinome. *Nat. Biotechnol.* **24,** 794–796.

Young, H., Baum, R., Cremerius, U., Herholz, K., Hoekstra, O., Lammertsma, A. A., Pruim, J., and Price, P. (1999). Measurement of clinical and subclinical tumour response using [18F]-fluorodeoxyglucose and positron emission tomography: Review and 1999 EORTC recommendations. European organization for research and treatment of cancer (EORTC) PET study group. *Eur. J. Cancer* **35,** 1773–1782.

Zhang, H., Chung, D., Yang, Y. C., Neely, L., Tsurumoto, S., Fan, J., Zhang, L., Biamonte, M., Brekken, J., Lundgren, K., and Burrows, F. (2006). Identification of new biomarkers for clinical trials of Hsp90 inhibitors. *Mol. Cancer Ther.* **5,** 1256–1264.

Biomarker Assay Translation from Discovery to Clinical Studies in Cancer Drug Development: Quantification of Emerging Protein Biomarkers

Jean W. Lee,* Daniel Figeys,[†] and Julian Vasilescu[†]

**Amgen Inc., Thousand Oaks, California 91320;*
[†]*Institute of Systems Biology, University of Ottawa, Ottawa K4A 4N2, Canada*

I. Introduction
II. Biomarkers and Cancer Drug Development
III. Biomarker Research Challenges: Technology Translation
  A. Considerations Prior to Translation
  B. Technology Translation
  C. Methods of Choice
IV. Biomarker Research Challenges: Cultural and Process Translation
  A. Diversity of Biomarker Methods and Data Types
  B. Translation from Research to Bioanalytical Environment: Preanalytical Considerations
  C. Method Validation: A Continual Process of Assay Refinement for the Intended Application
V. Clinical Qualification
  A. Clinical Trials
  B. Biomarker Assay Commercialization
VI. Conclusions and Perspectives
References

Many candidate biomarkers emerging from genomics and proteomics research have the potential to serve as predictive indexes for guiding the development of safer and more efficacious drugs. Research and development of biomarker discovery, selection, and clinical qualification, however, is still a relatively new field for the pharmaceutical industry. Advances in technology provide a plethora of analytical tools to discover and analyze mechanism-and-disease-specific biomarkers for drug development. In the discovery phase, differential proteomic analysis using mass spectrometry enables the identification of candidate biomarkers that are associated with a specific mechanism relevant to disease progression and affected by drug treatment. Reliable bioanalytical methods are then developed and implemented to select promising biomarkers for further studies in animals and humans. Quantitative analytical methods capable of generating reliable data constitute a solid basis for statistical assessment of the predictive utility of biomarkers.

Advances in CANCER RESEARCH
0065-230X/07 $35.00
DOI: 10.1016/S0065-230X(06)96010-2

Biomarker method validation is diverse and for purposes that are very different from those of drug bioanalysis or diagnostic use. Besides being flexible, it should sufficiently demonstrate the method's ability to meet the study intent and the attendant regulatory requirements. Several papers have been published outlining specific requirements for successful biomarker method development and validation using a "Fit-for-Purpose" approach. Many of the challenges faced during biomarker discovery as well as during technology and process translation are discussed in this chapter, including preanalytical planning, assay development, and preclinical and clinical validation. Specific references to protein biomarkers for cancer drug development are also discussed.

I. INTRODUCTION

In the post-genomics era, multiple disciplines are contributing to the success of rational drug development strategies including proteomics, metabolomics, bioinformatics, *in silico* simulation, and computational chemistry of both small and macromolecules. Innovative tools enable researchers toward the ultimate goal of personalized medicine. Over the last decade, the number of new drug targets has grown significantly without a corresponding increase in the number of new drug approvals (DiMasi *et al.*, 2003; Reichert, 2003). In the United States for example, there are only about 90 oncology drugs currently in the market (Booth *et al.*, 2003). Although more than 500 anticancer drugs are in development, only a few will ultimately achieve regulatory approval. The primary reason for this is because late-phase clinical trials often reveal unexpected toxicities or side effects. Notable examples include the anticancer drugs Gefitinib (Iressa; AstraZeneca) and Cetuximab (Erbitux; Imclone). Even more problematic are the recent cases of approved cyclooxygenase-2 (COX-2) inhibitor drugs that have been taken off the market or required additional risk warning labels. To address these critical issues, the US Food and Drug Administration (FDA) has proposed several new toolkits in a Critical Path of Innovation document (FDA Document, 2004; www.fda.gov/opacom/hpview.html). The use of biomarkers is among these to aid in drug candidate selection, attrition, optimization, and mechanism confirmation.

According to the Biomarkers Definitions Working Group (2001), a biomarker is defined as a characteristic that is objectively measured and evaluated as an indicator of a biological response to a therapeutic intervention. Biomarkers often fit into the cascade of pathological events that underlie a disease and as a result may serve as a surrogate endpoint[1] during clinical trials. Biomarkers may be categorized into four groups: those of unknown or uncharacterized mechanism; with a demonstrated mechanism but lacking a

[1] A *clinical endpoint* quantifies a characteristic of a patient's condition (i.e., how they feel or function, or the survival rate of a population) and is used to determine the outcome of a clinical trial. A *surrogate endpoint* predicts the safety and efficacy of a drug based on subjective and quantitative data and can be used as a substitute for the clinical endpoint.

dose–response relationship; with a well-characterized dose–response relationship; and those with a proven mechanism, which may serve as a surrogate endpoint (Bjornsson, 2005). Many biomarkers emerging from genomics and proteomics research have been closely linked to specific signaling pathways within cancer cells and may serve as drug target and/or predictive indexes to guide successful drug development. However, the analysis of novel biomarkers often requires sophisticated technologies that are not yet widely available (Lee *et al.*, 2005). Reliable methods are instrumental to generate quantitative data, and to statistically establish the predictive utility of a novel biomarker as a surrogate endpoint or "valid marker" through clinical qualification. During the course of a drug development program, both well-established and emerging novel biomarkers are used for Proof of Biology (PoB), correlation of dose–response relationship, and other purposes from discovery to the post-approval phase (shown by the horizontal progression in Fig. 1). The development process of a novel biomarker (depicted in the vertical progression) is intertwined with the drug development processes. Biomarker development may happen concurrently

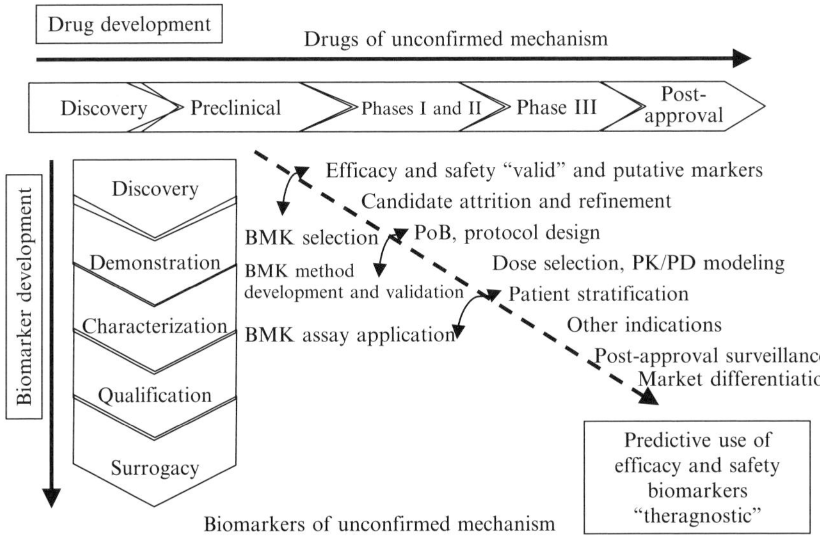

Fig. 1 The intertwined processes of drug development and biomarker development. The horizontal blocks depict the progression of drug development of a new chemical or biological entity with unconfirmed mechanism of action. The drug development uses multiple biomarkers in various purposes from efficacy/safety assessment, down to market differentiation. The vertical blocks depict the developmental processes of moving a novel biomarker of unconfirmed mechanism to proof of biology, and to surrogacy. The processes include biomarker (BMK) selection of on- and off-target markers, method development, validation, and application. The intertwined processes lead to ultimate application of biomarkers in "theragnostics" (diagonal dashed arrow).

with the development of a single or multiple drug candidates, a refined new chemical or biological entity, or for extended indications and/or additional mechanisms.

Biomarkers are becoming essential to the drug development process from preclinical, to clinical, and for post-approval therapeutics monitoring (such as conditional approval with additional efficacy-safety data) and diagnostic/ treatment decisions. In cancer drug development, biomarkers are used to help monitor drug effects and their data for early decision on candidate selection. The level of confidence in the biomarker data, and thus its role in making the right decision, is dependent on the amount and quality of information available about the biomarker. While some of the technologies in the field of biomarker discovery might be suitable for diagnostic purpose, they may not be suitable for preclinical and clinical application that often requires reliable quantitative data. Unfortunately, there is significant confusion surrounding selection of technology and the validation of method for biomarker discovery, and the translation of methods from discovery through the subsequent phases of drug development. The intent of this chapter is to help clarify the technical process of cancer biomarker discovery and assay translation.

II. BIOMARKERS AND CANCER DRUG DEVELOPMENT

Data of gene mutation and aberrant control of gene expression, together with differential proteomics have the potential to provide a better understanding of the mechanisms of tumor initiation, progression, the effects of therapeutic intervention, and drug resistance in nonresponders or relapse. Genomic and protein biomarkers are being introduced in the new drug development process such as for candidate attrition and refinement, PoB, confirmation of efficacy/safety, dose selection from drug exposure–response relationship, and patient selection (Fig. 1). The on-target effect of a new drug can be measured through novel biomarkers associated with the target (proximal biomarkers). However, since cancer is a complex disease with multiple contributors to the disease progression, off-target effects of connected pathways should also be monitored with other sets of biomarkers (distal biomarkers) to track the biological effects leading to disease intervention (efficacy, benefits) and/or adverse reactions (safety, risks).

Biomarker investigations can serve as early filters of go/no-go decisions for drug candidates and are important for lead optimization. Pharmacodynamic (PD) biomarkers provide mechanistic and efficacy information about an investigative compound. Proximal biomarkers are a subset of PD biomarkers that reflect drug action on the specific target, while distal biomarkers reflect downstream actions that lead to disease progression (Wagner, 2002). Data from both proximal and distal PD biomarkers enable tracking of the *in vivo* sequence of events with respect to drug treatment. The relationship between the pharmacokinetic (PK) data of drug dose exposure and the PD biomarker concentration (the PK/PD relationship) can be used to provide an initial assessment of drug absorption, distribution, metabolism and elimination, efficacy, and toxicity. For example, time course analysis of proximal biomarkers tends to closely follow the drug concentrations in plasma. Therefore, this data may be used to establish effective dose ranges and allow for patient–dose titrations. This represents an important step toward the goal of personalized medicine, where optimal treatment is selected for patients that ensures maximum therapeutic response with minimal side effects. Biomarkers that can predict the clinical outcome (a surrogate endpoint or "valid biomarker") would provide better assessment and improved treatment plan with benefits that far outweigh risks. This is especially important for most cancer patients since early diagnosis and correct treatment is key to survival.

The establishment of a biomarker as a surrogate endpoint involves iterative processes and multiple test systems to study the cause-and-effect relationship in *in vitro* cell lines, animal models, and patient clinical trials (Fig. 2). Cancer cell lines with genetic aberrances enable the discovery of protein biomarkers that are up- or down-regulated with respect to disease progression. Often, these cell lines are used for high-throughput screening of drug candidates and hundreds of gene and protein profiles are simultaneously evaluated (Weinstein, 2004). Unlike the three-dimensional (3D) scaffold structures that are present in tissue, *in vitro* test systems often consist of homogeneous cell populations that are spread out in layers. Therefore, signaling pathways and cell-to-cell communication may differ significantly compared to what normally occurs in animal models and in humans. As well, *in vitro* test systems cannot provide information on the effects of a drug in the microenvironment of an organ or other organs that affect paracrine pathways, vascular escape, inflammation, invasion, and metastasis. Animal models of xenografts, transgenic, and knockouts are therefore used to provide the transition from *in vitro* to *in vivo* testing of candidate biomarkers. Preclinical studies with animal models are also used to evaluate if candidate biomarkers should be selected for further clinical investigation.

Evidence for the role of potential biomarkers in cancer progression, relapse, and the linkage of biological pathways leading to regression or side effects is gathered in the *in vivo* system. The usefulness of the biomarkers is then evaluated with respect to drug treatment in early phase II clinical trials with a small human population. Only a subset of the putative biomarkers (usually less than 10) with promising results and biologically relevant mechanisms are then selected for late-phase clinical trials. Large

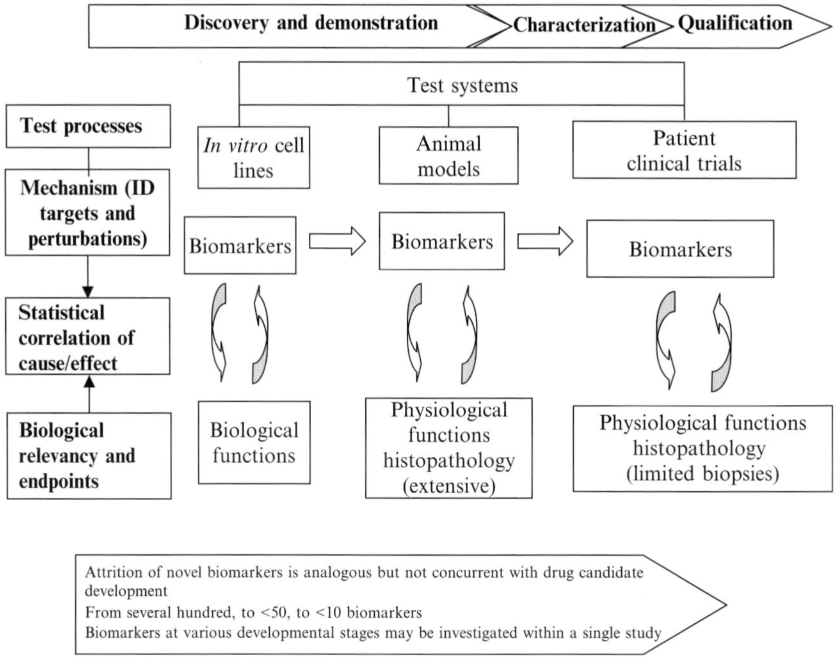

Fig. 2 Development of novel biomarker. Discovery, demonstration, characterization, and qualification may take place in one of the test systems of *in vitro* cell lines, animal models, and patient clinical trials. The test processes are to measure the biomarker perturbations, identify the causal mechanism of perturbation, and to correlate the data with biological and clinical endpoints to find statistical and clinical meaning biomarkers for predictive use.

datasets from phase III and post-approval clinical trials are subsequently used to build a statistical correlation to the clinical outcome and to determine the predictive utility of the biomarkers (clinical qualification). Finally, the same subset of biomarkers may be further evaluated with respect to other drug treatments of the same or similar mechanism. Investigations may also be extended to other diseases that overlap or share similar characteristics.

The utility of a potential biomarker must be demonstrated by statistical correlation to the disease clinical endpoint. It is important to note that clinical and biological endpoints differ at each of the systems depicted in Fig. 2; they may also vary with the target mechanism in question. For example, the biological endpoint of a tyrosine kinase inhibitor may simply be cell death, while the relevant biomarker signal readout could be the phosphorylation activity of the tyrosine kinase. However, the clinical endpoint of an animal model is more complex than that of a cell line because other factors such as metabolic rate and tumor size would be evaluated. For human studies, survival is the ultimate clinical endpoint for a disease such as cancer, which requires a considerable amount of time for data collection. Therefore, earlier clinical endpoints, such as time to tumor progression and metastasis (or disease stabilization), may be used and additional data related to survival and quality of life would be collected during the post-approval phase. In addition, retroactive samples from patients with survival data could be used to add to the statistical correlation of the novel biomarker to early and morbid clinical endpoints.

Biological variability is an important factor to consider when a statistical correlation to a clinical endpoint is determined for a candidate biomarker. This is because the overall noise of sample is the sum of both analytical and biological variability. Analytical variability can often be measured and controlled, while the biological variability is a lot more difficult to assess and control. For example, a proteomic or bioanalytical analysis can be broken down into several steps where tests can be designed to measure the noise from each of the steps. Controls can then be put in place to monitor variation and minimize noise by a normalization process. These controls also provide information on experimental variations, which are necessary when deciding to include different datasets in the overall analysis. Without the establishment of appropriate controls, the rejection of a particular dataset becomes arbitrary. Once a well-characterized analytical process has been put in place, standard statistical tools can be used to evaluate the quality of the different datasets.

The progress of a potential biomarker does not always coincide with or parallel to that of a new drug development, which begins in the discovery phase. Therefore, a 2D development matrix (Fig. 1) should be considered, one for biomarker development and the other for drug candidate development. To avoid confusion, a detailed work plan should be prepared that includes study objectives for each potential biomarker. Innovative companies often organize biomarker work groups to facilitate timely input and communication among therapeutic areas and supporting teams. The time required for biomarker assay development and method validation, the operational and logistical issues including preanalytical factors, and the limitations in data interpretation are key elements for careful planning and executions to support biomarker development and their applications.

III. BIOMARKER RESEARCH CHALLENGES: TECHNOLOGY TRANSLATION

The selection and implementation of technologies throughout the different stages of biomarker research (Fig. 3) must be carefully planned. Several factors should be considered including sample requirements, sensitivity and

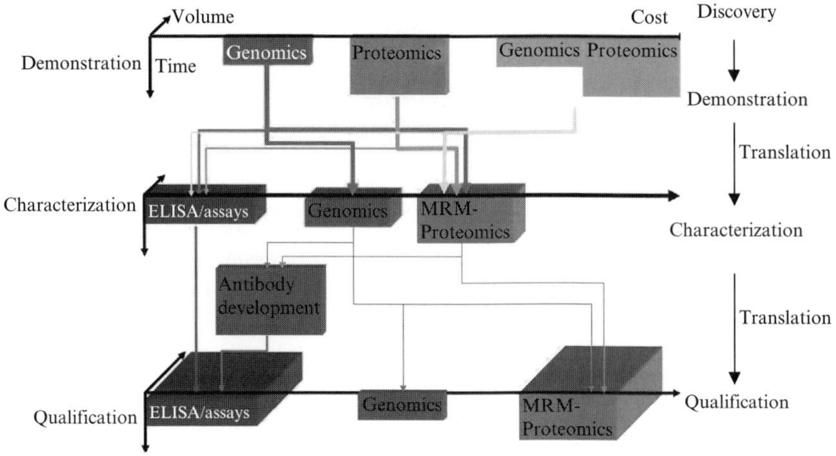

Fig. 3 Technology consideration, translation, and implementation at different stages of biomarker research. Three major considerations are depicted in the three dimensions of sample volume, cost, and time of method development and assay. The technologies suitable for each stage are shown in the color blocks. The connecting down arrows represents the translational processes from one phase into the next. (See Color Insert.)

assay range, analysis time, and cost. Challenges in technology translation and methodology limitation may be encountered during each stage of biomarker research, particularly during the discovery phase. For example, if a putative phosphorylated biomarker is identified and a phospho-specific antibody is not available or cannot be produced, development of a mass spectrometry (MS)-based approach capable of detecting its phosphorylation state may be required for further studies.

The use of genomic and proteomic technologies for biomarker discovery has attracted considerable interest over the past few years. Although gene expression analysis is currently less expensive and more rapid, the type of biological samples that can be used often limits its applicability. In contrast, proteomics enables screening for biomarkers in readily available biological fluids such as serum and plasma. A variety of proteomic platforms are available for biomarker discovery, such as 1D or 2D gel electrophoresis combined with mass spectrometry (Lambert *et al.*, 2005), gel-free liquid chromatography coupled to mass spectrometry (LCMS) (Lambert *et al.*, 2005), and gel-free multidimensional liquid chromatography coupled to mass spectrometry (MudPIT) (Washburn *et al.*, 2001). Differential proteomic analysis using isotope-coded affinity tags (i.e., ICAT or iTRAC reagents) and/or specialized software applications are available options. More recently, matrix assist laser desorption ionization (MALDI)-based imaging of tissue has been introduced as an approach for the discovery of potential biomarkers associated with drug treatment in breast cancer (Reyzer *et al.*, 2004). MALDI imaging is a promising new approach to study tissue slides and analyze small molecules, their metabolites, as well as changes in protein expressions. It is still limited in terms of spatial resolution and the number of protein identified. However, it is becoming an area of intense technological research, which will lead to improved techniques.

The selection and implementation of the appropriate technologies for biomarker discovery generates experimental data to be translated into useful information to guide subsequent studies. When evidence for a candidate biomarker is only provided by a single type of technology in the discovery phase, this may seriously limit the use of secondary technologies down the road. For example, a peak generated during a surface-enhanced laser desorption ionization (SELDI)-MS experiment is often considered as sufficient evidence for the identification of a candidate biomarker. However, this information cannot be utilized for subsequent experimentation with an antibody-based technology, such as enzyme-linked immunosorbent assay (ELISA), and will likely require the use of SELDI-MS throughout the remaining stages of biomarker research. Recent work has demonstrated that a peak detected by SELDI-MS can potentially be identified by affinity purification coupled with gel electrophoresis and mass spectrometry (Paradis *et al.*, 2005). However, so far the identification of SELDI-MS peak as proteins has been more the exception than the rule.

A. Considerations Prior to Translation

The list of potential biomarkers derived from high-throughput discovery technologies, such as proteomics, can be extensive. It would not be practical to move all the candidate biomarkers into the demonstration phase or possible to develop assays for all of them. Therefore, it is important to define parameters for the rational, statistical, and evidence-based selection and rejection of candidate biomarkers (Hunt *et al.*, 2005; Listgarten and Emili, 2005). Many parameters will be question specific and will have to be defined by the results of individual experiments. A minimum list of parameters, such as the following, is vital to avoid haphazard evaluation of moving the candidate biomarkers forward into preclinical and clinical trials.

1. Statistical figure of merit: Most proteomic approaches provide a differential value, that is, protein x has a ratio of y between two samples (experimental and control). The following questions regarding the statistical robustness of the result must be answered as a minimal requirement:

Is the ratio statistically different than 1:1? Was the measurement repeated on multiple biological samples? What are the standard deviations on the ratio measured from the same sample and measured from samples to samples? Is the study design sound (e.g., were there appropriate controls to filter out noises)?

2. The absolute or normalized level of the candidate biomarker: Most proteomic approaches provide a differential measurement and not an absolute measurement. In some instances, the candidate biomarker might be a protein that is already present in normal biological fluids. In these cases, it is important to know the protein concentration and its fluctuations between different biological fluids in the normal population. For a novel biomarker, an absolute measurement would not be possible. Normalization against a known control (e.g., a consistent protein as an internal standard or an added protein as an external standard) could be used to provide reference points for comparisons from experiment to experiment. The candidate biomarker might be rejected due to wide biological variability or time fluctuation that is difficult to control in the clinical phase.

3. The biological annotation surrounding the candidate biomarkers: Can the biomarker be mechanistically associated to the study (i.e., disease, drug treatment)? If the answer is yes, then this reinforces the position of the candidate biomarker.

4. The tissue distribution of the biomarker: Is there information available on the expression of the candidate biomarker in different tissues, diseases, or treatment? This information can be used to reject candidate biomarkers due to their widespread expression or due to their lack of specificity as indicated by their differential expression.

5. The translation of technologies: Is the candidate biomarker independent of a technology? If not, is the technology amenable technically and statistically for high-throughput studies? (See the following paragraph.)

6. Are reagents available for method development and validation for this biomarker?

The candidate biomarkers that satisfy these parameters as well as experimentally specific parameters can then be moved forward to the demonstration stage. The design of the demonstration stage depends on the availability of biological materials and the intended application of the biomarker. Biomarkers for clinical application require broader validations (tissue specificity, disease specificity, and so on) than a biomarker intended only for cell culture. However, basic technologies focused on protein measurement can be used during the demonstration stage of novel protein biomarkers (Fig. 3). These technologies can be used individually, or in combination to qualify and to eliminate candidate biomarkers. The degree of validation will depend on the ultimate application for the biomarker.

B. Technology Translation

The translation of technologies from discovery research to preclinical/clinical laboratories is a serious undertaking. Many parameters that are often overlooked in discovery research should be considered before approaching the transfer of technology to preclinical/clinical laboratories. The following are some of the important points that need to be considered:

1. Robustness of the technology: Often technologies used in research phase are not robust enough to sustain the demand of high-throughput preclinical/clinical laboratory setting. Lack of acceptance criteria, high variability, and high failure rate may exist in a research environment. However, these would not be acceptable in a high-throughput production environment; there is a required shift in culture and the technology (Section IV).
2. Is the procedure simple or can it be easily simplified? Some proteomics discovery approach can be very complex with multiple steps. The complex multistep approaches often have poor reproducibility in a high-throughput production environment.
3. Is the procedure under a standard operating procedure (SOP)? The standardization of the operating procedure is a requirement for preclinical/clinical laboratory setting. Once the method is validated, an SOP should be written and the assays carried out according to the SOP with detailed documentation. This is a marked difference from the research environment.
4. Are the different contributions to noise identified and characterized at the different steps in the procedure?
5. Are controls in place to assess the performance of the overall assay and different steps in the procedure?
6. Is the cost per analysis reasonable? This is a very important factor in application of the technology in a preclinical/clinical bioanalytical laboratory. Most proteomic approaches used in discovery phase are way too expensive per analysis to be used in bioanalytical laboratories.

C. Methods of Choice

Discovery of candidate biomarkers by proteomic approaches usually deliver a list of proteins that had statistically significant changes in expression across different cellular states, time, and drug responses. However, the "quantitation" is typically of a relative trend (A versus B or A versus B, C, D . . .), which is insufficient for preclinical/clinical application to study dose–response relationship. Furthermore, in some instances such as proteomic measurement that are focused on changes in the level of posttranslational modification often do not quantify changes in protein expression, which could have a significant impact on the relative level of posttranslational modification observed. Quantification of protein expression is important when considering the translation of candidate protein biomarkers. Therefore, when changes in posttranslational modification are considered as a potential biomarker, it would be preferable to be able to quantify levels of both protein expression and posttranslational modification.

Established technologies such as ELISA, enzymatic assays, LCMS in multiple reaction mode (MRM-LCMS), protein assay, and real-time polymerase chain reaction (RT-PCR) genomics are commonly used for biomarker analysis. ELISA can be performed in high throughput at a relatively low cost per samples. For example, the average price per biomarker analysis in preclinical/clinical samples is usually less than US$50 per analyte for an ELISA method. However, it requires the availability of antibodies against the candidate biomarker. For novel biomarkers at the preclinical phase, purified reference standard of the biomarker and antibodies are usually not available. Typically, the purification of the biomarker protein and the generation of antibodies is time consuming and expensive. Moreover, usually it takes at least 3 months to develop an ELISA per analyte. Most ELISA methods are developed for one analyte at a time. Technology for multiplexing immunoaffinity methods exists, such as Luminex LabMAP, which uses unique fluorescent labels on each analyte and flow cytometry for simultaneous detection of multiple analytes. However, quantitative method development and sample analysis are complicated due to the disproportionately variable biological ranges of the analytes and nonlinearity of the assays (Ray *et al.*, 2005). Thus, immunoassays may not be a desirable option at the exploratory preclinical stage because enormous resources would be required for assay development of the large number of emerging biomarkers from genomics and proteomics. The other options are to either perform MRM-LCMS protein assay or gene expression analysis.

MRM-LCMS protein assays measures the specific peptides that have been previously identified for the candidate protein biomarkers. Although the response curve (signal versus concentration) of the mass spectrometer is specific for individual peptides, absolute quantitation is possible by utilizing a synthetic isotopically labeled version of the peptides. The heavy isotope-labeled peptide can be used as a calibrator or as an internal standard to obtain quantitative measurements of the protein concentration. Typically, the protein sample of interest is digested with trypsin, and the isotope-labeled control peptides added to the mixture. The digest is then separated online by HPLC-ESI-MS/MS. About 100 MRM measurements can be performed per analysis considering the elution time and masses of the different peptides of interest. This is readily feasible for candidate biomarkers that were derived from a similar discovery proteomic technology. In the instance of genomic candidate biomarkers, software can be used to predict the elution times of the tryptic digest peptides that will be selected for the specific MRM-LCMS analysis. For preclinical studies involving disease models of several animal species, the homologue sequences in a different species can also be synthesized and tested. Overall, the MRM-LCMS approach can be generalized for multiple species and multianalyte assays. The time saving and relative cost of the MRM-LCMS approach are well suited for preclinical studies. After the candidate biomarkers have been chosen to enter clinical trials, resources can then be focused on the fewer chosen biomarkers for protein purification, antibodies production, and ELISA methods development for late-phase clinical application.

IV. BIOMARKER RESEARCH CHALLENGES: CULTURAL AND PROCESS TRANSLATION

The lack of stringent controls and understanding of statistical requirements for quantitative methods are often the pitfalls in many biomarker applications at the preclinical and clinical phases. The development of novel approaches such as proteomics and its associated technologies has spurred a push for the discovery of protein/peptide biomarkers in biological fluids. Unfortunately, the technology being used and the associated discoveries are often performed with little regard to the subsequent steps of biomarker discovery, that is, its translation to a preclinical/clinical environment. Therefore, bridging the gap in translation from the discovery phase in a research environment to the production environment of preclinical/clinical laboratories would remove a roadblock to bring success to the later phases. Such translation often involves changes of: (1) personnel from a creative flexible mode to that of production and control and (2) processes from meeting study objectives of differentiating marked effects to quantifying graded drug dose effects. Table I compares the different objectives and environments between the discovery and post-discovery phases in biomarker research. The demand for more rigorous and robust method validation increases as the biomarker progresses from demonstration, characterization, and qualification (vertical processes in Fig. 1). Often the transition involves different teams (such as discovery, preclinical and clinical bioanalytical), and thus communications throughout these teams are vital to the success of technological translation.

A. Diversity of Biomarker Methods and Data Types

Lee *et al.* (2003) discussed the complex processes in method validation and validation for biomarkers due to the inherent diversity of assays. Unlike drug

Table I Different Objectives and Environments of Biomarker Research During the Discovery and Post-Discovery Phases

	Discovery	Preclinical/clinical
Objectives	Differentiating marked effects	Graded drug dose effects
Analytical controls	Usually less defined or lacking, that is, normalized against a constant component. Usually no QCs	Use internal or external standards to minimize variance. QC samples to monitor assay performance in every analytical batch
Biological controls	Cell homogeneity, inactivated state as control	Diseased versus healthy (control) states, pre- (control) and postdrug treatments, drug versus placebo
Method validation requirements	Minimal	Increase rigor and robustness with program
Processes of operation	Usually no SOP, minimal documentation	SOPs and documentation for traceable sample custody and data generation
Personnel orientation	Creative mode	Production and control mode

assays, where samples are quantified against calibrators prepared from a highly purified and well-defined reference standard, biomarker assays differ considerably depending on the type of analytical measurement, the type of analytical data that arises from the assay, and the intended use of the data (Table II). In general, the intended application dictates the rigor of method validation required. For example, numerous candidate biomarkers can be obtained during discovery phases. For internal use, the extent of analytical validation can be limited to a few basic components to expedite the process of initial exploratory demonstration (Lee *et al.*, 2006). However, their use for characterization and qualification studies would require more extensive method validation and documentation. In addition, specific biomarker assay validation will generate different data type that may require special considerations and cautions during method validation. The diversity of method categories and the types of data generated in biomarker research are depicted in Fig. 4 and discussed in the following paragraphs.

A *definitive quantitative assay* uses a well-characterized reference standard that is fully representative of the endogenous biomarker. Absolute quantitative values for unknown samples are calculated from a regression function. Such assays are common for drug analysis for PK studies but only applicable to a small fraction of biomarkers such as small molecule bioanalytes (e.g., steroids). Instead, most biomarker assays are *relative quantitative assays* where the reference standard is not well characterized, not

Table II Application of Biomarkers from Drug Pharmacokinetic (PK) Analysis and Clinical Diagnosis

	Pharmacokinetic study	Biomarker for drug development study	Biomarker for diagnosis
Intended application	PK parameters of BA and BE	PD—safety and efficacy	Distinguish diseased from healthy
Method types	Mostly definitive quantitation	All four types, relative quantitation predominant	All four types
Reference standard	Well characterized and pure	Many are not well characterized or pure. Research-grade standards often vary within and between vendors	Vendor consistent and well established
Analytes	Exogenous, well defined	Endogenous, less well defined	
Method and reagent source	Developed in-house	Developed in-house. Some sources from diagnostic research grade kits	Well established, from vendor
Calibrator matrix	Analyte-free biological matrix	Substituted matrix (buffer or depleted biological matrix)	
Validation sample and QC preparation	Spiked ref standard into bio matrix. Prepared in-house	Spiked ref standard and intended population samples. Often prepared in-house	From vendor, may not use the exact biological matrix, common pool among labs
Accuracy	Absolute accuracy	Mostly relative accuracy	QC assessment not performed in every run for acceptance
Assay acceptance criteria	4–6-X^a rule for each run	Confidence Interval or a variant of 4–6-X^a rule for each run	Westgard rule, CAP test for lab accreditation

[a] Out of six QCs, at least four must be within X% of the nominal or target value for the analytical run to be acceptable. The six QCs consist of two each at low, mid, and high concentration and at least one should be acceptable. If the batch is large, the number of QCs will be increased.

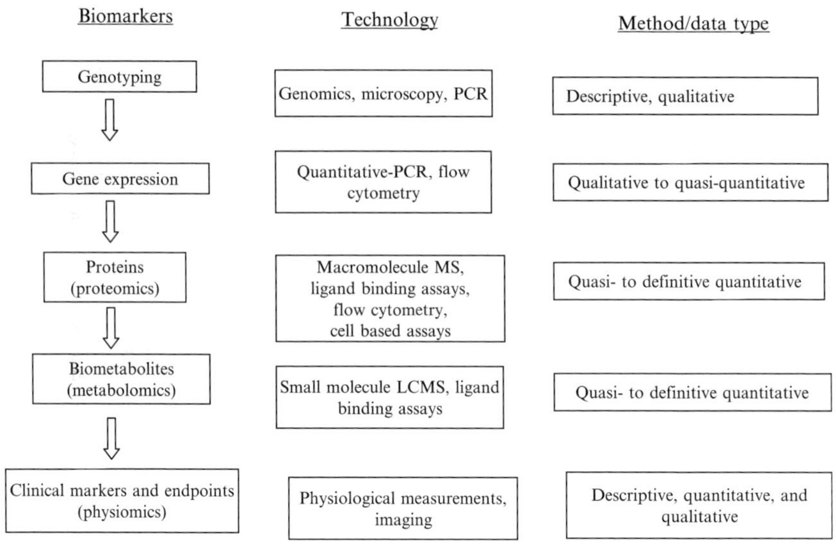

Fig. 4 Technology and method, data type of various biomarkers. The left column shows the diversity of biomarkers from gene expression to various clinical markers. The middle column shows the corresponding technologies commonly used for their measurements. The right column depict the method category and data type associated with these markers.

available in a purified form, or not fully representative of the endogenous form (e.g., cytokine immunoassays). Results from these assays are expressed in continuous numeric units of the relative reference standard. When no reference standard is available as calibrators, a *quasi-quantitative assay* is possible if the analytical response is continuous (numeric), with the analytical results expressed in terms of a characteristic of the test sample. Examples are antidrug antibody assays (where the readout is a titer or percentage bound), enzymatic assays (where activity might be expressed per unit volume), and flow cytometric assays (Mire-Sluis *et al.*, 2004).

In contrast to the "quantitative" categories, *qualitative assays* of biomarkers generate discrete (discontinuous) data, which are reported in either ordinal (low, medium, and high, scores 1–5) or nominal (yes/no, positive/negative) formats. In general, qualitative methods are suitable for differentiating marked effects such as all or none effect of gene expression, activation or inhibition in relatively homogenous cell populations in discrete scoring scales (1–5; −, +, +++, and so on) such as immunohistochemical assays.

Protein biomarkers are often heterogeneous, existing in multiple forms. The concentration of each component and its biological activities *in vivo* are unknown and may vary within individual (health status), time (diurnal, seasonal), and between individuals (gender, age, polymorphism). Longitudinal study design may benefit from well-defined target and control populations, which would require fewer subjects than that of a random population design. Instead of absolute quantification against a well-defined reference standard, greater emphasis would be placed on dose–exposure effect and temporal changes in biomarker concentrations of samples before and after drug treatment of various doses. Additional experimental controls (such as normal versus disease, placebo versus dosed) in the study design are useful to provide the appropriate comparison in a study using relative quantitative assays.

Usually, the emerging biomarkers evolved from discovery have not been purified or characterized and thus are not available as a reference standard for definitive or relative quantitative assay; the data type would be either quasi-quantitative or qualitative. These types of assay are suitable to differentiate marked effects. As the candidate biomarker proceeds to clinical phases the study objectives would shift to obtain quantitative data to characterize the drug exposure–response relationship. This often involve the correlation of the biomarkers with disease progression or treatment effect by PK/PD modeling, the choices of biomarker assays should be those most amenable to quantitative analysis. The translation from discovery to subsequent development would require method modification from a qualitative or quasi-quantitative to a relative quantitative assay and with the understanding of the possible limitations impacting the confidence in the bioanalytical measurements. Biomarkers are used to support decisions that impact the fate of a drug development program. Therefore, it is essential that the assay be based on reliable, quantitative, and objective data.

In addition, clinical physiological markers are often used in trials and are usually based on subjective, qualitative measurements (such as pain scores). However, attempts have been made to obtain quantitative measurements that will improve the confidence of statistical analyses with reasonable sampling size. For example, electrocardiogram (EKG) patterns, tissue immunostaining, and imaging have been digitalized to enable quantitative measurements (Leong and Leong, 2004). Furthermore, imaging such as magnetic resonance imaging (MRI) and positron emission tomography (PET) have provided useful data for clinical physiological markers (Dubey *et al.*, 2003).

B. Translation from Research to Bioanalytical Environment: Preanalytical Considerations

1. RESEARCH WORK PLAN

The development of a biomarker research work plan is an important element in a biomarker study. The plan should include clear definition of the control and experimental sample sets, the population in which the marker will be measured, the rigor of method validation, and an initial tolerance limit for method variability (acceptance criteria) required to meet study objectives. For example, the plan could include the expected clinical differentiation of patients versus the control set as well as the drug effect. It is important to consider that longitudinal analysis of the biomarker in the targeted patient population may be needed to differentiate the effects of the drug from other complex effectors.

There are three major factors that help define and set limits on assay tolerance of imprecision: (1) the intended use of the data during various stages of drug development, (2) the nature of the assay methodology and the different types of data that they provide, and (3) the biological variability of the biomarker that exists within and between populations. The first factor helps shape the assay tolerance or acceptance criteria for biomarkers. Contrary to the general perception held by the analytical laboratories, the required assay acceptance criteria do not necessarily depend on the method deliverable, but rather on the intended application. Once the study purpose is identified, the method requirements become the basis of method development choices and validation rigor to arrive at a suitable method to meet the study objectives.

2. METHODS SELECTION AND FEASIBILITY TEST

The ultimate goal of a biomarker assay is to evaluate the impact of a drug on the activity of biomarkers *in vivo*. Therefore, a method that measures the intended biological activity should be most desirable. However, activity assays using cell-based and enzyme activity methods do not generally have adequate sensitivity and precision. Moreover, they are usually laborious, low throughput, and often lack definitive reference standards. In contrast, physicochemical (such as chromatographic methods) or biochemical (such as immunoassays) methods provide higher sensitivity and more precise quantitative data to enable drug exposure–response statistical correlation. Thus, these methods are often chosen for biomarker analysis over activity measurements. The data user should be aware that the application rests on the assumption that these measurements are reflective of the biological activities. This tacit assumption might not always hold true. Therefore, complementary activity assays using higher concentrations, and/or on smaller sets of samples are sometimes used to correlate and confirm that the physicochemical or biochemical characteristics of the biomarker are reflective of biological activities.

As discussed in the previous section, the method used in discovery can be translated into a bioanalytical environment. Feasibility tests and exploratory validation are carried out to assess if the method is adequate in assay performance with respect to accuracy, precision, selectivity, dynamic range, sample integrity in biological matrix, and dilutional linearity.

3. SAMPLE INTEGRITY

Results from biomarker assays are only valid if sample integrity has been maintained. Sample integrity includes stability of the analyte in the biological matrix throughout variable environments spanning from sample collection, storage, shipping, and further storage up to the last sample analysis. Sample integrity can be compromised by many factors such as inappropriate sample collection and storage, lack of controls, as well as insufficient training of the clinical staff. The impact of sample collection and storage on individual technique can be investigated as part of the experimental plan. In some instance, genomic and proteomic functional markers can be used for the assessment of sample integrity. It may be prudent to perform extensive sample integrity investigation for a definitive study to cover unduly stressful environment. The study protocol should include procedures for sample collection, shipping and analysis, as well as sample storage and disposal. If the procedures involve complex processing, on-site training of the clinical staff may be warranted. Control and standardization of sample collection is necessary to minimize variability from multiple clinical sites. Documentation on the sample chain of custody should be similar to that of a clinical study for PK purpose. In late phase drug development with multiple clinical sites, central sample repository provides a uniformed collection, storage, and shipping process.

Serum is a cleaner matrix and more preferable than plasma and pose less problem on lab automation. However, labile biomarkers may not tolerate the coagulation process. Biomarkers involved in the coagulation pathway or in platelet activation can only be quantified accurately in plasma. Sample collection and handling present microenvironmental changes on dynamic cell systems such as whole blood and tissue samples. The possible effect on the subsequent biomarker assay must be considered to minimize artifacts. For example, for flow cytometric assays, inappropriate handling of the blood could activate certain cells, thus complicating the quantification of cells, cell surface markers, or the expression of blood cell products. The collection and processing of human tissue biopsies has historically focused on obtaining clinical data (e.g., cancer diagnosis and staging). Until recently, little emphasis has been put on the development of appropriate sampling handling techniques (e.g., cryopreservation), which are essential for tissue-based biomarker analyses. For tumor marker analysis, it is necessary to freeze tissue specimens on dry ice in the pathology suite within 20–30 min of collection or immediately immersed in ice and rapidly transported to a site for cryopreservation and storage at $-80\,^{\circ}C$. Minimizing heterogeneity during sampling is also important. Technologies such as flow cytometry and laser capture microdissection can differentiate and separately collect cancerous and noncancerous cells, where the latter can be used as controls.

Normalization can improve the consistency of a tissue biomarker. An appropriate normalization factor (e.g., weight, protein content, precursor protein, cell counts of the target cell type) should be in place. For example, the total kinase content can be used to normalize phosphorylated kinases. The sample integrity of the biomarker and the normalization factor should both be maintained.

C. Method Validation: A Continual Process of Assay Refinement for the Intended Application

1. REFERENCE STANDARD AND CALIBRATORS

A common problem of novel biomarker assays is the lack of well-characterized and standardized reference materials. Initially, an emerging novel biomarker prohibits the development of an "official" standard. Eventually, when the novel biomarker becomes well established in the research community, reagent standardization can be addressed more effectively (Section V.B on commercialization). Most biomarkers are endogenous compounds with measurable levels in the biological matrix of interest. To prepare calibrators in an analyte-free biological matrix, an altered substitute matrix is often used. The substitute may be a stripped matrix with the analyte depleted by charcoal or an affinity solid phase, other species matrix, a buffer solution, or a matrix treated to degrade the analyte of interest. The basic principle of the quantitation relies on the similarity of the response of the analyte in the substituted matrix and in the biological matrix. The search for a suitable substituted matrix is often a challenge for the development of analytical methods for biomarker.

2. ASSAY DYNAMIC RANGE

The levels of the biomarker in the targeted and normal populations should be used to plan the assay dynamic range, from lower to upper limits of quantification (LLOQ to ULOQ) with expansion to accommodate high concentration samples. Ideally, the assay range of the method should commensurate with the biological range. However, this is not often the case. For example, the biological range of a proinflammatory cytokine can be in the low pg/ml in sera from asthmatic and rheumatoid arthritic patients, increasing for certain patients with cancer and cardiovascular diseases, and to very high concentrations in ng/ml range in the case of acute infection. In addition, sample concentrations are matrix dependent. In general, biological fluids in the circulation have lower concentrations than that of the target organ and peripheral tissues. Usually, the calibrator range of an ELISA method would not cover the entire biological spectrum. Consequently, high concentration samples are diluted to extend the assay range. However, low concentration clinical samples may fall beyond the LLOQ of the method, that is, the method lacks the required sensitivity. In such case, the project team needs to understand the sensitivity limitation and decide on the appropriate action. For example, the investigator may use values below the LLOQ but above the limit of detection (LOD) to evaluate the changes, taking the risk of the higher variability in that region.

If diurnal variability is expected, it is prudent to pool samples over 24 h or to always collect samples at the same time of the day. The initial survey of analyte range in the intended study population should also provide information on assay variability versus biological (within and between individual) variability, which help to design appropriate clinical and assay controls to produce unbiased clinical answers. For cancer studies, there may not be placebo or baseline samples available to differentiate the true drug effect from the nonspecific variability of the biomarker expression or measurement variability. In such case, the dose–response relationship data of the biomarker and the correlation to clinical markers will be critical to demonstrate the selective drug effect on the biomarker and its biological significance.

3. PARALLELISM AND DILUTION LINEARITY TO EVALUATE MATRIX EFFECT

The calibrator preparation in an analyte-free substituted matrix instead of the intended sample matrix is one major difference of biomarker assays from that of drug compounds. It is crucial to demonstrate that the concentration–responses relationship in the sample matrix is similar to that of the substituted matrix. Spike recovery experiments of the reference standard may be inadequate to evaluate the matrix effect, as the reference standard may not fully represent the endogenous analyte. Instead, parallelism experiments should be performed through serial dilution of a high concentration sample with the calibrator matrix. Multiple individual matrix lots should be tested to compare lot-to-lot consistency. In the instance that limited amounts of sample are available, a pooled matrix strategy can be used with caution as discussed by Lee *et al.* (2006).

When parallelism cannot be performed due to unavailability of samples with sufficiently high analyte concentrations of analyte, method selectivity may be assessed using spiked recovery and dilution linearity of spiked samples. If there is an interference or matrix effect, the sample could be diluted to reduce the background. The minimal required dilution should be determined and the same dilution applied to all samples to remove the matrix effect (DeSilva *et al.*, 2003).

4. VALIDATION SAMPLES AND QUALITY CONTROLS

Validation samples are used in prestudy validation to define intra- and inter-run accuracy/precision and stability, providing data to demonstrate the assay's ability to meet study requirement for its intended application. On the other hand, quality control (QC) samples play a role for determining run-acceptance during specimen analysis. The validation samples can be employed as QCs after prestudy validation. Validation samples and QC should be as closely related to the study samples as possible. They can be prepared from a pool of low-level samples after the initial screen of multiple matrix lots. If a reference standard is available, known amounts of the standard should be added to the low-level pool to prepare higher levels of QC. Alternatively, lots with higher level could be pooled to prepare the high concentration QCs for quasi-quantitative assays, and in some cases, also for relative quantitative assays. The importance is to determine the values of the validation samples and QC levels during prestudy validation, and then use the determined values as targets for in-study assay performance assessment for relative accuracy, precision, and stability. Preparation of validation samples and QC in the substitute matrix, such as a protein-buffer, is not recommended (Lee *et al.*, 2006). In the case of rare matrix without sufficient volume to prepare multiple levels, attempts should be made to prepare at least one level of QC in that matrix.

The intended purposes and the expected method performance requirements should be predefined in a biomarker work plan. Many biomarker assays are used for quantitative comparisons of dosing (or regiment treatment) effects. The precision evaluation of a biomarker assay lends support to the statistical significance of the results. Similar to drug/metabolite bioanalysis, precision is normally evaluated as inter- and intra-assay variations with validation samples during method validation and QCs during clinical study sample analysis. Recommendations for method development, validation and sample analysis of biomarkers has been presented by Lee *et al.* (2006).

5. REGULATORY ISSUES

Biomarker development during discovery is not subjected to regulatory control by government agencies or by the company's internal Quality Assurance (QA) units. In general, the objective is to detect dramatic effects. During preclinical and clinical phases, the objectives shift to differentiate graded trends and quantifiable changes to provide dose–exposure and response calculations. In general, data in exploratory and/or early phase would not be used for submission to government agencies to support safety and efficacy. If the data are used for important business decision-making, they could be reviewed by the internal QA to assure that the data are reliable for the important decision-making. On the other hand, biomarkers to support drug safety should be performed under good laboratory practice (GLP).

There is a general lack of regulatory guidance on what needs to be done to validate a biomarker assay. Routine and novel biomarkers are often performed in both bioanalytical and clinical laboratories. Bioanalytical methods for PK studies follow the FDA guidance published in May 2001, which is often referred to as "GLP-compliant" for convenience (FDA, 2001). Although biomarker laboratory analyses have many similarities to those used in toxicology and PK studies, the variety of novel biomarkers and the nature of their applications often preclude the use of previously established bioanalytical validation guidelines. Assays for diagnostic and/or treatment decisions follow the standards, guidelines, and best practices required in the Clinical Laboratory Improvement Amendments of 1988 (CLIA), developed and published by Clinical and Laboratory Standards Institute [CLSI, formerly the National Committee for Clinical Laboratory Standards (NCCLS)] (FDA 2001; National Committee for Clinical Laboratory Standards, 1999). The CLSI/CLIA acceptance criteria are defined by the proven performance of the method and adjusting them over time using a confidence limit-based approach (Westgard and Klee, 1996).

For novel biomarker assays in early clinical phase development where attrition of the clinical candidate and associated biomarker assays is high, the assay may only be used for a short period of time to preclude the establishment of appropriate confidence limits. In addition, the intended applications of novel biomarkers differ in a pilot PoB study from that of a confirmatory purpose. The rigor of the method validation usually increases from the exploratory phase to the definitive studies. Therefore, current regulatory guidelines are not suitable. Biomarker method development and validation should be "Fit-for-Purpose" so that assays could be successfully applied to meet the intended purpose of the study, instead of following a one-size-fit all guideline as discussed in the conference report of the AAPS and Clinical Ligand Assay Society (CLAS) Biomarker Method Validation Workshop in 2003 (Lee *et al.*, 2005).

V. CLINICAL QUALIFICATION

A. Clinical Trials

Clinical qualification of a biomarker's predictive use in drug development is a graded evidentiary process of linking a biomarker with biological processes, events, and clinical endpoints. Each study in a particular phase has special objectives that should be defined in a research plan. During the clinical qualification, it is important to establish if the sensitivity and specificity of the biomarker to detect and assess the drug effect on the disease can also be demonstrated in human. The ultimate goal is to determine if the novel biomarker or a panel of biomarkers can substitute for a clinical endpoint (surrogacy) base on evidences from comprehensive datasets. The designation of surrogate endpoint or "valid biomarkers" would require agreement with regulatory authorities.

The research plan for the clinical qualification of a novel biomarker is often linked to more than one drug development programs of a similar mechanism of action. Both novel and routine biomarkers are included in a research plan of a target disease. Exploratory and confirmatory data of the novel biomarker on the target hit will be accumulated and correlated (or compared) to the established routine biomarkers, as well as the clinical endpoints. Overall, the research plan should have sufficient details in the design, execution, and data interpretation as well as the technical performance for the candidate biomarker. The plan should include a section on regulatory-related objectives such as confirmatory evidence, support of a claim, replacement of the routine conventional measures, or voluntary data submission. The development plan should specify the sampling size required and criteria in statistical methods (e.g., receiver operation curve, principal component analysis, Bayesian and artificial neural network, and so on) to establish correlation, predictive utilities, and ultimately surrogacy (Antal *et al.*, 2004; Crawford, 2003; Liggett *et al.*, 2004).

Survival is the standard clinical endpoint used in cancer drug approval. However, it represents a pretty high bar for patients with refractory and progressive tumors. The utilization of other clinical endpoints has not been consistent. For example, time to progression and/or relapse, disease stabilization, tumor size reduction, or other imaging clinical markers have been considered as clinical endpoints. It is prudent to discuss with the regulatory agency during preparation of the clinical trials on the selection of the intended clinical markers and other clinical endpoints. Considerations should be given to phase II trial designs that are more predictive of phase III success. In addition, the clinical protocol design should allow evaluation of biological heterogeneity for multicomponent statistical analysis.

Traditional measurements to assess clinical endpoints are often qualitative (e.g., pain, quality of life, or tumor size in nominal scores; pathological staining in ordinal scores). For effective statistical correlations of biochemical markers and clinical markers, it would be ideal if both sets of data were quantitative. Technological advances offer unprecedented possibilities of quantitative physiological measurements. For example, imaging is a relatively noninvasive clinical marker that can now be used in a quantitative manner. Clearly, there is an increasing need for cheaper and quantitative imaging

(Leong and Leong, 2004). Imaging can become a clinical or surrogate endpoint that provides earlier correlation than morbidity and mortality. Digital imaging has become common practice in anatomical pathology to replace photographic prints for reporting. More advanced computerized imaging systems provide greater versatility, speed of turnaround, as well as lowering the cost for incorporating macroscopic and microscopic pictures into pathology reports. It allows the transmission to remote sites via the Internet for consultation, quality assurance, and educational purposes, and can be stored on and disseminated by CD-ROM. In addition, 3D images of gross specimens can be assembled with the use of positron-emitting radionuclides (^{11}C, ^{13}N, ^{15}O, and ^{18}F) or fluorescent molecular probes of the drug, tumor-specific ligand or metabolite. Their applications in research allow more objective and automated quantitation of a variety of morphological and immunohistological parameters.

B. Biomarker Assay Commercialization

The translation of a novel biomarker methodology to an approved clinical diagnostic test can help in post-approval surveillance of drug safety/efficacy using the dataset from large patient populations. Therefore, the commercialization of novel biomarkers becomes a necessity. Initially, diagnostic kits and components are often "borrowed" for biomarker application during drug development. These include FDA-approved products, kits for "research only" (or in-house developed assays) and other commercial reagents. Because the intended use for drug development is different from the diagnostic purposes, the laboratory must define the intended application and carry out the appropriate validation. The shifted application of an assay system from drug development to post-approval patient monitoring would require 510(K) clearance from Office of *In Vitro* Diagnostics Device Evaluation and Safety of the FDA for clinical use. Several aspects should be considered for commercialization:

1. Standardization of reference material: Most protein biomarkers are heterogeneous in nature; the endogenous form or forms may change dependent on the health status. There is a general lack of standard reference material in most kits for research use. For example, the commercial supplier may assign a numerical value to the reference standard according to activity results from its bioassays. As a result, it is not uncommon that the standard material values may differ substantially between manufacturers or even between lots from one manufacturer (Sweep *et al.*, 2003). It is difficult to define which form or combination of forms of the biomarker should be used as the standard reference material. There have been continuing efforts to

establish "gold" standards for tumor markers that have been deemed as predictive to provide diagnostic and prognostic assessment for patients (Barker, 2003; Basuyau *et al.*, 2003; Hammond *et al.*, 2003; Rafferty *et al.*, 2000). For novel biomarkers, similar collaborative efforts from the diagnostic and pharmaceutical industries would be needed to render gold standards as reference materials.

2. Standardization of QCs: For biomarker characterization and evaluation the bioanalytical laboratories often use QCs prepared in house to assess the assay performance (Table I). However, if the assay is going to be used in a diagnostic laboratory, standardized QCs with the defined target values must be available from a repository or commercial sources for clinical laboratory certification and tracking as well as providing the ability to pool the statistics within and among laboratories (Westgard and Klee, 1996; Westgard *et al.*, 1981).

3. Information sharing: The following information should be available to facilitate commercialization: the expected biomarker ranges in normal individuals and targeted populations, stability information of the analyte in biological matrices, and the conditions for sample collection and storage to preserve analyte integrity.

4. Codevelopment of FDA approved commercial test by diagnostic and pharmaceutical companies: For example, HercepTest® was codeveloped with Herceptin for patient entry criteria (DAKO HercepTest). In addition, an immunoassay kit for the extracellular domain (p97–115 KD) of the HER-2/neu receptor in human serum was developed for clinical patient monitoring (Carney *et al.*, 2003).

VI. CONCLUSIONS AND PERSPECTIVES

Genomic alteration at significant frequencies occurred in major cancer types, which provides a starting point of investigative experiments on gene expressions in mRNA and protein alteration to identify drug targets (Futreal *et al.*, 2004). Modulation of transcription at the protein level by the use of perturbing molecules that mediate between activation domains can be studied *in vitro*, and further extend into the human system. There are a lot more to learn about the genetics–chemistry–biology interplay for cancer progression and intervention in the biomarkers' translational processes from *in vitro* and animal models to clinical application. More than ever, collaborative efforts to pool, organize, and disseminate data on biomarkers into useful knowledge must happen. One example is the volunteer pharmacogenomics data submission encouraged by the FDA (FDA, 2005; Lesko and Woodcock, 2004). Pharmacogenomics data play an important role in cancer drug development and personalized treatment. The issues and complexity of biomarker data are reflected in the challenges of pharmacogenomics data generation and application. This includes assay validation and standardization; uniform terminology of analytes (genes and gene products); interpretation of the biology and linkage to clinical significance within the right context and with proper controls; and the standards for transmission, processing, and storage of multidimensional data. At the same time, important issues and concerns within the scientific and medical communities, such as patient privacy, intellectual property, data, and process control, will have to be addressed.

The importance of biomarker data to help moving the right drug candidates forward from discovery to preclinical and clinical and for post-approval surveillance is increasing. This is in part due to the increased costs and productivity pressure on pharmaceutical companies and the intensified concern of drug safety during post-approval applications. Therefore, there is a race for the discovery of novel biomarkers as well as novel analytical methodologies for biomarker measurements. Genomic, proteomic, and metabolomic research discover many individual or groups of putative biomarkers. These emerging biomarkers lengthen the list of biomarkers already implicated by other approaches (e.g., epidemiology). Hundreds of candidate biomarkers are being identified in their causal relationship to disease progression and drug intervening mechanism. Both on- and off-target biomarkers are studied to investigate the drug effect and the data used for drug candidate selections. Bioinformatic tools with sound statistical treatments are crucial for choosing the panels of biomarkers to be studies as well as making the right decision to choose the right drug candidate to move forward in drug development.

Successful identification and characterization of novel biomarkers will improve the effectiveness of drug development. Technological translation from frontier technologies used in discovery to preclinical and clinical environment should be vigorously pursued to increase method throughput and robustness for the application. Process and technological translation of the candidate biomarkers from discovery into preclinical and clinical phases demand careful planning and execution. The complexity and level of uncertainty increase from the cell systems in discovery to preclinical animal models and to human. In general, variability from the methodology is less than that from the biological sources. Therefore, statistical tools must be used to assess the variability from the different components for appropriate data treatment and correlations. Judicious controls in clinical studies and technological steps are necessary to tease out the true effects of the drug treatment. A well-defined research plans in novel biomarker development which are carried out with properly controlled experiments in clinical trials will lead to successful biomarkers selection and application.

Technological advances that provide improved sensitivity and selectivity are fundamental for the success of biomarker discovery. In particular, technology integration of laser-microdissected cryostat sectioning, ProteinChip, gene microarray, immunohistochemistry, multiplex binding assays, and hyphenated MS methods (e.g., FACS-MS, MALDI-MS, and affinity-MS) will likely brighten the future of biomarker discovery and translation through the processes. Because these technologies evolved in a research environment, translation for the application of preclinical and clinical samples requires the cooperation of the scientists from discovery and clinical realms. A consortium of biomarker has been proposed and finally formed in October 2006 by NIH, the FDA, and Pharmaceutical Research and Manufacturers of America (PhRMa) to deposit information, share knowledge and resources, and build consensus (MacGregor, 2004; Srivastava and Gopal-Srivastava, 2002). Standardization of methods, reagents, and QCs could also be included in the organization goals. It takes concerted efforts and commitment to materialize the potential of biomarkers to efficiently develop safer and more efficacious pharmaceutical products, and provide clinical practitioners with sensitive and specific biomarkers for cancer screening, patient monitoring, and choice of therapy.

ACKNOWLEDGMENTS

The authors are grateful to ideas and collaborations from the Biomarker Subcommittee in the Ligand Binding Assay Bioanalytical Focus Group of the American Association of Pharmaceutical Scientists and our former colleague, Robert Masse, Ph.D. (MDS Pharma Services), which contribute to the content in this chapter.

REFERENCES

Antal, P., Fannes, G., Timmerman, D., Moreau, Y., and De Moor, B. (2004). Using literature and data to learn Bayesian networks as clinical models of ovarian tumors. *Artif. Intell. Med.* **30**, 257–281.

Barker, P. E. (2003). Cancer biomarker validation: Standards and process: Roles for the National Institute of Standards and Technology (NIST). *Ann. NY Acad. Sci.* **983**, 142–150.

Basuyau, J. P., Blanc-Vincent, M. P., Bidart, J. M., Daver, A., Deneux, L., Eche, N., Gory-Delabaere, G., Pichon, M. F., and Riedinger, J. M. (2003). Summary report of the standards, options and recommendations for the use of serum tumour markers in breast cancer: 2000. *Br. J. Cancer* **89**(Suppl. 1), S32–S34.

Biomarkers Definitions Working Group. (2001). Biomarkers and surrogate endpoints: Preferred definitions and conceptual framework. *Clin. Pharmacol. Ther.* **69**, 89–95.

Bjornsson, T. D. (2005). Biomarkers applications in drug development. *Eur. Pharm. Rev.* **1**, 17–21.

Booth, B., Glassman, R., and Ma, P. (2003). From the Analyst's couch. Oncology's trials. *Nat. Rev.* **2,** 609–610.

Carney, W. P., Neumann, R., Lipton, A., Leitzel, K., Ali, S., and Price, C. P. (2003). Potential clinical utility of serum HER-2/neu oncoprotein concentrations in patients with breast cancer. *Clin. Chem.* **49,** 1579–1598.

Crawford, E. D. (2003). Use of algorithms as determinants for individual patient decision making: National comprehensive cancer network versus artificial neural networks. *Urology* **62,** 13–19.

DeSilva, B., Smith, W., Weiner, R., Kelley, M., Smolec, J., Lee, B., Khan, M., Tacey, R., Hill, H., and Celniker, A. (2003). Recommendations for the bioanalytical method validation of ligand-binding assays to support pharmacokinetic assessments of macromolecules. *Pharm. Res.* **20,** 1885–1900.

DiMasi, J. A., Hansen, R. W., and Grabowski, H. G. (2003). The price of innovation: New estimates of drug development costs. *J. Health Econ.* **22,** 151–185.

Dubey, P., Su, H., Adonai, N., Du, S., Rosato, A., Braun, J., Gambhir, S. S., and Witte, O. N. (2003). Quantitative imaging of the T cell antitumor response by positron-emission tomography. *Proc. Natl. Acad. Sci. USA* **100,** 1232–1237.

FDA(2001). Guidance for industry. Bioanalytical method validation: Availability. *Fed. Regist.* **66,** 28526–28527.

FDA Document(2004). Innovation or stagnation: Challenge and opportunity on the critical path. http://www.fda.gov/oc/initiatives/criticalpath/

FDA(2005). Guidance for industry. Pharmacogenomic data submissions.

Futreal, P. A., Coin, L., Marshall, M., Down, T., Hubbard, T., Wooster, R., Rahman, N., and Stratton, M. R. (2004). A census of human cancer genes. *Nat. Rev. Cancer.* **4,** 177–183.

Hammond, M. E., Barker, P., Taube, S., and Gutman, S. (2003). Standard reference material for Her2 testing: Report of a National Institute of Standards and Technology-sponsored Consensus Workshop. *Appl. Immunohistochem. Mol. Morphol.* **11,** 103–106.

Hunt, S. M., Thomas, M. R., Sebastian, L. T., Pedersen, S. K., Harcourt, R. L., Sloane, A. J., and Wilkins, M. R. (2005). Optimal replication and the importance of experimental design for gel-based quantitative proteomics. *J. Proteome Res.* **4,** 809–819.

Lambert, J. P., Ethier, M., Smith, J. C., and Figeys, D. (2005). Proteomics: From gel based to gel free. *Anal. Chem.* **15,** 3771–3787.

Lee, J. W., Smith, W. C., Nordblom, G. D., and Bowsher, R. R. (2003). Validation of assays for the bioanalysis of novel biomarkers. *In* "Biomarkers in Clinical Drug Development" (J. C. Bloom and R. A. Dean, Eds.), pp. 119–149. Marcel Dekker, New York.

Lee, J. W., Weiner, R. S., Sailstad, J. M., Bowsher, R. R., Knuth, D. W., O'Brien, P. J., Fourcroy, J. L., Dixit, R., Pandite, L., Pietrusko, R. G., Soares, H. D., Quarmby, V., *et al.* (2005). Method validation and measurement of biomarkers in nonclinical and clinical samples in drug development. *Pharm. Res.* **22,** 499–511.

Lee, J. W., Devanarayan, V., Barrett, Y. C., Allinson, J., Fountain, S., Keller, S., Weinryb, I., Green, M., Duan, L., Rogers, J. A., Millham, R., O'Brien, R., *et al.* (2006). Fit-for-Purpose method development and validation for successful biomarker measurement. *Pharm. Res.* **23,** 312–328.

Leong, F. J., and Leong, A. S. (2004). Digital photography in anatomical pathology. *J. Postgrad. Med.* **50,** 62–69.

Lesko, L. J., and Woodcock, J. (2004). Translation of pharmacogenomics and pharmacogenetics: A regulatory perspective. *Nat. Rev. Drug Discov.* **3,** 763–769.

Liggett, W. S., Barker, P. E., Semmes, O. J., and Cazares, L. H. (2004). Measurement reproducibility in the early stages of biomarker development. *Dis. Markers* **20,** 295–307.

Listgarten, J., and Emili, A. (2005). Statistical and computational methods for comparative proteomic profiling using liquid chromatography-tandem mass spectrometry. *Mol. Cell. Proteomics* **4,** 419–434.

MacGregor, J. T. (2004). Biomarkers of cancer risk and therapeutic benefit: New technologies, new opportunities, and some challenges. *Toxicol. Pathol.* 32(Suppl. 1), 99–105.

Mire-Sluis, A. R., Barrett, Y. C., Devanarayan, V., Koren, E., Liu, H., Maia, M., Parish, T., Scott, G., Shankar, G., Shores, E., Swanson, S. J., Taniguchi, G., *et al.* (2004). Recommendations for the design and optimization of immunoassays used in the detection of host antibodies against biotechnology products. *J. Immunol. Methods* **289,** 1–16.

National Committee for Clinical Laboratory Standards (1999). Document EP5-A: Evaluation of Precision Performance of Clinical Chemistry Devices: Approved Guideline.

Paradis, V., Degos, F., Dargere, D., Pham, N., Belghiti, J., Degott, C., Janeau, J. L., Bezeaud, A., Delforge, D., Cubizolles, M., Laurendeau, I., and Bedossa, P. (2005). Identification of a new marker of hepatocellular carcinoma by serum protein profiling of patients with chronic liver diseases. *Hepatology* **41,** 40–47.

Rafferty, B., Rigsby, P., Rose, M., Stamey, T., and Gaines Das, R. (2000). Reference reagents for prostate-specific antigen (PSA): Establishment of the first international standards for free PSA and PSA (90:10). *Clin. Chem.* **46,** 1310–1317.

Ray, C. A., Bowsher, R. R., Smith, W. C., Devanarayan, V., Willey, M. B., Brandt, J. T., and Dean, R. A. (2005). Development, validation, and implementation of a multiplex immunoassay for the simultaneous determination of five cytokines in human serum. *J. Pharm. Biomed. Anal.* **36,** 1037–1044.

Reichert, J. M. (2003). Trends in development and approval times for new therapeutics in the United States. *Nat. Rev. Drug Discov.* **2,** 695–702.

Reyzer, M. L., Caldwell, R. L., Dugger, T. C., Forbes, J. T., Ritter, C. A., Guix, M., Arteaga, C. L., and Caprioli, R. M. (2004). Early changes in protein expression detected by mass spectrometry predict tumor response to molecular therapeutics. *Cancer Res.* **64,** 9093–9100.

Srivastava, S., and Gopal-Srivastava, R. (2002). Biomarkers in cancer screening: A public health perspective. *J. Nutr.* **132,** 2471S–2475S.

Sweep, F. C., Fritsche, H. A., Gion, M., Klee, G. G., and Schmitt, M. (2003). Considerations on development, validation, application, and quality control of immuno(metric) biomarker assays in clinical cancer research: An EORTC-NCI working group report. *Int. J. Oncol.* **23,** 1715–1726.

Wagner, J. A. (2002). Overview of biomarkers and surrogate endpoints in drug development. *Dis. Markers* **18,** 41–46.

Washburn, M. P., Wolters, D., and Yates, J. R.r. (2001). Large-scale analysis of the yeast proteome by multidimensional protein identification technology. *Nat. Biotechnol.* **19,** 242–247.

Weinstein, J. N. (2004). Integromic analysis of the NCI-60 cancer cell lines. *Breast Dis.* **19,** 11–22.

Westgard, J. O., and Klee, G. G. (1996). Quality management. *In* "Fundamentals of Clinical Chemistry" (C. Burtis, Ed.), pp. 211–223. WB Saunders, Philadelphia.

Westgard, J. O., Barry, P. L., Hunt, M. R., and Grove, T. (1981). A multi-rule Shewhart chart for quality control in clinical chemistry. *Clin. Chem.* **27,** 493–501.

Molecular Optical Imaging of Therapeutic Targets of Cancer

Konstantin Sokolov,[*] Dawn Nida,[†] Michael Descour,[‡] Alicia Lacy,[§] Matthew Levy,[¶] Brad Hall,[¶] Su Dharmawardhane,[||] Andrew Ellington,[¶] Brian Korgel,[**] and Rebecca Richards-Kortum[†]

[*]*Department of Imaging Physics, MD Anderson Cancer Center, Houston, Texas 77030;*
[†]*Department of Bioengineering, Rice University, Houston, Texas 77030;*
[‡]*Optical Sciences Center, The University of Arizona, Tucson, Arizona 85724;*
[§]*Department of Biomedical Engineering, The University of Texas at Austin, Austin, Texas 78712;*
[¶]*Department of Chemistry and Biochemistry, The University of Texas at Austin, Austin, Texas 78712;*
[||]*Department of Anatomy and Cell Biology, Universidad Central del Caribe, School of Medicine, Bayamon, Puerto Rico 00960; and*
[**]*Department of Chemical Engineering, The University of Texas at Austin, Austin, Texas 78712*

I. Introduction
II. Role of Molecular-Specific Optical Imaging of Carcinogenesis
  A. Monitoring Effectiveness of Therapy
  B. Rational Selection of Targeted Therapeutics
III. Cancer Biomarkers and Therapeutic Targets
  A. Epidermal Growth Factor Receptor
  B. VEGF and Its Receptors
  C. Matrix Metalloproteinases
IV. Optical Technologies
  A. Widefield Microscopy
  B. High-Resolution Microscopy
  C. Fiber Optic Microscopes
V. Optically Active Contrast Agents
  A. Metal Nanoparticles
  B. Quantum Dots
  C. Smart Contrast Agents
VI. Delivery of Contrast Agents *In Vivo*
  A. Topical Delivery
  B. Systemic Delivery
VII. Molecular-Specific Optical Imaging
  A. Metal Nanoparticles
  B. Quantum Dots
VIII. Future Directions: "Smart" Contrast Agents
IX. Conclusions
References

Advances in CANCER RESEARCH

0065-230X/07 $35.00
DOI: 10.1016/S0065-230X(06)96011-4

Recent progress in discerning the molecular events that accompany carcinogenesis has led to development of new cancer therapies directly targeted against the molecular changes of neoplasia. Molecular-targeted therapeutics have shown significant improvements in response rates and decreased toxicity as compared to conventional cytotoxic therapies which lack specificity for tumor cells. In order to fully explore the potential of molecular-targeted therapy, a new set of tools is required to dynamically and quantitatively image and monitor the heterogeneous molecular profiles of tumors *in vivo*. Currently, molecular markers can only be visualized *in vitro* using complex immunohistochemical staining protocols. In this chapter, we discuss emerging optical tools to image *in vivo* a molecular profile of risk-based hallmarks of cancer for selecting and monitoring therapy. We present the combination of optically active, targeted nanoparticles for molecular imaging with advances in minimally invasive optical imaging systems, which can be used to dynamically image both a molecular and phenotypic profile of risk and to monitor changes in this profile during therapy.

I. INTRODUCTION

Cancer is a major public health problem. Worldwide, more than 6 million people die from cancer each year and more than 10 million new cases are detected. In the last decade, the enormous progress made in discerning the molecular events that accompany carcinogenesis has led to development of new cancer therapies directly targeted against the molecular changes of neoplasia. In preclinical and early clinical trials, molecular-targeted therapeutics (alone and in combination) have shown impressive response rates and less toxicity than conventional cytotoxic therapies, which lack specificity for tumor cells. For example, drugs such as Erlotinib and Erbitux have increased responsiveness to chemotherapy and have shown responses in patients who progress on conventional therapy. *However, in order to achieve the promise of molecular-targeted cancer therapeutics, a new set of tools is required to dynamically and quantitatively image and monitor the heterogeneous molecular profiles of tumors in vivo.* This new set of tools can: (1) enable rational selection of a regimen of targeted therapies, given a quantitative image of the molecular profile of a patient's tumor; (2) provide a much more sensitive and rapid way to monitor the efficacy of targeted therapeutics across an entire tumor, enabling changes to be made quickly before a tumor has advanced to a clinically significant degree; (3) allow design of more appropriate clinical trials of new molecular agents based on alteration in molecular profiles of risk rather than maximal tolerated dose; and (4) provide a dynamic approach to guide conventional tumor ablation, based on a molecular profile of risk.

The traditional classification of cancer and its precursors is based on a phenotypic profile of risk, including macroscopic parameters such as tumor size, extent of invasion, and presence of metastases and microscopic parameters such as nuclear to cytoplasmic (N/C) ratio, differentiation, and depth of epithelial involvement. Increasingly, molecular parameters of risk, such as the status of the epidermal growth factor receptor (EGFR), estrogen/progesterone receptor, Her-2/neu, and c-kit, are being incorporated into clinical therapeutic paradigms. Currently, molecular markers can only be visualized *in vitro* using complex immunohistochemical staining protocols. *In this chapter, we will discuss emerging tools to image in vivo a molecular profile of risk-based hallmarks of cancer, for selecting and monitoring therapy.* Imaging the molecular features of cancer requires molecular-specific contrast agents which can safely be used *in vivo* as well as cost-effective imaging systems to rapidly and noninvasively image the uptake, distribution, and binding of these agents (Fig. 1). Here, we present the combination of optically active, targeted nanoparticles for molecular imaging with advances in minimally invasive optical imaging systems, which provide tools to dynamically image both a molecular and a phenotypic profile of risk and to monitor changes in this profile during therapy. Images obtained in real time through low-magnification optical microscopes can rapidly characterize tumor size and distribution, margin location, and molecular heterogeneity across a large field of view, based on a molecular profile of risk. Then fiber optic endoscopes can be used to quantitatively image areas of potential risk in 3D with subcellular resolution, in real time. The combination of 2D surface imaging and 3D high-resolution imaging affords the ability to profile a tumor and provide precise molecular information with histologic quality spatial resolution. Further, optical imaging

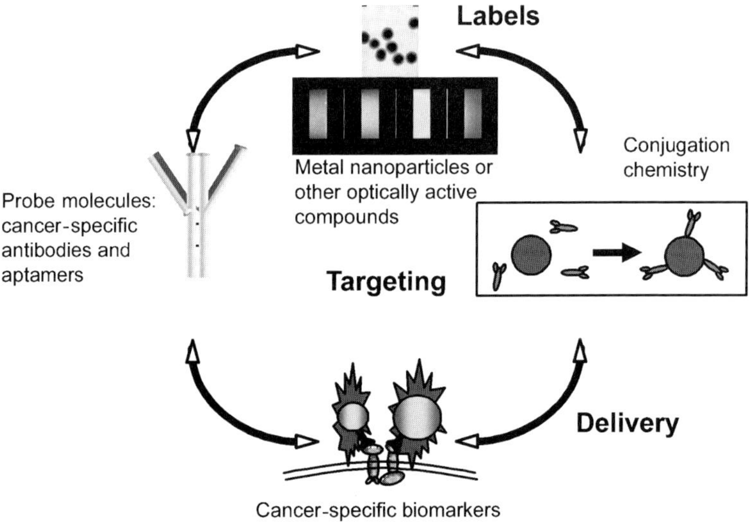

Fig. 1 Development of molecular-specific contrast agents for *in vivo* optical imaging of carcinogenesis. (See Color Insert.)

systems are inexpensive, robust, and portable because of advances in computing, fiber optics, and semiconductor technology. Thus, optical imaging systems are ideally suited for minimally invasive, real-time selection, and assessment of response to targeted therapeutics.

II. ROLE OF MOLECULAR-SPECIFIC OPTICAL IMAGING OF CARCINOGENESIS

Rapid and noninvasive imaging of molecular features of cancer can provide unprecedented ability to study the molecular processes associated with carcinogenesis *in vivo* in humans. In particular, they can enable direct imaging of the biology of invasion and monitoring of host response serially over time. In this section, we give examples of the impact that molecular optical imaging can play in advancing molecular therapeutics into future clinical practice.

A. Monitoring Effectiveness of Therapy

When choosing therapy, it is important to quickly determine whether the patient is responding, to avoid progression of a nonresponsive tumor. There is no reliable method to quickly determine the response of an individual's tumor to nonsurgical therapy. Determination of tumor response is routinely assessed 8 weeks (chemotherapy) or 12 weeks (radiation) after initiation of therapy by physical examination and radiographic scans. These methods are grossly inadequate, relying predominately on lesion size. Even positron emission tomography (PET) scans have limited applicability in this situation. While an invasive biopsy can be performed during treatment, it only removes a small portion (millimeters) of the area of interest and may not accurately reflect overall treatment effectiveness. Furthermore, biopsy can be very painful due to mucositis and inflammation associated with therapy, and may lead to problems with wound healing. It is even more difficult to assess response of tumors to new, molecularly targeted therapy. For example, with antiangiogenesis therapy, there is typically no major change in the gross size of tumors, despite effective blockage of tumor vascular neoangiogenesis.

Thus, advancements that improve the ability to quickly and noninvasively assess the response of a tumor to therapy have important benefits. Optical molecular imaging can provide an important new tool to accurately assess therapeutic response, bridging the gap between clinical exam and histologic indicators of response. Because optical molecular imaging can assess cell morphology, tissue architecture, and cancer-related biomarkers noninvasively, it provides a tool to assess response to therapy across an entire tumor. Thus, it could enable clinicians to avoid unnecessary delay in identifying most effective therapy before a cancer has progressed to a stage where it is difficult or impossible to treat successfully.

B. Rational Selection of Targeted Therapeutics

The development of cancer is a multistep process of genetic, epigenetic, and metabolic changes resulting from exposure to carcinogens. For example, in the progression of oral carcinogenesis, an accumulating series of molecular alterations leads to EGFR activation, telomerase activation, up-regulation of cyclooxygenase-2 (COX-2), and overexpression of cyclin D1 (Ang *et al.*, 2002; Califano *et al.*, 1996; Hanahan and Weinberg, 2000). Elucidation of signaling pathways involved in initiation and progression of malignancy has facilitated development of novel, targeted approaches to cancer treatment. Signaling pathways targeted by these agents include EGFR, vascular endothelial growth factor (VEGF), Ras, COX-2, and p53 (Lippman *et al.*, 2005; Rhee *et al.*, 2004). Understanding this progression has provided a wealth of targeted therapeutics, including COX-2 inhibitors (e.g., celecoxib, sulindac), tyrosine kinase inhibitors (e.g., Iressa), and vectors that cause lysis of cells with deficient p53 activity (e.g., Ad-p53). These agents represent a fundamental new approach to cancer therapy. Current cancer therapeutics rely on cytotoxic drugs that lack specificity for tumor cells; targeted therapeutics may provide greater efficacy with reduced side effects (Holsinger *et al.*, 2003). The success of targeted therapeutics for chronic myeloid leukemia (CML) and gastrointestinal (GI) stromal tumors illustrates the unique potential of these approaches. More recent work has shown that combination therapies combining targeted agents can have substantially greater response (Torrance *et al.*, 2000). Many clinical trials are currently underway with agents targeting EGFR and other tyrosine kinases (ZD1839-Iressa, C225-Cetuximab), Ras signaling via inhibition of farnesyltransferase (SCH6636), p53 gene pathways via gene transduction with RPR/INGN 201, and COX-2 signaling with celecoxib. However, little is known about how to rationally select the combination of agents for a patient and how to rationally design dosing schedules and monitoring strategies for future clinical trials.

Technological advancements in optical molecular imaging can provide the ability to dynamically and quantitatively image the molecular profile of a tumor. This can open the possibility for rational selection of a regimen of targeted therapeutics, based on a molecular profile of risk and can provide a tool to monitor whether the targeted therapy is having the desired effect, enabling clinicians to avoid unnecessary delay in identifying the most effective therapy before a cancer has progressed to a stage where it is difficult or impossible to successfully treat.

III. CANCER BIOMARKERS AND THERAPEUTIC TARGETS

Cancer results from accumulation of a series of key mutations in expanding clones of cells. It has been shown that these molecular events play a major role in cancer progression and can be valuable biomarkers for cancer diagnosis, grading, and prognosis (Ang *et al.*, 2002; van de Vijver *et al.*, 2002). It has been suggested that there are six acquired capabilities shared by most human cancers which collectively dictate malignant growth (Hanahan and Weinberg, 2000) (Fig. 2). In this chapter, we consider biomarkers associated with four of these: (1) self-sufficiency in growth signals (EGFR), (2) evasion of programmed cell death (anionic phospholipids), (3) tissue invasion and metastasis: matrix metalloproteinases (MMPs), and (4) sustained angiogenesis: VEGF and its receptors (Fig. 2).

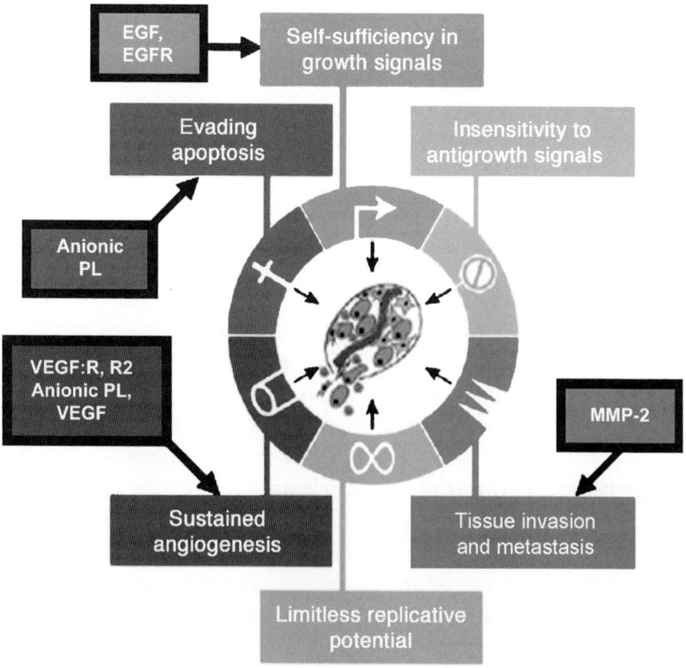

Fig. 2 Strategy to develop contrast agents to monitor hallmarks of cancer (Califano *et al.*, 1996). Targets for contrast agents are shown in boxes outlined with dark black lines. Reprinted from Cell, 100, Hanahan, D., Weinberg, R. A., The Hallmarks of Cancer, p. 57–70, 2000, with permission from Elsevier. (See Color Insert.)

A. Epidermal Growth Factor Receptor

EGFR overexpression occurs in 80–100% of precancerous and cancerous lesions of the oral cavity (Lippman and Hong, 2001). EGFR is also frequently overexpressed in non-small cell lung cancer (NSCLC) (Hirsch *et al.*, 2003) and is a target of novel therapeutics. There are seven known EGFR ligands, including EGF and transforming growth factor-α. Ligand binding leads to receptor dimerization and autophosphorylation of the receptor, which in turn activates signal transduction pathways leading to increased cell proliferation and survival.

B. VEGF and Its Receptors

VEGF is a key angiogenic factor that stimulates blood vessel growth in normal and neoplastic tissues (Zhang *et al.*, 2002). VEGF is expressed at high levels in malignant tissues in many organ sites. There are two high-affinity receptors for VEGF—R1 and R2. VEGFR2 is thought to be the major initiator of angiogenesis in tumors. VEGFR1 and VEGFR2 are both overexpressed in tumor cells and tumor endothelium, due to both increases in ligand concentration and hypoxia (Brekken and Thorpe, 2001).

C. Matrix Metalloproteinases

At late stages of tumor progression, tumor cells invade adjacent tissue and travel to distant sites to form metastases. Metastases cause 90% of human cancer death (Sporn, 1996). MMPs enable invasion and metastasis and are implicated in other activities important for tumor growth, such as angiogenesis and growth signaling (Hanahan and Weinburg, 2000). For example, it has been demonstrated in many studies that MMPs (MMP-1, -2, -3, -9, -10, -11, -13, MT1-MMP) are all expressed in oral cancer and have roles in tumor progression (Thomas *et al.*, 1999 and references therein). MMP-2 and -9 are specific for collagen IV—these are the enzymes that allow for invasion through the basement membrane into blood vessels/lymphatics. MMP staining has been shown to progressively increase as the grade of the lesion or stage of the tumor increased. Sienel *et al.* showed that homogeneous expression of MMP-9 and -2 can serve as significant indicators of poor patient survival in lung carcinomas, as well (Passlick *et al.*, 2000; Sienel *et al.*, 2003). MMP-9 was overexpressed by a factor of 2, while MMP-2 activity was increased by a factor of 17 in 36 lung cancer patients (Hrabec *et al.*, 2002).

IV. OPTICAL TECHNOLOGIES

Molecular imaging requires two components: a molecular-specific source of signal (typically provided through a contrast agent) and an imaging system to detect this signal. In recent years, high-resolution micro-PET, MRI, and ultrasound have shown promise for molecular imaging in animal studies (Pomper, 2001). However, these systems are expensive and do not provide sufficient resolution to image subcellular detail in real time. An alternative is optical imaging. Optical imaging can be carried out noninvasively in real time, yielding unprecedented spatial resolution (less than 1 μm lateral resolution). Optical imaging systems are inexpensive, robust, and portable. Confocal microendoscopes which image near-infrared (NIR) reflected light have been used to image subcellular features in epithelial tissue at video rate to depths exceeding 400 μm (Collier *et al.*, 1998, 2000, 2002; González *et al.*, 1999a–e; Huzaira *et al.*, 2001; Langley *et al.*, 2001; Rajadhyaksha, 2001; Rajadhyaksha *et al.*, 1995, 1999a,b, 2001; Selkin *et al.*, 2001; White *et al.*, 1999). Optical coherence tomography (OCT) has been used to profile NIR reflectance, assessing tissue architecture to a depth of 2 mm, yielding information about the transition from *in situ* carcinoma to microinvasive carcinoma (Bouma *et al.*, 2000; Das *et al.*, 2001; Jesser *et al.*, 1999; Li *et al.*, 2000; Pitris, 1998; Pitris *et al.*, 1999). Optical imaging systems that detect tissue autofluorescence can give important molecular information, mapping the intracellular concentrations of NADH and FAD, for example (Richards-Kortum and Sevick-Muraca, 1996). We developed two types of optical systems to image tissue noninvasively: (1) *widefield microscopy*, to image large fields of view with limited spatial resolution and (2) *confocal microscopy*, to image tissue noninvasively with subcellular spatial resolution. Over the last 15 years, we have intensively tested these systems to aid in the early detection of cervical and oral neoplasia. Later in this chapter, we demonstrate the integration of optical imaging devices with molecular-specific contrast agents for monitoring of carcinogenesis. Here, we describe in detail real-time optical imaging technologies that allow imaging of morphologic and molecular features of neoplasia at two length scales. In the first, low-resolution, widefield microscopes, capable of imaging areas with a large field of view (5–15 cm), are used to identify areas suspicious for neoplasia. In the second approach, high-resolution microscopes, operating near the diffraction resolution limit, are used to image the morphologic and molecular characteristics of neoplastic lesions.

A. Widefield Microscopy

In widefield microscopy, the tissue surface is illuminated with light; light remitted from the surface is collected through an objective lens and directed to a detector to form a 2D widefield image (Fig. 3). We developed simple,

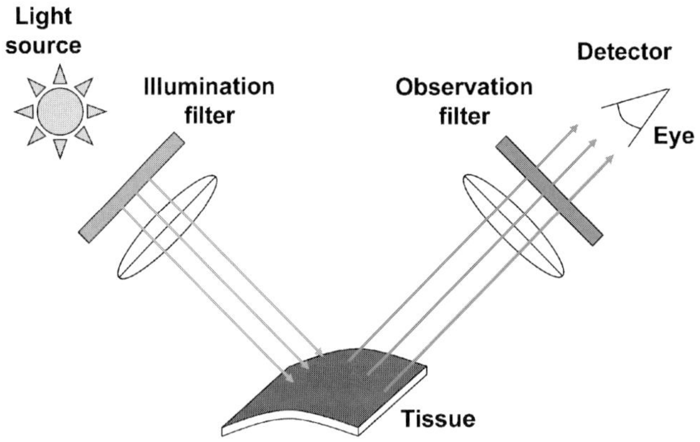

Fig. 3 *In vivo* widefield microscopy. (See Color Insert.)

inexpensive systems to image tissue at video rate *in vivo*, providing information to guide placement of higher resolution imaging systems. In one approach, a simple colposcope (a low-power microscope used to view the cervix) was modified to enable collection of quantitative images of tissue autofluorescence—the resulting device is a multispectral digital colposcope (MDC). In constructing the MDC, we used an inexpensive (<US$500), commercially available, video-rate, color CCD camera to capture autofluorescence images at video rate *in vivo* (Benavides *et al.*, 2003; Park *et al.*, 2005). Figure 4 illustrates typical results to image autofluorescence and reflectance of the cervix. In colposcopy, a health care provider visually examines pattern of white light reflected from the cervix before and after application of acetic acid. We extended the ability of the colposcope to measure tissue autofluorescence, which has the potential to increase both specificity and sensitivity of the procedure and to guide where higher resolution images should be obtained (Park *et al.*, 2005; Svistun *et al.*, 2004). We explored the use of nonspecific contrast agents such as acetic acid with the MDC to indicate regions of tissue to be probed in greater detail.

We also applied widefield microscopy to demonstrate the feasibility of detecting tumor margins in real time in the oral cavity. To evaluate the concept of multispectral imaging of oral tissue, we investigated the autofluorescence of *ex vivo* oral cancer lesions at several wavelength combinations. Fluorescence of oral tissue specimens was observed through a widefield multispectral imaging system. Figure 5 shows typical results. The system consisted of a xenon light source and a CCD camera. The tissue was excited at 340, 380, 400, and 440 nm, and tissue autofluorescence was collected at 510 and 530 nm. The specimen photographs were handed to an experienced head and neck surgeon, blinded to the histopathological diagnosis of the

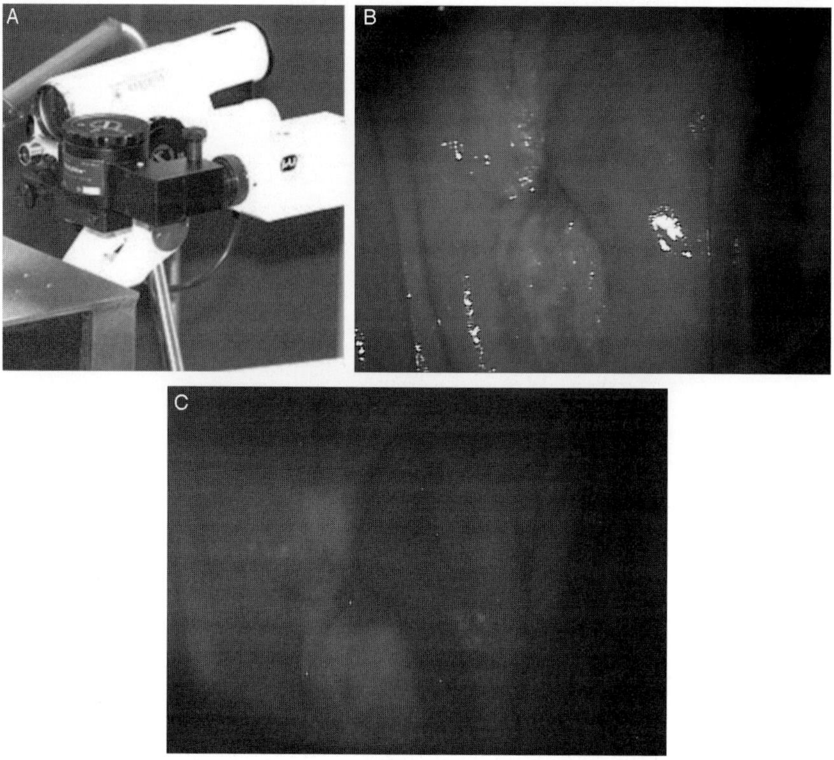

Fig. 4 Multispectral digital microscope (top) and *in vivo* images of cervical tissue obtained with white light (middle) and 340 nm excitation (bottom). Published with permission from Optical systems for *in vivo* molecular imaging of cancer, Technology in Cancer Research and Treatment, 2, 491–504, 2003, TCRT. http://www.tcrt.org. (See Color Insert.)

biopsies. Suspected neoplastic areas were encircled on photos, and compared to histologic results. Best results were achieved at 400 nm excitation and 510 nm detection (Svistun *et al.*, 2004) and indicated that oral cavity autofluorescence can easily enhance a clinician's visual assessment of normal and neoplastic mucosa in the oral cavity.

These systems can be also used to image fluorescence in intact small animal models of neoplasia. Figure 6 shows white light and fluorescence images of subcutaneous tumors from MDA-MB-435 human breast cancer cells in a nude mouse model. Tumors formed with red fluorescence protein (RFP) expressing cells show easily detectable autofluorescence as early as 2 weeks following injection of the cancer cells, while RFP negative control tumors do not show detectable fluorescence. Micrometastases in the excised lung (Fig. 7) can easily be detected via the RFP fluorescence.

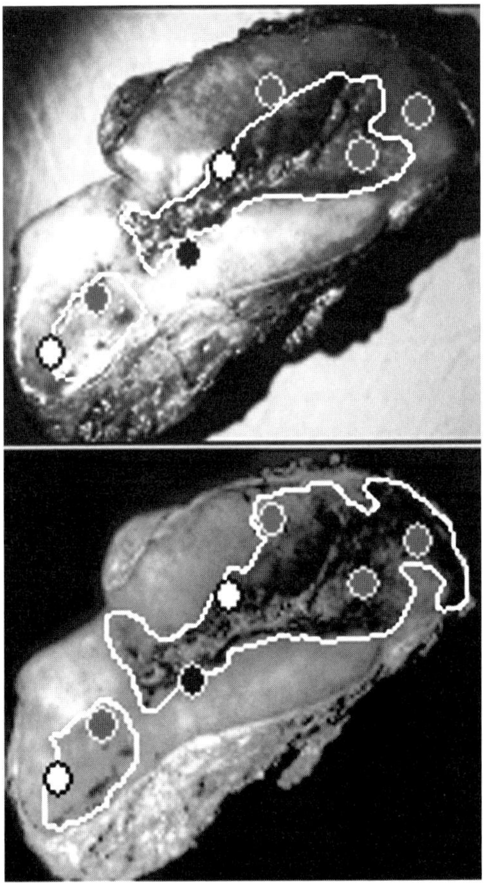

Fig. 5 Images of resected oral cancer obtained with white light illumination (top) and 440 nm excitation, 530 nm emission (bottom). Yellow lines represent surgeon's assessment of the margins of abnormal tissue. Circles are biopsy sites: red = cancer, white = dysplasia, blue = normal. Published with permission from Vision enhancement system for detection of oral cavity neoplasia based on autofluorescence, Head & Neck, 26, 205–215, 2004, Wiley. (See Color Insert.)

B. High-Resolution Microscopy

In order to image the morphologic and molecular changes of neoplasia with cellular and subcellular detail, higher resolution microscopes are required. *In vivo* confocal microscopes can provide detailed images of tissue architecture and cellular morphology in living tissue in near real time. In epithelial tissue, 1 μm resolution has been achieved with a 200–400 μm field of view and penetration depth up to 500 μm (Collier *et al.*, 2002; Delaney and Harris,

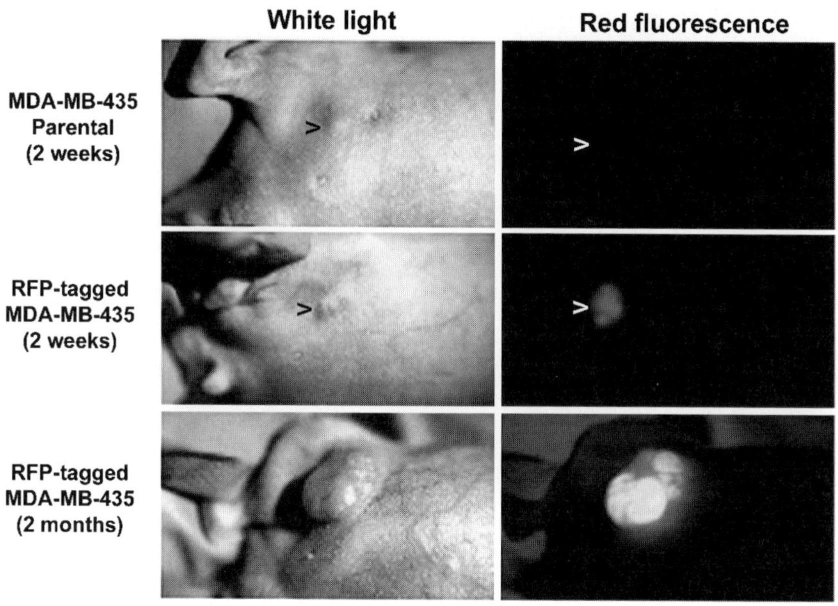

Fig. 6 Multispectral images of nude mice with subcutaneous tumors (MDA-MB-435 cells). The top row shows white light and fluorescent images of tumors formed with the parental cell line, 2 weeks following injection. The middle row shows white light and fluorescent images of tumors formed with RFP-expressing cells 2 weeks after injection, while the bottom row shows images 2 months after injection. (See Color Insert.)

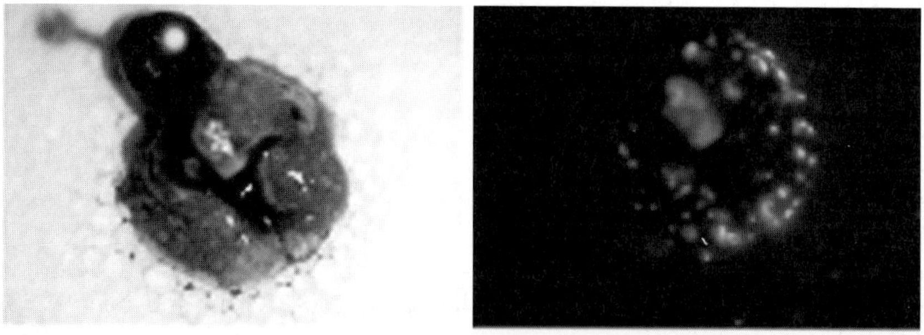

Fig. 7 Micrometastases in the excised lung. Left image (white light) and right image (red fluorescence). (See Color Insert.)

1995; Drezek *et al.*, 2000; Rajadhyaksha, 2001; Rajadhyaksha, *et al.*, 1995, 1999a,b, 2001; Selkin *et al.*, 2001; White *et al.*, 1999). A nonfiber optic version of a real-time reflectance-based confocal microscope showed promise to detect changes associated with cervical precancer (Collier *et al.*, 2002; Smithpeter *et al.*, 1998). Figure 8 shows image pairs of normal (left) and abnormal (right) biopsies from several patients and the corresponding histology. In each pair, the difference in nuclear density and area is evident between the normal and abnormal biopsy. Architectural similarities are also evident between confocal and histologic images. The N/C ratio extracted from confocal images gave best discrimination between high-grade lesions and non-high-grade lesions.

C. Fiber Optic Microscopes

Fiber optic confocal microscopes are needed to obtain images clinically (Minsky, 1955). These instruments are designed to be inserted through a speculum, catheter, large-bore needle, or biopsy channel of an endoscope. An example of a fiber optic reflectance confocal microscope with a miniature, plastic injection-molded objective lens is shown in Fig. 9 (Liang *et al.*, 2002; Sung *et al.*, 2002a,b). This system has a lateral resolution of $\sim$2 μm, an axial resolution of $\sim$5 μm, and can obtain images throughout the entire epithelial thickness in cervical tissue.

Later in this chapter, we discuss the combination of these optical imaging technologies with optical contrast agents specific for cancer biomarkers.

V. OPTICALLY ACTIVE CONTRAST AGENTS

Luminescent organic dyes and fluorescent proteins have been traditional optical contrast agents and their applications in biology and medicine are described in a number of excellent reviews and books (Achilefu, 2004; Geddes, 2004; Hassan and Klaunberg, 2004; Shah and Weissleder, 2005; Tsien, 2005). Recently, a new paradigm has emerged in development of optically active compounds which is based on nanoparticles with unique size-dependent optical properties (Alivisatos, 2004; Chan and Nie, 1998; Michalet *et al.*, 2005; Santra *et al.*, 2004). The nanoparticle platform offers a number of advantages over the traditional fluorescent dyes including greatly improved photostability, bright signal, and simple tunability of optical properties (Alivisatos, 2004). Nanoparticles also provide high surface area which can be easily modified using a variety of delivery, targeting, and therapeutic moieties that enable the use of nanoparticles as a common platform for multiple applications (Michalet *et al.*, 2005).

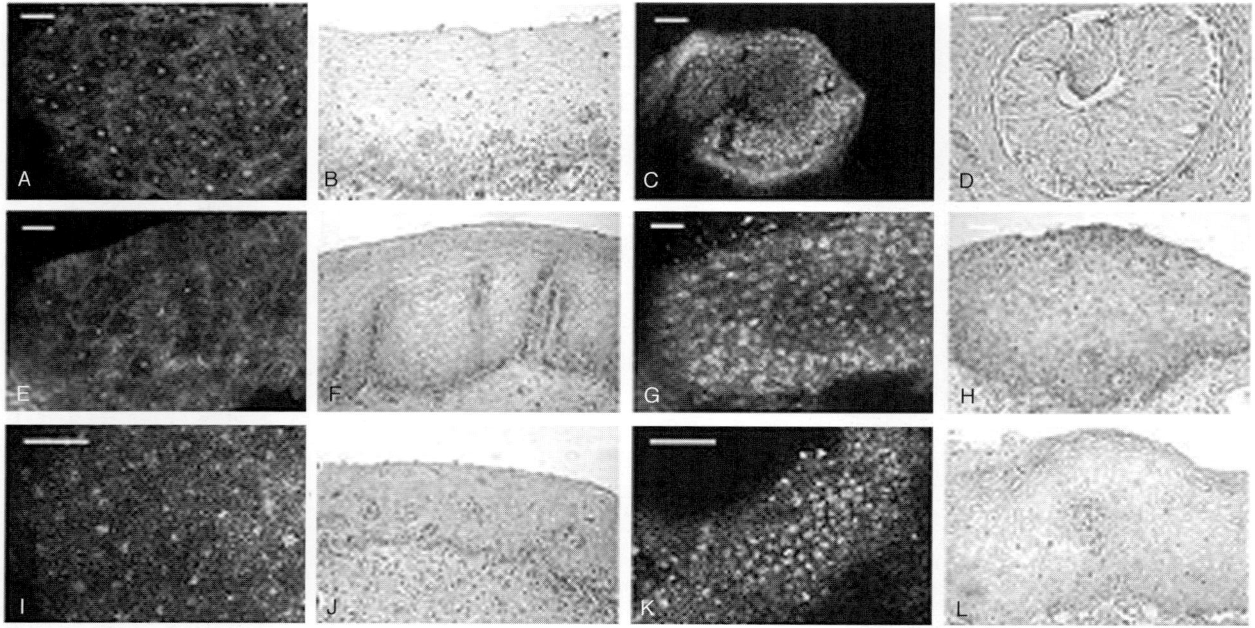

Fig. 8 Reflectance confocal images from the normal (left two columns) and abnormal (right two columns) biopsy pairs. Increased nuclear density can be seen in the confocal images of the abnormal samples (C, G, and K). The confocal images were taken 50 μm below the surface. The histologic images were classified as normal (B, F, and J), CIN II/III (D and H) and cancer (L). Scale bars in the confocal images are 50 μm and 100 μm in the histology images. Reprinted from Academic Radiology, Vol. 9, Collier, *et al.*, Near real time confocal microscopy of amelanotic tissue: detection of dysplasia in *ex vivo* cervical tissue.

Fig. 9 Miniature injection-molded 3.3X/1.0 NA microscope objective for reflectance confocal microscopy. Published with permission from *In vivo* fiber optic confocal reflectance microscope with an injection-molded plastic miniature objective lens, Applied Optics, 44, 1792–97, 2005. (See Color Insert.)

In this chapter, we outline research activities in development of two emerging types of optically active contrast agents: those based on metal nanoparticles, which give a strong source of reflected light, and those based on quantum dots, which are a strong source of luminescence. We also discuss how the unique nanoscale sensitivity of quantum dots and gold nanoparticles can be used to design smart contrast agents, which are dark in the absence of a target ligand but become brightly luminescent on exposure to the ligand to select and monitor targeted therapies. These smart contrast agents rely on the strong ability of gold to quench the luminescence of quantum dots when in close proximity.

Our strategies in the development of the optically active contrast agents are outlined in Fig. 10. The agents consist of three parts: (1) a probe molecule which provides molecular-specific recognition of cancer biomarkers conjugated to (2) a nanoscale optically interrogatable label in (3) a permeation-enhancing delivery formulation. We use three types of molecular probes: monoclonal antibodies against cancer-specific biomarkers, peptides selectively cleaved by cancer-related enzymes, and aptamers which undergo conformational change on selective binding to cancer-related growth factors. Together, this cocktail of contrast agents can constitute a powerful approach to image the hallmarks of cancer at the molecular level and to monitor the effects of targeted therapies on multiple crucial signaling pathways, alone and in combination.

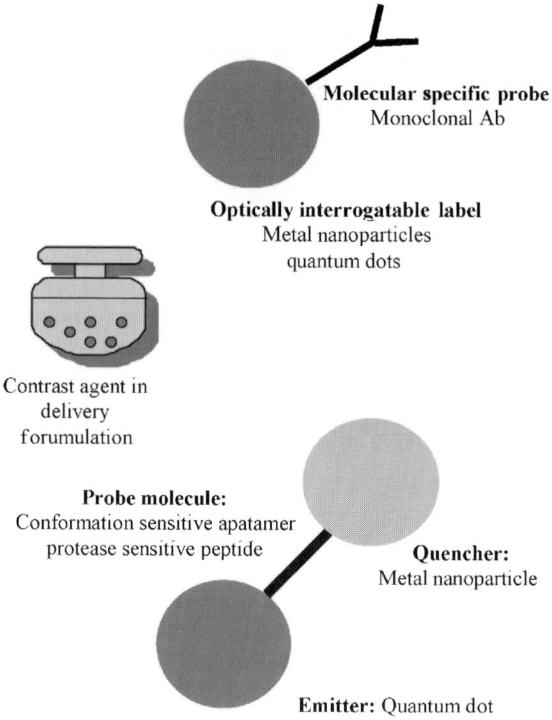

Fig. 10 Contrast agent strategy. Published with permission from Optical Systems for *in vivo* molecular imaging of cancer, Technology in Cancer Research and Treatment, 2, 491–504, 2003, TCRT. http://www.tcrt.org.

A. Metal Nanoparticles

1. OPTICAL PROPERTIES

Gold nanoparticles have been extensively used as stains in electron microscopy of Horisberger (1981) and Geoghegan and Ackerman (1977). As a result, the fundamental principle of interactions between gold particles and biomolecules, especially proteins, have been thoroughly studied. However, colloidal gold nanoparticles exhibit beautiful and intense colors in the visible and NIR spectral regions. These colors are the result of excitation of surface plasmon resonances and are extremely sensitive to the sizes, shapes, and aggregation state of the particles, to the dielectric properties of the surrounding medium, and to adsorption of ions on the particle surface (Mulvaney, 1996).

The tremendous potential of using metal nanoparticles as optically interrogatable biological labels has recently been recognized, leading to development of a variety of novel applications in bioanalytical chemistry with unprecedented sensitivity (Elghanian *et al.*, 1997; Kneipp *et al.*, 2002; Nabiev *et al.*, 1991; Nie and Emory, 1997; Schultz *et al.*, 2000; Yasuda *et al.*, 2001; Yguerbide, 2001). The ability to resonantly scatter light at frequencies coinciding with the particles' surface plasmon resonances is still to be explored for *in vivo* biological applications. Recently, it has been demonstrated that this property can be used in the development of contrast agents for *in vivo* reflectance (Sokolov *et al.*, 2003a,b). The scattering cross section of gold nanoparticles is extremely high compared to polymeric spheres of the same size (Fig. 11), especially in the red. This property is crucial for development of contrast agents for optical imaging in living organisms because light penetration depth in tissue dramatically increases toward the red and NIR optical region. Another interesting optical property of metal nanoparticles is the increase in scattering cross section per particle when the particles aggregate (Fig. 11, right). These changes produce a large optical contrast, in both the scattering cross section and the wavelength dependence of the scattering, between isolated gold particles and assemblies of gold particles. This increase in contrast improves the ability to image

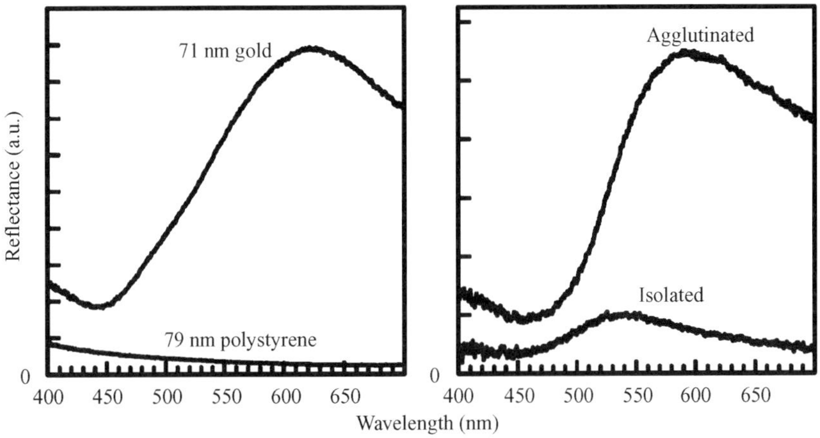

Fig. 11 Light scattering by suspensions of polystyrene spheres and gold nanoparticles with the same concentration (left). Comparison of scattering of isolated and agglutinated conjugates of 12 nm gold particles with EGFR antibodies (right). Published with permission from Real-Time Vital Optical Imaging of Precancer Using Anti-Epidermal Growth Factor Receptor Antibodies Conjugated to Gold Nanoparticles, Sokolov, K., Follen, M., Aaron, J., Pavlova, I., Malpica, A., Lotan, R., and Richards-Kortum, R., Cancer Research, 63:1999–2004, May 1, 2003.

markers which are not uniquely expressed in diseased tissue, but are expressed at higher levels relative to normal tissue, and to develop sensitive labeling procedures which do not require washing steps to remove single unbound particles.

2. SYNTHESIS

We explore vital reflectance imaging using both gold and silver nanoparticles. The major focus is on development of agents based on gold because it is widely recognized that gold is biocompatible and can be used directly for *in vivo* applications. However, silver particles exhibit higher extinction coefficients and provide higher enhancement of the local electromagnetic field and of other effects associated with optical excitation of surface plasmon resonances (Zeman and Schatz, 1987). Silver particles are not as stable or biocompatible as gold nanoparticles. This issue can be addressed by encapsulating silver particles inside an inert material such as silica coating (Sokolov *et al.*, 1998). This coating stabilizes particles and provides a well-characterized surface for chemical immobilization of biomolecules.

Gold and silver colloids can be prepared from chloroauric acid ($HAuCl_4$) and silver nitrate ($AgNO_3$), respectively, by using a variety of reducing agents (Cotton, 1988; Frens, 1973; Hildebrandt and Stockburger, 1984; Lee and Meisel, 1982; Schneider, 1988; Wilenzick *et al.*, 1967). Preparation of highly uniform gold colloids with particle sizes ranging from about 10 nm to ca. 100 nm was demonstrated using sodium citrate reduction of chloroauric acid (Frens, 1973). This colloid exhibits a single extinction peak ranging from 500 nm to about 540 nm depending on particle size. Sodium citrate reduction of silver nitrate results in a colloidal solution with about 35-nm-diameter silver particles and a single peak at 410 nm (Hildebrandt and Stockburger, 1984; Lee and Meisel, 1982). The distribution of silver particles is significantly broader; however, the procedure is highly reproducible. Silver particles with narrower distributions and different diameters can be prepared using a starter hydrosol (Schneider, 1988).

Scattering properties of metal nanoparticles strongly depend on their size and shape. By changing sizes, metal nanoparticles that exhibit different colors in reflected light can be produced (Fig. 12). The strength and wavelength dependence of this scattering can be predicted using Mie theory. Additionally, the position of the surface plasmon resonances of gold and silver can be significantly altered by using nanocomposite materials described by Sershen *et al.* (2000) and Averitt *et al.* (1999). These materials consist of a dielectric optically inert core particle and an optically active gold shell and can be prepared in a variety of sizes. Another approach to tune optical properties of metal nanoparticles is based on template synthesis (Haes and Van Duyne, 2002). In one method, metal nanoparticles of

Fig. 12 Scattering of silver nanoparticles with different sizes. Similar tuning of optical properties can be achieved with gold. (See Color Insert.)

pyramidal shape with different sizes can be synthesized inside cavities formed by a dense monolayer of polystyrene beads on a flat substrate. After synthesis, the nanoparticles can be removed from the surface by a simple one-step procedure. The available variety of the synthetic methods allow optimization of scattering properties of contrast agents to take advantage of optical regions where tissue is most transparent depending on the degree of tissue penetration required. Another important venue is to use particles of different sizes conjugated to different probe molecules to achieve multicolor labeling.

3. CONJUGATION

There are a variety of possible strategies for preparation of conjugates of gold particles with cancer-specific probe molecules. Well-characterized conjugation protocols have been developed to prepare gold immunostains for electron microscopy (Geoghegan and Ackerman, 1977; Horisberger, 1981). Briefly, the procedure is based on noncovalent binding of proteins at their isoelectric point (point of zero net charge) to gold particles. The complex formation is irreversible and very stable. In fact, the shelf life of the conjugates is so long that the most commonly used gold immunostains can be routinely purchased from major biochemical companies. Moreover, a variety of thiol-terminated heterogeneous cross-linkers can be also used to covalently attach monoclonal antibodies to a gold surface. This approach is preferable for *in vivo* applications because it allows coadsorption of other thiol-terminated molecules, such as polyethylene glycol (PEG), which is important to increase circulation of the contrast agents in the body after systemic delivery. For vital imaging with contrast agents based on metal nanoparticles, it is imperative to develop bioconjugates that have very low nonspecific binding and are not accumulated by the reticuloendothelial system. This issue is usually addressed by coadsorbing PEG and probe molecules on the surface on nanoparticles. This strategy has been recently demonstrated in experiments on *in vivo* molecular-specific imaging of embryogenesis using quantum dots (Dubertret *et al.*, 2002) and in colloidal gold drug delivery system in live mice (Paciotti *et al.*, 2001).

B. Quantum Dots

1. OPTICAL PROPERTIES

Semiconductor nanocrystals with diameters smaller than the Bohr exciton diameter—on the order of 1–10 nm—exhibit size-dependent optical properties due to quantum confinement of electrons and holes and are often referred to as "quantum dots." Fluorescence emission from the quantum dots can be tuned by size and composition to range from 400 nm to 2 μm with very narrow emission bandwidths of approximately 20–30 nm (Banin *et al.*, 1998; Murray *et al.*, 1993). Quantum dots of different size that emit fluorescence at different wavelength can be excited at a single wavelength greater than their respective absorption edges. This provides a unique opportunity to do multicolor imaging experiments with a single excitation wavelength (Lacoste *et al.*, 2000). Despite the obvious advantages of quantum dots as compared to conventionally used fluorescence labels, their biological applications have been hampered by the low solubility of semiconductor materials which comprise the dots. Recently, new chemical strategies were proposed to make water-soluble quantum dots that immediately resulted in exciting applications of quantum dots for biological imaging (Akerman *et al.*, 2002; Bruchez *et al.*, 1998; Chan and Nie, 1998; Chan *et al.*, 2002; Dubertret *et al.*, 2002; Parak *et al.*, 2002). Comparison of quantum dots to one of the brightest fluorescent molecules—Rhodamine 6G (R6G)—showed that the quantum dots are 20 times as bright, 100 times as stable against photobleaching, and one-third as wide in spectral linewidth (Chan and Nie, 1998).

A variety of different types of quantum dots have been used for biological labeling: CdS for UV-blue, CdSe for the bulk of the visible spectrum, and CdTe for the far red and NIR. Although the semiconductor dictates the spectral region where the emission occurs, the size of the particle can shift the wavelength—the absorption onset and emission shift to larger energy with decreasing size (Bruchez *et al.*, 1998). Our research has recently focused on CdTe quantum dots because the emission wavelengths fall within 600–1300 nm, the spectral region which has been demonstrated as best suited for biological imaging due to an increase in the penetration depth of light in tissue (Peng and Peng, 2001).

2. SYNTHESIS AND CONJUGATION STRATEGIES

A variety of semiconductor nanocrystals, or quantum dots, with relatively high-quality optical properties can be produced using solution-phase methods (Holmes *et al.*, 2001; Korgel and Fitzmaurice, 1999; Murray *et al.*, 1993). The nanocrystal preparations must yield nanocrystals with a tight size distribution (to eliminate inhomogeneous broadening of optical properties), crystalline cores with few compositional and structural defects, and well-passivated surfaces. The most successful route to synthesizing semiconductor nanocrystals has been through arrested precipitation with subsequent size-selective precipitation (Korgel and Fitzmaurice, 1999; Murray *et al.*, 1993). Arrested preparation methods rely on binding bulky "inert" ligands to the particle surfaces during growth. Thiols have been used as capping ligands in a relatively general way since they adsorb to a wide variety of semiconductor materials. Other capping ligands include phosphines, amines, and carboxyl groups, depending on the chemistry of the inorganic material. The ligand extending away from the particle surface determines the particle solubility. Particles can be functionalized with either hydrophobic (i.e., alkanes) or hydrophilic (e.g., carboxyl or amine groups) moieties. The nanocrystals are sufficiently stable that chemistry can be done to their surfaces.

CdTe quantum dots are synthesized using a modification of the organometallic method developed by Peng and Peng (2001). Briefly, cadmium oxide (CdO) and *n*-tectradecylphosphonic acid (TDPA) are loaded with trioctylphosphine oxide (TOPO) into a reaction flask. The mixture is heated to 340 °C under nitrogen to form a Cd-TDPA complex. As the mixture cools, a tellurium-trioctylphosphine (TOP) complex is injected. Spherical nanocrystals form quickly, resulting in monodisperse (Fig. 13), highly luminescent CdTe nanocrystals. The mixture is cooled and chloroform is injected to quench the reaction. Nanoparticles are isolated and cleaned by precipitation with ethanol. We showed that the size of the resulting nanocrystals is dependent on the temperature at which the TOP/tellurium is injected into the cadmium/hexylphosphonic acid (HDPA)/trioctylphosphine oxide (TOPO); as temperature of injection increases, the size of the nanocrystal increases but with a loss of quantum yield. The TOPO/TOP capping renders the quantum dots soluble in organic solvents, but not in aqueous environment. A post-synthesis ligand exchange is performed with mercaptopropionic acid (MPA) to render the quantum dots water soluble, and the carboxyl group extends into the solvent for use in subsequent linking to a targeting biomolecule using cross-linking agents similar to procedure described by Chan and Nie (1998). In another approach, we carried out a partial ligand exchange with biotinylated PEG thiol; the avidin-biotin reaction can be used to add biotinylated targeting moieties (Fig. 14).

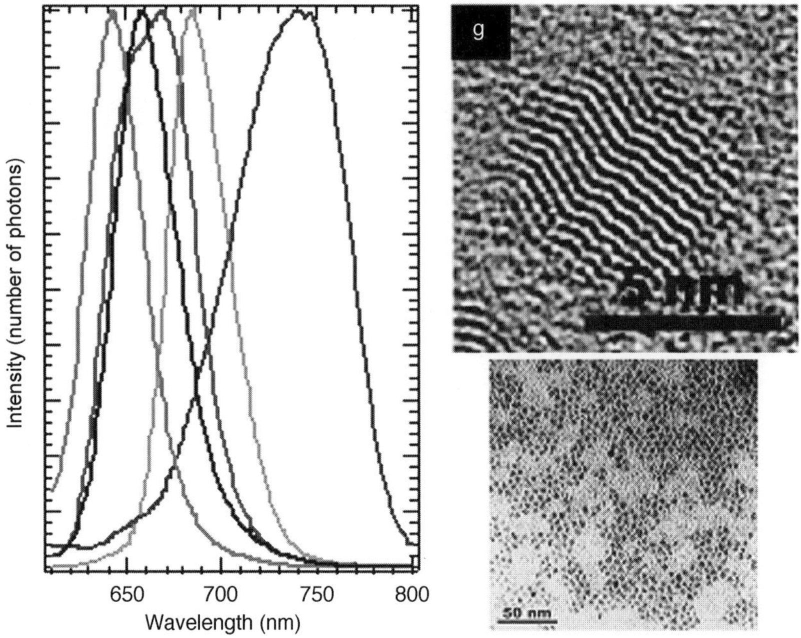

Fig. 13 PL emission spectra of CdTe nanocrystals of increasing diameter (from ~2 to 6 nm) left to right. The excitation wavelength was 540 nm for all of the nanocrystals. (Right) TEM image of a CdTe nanocrystals showing its internal crystallinity and spherical shape (Shieh, Saunders, and Korgel, unpublished data). (See Color Insert.)

Alternatively, quantum dots can be encapsulated by silica and, then, a variety of silanization reagents can be used to introduce functional groups to silica surfaces for subsequent protein immobilization (Bhatia, 1989).

C. Smart Contrast Agents

A number of enzyme-activatable fluorescent probes have been developed and tested for use in optical molecular imaging of cancer. In particular, a number of protease-specific imaging probes have been developed to image cathepsin-B (Bogdanov *et al.*, 2002; Bremer *et al.*, 2002, 2005), MMP-2 (Bremer *et al.*, 2001; Funovics *et al.*, 2003), and caspase-3 (Chiang and Truong, 2005). In this approach, multiple residues of an NIR fluorophore, typically Cy5.5, are coupled via a long circulating graft copolymer containing a cleavable peptide spacer that serves as cleavage sites for recognizing proteases. When the fluorophores are in close proximity, mutual energy

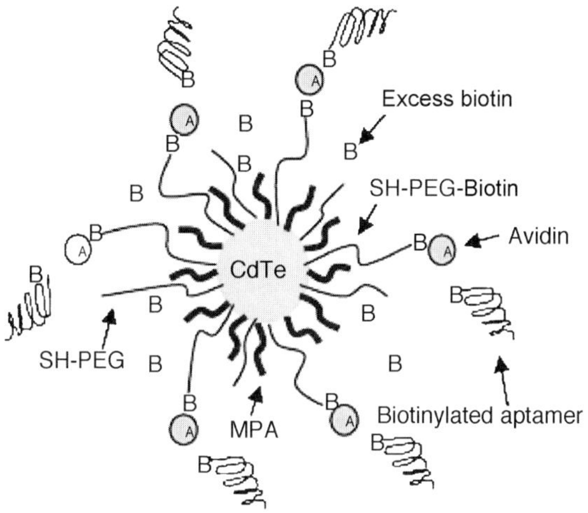

Fig. 14 Capped, aptamer targeted quantum dot. (See Color Insert.)

transfer results in strong quenching of fluorescence. However, following enzymatic cleavage of the backbone, the fluorophores are released, resulting in strong detectable fluorescence. These enzyme-sensing molecular beacons were used to image the presence of cathepsin-B in dysplastic adenomatous polyps in a mouse model (Marten *et al.*, 2002) and MMP-2 via a synthetic MMP-2 substrate peptide spacer (Gly-Pro-Leu-Gly-Val-Arg-Gly-Lys) (Bremer *et al.*, 2001). *In vitro* cleavage of the peptide spacer by MMP-2 releases dye molecules; fluorescence intensity increases by 850% due to dequenching (Bremer *et al.*, 2001). Activation could be blocked by MMP-2 inhibitors. Activated probe could be detected in MMP-2 positive tumors implanted in nude mice 1 h following i.v. injection with a twofold increase in fluorescence compared to MMP-2 negative tumors.

Molecular beacons have been used to probe mRNA expression in living cells (Tsourkas *et al.*, 2002). This approach uses a single beacon consisting of an oligonucleotide probe capable of forming a stem-loop hairpin structure with a reporter fluorescent dye at one end and a quencher at the other end. In the absence of the target, the beacon is dark. When target RNA is present, the beacon opens and hybridizes; the quencher moves sufficiently far from the fluorophore resulting in bright emission (Tsourkas *et al.*, 2002). While this approach provides a tool to probe gene expression in living cells, background fluorescence due to incomplete quenching limits sensitivity. A more sensitive approach uses a pair of molecular beacons, one with a donor and the other with an acceptor fluorophore, that hybridize to adjacent regions on a single mRNA target (Santangelo *et al.*, 2004).

Thus, incomplete quenching limits the sensitivity of current molecular beacons for molecular imaging. Recent studies show that gold metal efficiently quenches the fluorescence of many fluorophores (Dulkeith *et al.*, 2002), and this approach has been used to provide efficient quenching in molecular beacons recently (Dubertret *et al.*, 2001). Gold nanoparticles quench the fluorescence of cationic polyfluorene with Stern–Volmer constants near 10^{11} M^{-1}, which is 9–10 orders of magnitude larger than small molecule dye-quench pairs. In addition, gold is immune to photobleaching problems associated with dye-based systems (Dulkeith *et al.*, 2002). Dulkeith *et al.* (2002) examined these effects for quenching lissamine fluorescence via colloidal gold as a function of particle diameter ranging from 2 to 60 nm. Multiple lissamine molecules were covalently attached to the gold via a 1 nm thioester spacer. Quenching of greater than 99% of fluorescence was observed with 30-nm diameter gold particles.

VI. DELIVERY OF CONTRAST AGENTS *IN VIVO*

Different delivery formulations may be required for different clinical applications, and optimizing these formulations is an important part of the design of a successful imaging strategy. To select therapy, to guide resection, and to monitor response to targeted therapies in epithelial malignancies, topical application may be the route of choice, since the contrast agent can be applied directly to the tumor, the margin, and the tumor bed. To monitor molecular therapeutics of deep tumors and metastases, greater penetration may be required and thus systemic injection may be the optimal approach.

A. Topical Delivery

Systemic delivery of imaging nanoparticles such as quantum dots have been recently described in a number of reports (Akerman *et al.*, 2002; Kim *et al.*, 2004; Larson *et al.*, 2003). However, less data are available on topical delivery of contrast agents (Sokolov *et al.*, 2003b). Here, we describe topical delivery formulations which combine mucoadhesive polymers and penetration enhancers. The goal is to (1) increase mucoadhesion of the nanoparticles, (2) increase penetration into the epithelial layer, and (3) to preserve the specificity of the contrast agents for cancer biomarkers. Below, we give three examples of polymer formulations that appear promising for topical delivery of contrast agents.

Polycarbophil–PVP: It is based on a combination of mucoadhesive polycarbophil (1–3% w/v, B. F. Goodrich) and 1–3% polyvinylpyrrolidone (PVP K-90, BASF). Polycarbophil is a water insoluble (but swellable) polymer that is highly mucoadhesive and has been used in vaginal and topical drug delivery (Bernkop-Schnurch *et al.*, 2000; Chang *et al.*, 2002a,b; Eouani *et al.*, 2001; Kerec *et al.*, 2002; Lehr *et al.*, 1992, 1994; Luessen *et al.*, 1995), specifically for the delivery of progesterone (e.g., Replens, Columbia Laboratories Inc.). PVP is a widely used excipient in topical formulations (e.g., Povidone) and could enhance the permeability of the mucosal surface.

Carbopol–PVP: Carbopol (Noveon Inc., or Polymer Sciences, NJ), a water-soluble polymer, along with PVP as a permeation enhancer. Carbopol (polyacrylic acid) is a highly bioadhesive polymer that has been shown to adhere strongly with different mucosal surfaces (Chun *et al.*, 2002; Lee and Chien, 1996; Oechsner and Keipert, 1999; Park and Robinson, 1987; Rossi *et al.*, 1999; Ugwoke *et al.*, 2000; Warren and Kellaway, 1998). Low concentration should be used to ensure that the formulation can be washed away before imaging and to ensure adequate diffusion of particles out of the formulation and into the mucosa.

Chitosan–Carboxymethylcellulose: Chitosan is a biocompatible polymer that has been widely used in drug delivery and topical wound-healing applications. Chitosan, a highly mucoadhesive polymer (Bernkop-Schnurch, 2000; Ferrari *et al.*, 1997; Filipovic-Grcic *et al.*, 2001; Singla and Chawla, 2001), increases paracellular permeation of drugs, peptides, and proteins across the mucosal epithelium (Dodane *et al.*, 1999; Kotze *et al.*, 1997, 1999; Senel and Hincal, 2001; Thanou *et al.*, 2000, 2001a,b) and is widely used in nasal and oral delivery as well as in some studies for vaginal delivery (Genta *et al.*, 1999; Kast *et al.*, 2002). A 0.1–0.5% solution of chitosan (MW ~290,000) containing 5% carboxymethyl cellulose (a viscosity enhancer) would provide an appropriate solution for the topical application of contrast agents.

Both carbopol and polycarbophil are widely used in mucosal formations; however, neither has been previously used for nanoparticle delivery. The third approach differs from the first two because of the permeation-enhancing properties of chitosan. These delivery formulations have to be still thoroughly tested in tissue cultures and animal experiments before transition to clinical testing.

B. Systemic Delivery

One of the promising approaches to formulating nanoparticles for systemic delivery is the use of PEG-functionalized nanoparticles where the distal end of the PEG is conjugated to the targeting ligand. This design ensures appropriate targeting of the nanoparticles without steric hindrance while at the same time providing "stealth" properties to the particles for prolonged systemic bioavailability. This can be achieved using a heterobifunctional PEG carrying a thiol group at one end and an N-hydroxysuccinimide (NHS) ester at the other end (SH-PEG-NHS, Nektar Therapeutics, MW 3400). First, the SH-PEG-NHS is reacted with the targeting antibody or aptamer (carrying a primary amine end group) under reducing condition (i.e., presence of 0.1 mM dithiotrietol, pH 8.5, phosphate buffer) to generate SH-PEG-ligand.This ensures that the SH groups do not form disulfide bonds during the reaction. Then, SH-PEG-ligand is conjugated directly to the gold nanoparticles or is attached to the surface of quantum dots using a ligand exchange reaction. Another approach has been reported by Nie (Gao *et al.*, 2004) where amphiphilic triblock copolymers (polybutacrylate–polyethacrylate–polymethacrylic acid) are used as a micellar coating on quantum dots for *in vivo* delivery.

VII. MOLECULAR-SPECIFIC OPTICAL IMAGING

A. Metal Nanoparticles

First applications of silver and gold nanoparticles as scattering contrast agents for *in vitro* biological assays were reported by Schultz *et al.* (2000) and Yguerabide *et al.* (2001). It was demonstrated that scattering signal from the single nanoparticles can be millions of times brighter than signal from individual fluorescent molecules that can enable development of ultrasensitive immuno and DNA probe assays. Our group has been working on development of contrast agents based on gold nanoparticles for detection of cancer biomarkers in living cells (Sokolov *et al.*, 2003a,b). In this section, we demonstrate molecular-specific imaging of EGFR using gold bioconjugates with anti-EGFR monoclonal antibodies. In these studies, gold nanoparticles with ca. 12 nm in diameter were used. Our data show that gold conjugates are stable in biological samples and do not exhibit nonspecific aggregation.

1. METAL NANOPARTICLES AND EGFR

Figure 15 shows confocal reflectance and combined transmittance/reflectance images of SiHa (human cervical cancer cell line) cells labeled with anti-EGFR/gold conjugates. Labeling predominately occurs on the surface of the cytoplasmic membrane, specifying the transmembrane localization of EGFR. The intensity of light scattering from the labeled cells is ca. 50 times higher than from unlabeled cells. Therefore, unlabeled cells cannot be

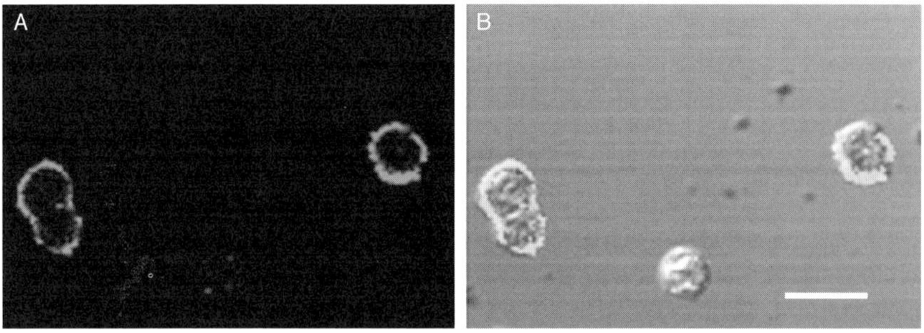

Fig. 15 Confocal reflectance (A) and combined reflectance/transmittance (B) images of labeled SiHa cells. Scattering from gold conjugates is false-colored red (scale bar: ca. 20 μm). Published with permission from Real-Time Vital Optical Imaging of Precancer Using Anti-Epidermal Growth Factor Receptor Antibodies Conjugated to Gold Nanoparticles, Sokolov, K., Fallen, M., Aaron, J., Pavlova, I., Malpica, A., Lotan, R., and Richards-Kortum, R., Cancer Research, 63: 1999–2004, May 1, 2003. (See Color Insert.)

resolved on the dark background. In fact, light scattering from the labeled cells is so strong that it can be easily observed using low-magnification optics and an inexpensive light source such as a laser pointer (Sokolov *et al.*, 2003b). No labeling was observed with nonspecific monoclonal antibodies. We observed heterogeneous labeling of SiHa cells in suspension. Heterogeneity of protein expression in cell lines is not uncommon and has been described before in the case of EGFR (Monaghan *et al.*, 1990). We determined that ca. 5×10^4 gold conjugates are bound per cell. This number correlates very well with our measurements of the amount of EGFR per SiHa cell using flow cytometry that yielded 5×10^4 receptors per cell.

We extended our work using anti-EGFR gold conjugates to label organ cultures of normal and neoplastic cervical tissue. Cultures were incubated in a solution containing contrast agent, then the excess of the contrast agents was removed and the specimens were imaged with reflectance-based confocal microscopy. The bright "honeycomb" like structure of labeled cellular cytoplasmic membranes of closely spaced cells can be easily seen in a section of an abnormal biopsy (Fig. 16, top A). No labeling of the normal biopsy can be seen when the sample is imaged under the same acquisition conditions (Fig. 16, top B). Anti-EGFR/gold conjugates did not bind to the stromal layer of cervical biopsies.

2. COMPARISON OF GOLD AND FLUORESCENT LABELING

We conducted studies of EGFR labeling in cervical biopsies using gold anti-EGFR conjugates (Sokolov *et al.*, 2003b) and in oral cavity biopsies using fluorescent dyes (Hsu *et al.*, 2004). The cervix and oral cavity are both

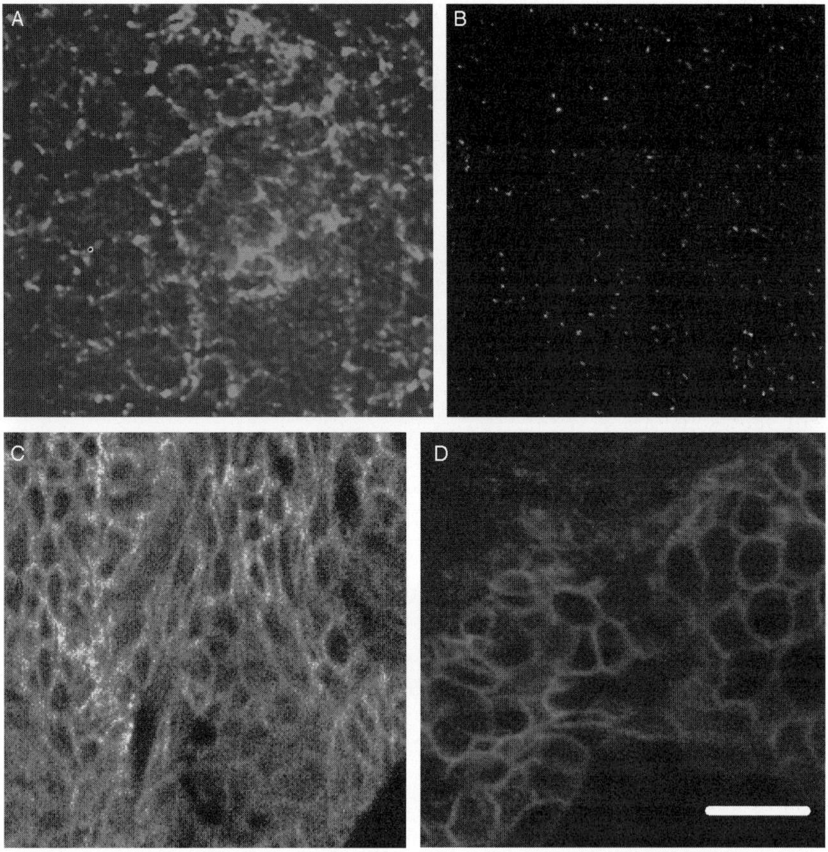

Fig. 16 Squamous cervical epithelium labeled with anti-EGFR antibodies/gold nanoparticles conjugates: (A) abnormal and (B) normal epithelium obtained under the same acquisition conditions. Published with permission from Real–Time Vital Optical Imaging of Precancer Using Anti-Epidermal Growth Factor Receptor Antibodies Conjugated to Gold Nanoparticles, Sokolov, K., Follen, M., Aaron, J., Pavlova, I., Malpica, A., Lotan, R., and Richards-Kortum R., Cancer Research, 63:1999–2004, May 1, 2003. Squamous oral epithelium labeled with anti-EGFR biotinylated antibodies and Alexa Fluor 660 fluorescent dye/streptavidin complex: (C) clinically abnormal (moderate dysplasia), (D) clinically normal. (See Color Insert.)

covered by a squamous epithelium which undergoes very similar progressions from normal to malignant stages. The level of EGFR expression of cancerous cells is very similar in both epithelia (Carpenter, 1987; King and Sartorelli, 1989). Therefore, the labeling results obtained in these two studies can be compared. Figure 16 shows images of abnormal (moderate dysplasia)/normal pairs of human squamous epithelium labeled using anti-EGFR monoclonal antibodies conjugated with 12 nm gold nanoparticles

(Fig. 16, top) and anti-EGFR monoclonal antibodies complexes with Alexa Fluor 660 streptavidin (Fig. 16, bottom). The ratio of signal intensity between the normal and the abnormal biopsies is ca. 15 and ca. 3 in the case of gold nanoparticles and the fluorescent dye, respectively. It is important to note that two washing steps in 1× phosphate buffered saline (PBS) had been carried out before the fluorescent images were obtained. No washing steps were required for the reflectance imaging with gold nanoparticles.

High-contrast imaging with gold nanoparticles without intermediate washing steps is possible because nanoparticles undergo significant changes in optical properties when they form closely spaced assemblies on the surface of malignant cells. In assemblies, the total scattering cross-section scales with N^2, where N is the number of nanoparticles in the assembly. In addition, electrodynamic coupling between the scattered fields of adjacent nanoparticles results in a simultaneous red shift of the resonance scattering spectra (Sokolov *et al.*, 2003b; Fig. 11, right). This shift can be used to tune the excitation wavelength for detection of labeled cells in the presence of single unbound gold bioconjugates.

3. SIGNAL INTENSITY: GOLD NANOPARTICLES VERSUS FLUORESCENT DYES

The signal intensity of the abnormal biopsy labeled using gold bioconjugates (Fig. 16, top A) is ca. 80 times higher than the signal collected from the fluorescently labeled specimen (Fig. 16, bottom). It has been reported that the cross section of scattering from a 50 nm gold particle is approximately a millionfold larger than the absorption or emission cross sections of the brightest known organic molecules or even quantum dots (Alivisatos, 2004). The scattering signal can be dramatically increased because the scattering cross section of gold nanoparticles increases as the sixth power of their radius. This enables a very simple way of changing signal strength which is not that simple in the case of fluorescent dyes or quantum dots. Gold nanoparticles are also immune to photobleaching and do not undergo any photochemical reactions which can be potentially toxic.

4. IMAGING OF METAL NANOPARTICLES

Imaging of cancer cells labeled with gold nanoparticles can be performed using microscopic detection with widefield illumination (not shown) or with scanning confocal microscope (Figs. 15 and 16). The detection can be also carried out using macroscopic widefield of view CCD cameras with a broad band white light source. The macroscopic detection can be optimized using illumination through band-pass filters in red optical region where the labeled cells have increased scattering.

5. SAFETY OF METAL NANOPARTICLES FOR *IN VIVO* USE

Contrast agents based on gold nanoparticle antibody conjugates have the potential for *in vivo* use, with topical or systemic delivery. The inherent biocompatibility of gold implies that they can be used directly *in vivo* without the need for protective layer growth. In fact, long-term treatment of rheumatoid arthritis utilizes gold (Abrams and Murrer, 1993) (up to a cumulative dose of 1.2–1.8 g/year for up to 10 years). It has been shown that prolonged treatment with gold salts leads to formation of gold crystals and their deposition in lysosomes of macrophages (Beckett *et al.*, 1982; Smith *et al.*, 1995; Yun Patricia *et al.*, 2002). This condition is called chrysiasis and it was first described in 1928. Chrysiasis is the development of a blue–gray pigmentation in skin. The effects of this condition have no serious pathological significance and are considered to be purely cosmetic. Chrysiasis can develop only after a threshold, equivalent to 20 mg/kg gold content or after few grams of gold are administered through i.v. or other type of treatment (Smith *et al.*, 1995). We anticipate that between a few hundred micrograms to a few milligrams of gold would be required for a diagnostic procedure. This is thousands times less than the threshold for any types of side effects detected in clinical practice. Humanized antibodies, where a mouse antibody-binding site is transferred to a human antibody gene, are much less immunogenic in humans (Holliger and Bohlen, 1999), and many humanized antibodies are currently in clinical trials. Since 1997, the Food and Drug Administration (FDA) has approved more than 10 monoclonal antibody-based drugs, including Herceptin for metastatic breast cancer therapy (Holliger and Bohlen, 1999; Weiner, 1999). Gold nanoshells are also a focus of a new minimally invasive thermal treatment of cancer (Hirsch *et al.*, 2003). Preclinical studies of the technology in animals have shown promising therapeutic effect as well as biocompatibility and lack of any cytotoxicity associated with the particles.

In summary, metal nanoparticles are a new class of very bright contrast agents which exhibit a number of unique optical properties including dramatic nonlinear increase in scattering cross section per particle and red shift in plasmon resonance frequency (color change) when the particles form closely spaced assemblies. These properties account for the higher contrast observed in the imaging of normal/abnormal human tissue using gold bioconjugates as compared to fluorescent labeling (Fig. 16).

B. Quantum Dots

Recently, the first *in vivo* applications of quantum dots were demonstrated (Akerman *et al.*, 2002; Dubertret *et al.*, 2002; Michalet *et al.*, 2005).

In a study by Akerman *et al.* (2002), quantum dots conjugated with peptides specific for normal lung or tumor blood vessels, or for tumor lymphatic vessels were i.v. injected into tumor-bearing mice. Specific targeting of lung or tumor vasculature using peptide-coated quantum dots was demonstrated in this study. No acute toxicity associated with i.v. administration of the quantum dots was observed even after 24 h of circulation. In another study, quantum dots were solubilized for biological *in vivo* imaging by encapsulation inside phospholipid block-copolymer micelles (Dubertret *et al.*, 2002). The encapsulated dots were microinjected into individual cells at early stage *Xenopus* embryos. The authors followed development of the labeled cells because the dots were confined to progeny of the injected cells. The encapsulated quantum dots do not exhibit any biological activity and are nontoxic for embryos at the levels of ca. 2×10^9 dots per cell. These studies suggest that the encapsulated nanoparticles are stable *in vivo*.

1. APTAMER-TARGETED QUANTUM DOTS

Recently, specific labeling of prostate cancer cells using aptamer targeting of CdTe quantum dots was demonstrated (Shieh *et al.*, 2005). CdTe 3-nm-diameter nanocrystals were synthesized, then rendered water soluble and biocompatible by ligand exchange, first with MPA and then partial ligand exchange with thiolated and biotinylated PEG. The biotinylated PEG–quantum dot complex was then incubated with avidin. An anti-prostate-specific membrane antigen (PSMA) aptamer was biotinylated and affixed to the avidin-coated quantum dots. The aptamer-quantum dot conjugates were then used to label either LNCaP cells, a prostate tumor line that overexpresses PSMA on its surface, or PC3 cells, a line that has little PSMA on its surface. As can be seen in Fig. 17, the quantum dot conjugates specifically light up the LNCaP cells. No emission was seen in PC3 cells.

2. BIOCOMPATIBILITY AND CYTOTOXICITY OF QUANTUM DOTS

There is significant concern and uncertainty about the biocompatibility and cytotoxicity of semiconductor nanocrystals. In bulk form, the Cd-based Groups II–VI semiconductors, CdE (where E is S, Se, and Te) and the Groups III–V semiconductors (such as InAs and InP) are well known to be acutely toxic and carcinogenic (Gottschling *et al.*, 2001; Morgan *et al.*, 1997; Tanaka, 2004). However, nanocrystals of these materials are coated with organic ligands and often with inorganic shell materials like silica. Some researchers have not observed any indications of toxicity in live cell-imaging studies and have suggested that semiconductor nanocrystals will

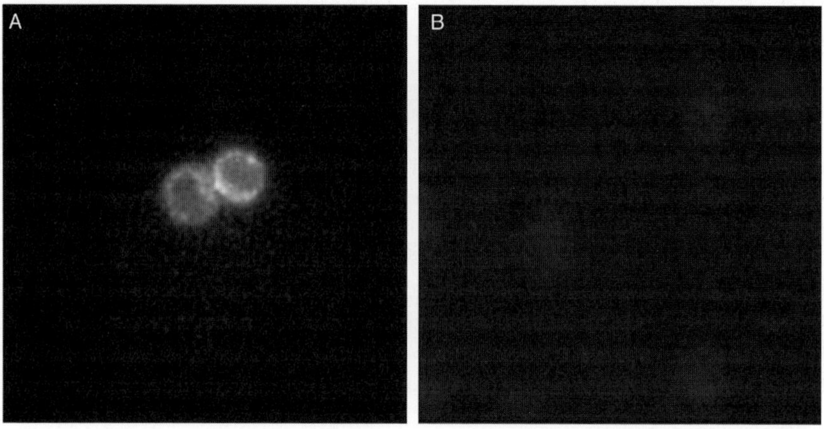

Fig. 17 Quantum dot conjugates targeted against PSMA (A) and untargeted (B). (See Color Insert.)

not be toxic (Akerman *et al.*, 2002; Michalet *et al.*, 2005; Voura *et al.*, 2004). Recent studies, however, showed that the nanocrystals can indeed be acutely toxic under certain situations, primarily as a result of heavy metal (i.e., Cd^{2+}) leaching from the nanocrystal core (Derfus *et al.*, 2004a; Kirchner *et al.*, 2005). In long-term *in vivo* applications, the toxicity limits become more important and more rigorous as exposure times are extended. The inherent toxicity is still somewhat of an open question, as both Derfus *et al.* (2004b) and more recently Kirchner *et al.* (2005) have shown that under certain circumstances, effective passivation of the nanocrystals can indeed render semiconductor nanocrystals nontoxic. The situation is somewhat complicated, however, as Kirchner *et al.*'s (2005) recent study proved that the capping ligand chemistry and cell-binding efficiency can be equally as important as the inorganic core chemistry in terms of cytotoxicity or biocompatibility. These results are consistent with those of Dubertret *et al.* (2002), who observed robust biocompatibility with ZnS-overcoated CdSe nanocrystals encapsulated in phospholipids micelles when injected into *Xenopus* embryos for lineage-tracking experiments in embryogenesis.

The general guidelines for reduced cytotoxicity that have emerged from recent studies are: (1) inorganic shells like silica or ZnS can greatly reduce cytotoxicity, (2) PEGylation can greatly reduce cytotoxicity by inhibiting intracellular nonspecific uptake, and (3) particle aggregation on the cell surface leads to rapid cell poisoning (Kirchner *et al.*, 2005). These considerations must be taken into account in the design of quantum dots for *in vivo* molecular imaging.

VIII. FUTURE DIRECTIONS: "SMART" CONTRAST AGENTS

The contrast agents that we described above rely on the optical signature of metal nanoparticles and quantum dots to provide strong sources of optical signal for molecular detection. However, the unique nanoscale sensitivity of quantum dots and gold nanoparticles can be exploited to design smart contrast agents, which are dark in the absence of a target ligand but become brightly luminescent on exposure to the ligand. These smart contrast agents rely on the strong ability of gold to quench the luminescence of quantum dots when in close proximity. Here, we discuss two approaches to the development of these promising "smart" contrast agents.

Protease-sensitive nanoparticle contrast agents: We propose design of a combination contrast agent in which a quantum dot is closely linked to a gold nanoparticle using a peptide spacer which is specifically cleaved by tumor-specific proteases. In the absence of protease, the gold will quench the quantum dot luminescence; however, in the presence of protease, cleavage of the peptide will result in release of quantum dots and a bright luminescent signal. Such agents can be used to report on the activity of MMPs, for example, MMP-2.

The proposed design couples the advantages of quantum dots (strong fluorescence, immunity to photobleaching) with those of colloidal gold (strong quenching, immunity to photobleaching) to develop sensitive, photostable molecular beacons sensitive to the presence of MMPs. One possible implementation of this design is illustrated in Fig. 18. Multiple PEGylated CdTe quantum dots are attached to a 30-nm-diameter gold particle via an MMP-2 substrate (Gly-Pro-Leu-Gly-Val-Arg-Gly-Lys) with an appropriate spacer. A number of exploratory studies need to be carried out in order to develop these contrast agents. It is essential to explore the relative fluorescence intensity of the beacon in the presence and absence of physiologically relevant concentrations of MMP-2. Also the ratio of signal in the "on" and "off" conformations should be optimized as the size of the gold nanoparticles and the separation distance between the gold and the quantum dot are varied.

Reporters of the molecular hallmarks of self-sufficiency in growth signals: In this design, quantum dots and gold nanoparticles can be linked using nucleic acid aptamers which undergo conformational change on binding their target ligand. The aptamer is selected so that the target-dependent modulation of the aptamer conformation leads to dequenching, resulting in "aptamer beacons" that can directly signal the presence of cognate ligands in solution, without the need for immobilization, washing, or other

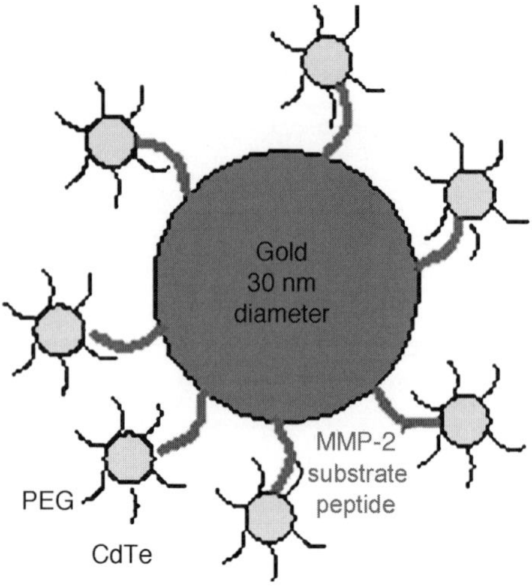

Fig. 18 Schematic of a smart contrast agent. (See Color Insert.)

processing steps. We are pursuing aptamers which undergo conformational change on exposure to soluble growth factors, including VEGF and EGF.

The use of affinity reagents, such as aptamers and antibodies, for targeting nanoparticle-based contrast agents is described in Section VII of this chapter. However, beyond merely acting as targeting agents aptamers can participate directly in the function of contrast agents. The fact that aptamers can undergo ligand-dependent conformational changes can be used to develop unique imaging technologies. In particular, aptamers were previously adapted to function as conformation-switching beacons. In general, the native secondary structure of an aptamer can be perturbed by the addition of oligonucleotide sequences in *trans* or in *cis* (shown in Fig. 19); in the absence of ligand, this is the "off" conformation, just as with a sequence-sensing molecular beacon. However, in presence of cognate ligand, the native structure is stabilized, and it assumes the "on" conformation, allowing signaling.

Aptamer beacon technology can be adapted to both organic fluorophores and more complex signaling reagents, such as quantum dots. In the example shown (Fig. 20), multiple aptamers (ca. 40) were mounted on the surface of a quantum dot; the aptamers served as platforms for hybridization of multiple quenchers to the dot. In the presence of either an antisense sequence or cognate protein (thrombin) the quencher was displaced, resulting in dequenching. A noncognate, quite basic protein (lysozyme) did not bind to the beacon in a way as to lead to dequenching.

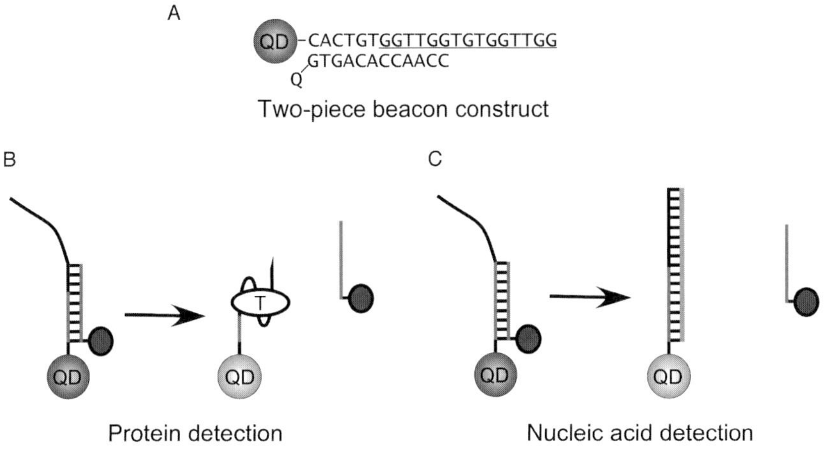

Fig. 19 Aptamer-based smart contrast agents (QD, quantum dot; T, thrombin).

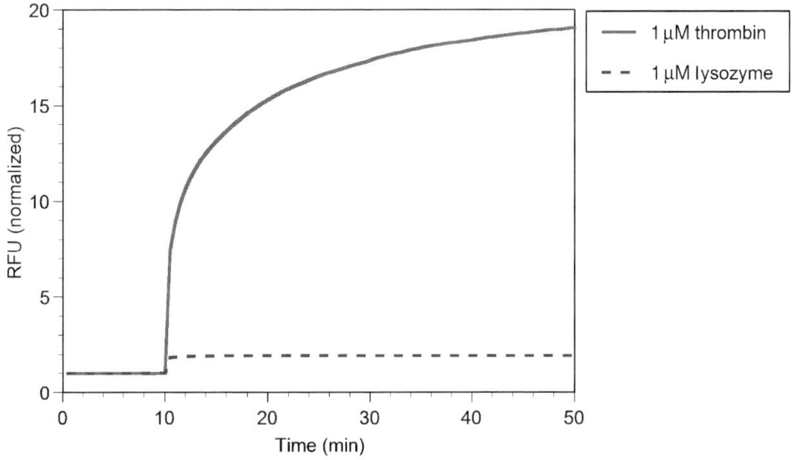

Fig. 20 Emission from aptamer-based smart contrast agent. As thrombin is added, the aptamer undergoes a conformational change, displacing the quencher. (See Color Insert.)

One interesting direction that we are pursuing is development of "one-piece" beacons in which the aptamer is covalently attached to both the quantum dot and the quencher as opposed to the "two-piece" quantum dot beacons described above. For this purpose, well-known RNA and modified RNA aptamers against the cytokines EGF and VEGF can be used

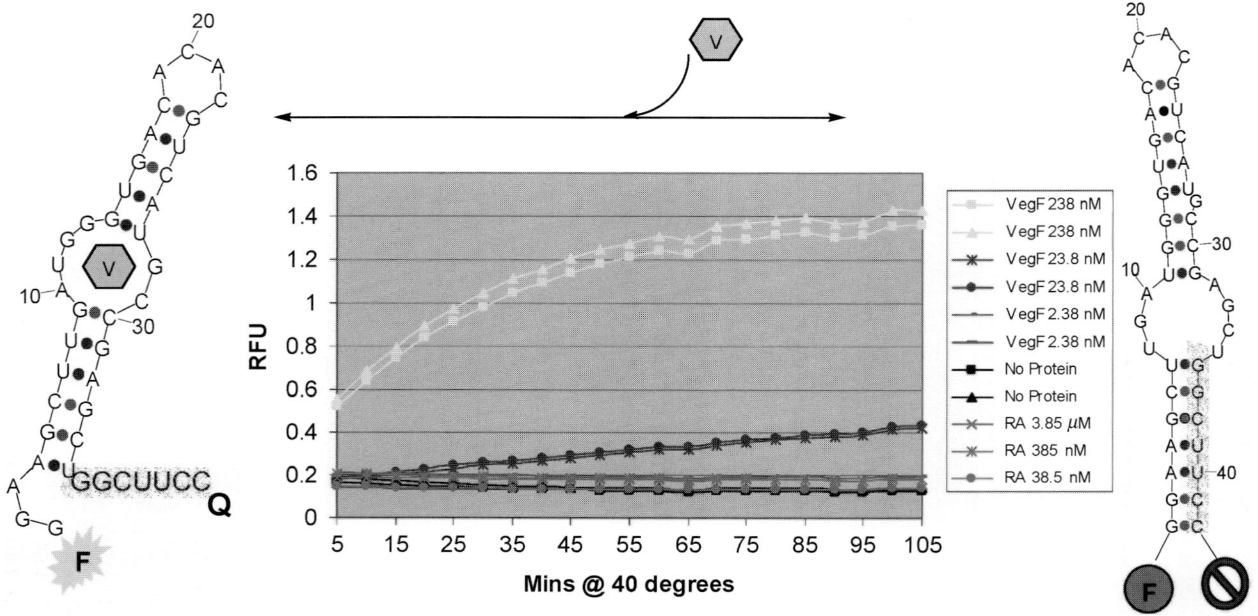

Fig. 21 VEGF-sensitive smart contrast agent. Right: beacon in the "off" configuration. Aptamer hybridizes so that qdot (F) and quencher are close to one another. Left: beacon binds VEGF (V). Aptamer hybridization changes so that qdot and quencher (Q) move apart to allow dequenching. Graph shows time dependence of fluorescence signal at different concentrations of VEGF.

to generate new quantum dot beacons. An example of a one-piece beacon to detect nanomolar VEGF is shown in Fig. 21. One-piece beacons are preferable to two-piece beacons because they reversibly signal the presence of a cognate ligand, and because their performance will not degrade on dilution in the bloodstream (a two-piece beacon will slowly lose its quencher on dilution). In addition, because synthesis of RNA and modified RNA molecules is both relatively inefficient and expensive, it is also important to focus on selection of new DNA aptamers that can bind to VEGF and EGF. Aptamer configurations which give optimal ratio of "on" to "off" signal should be explored as the length of the distal sequence adjacent to the quencher is varied.

The use of aptamer beacons in dynamic contrast agents should be generalizable to other nanoparticles, as well. As an initial step, the sensitivity of quantum dot beacons can be increased by using a more efficient quencher, a gold nanoparticle, rather than an organic dye. Aptamer beacons could be simultaneously conjugated to the surface of the quantum dot (at their 5′ ends) and to the surface of the gold nanoparticle (at their 3′ ends). Binding of the cognate ligand (VEGF or EGF) to the aptamer would stabilize the binding conformation and change the apposition of the quantum dot and gold nanoparticles relative to one another. By engineering the sequence and secondary structure of the aptamer beacon, it should be possible to modulate the separation of the quantum dot and gold nanoparticle, both in the quenched and unquenched states, and thus to modulate the signal: background ratio that is exhibited by the dynamic contrast reagents.

IX. CONCLUSIONS

Development of optical molecular imaging approaches for *in vivo* monitoring of therapeutic target of cancer is a multidisciplinary problem that requires the combination of expertise in the fields of biomedical optics, nanotechnology, molecular-specific targeting, *in vivo* delivery, and cancer molecular biology. In this chapter, we presented the synergy of these diverse scientific areas in development of molecular-specific imaging which is based on two emerging types of optically active contrast agents—metal nanoparticles and quantum dots. The cocktail of presented contrast agents can provide the ability to monitor major hallmarks of cancer including: (1) self-sufficiency in growth signals (EGFR and EGF), (2) evasion of programmed cell death (anionic phospholipids), (3) tissue invasion and metastasis (MMPs), and (4) sustained angiogenesis (VEGF and its receptors).

REFERENCES

Abrams, M. J., and Murrer, B. A. (1993). Metal compounds in therapy and diagnosis. *Science* **261**(5122), 725–730.

Achilefu, S. (2004). Lighting up tumors with receptor-specific optical molecular probes. *Technol. Cancer Res. Treat.* **3**(4), 393–409.

Akerman, M. E., Chan, W. C., Laakkonen, P., Bhatia, S. N., and Ruoslahti, E. (2002). Nanocrystal targeting *in vivo*. *Proc. Natl. Acad. Sci. USA* **99**(20), 12617–12621.

Alivisatos, P. (2004). The use of nanocrystals in biological detection. *Nat. Biotechnol.* **22**(1), 47–52.

Ang, K. K., Berkey, B., Tu, X., Zhang, H.-Z., Katz, R., Hammond, E., Fu, K. K., and Milas, L. (2002). Impact of epidermal growth factor receptor expression on survival and pattern of relapse in patients with advanced head and neck carcinoma. *Cancer Res.* **62**(24), 7350–7356.

Averitt, R. D., Westcott, S. L., and Halas, N. J. (1999). The linear optical properties of gold nanoshells. *J. Opt. Soc. Am. B* **16,** 1824–1832.

Banin, U., Lee, C. J., Guzelian, A. A., Kadavanich, A. V., Alivisatos, A. P., Jaskolski, W., Bryant, G. W., Efros Al, L., and Rosen, M. (1998). Size-dependent electronic level structure of InAs nanocrystal quantum dots: Test of multi-band effective mass theory. *J. Chem. Phys.* **109**(6), 2306–2309.

Beckett, V. L., Doyle, J. A., Hadley, G. A., and Spear, K. L. (1982). Chrysiasis resulting from gold therapy in rheumatoid arthritis: Identification of gold by X-ray microanalysis. *Mayo Clin. Proc.: Mayo Clin.* **57**(12), 773–777.

Benavides, J. M., Chang, S., Park, S. Y., MacKinnon, N., MacAulay, C., Milbourne, A., Malpica, A., Follen, M., and Richards-Kortum, R. (2003). Multispectral digital colposcopy for *in vivo* detection of cervical cancer. *Opt. Express* **11,** 1223–1236.

Bernkop-Schnurch, A. (2000). Chitosan and its derivatives: Potential excipients for peroral peptide delivery systems. *Int. J. Pharm.* **194**(1), 1–13.

Bernkop-Schnurch, A., Scholler, S., and Biebel, R. G. (2000). Development of controlled drug release systems based on thiolated polymers. *J. Control. Rel.* **66**(1), 39–48.

Bhatia, S. N. (1989). Use of thiol-terminal silanes and heterobifunctional crosslinkers for immobilization of antibodies on silica surfaces. *Anal. Biochem.* **178,** 408–413.

Bogdanov, A. A., Lin, C. P., Simonova, M., Matuszewski, L., and Weissleder, R. (2002). Cellular activation of the self-quenched fluorescent reporter probe in tumor microenvironment. *Neoplasia* **4**(3), 228–236.

Bouma, B. E., Tearney, G. J., Compton, C. C., and Nishioka, N. S. (2000). High-resolution imaging of the human esophagus and stomach *in vivo* using optical coherence tomography. *Gastrointest. Endosc.* **51**(4, Part 1), 467–474.

Brekken, R. A., and Thorpe, P. (2001). VEGF-VEGF receptor complexes as markers of tumor vascular endothelium. *J. Control. Rel.* **74**(1–3), 173–181.

Bremer, C., Bredow, S., Mahmood, U., Weissleder, R., and Tung, C. (2001). Optical imaging of MMP-2 activity in tumors: Feasibility in a mouse model. *Radiology* **221,** 523–529.

Bremer, C., Tung, C. H., Bogdanov, A., and Weissleder, R. (2002). Imaging of differential protease expression in breast cancers for detection of aggressive tumor phenotypes. *Radiology* **222,** 814–818.

Bremer, C., Ntziachristos, V., Weitkamp, B., Theilmeir, G., Heindel, W., and Weissleder, R. (2005). Optical imaging of spontaneous breast tumors using protease sensing 'smart' optical probes. *Invest. Radiol.* **40**(6), 321–327.

Bruchez, M., Jr., Moronne, M., Gin, P., Weiss, S., and Alivisatos, A. P. (1998). Semiconductor nanocrystals as fluorescent biological labels. *Science* (Washington, DC) **281**(5385), 2013–2016.

Califano, J., van der Riet, P., Westra, W., Nawroz, H., Clayman, G., Piantadosi, S., Corio, R., Lee, D., and Greenbuerg, B. (1996). Genetic progression model for head and neck cancer: Implications for field cancerization. *Cancer Res.* **56**(11), 2488–2492.

Carpenter, G. (1987). Receptors for epidermal growth factor and other polypeptide mitogens. *Annu. Rev. Biochem.* **56,** 881–914.

Chan, W. C., and Nie, S. (1998). Quantum dot bioconjugates for ultrasensitive nonisotopic detection. *Science* **281**(5385), 2016–2018.

Chan, W. C., Maxwell, D. J., Gao, X., Bailey, R. E., Han, M., and Nie, S. (2002). Luminescent quantum dots for multiplexed biological detection and imaging. *Curr. Opin. Biotechnol.* **13**(1), 40–46.

Chang, J. Y., Oh, Y.-K., Choi, H.-G., Kim, Y. B., and Kim, C.-K. (2002a). Rheological evaluation of thermosensitive and mucoadhesive vaginal gels in physiological conditions. *Int. J. Pharm.* **241**(1), 155–163.

Chang, J. Y., Oh, Y.-K., Kong, H. S., Kim, E. J., Jang, D. D., Nam, K. T., and Kim, C.-K. (2002b). Prolonged antifungal effects of clotrimazole-containing mucoadhesive thermosensitive gels on vaginitis. *J. Control. Rel.: Off. J. Control. Rel. Soc.* **82**(1), 39–50.

Chiang, J. J., and Truong, K. (2005). Using co-cultures expressing fluorescence resonance energy transfer based protein biosensors to simultaneously image caspase-3 and Ca(2+) signaling. *Biotechnol. Lett.* **27**(16), 1219–1227.

Chun, M. K., Cho, C. S., and Choi, H. K. (2002). Mucoadhesive drug carrier based on interpolymer complex of poly(vinyl pyrrolidone) and poly(acrylic acid) prepared by template polymerization. *J. Control. Rel.* **81**(3), 327–334.

Collier, T., and Richards-Kortum, R. (1998). Fiber-optic confocal microscope for biological imaging. *In* "Conference on Lasers and Electro-Optics," Institute of Electrical and Electronics Engineers, San Francisco.

Collier, T., Shen, P., de Pradier, B., Sung, K.-B., Follen, M., Malpica, A., and Richards-Kortum, R. (2000). Near real time confocal microscopy of amelanotic tissue: Dynamics of acetowhitening enable nuclear segmentation. *Opt. Express* **6**(2), 40–48.

Collier, T., Lacy, A., Richards-Kortum, R., Malpica, A., and Follen, M. (2002). Near real-time confocal microscopy of amelanotic tissue: Detection of dysplasia in *ex vivo* cervical tissue. *Acad. Radiol.* **9**(5), 504–512.

Cotton, T. M. (1988). *In* "Spectroscopy of Surfaces" (R. J. H. Clark and R. E. Hester, Eds.), p. 91. Wiley, New York.

Das, A., Sivak, M. V., Jr., Chak, A., Wong, R. C., Westphal, V., Rollins, A. M., Willis, J., Isenberg, G., and Izatt, J. A. (2001). High-resolution endoscopic imaging of the GI tract: A comparative study of optical coherence tomography versus high-frequency catheter probe EUS. *Gastrointest. Endosc.* **54**(2), 219–224.

Delaney, P. M., and Harris, M. R. (1995). Fiber optics in confocal microscope. *In* "Handbook of Confocal Microscopy" (J. B. Pawley, Ed.), pp. 515–523. Plenum Press, New York.

Derfus, A., Chan, W., and Bhatia, S. (2004a). Intracellular delivery of quantum dots for live cell labeling and organelle tracking. *Adv. Mater.* **16,** 961–966.

Derfus, A., Chan, W., and Bhatia, S. (2004b). Probing the cytotoxicity of semiconductor quantum dots. *Nano. Lett.* **4,** 11–18.

Dodane, V., Amin Khan, M., and Merwin, J. R. (1999). Effect of chitosan on epithelial permeability and structure. *Int. J. Pharm.* **182**(1), 21–32.

Drezek, R. A., Collier, T., Brookner, C. K., Malpica, A., Lotan, R., Richards-Kortum, R. R., and Follen, M. (2000). Laser scanning confocal microscopy of cervical tissue before and after application of acetic acid. *Am. J. Obstet. Gynecol.* **182**(5), 1135–1139.

Dubertret, B., Calame, M., and Libchaber, A. J. (2001). Single-mismatch detection using gold-quenched fluorescent oligonucleotides. *Nat. Biotechnol.* **19**(4), 365–370.

Dubertret, B., Skourides, P., Norris, D. J., Noireaux, V., Brivanlou, A. H., and Libchaber, A. (2002). *In vivo* imaging of quantum dots encapsulated in phospholipid micelles. *Science* (Washington, DC) **298**(5599), 1759–1762.

Dulkeith, E., Morteani, A. C., Niedereichholz, T., Klar, T., Feldman, J., Levi, S., van Veggel, F., Reinhoudt, F., Moller, M., and Gittins, D. (2002). Fluorescence quenching of dye molecules near gold nanoparticles: Radiative and nonradiative effects. *Phys. Rev. Lett.* **2002**(89), 20.

Elghanian, R., Storhoff, J. J., Mucic, R. C., Letsinger, R. L., and Mirkin, C. A. (1997). Selective colorimetric detection of polynucleotides based on the distance-dependent optical properties of gold nanoparticles. *Science* **277**(5329), 1078–1080.

Eouani, C., Piccerelle, P., Prinderre, P., Bourret, E., and Joachim, J. (2001). *In vitro* comparative study of buccal mucoadhesive performance of different polymeric films. *Eur. J. Pharm. Biopharm.* **52**(1), 45–55.

Ferrari, F., Rossi, S., Bonferoni, M. C., Caramella, C., and Karlsen, J. (1997). Characterization of rheological and mucoadhesive properties of three grades of chitosan hydrochloride. *Farmaco* **52**(6–7), 493–497.

Filipovic-Grcic, J., Skalko-Basnet, N., and Jalsenjak, I. (2001). Mucoadhesive chitosan-coated liposomes: Characteristics and stability. *J. Microencapsul.* **18**(1), 3–12.

Frens, G. (1973). Controlled nucleation for the regulation of the particle size in monodisperse gold suspensions. *Nat. Phys. Sci.* **241**, 20–22.

Funovics, M., Weissleder, R., and Tung, C. (2003). Protease sensors for bioimaging. *Anal. Bioanal. Chem.* **377**, 956–963.

Gao, X., Cui, Y., Levenson, R. M., Chung, L. W. K., and Nie, S. (2004). *In vivo* cancer targeting and imaging with semiconductor quantum dots. *Nat. Biotechnol.* **22**(8), 969–976.

Geddes, C. L. (Ed.) (2004). "Reviews in Fluorescence 2004." Kluwer Academic/Plenum Publishers, New York.

Genta, I., Perugini, P., Pavanetto, F., Modena, T., Conti, B., and Muzzarelli, R. A. A. (1999). Microparticulate drug delivery systems. *EXS* **87**, 305–313.

Geoghegan, W. D., and Ackerman, G. A. (1977). Adsorption of horseradish peroxidase, ovomucoid and anti-immunoglobulin to colloidal gold for the indirect detection of concanavalin A, wheat germ agglutinin and goat anti-human immunoglobulin G on cell surfaces at the electron microscopic level: A new method, theory and application. *J. Histochem. Cytochem.: Off. J. Histochem. Soc.* **25**(11), 1187–1200.

González, S., Gonzalez, E., White, W. M., Rajadhyaksha, M., and Anderson, R. R. (1999a). Allergic contact dermatitis: Correlation of *in vivo* confocal imaging to routine histology. *J. Am. Acad. Dermatol.* **40**(5, Part 1), 708–713.

González, S., Rajadhyaksha, M., Gonzalez-Serva, A., White, W. M., and Anderson, R. R. (1999b). Confocal reflectance imaging of folliculitis *in vivo*: Correlation with routine histology. *J. Cutaneous Pathol.* **26**(4), 201–205.

González, S., Rajadhyaksha, M., Rubinstein, G., and Anderson, R. R. (1999c). Characterization of psoriasis *in vivo* by reflectance confocal microscopy. *J. Med.* **30**(5–6), 337–356.

González, S., Rubinstein, G., Mordovtseva, V., Rajadhyaksha, M., and Anderson, R. R. (1999d). *In vivo* abnormal keratinazation in Darier-White's disease as viewed by real-time confocal imaging. *J. Cutaneous Pathol.* **26**(10), 504–508.

González, S., White, W. M., Rajadhyaksha, M., Anderson, R. R., and Gonzalez, E. (1999e). Confocal imaging of sebaceous gland hyperplasia *in vivo* to assess efficacy and mechanism of pulsed dye laser treatment. *Lasers Surgery Med.* **25**(1), 8–12.

Gottschling, B., Maronpot, R. R., Hailey, J., Peddada, S., Moomaw, C., Klaunig, J., and Nyska, A. (2001). The role of oxidative stress in indium phosphide-induced lung carcinogenesis in rats. *Toxicol. Sci.* **64**, 28–40.

Haes, A. J., and Van Duyne, R. P. (2002). A nanoscale optical biosensor: Sensitivity and selectivity of an approach based on the localized surface plasmon resonance spectroscopy of triangular silver nanoparticles. *J. Am. Chem. Soc.* **124**(35), 10596–10604.

Hanahan, D., and Weinberg, R. A. (2000). The hallmarks of cancer. *Cell* **100**(1), 57–70.

Hassan, M., and Klaunberg, B. A. (2004). Biomedical applications of fluorescence imaging *in vivo*. *Comp. Med.* **54**(6), 635–644.

Hildebrandt, P., and Stockburger, M. (1984). Surface-enhanced resonance Raman spectroscopy of rhodamine 6G adsorbed on colloidal silver. *J. Phys. Chem.* **88**(24), 5935–5944.

Hirsch, L. R., Stafford, R. J., Bankson, J. A., Sershen, S. R., Rivera, B., Price, R. E., Hazle, J. D., Halas, N. J., and West, J. L. (2003). Nanoshell-mediated near-infrared thermal therapy of tumors under magnetic resonance guidance. *Proc. Natl. Acad. Sci. USA* **100**(23), 13549–13554.

Hirsch, F. R., Varella-Garcia, M., Bunn, P. A., Jr., di Maria, M. V., Veve, R., Bremnes, R. M., Baron, A. E., Zeng, C., and Franklin, W. A. (2003). Epidermal growth factor receptor in non-small-cell lung carcinomas: Correlation between gene copy number and protein expression and impact on prognosis. *J. Clin. Oncol.* **21**(20), 3798–3807.

Holliger, P., and Bohlen, H. (1999). Engineering antibodies for the clinic. *Cancer Metastasis Rev.* **18**(4), 411–419.

Holsinger, F. C., Doan, D. D., Jasser, S. A., Swan, E. A., Greenberg, J. S., Schiff, B., Bekele, B., Younes, M., Bucana, C. D., Fidler, I. J., and Myers, J. N. (2003). EGFR blockade potentiates apoptosis mediated by paclitaxel and leads to prolonged survival in a murine model of oral cancer. *Clin. Cancer Res.* **9**, 3183–3189.

Holmes, J. D., Ziegler, K. J., Doty, R. C., Pell, L. E., Johnston, K. P., and Korgel, B. A. (2001). Highly luminescent silicon nanocrystals with discrete optical transitions. *J. Am. Chem. Soc.* **123**(16), 3743–3748.

Horisberger, M. (1981). Colloidal gold: A cytochemical marker for light and fluorescent microscopy and for transmission and scanning electron microscopy. *Scanning Electron Microsc.* **2**, 9–31.

Hrabec, E., Strek, M., Nowak, D., Greger, J., Suwalski, M., and Hrabec, Z. (2002). Activity of type IV collagenases (MMP-2 and MMP-9) in primary pulmonary carcinomas: A quantitative analysis. *J. Cancer Res. Clin. Oncol.* **128**(4), 197–204.

Hsu, E. R., Anslyn, E. V., Dharmawardhane, S., Alizadeh-Naderi, R., Aaron, J. S., Sokolov, K. V., El-Naggar, A. K., Gillenwater, A. M., and Richards-Kortum, R. (2004). A far-red fluorescent contrast agent to image epidermal growth factor receptor expression. *Photochem. Photobiol.* **79**(3), 272–279.

Huzaira, M., Rius, F., Rajadhyaksha, M., Anderson, R. R., and Gonzalez, S. (2001). Topographic variations in normal skin, as viewed by *in vivo* reflectance confocal microscopy. *J. Invest. Dermatol.* **116**(6), 846–852.

Jesser, C. A., Boppart, S. A., Pitris, C., Stamper, D. L., Nielsen, G. P., Brezinski, M. E., and Fujimoto, J. G. (1999). High resolution imaging of transitional cell carcinoma with optical coherence tomography: Feasibility for the evaluation of bladder pathology. *Br. J. Radiol.* **72**(864), 1170–1176.

Kast, C. E., Valenta, C., Leopold, M., and Bernkop-Schnurch, A. (2002). Design and *in vitro* evaluation of a novel bioadhesive vaginal drug delivery system for clotrimazole. *J. Control. Rel.* **81**(3), 347–354.

Kerec, M., Bogataj, M., Mugerle, B., Gasperlin, M., and Mrhar, A. (2002). Mucoadhesion on pig vesical mucosa: Influence of polycarbophil/calcium interactions. *Int. J. Pharma.* **241**(1), 135–143.

Kim, S., Lim, Y. T., Soltesz, E. G., De Grand, A. M., Lee, J., Nakayama, A., Parker, J. A., Mihaljevic, T., Laurence, R. G., Dor, D. M., Cohn, L. H., Bawendi, M. G., *et al.* (2004). Near-infrared fluorescent type II quantum dots for sentinel lymph node mapping. *Nat. Biotechnol.* **22**(1), 93–97.

King, I., and Sartorelli, A. C. (1989). Epidermal growth factor receptor gene expression, protein kinase activity, and differentiation of human malignant epidermal cells. *Cancer Res.* **49**(20), 5677–5681.

Kirchner, C., Liedl, T., Kudera, S., Pellegrino, T., Javier, A., Gaub, H., Stolzle, S., Fertig, N., and Parak, W. (2005). Cytotoxicity of colloidal CdSe and CdSe/ZnS nanoparticles. *Nano. Lett.* **33,** 1–338.

Kneipp, K., Haka, A. S., Kneipp, H., Badizadegan, K., Yoshizawa, N., Boone, C., Shafer-Peltier, K., Motz, J. T., Dasari, R. R., and Feld, M. S. (2002). Surface-enhanced Raman spectroscopy in single living cells using gold nanoparticles. *Appl. Spectros.* **56**(2), 150–154.

Korgel, B. A., and Fitzmaurice, D. (1999). Small-angle X-ray-scattering study of silver-nanocrystal disorder-order phase transitions. *Phys. Rev. B: Condens. Matter Mater. Phys.* **59**(22), 14191–14201.

Kotze, A. F., Luessen, H. L., De Leeuw, B. J., De Boer, B. G., Verhoef, J. C., and Junginger, H. E. (1997). N-trimethyl chitosan chloride as a potential absorption enhancer across mucosal surfaces: *In vitro* evaluation in intestinal epithelial cells (Caco-2). *Pharm. Res.* **14**(9), 1197–1202.

Kotze, A. F., Thanou, M. M., Lueen, H. L., De Boer, A. G., Verhoef, J. C., and Junginger, H. E. (1999). Enhancement of paracellular drug transport with highly quaternized N-trimethyl chitosan chloride in neutral environments: *In vitro* evaluation in intestinal epithelial cells (Caco-2). *J. Pharm. Sci.* **88**(2), 253–257.

Lacoste, T. D., Michalet, X., Pinaud, F., Chemla, D. S., Alivisatos, A. P., and Weiss, S. (2000). Ultrahigh-resolution multicolor colocalization of single fluorescent probes. *Proc. Natl. Acad. Sci. USA* **97**(17), 9461–9466.

Langley, R. G., Rajadhyaksha, M., Dwyer, P. J., Sober, A. J., Flotte, T. J., and Anderson, R. R. (2001). Confocal scanning laser microscopy of benign and malignant melanocytic skin lesions *in vivo*. *J. Am. Acad. Dermatol.* **45**(3), 365–376.

Larson, D. R., Zipfel, W. R., Williams, R. M., Clark, S. W., Bruchez, M. P., Wise, F. W., and Webb, W. W. (2003). Water-soluble quantum dots for multiphoton fluorescence imaging *in vivo*. *Science* **300**(5624), 1434–1437(Washington, DC).

Lee, C. H., and Chien, Y. W. (1996). *In vitro* permeation study of a mucoadhesive drug delivery system for controlled delivery of nonoxynol-9. *Pharm. Dev. Technol.* **1**(2), 135–145.

Lee, P. C., and Meisel, T. J. (1982). Adsorption and surface-enhanced Raman of dyes on silver and gold soles. *J. Phys. Chem.* **86,** 3391–3395.

Lehr, C. M., Bouwstra, J. A., Kok, W., De Boer, A. G., Tukker, J. J., Verhoef, J. C., Breimer, D. D., and Junginger, H. E. (1992). Effects of the mucoadhesive polymer polycarbophil on the intestinal absorption of a peptide drug in the rat. *J. Pharm. Pharmacol.* **44**(5), 402–407.

Lehr, C. M., Lee, Y. H., and Lee, V. H. (1994). Improved ocular penetration of gentamicin by mucoadhesive polymer polycarbophil in the pigmented rabbit. *Invest. Ophthalmol. Vis. Sci.* **35**(6), 2809–2814.

Li, X. D., Boppart, S. A., Van Dam, J., Mashimo, H., Mutinga, M., Drexler, W., Klein, M., Pitris, C., Krinsky, M. L., Brezinski, M. E., and Fujimoto, J. G. (2000). Optical coherence tomography: Advanced technology for the endoscopic imaging of Barrett's esophagus. *Endoscopy* **32**(12), 921–930.

Liang, C., Sung, K.-B., Richards-Kortum, R., and Descour, M. (2002). Design of a high-numerical-aperture miniature microscope objective for an endoscopic fiber confocal reflectance microscope. *Appl. Opt.* **41**(22), 4603–4610.

Lippman, S. M., and Hong, W. K. (2001). Molecular markers of the risk of oral cancer. *N. Engl. J. Med.* **344**(17), 1323–1326.

Lippman, S. M., Sudbo, J., and Hong, W. K. (2005). Oral cancer prevention and the evolution of molecular-targeted drug development. *J. Clin. Oncol.* **23**(2), 346–356.

Luessen, H. L., Verhoef, J. C., Borchard, G, Lehr, C. M., de Boer, A. G., and Junginger, H. E. (1995). Mucoadhesive polymers in peroral peptide drug delivery. II. Carbomer and polycarbophil are potent inhibitors of the intestinal proteolytic enzyme trypsin. *Pharma. Res.* **12**(9), 1293–1298.

Marten, K., Bremer, C., Khazaie, K., Sameni, M., Sloane, B., Tung, C.-H., and Weissleder, R. (2002). Detection of dysplastic intestinal adenomas using enzyme-sensing molecular beacons in mice. *Gastroenterology* **122**(2), 406–414.

Michalet, X., Pinaud, F. F., Bentolila, L. A., Tsay, J. M., Doose, S., Li, J. J., Sundaresan, G., Wu, A. M., Gambhir, S. S., and Weiss, S. (2005). Quantum dots for live cells, *in vivo* imaging, and diagnostics. *Science* (Washington, DC) **307**(5709), 538–544.

Minsky, M. (1955). Microscopy apparatus. US Patent 3013467.

Monaghan, P., Ormerod, M. G., and O'Hare, M. J. (1990). Epidermal growth factor receptors and EGF-responsiveness of the human breast-carcinoma cell line PMC42. *Int. J. Cancer* **46**(5), 935–943.

Morgan, D., Shines, C., Jeter, S., Blazka, M., Elwell, M., Wilson, R., Ward, S., Price, H., and Moskowitz, P. (1997). Comparative pulmonary absorption, distribution, and toxicity of copper, gallium diselenide, copper indium diselenide and cadmium telluride in Sprague-Dawley rats. *Toxicol. Appl. Pharmacol.* **147,** 399–410.

Mulvaney, P. (1996). Surface plasmon spectroscopy of nanosized metal particles. *Langmuir* **12**(3), 788–800.

Murray, C. B., Norris, D. J., and Bawendi, M. G. (1993). Synthesis and characterization of nearly monodisperse CdE (E = sulfur, selenium, tellurium) semiconductor nanocrystallites. *J. Am. Chem. Soc.* **115,** 8706–8715.

Nabiev, I. R., Morjani, H., and Manfait, M. (1991). Selective analysis of antitumor drug interaction with living cancer cells as probed by surface-enhanced Raman spectroscopy. *Eur. Biophys. J.* **19**(6), 311–316.

Nie, S., and Emory, S. R. (1997). Probing single molecules and single nanoparticles by surface-enhanced Raman scattering. *Science* **275**(5303), 1102–1106.

Oechsner, M., and Keipert, S. (1999). Polyacrylic acid/polyvinylpyrrolidone bipolymeric systems. I. Rheological and mucoadhesive properties of formulations potentially useful for the treatment of dry-eye-syndrome. *Eur. J. Pharm. Biopharm.* **47**(2), 113–118.

Paciotti, G. F., Myer, L. D., Kim, T. H., Wang, S., Alexander, H. R., Weinreich, D., and Tamarkin, L. (2001). Colloidal gold: A novel colloidal nanoparticle vector for tumor-directed drug delivery. *In* "Proceedings of AACR-NCI-EORTC International Conference on Molecular Targets and Cancer Therapeutics: Discovery, Biology, and Clinical Applications." Miami Beach, Florida.

Parak, W. J., Boudreau, R., Le Gros, M., Gerion, D., Zanchet, D., Micheel, C. M., Williams, S. C., Alivisatos, A. P., and Larabell, C. (2002). Cell motility and metastatic potential studies based on quantum dot imaging of phagokinetic tracks. *Adv. Mater.* (Weinheim, Germany) **14**(12), 882–885.

Park, H., and Robinson, J. R. (1987). Mechanisms of mucoadhesion of poly(acrylic acid) hydrogels. *Pharm. Res.* **4**(6), 457–464.

Park, S. Y., Sokolov, K., Collier, T., Aaron, J., Markey, M., Mackinnon, N., MacAulay, C., Coghlan, L., Milbourne, A., Follen, M., and Richards-Kortum, R. (2005). Multispectral digital microscopy for *in vivo* detection of oral neoplasia in the hamster cheek pouch model of carcinogenesis. *Opt. Express* **13**, 749–762.

Passlick, B., Sienel, W., Seen-Hibler, R., Wocket, W., Thetter, O., Mutschler, W., and Pantel, K. (2000). Overexpression of matrix metalloproteinase 2 predicts unfavorable outcome in early-stage non-small cell lung cancer. *Clin. Cancer Res.: An Off. J. Am. Assoc. Cancer Res.* **6**(10), 3944–3948.

Peng, Z., and Peng, X. (2001). Formation of high-quality CdTe, CdSe, and CdS nanocrystals using CdO as precursor. *J. Am. Chem. Soc.* **123,** 183–184.

Pitris, C. (1998). Cellular and neoplastic tissue imaging with optical coherence tomography. *In* "Conference on Lasers and Electro-Optics Europe." Institute of Electrical and Electronics Engineers, San Francisco.

Pitris, C., Goodman, A., Boppart, S. A., Libus, J. J., Fujimoto, J. G., and Brezinski, M. E. (1999). High-resolution imaging of gynecologic neoplasms using optical coherence tomography. *Obstet. Gynecol.* **93**(1), 135–139.

Pomper, M. G. (2001). Molecular imaging: An overview. *Acad. Radiol.* 8(11), 1141–1153.

Rajadhyaksha, M. (2001). Confocal cross-polarized imaging of skin cancers to potentially guide Mohs micrographic surgery. *Opt. Photonics News* **12,** 30.

Rajadhyaksha, M., Grossman, M., Esterowitz, D., Webb, R. H., and Anderson, R. R. (1995). *In vivo* confocal scanning laser microscopy of human skin: Melanin provides strong contrast. *J. Invest. Dermatol.* **104**(6), 946–952.

Rajadhyaksha, M., Anderson, R. R., and Webb, R. H. (1999a). Video-rate confocal scanning laser microscope for imaging human tissues *in vivo*. *Appl. Opt.* **38**(10), 2105–2115.

Rajadhyaksha, M., Gonzalez, S., Zavislan, J. M., Anderson, R. R., and Webb, R. H. (1999b). *In vivo* confocal scanning laser microscopy of human skin II: Advances in instrumentation and comparison with histology. *J. Invest. Dermatol.* **113**(3), 293–303.

Rajadhyaksha, M., Menaker, G., Flotte, T., Dwyer, P. J., and Gonzalez, S. (2001). Confocal examination of nonmelanoma cancers in thick skin excisions to potentially guide mohs micrographic surgery without frozen histopathology. *J. Invest. Dermatol.* **117**(5), 1137–1143.

Rhee, J. C., Khuri, F. R., and Shin, D. M. (2004). Emerging drugs for head and neck cancer. *Expert Opin. Emerg. Drugs* **9**(1), 91–104.

Richards-Kortum, R., and Sevick-Muraca, E. (1996). Quantitative optical spectroscopy for tissue diagnosis. *Annu. Rev. Phys. Chem.* **47,** 555–606.

Rossi, S., Bonferoni, M. C., Ferrari, F., and Caramella, C. (1999). Drug release and washability of mucoadhesive gels based on sodium carboxymethylcellulose and polyacrylic acid. *Pharm. Dev. Technol.* 4(1), 55–63.

Santra, S., Xu, J., Wang, K., and Tan, W. (2004). Luminescent nanoparticle probes for bioimaging. *J. Nanosci. Nanotechnol.* **4**(6), 590–599.

Santangelo, P. J., Nix, B., Tsourkas, A., and Bao, G. (2004). Dual FRET molecular beacons for mRNA detection in living cells. *Nucleic Acids Res.* **32**(6), e57/1–e57/9.

Schneider, S. (1988). Reproducible preparation of silver sols with uniform particle size for application in surface-enhanced Raman spectroscopy. *Photochem. Photobiol.* **60**(6), 605–610.

Schultz, S., Smith, D. R., Mock, J. J., and Schultz, D. A. (2000). Single-target molecule detection with nonbleaching multicolor optical immunolabels. *Proc. Natl. Acad. Sci. USA* **97**(3), 996–1001.

Senel, S., and Hincal, A. A. (2001). Drug permeation enhancement via buccal route: Possibilities and limitations. *J. Control. Rel.* 72(1–3), 133–144.

Shah, K., and Weissleder, R. (2005). Molecular optical imaging: Applications leading to the development of present day therapeutics. *NeuroRx* **2**(2), 215–225.

Sienel, W., Hellers, J., Morrese-Hauf, A., Lichtinghagen, R., Mutschler, W., Jochum, M., Klein, C., Passlick, B., and Pantel, K. (2003). Prognostic impact of matrix metalloproteinase-9 in operable non-small cell lung cancer. *Int. J. Cancer* **103**(5), 647–651.

Selkin, B., Rajadhyaksha, M., Gonzalez, S., and Langley, R. G. (2001). *In vivo* confocal microscopy in dermatology. *Dermatol. Clin.* **19**(2), 369–377.

Sershen, S. R., Westcott, S. L., Halas, N. J., and West, J. L. (2000). Temperature-sensitive polymer-nanoshell composites for photothermally modulated drug delivery. *J. Biomed. Mater. Res.* **51**(3), 293–298.

Shieh, F., Lavery, L., Chu, C. T., Richards-Kortum, R., Ellington, A. D., and Korgel, B. A. (2005). Semiconductor nanocrystal-aptamer bioconjugate probes for specific prostate carcinoma cell targeting Proceedings of SPIE—The International Society for Optical Engineering, Volume 5705, Nanobiophotonics and Biomedical Applications II159–165.

Singla, A. K., and Chawla, M. (2001). Chitosan: Some pharmaceutical and biological aspects—an update. *J. Pharm. Pharmacol.* **53**(8), 1047–1067.

Smith, R. W., Leppard, B., Barnett, N. L., Millward-Sadler, G. H., McCrae, F., and Cawley, M. I. (1995). Chrysiasis revisited: A clinical and pathological study. *Br. J. Dermatol.* **133**(5), 671–678.

Smithpeter, C. L., Dunn, A., Drezek, R., Collier, T., and Richards-Kortum, R. (1998). Near real time confocal microscopy of cultured amelanotic cells: Sources of signal, contrast agents, and limits of contrast. *J. Biomed. Opt.* **3**(4), 429–436.

Sokolov, K., Chumanov, G., and Cotton, T. M. (1998). Enhancement of molecular fluorescence near the surface of colloidal metal films. *Anal. Chem.* **70**(18), 3898–3905.

Sokolov, K., Aaron, J., Hsu, B., Nida, D., Gillenwater, A., Follen, M., MacAulay, C., Adler-Storthz, K., Korgel, B., Descour, M., Pasqualini, R., Arap, W., *et al.* (2003a). Optical systems for *in vivo* molecular imaging of cancer. *Technol. Cancer Res. Treat.* **2**(6), 491–504.

Sokolov, K., Follen, M., Aaron, J., Pavlova, I., Malpica, A., Lotan, R., and Richards-Kortum, R. (2003b). Real time vital imaging of pre-cancer using anti-EGFR antibodies conjugated to gold nanoparticles. *Cancer Res.* **63**(9), 1999–2004.

Sporn, M. B. (1996). The war on cancer. *Lancet* **347**(9012), 1377–1381.

Sung, K.-B., Liang, C., Descour, M., Collier, T., Follen, M., and Richards-Kortum, R. (2002a). Fiber-optic confocal reflectance microscope with miniature objective for *in vivo* imaging of human tissues. *IEEE Trans. Biomed. Eng.* **49**(10), 1168–1172.

Sung, K-.B., Liang, C., Descour, M., Collier, T., Follen, M., Malpica, A., and Richards-Kortum, R. (2002b). Near real time *in vivo* fibre optic confocal microscopy: Sub-cellular structure resolved. *J. Microsc.* **207**(Part 2), 137–145.

Svistun, E., Alizadeh-Naderi, R., El-Naggar, A., Jacob, R., Gillenwater, A., and Richards-Kortum, R. (2004). Vision enhancement system for detection of oral cavity neoplasia based on autofluorescence. *Head Neck* **26**(3), 205–215.

Tanaka, A. (2004). Toxicity of indium arsenide, gallium arsenide, and aluminium gallium arsenide. *Toxicol. Appl. Pharmacol.* **198**(3), 405–411.

Thanou, M., Florea, B. I., Langemeyer, M. W. E., Verhoef, J. C., and Junginger, H. E. (2000). N-trimethylated chitosan chloride (TMC) improves the intestinal permeation of the peptide drug buserelin *in vitro* (Caco-2 cells) and *in vivo* (rats). *Pharma. Res.* **17**(1), 27–31.

Thanou, M., Verhoef, J. C., and Junginger, H. E. (2001a). Oral drug absorption enhancement by chitosan and its derivatives. *Adv. Drug Deliv. Rev.* **52**(2), 117–126.

Thanou, M., Verhoef, J. C., and Junginger, H. E. (2001b). Chitosan and its derivatives as intestinal absorption enhancers. *Adv. Drug Deliv. Rev.* **50**(Suppl. 1), S91–S101.

Thomas, G. T., Lewis, M. P., and Speight, P. M. (1999). Matrix metalloproteinases and oral cancer. *Oral Oncol.* **35**(3), 227–233.

Torrance, C. J., Jackson, P. E., Montgomery, E., Kinzler, K. W., Vogelstein, B., Wissner, A., Nunes, M., Frost, P., and Discafani, C. M. (2000). Combinatorial chemoprevention of intestinal neopalsia. *Nat. Med.* **6**(9), 1024–1028.

Tsien, R. Y. (2005). Building and breeding molecules to spy on cells and tumors. *FEBS Lett.* **579**(4), 927–932.

Tsourkas, A., Behlke, M. A., and Bao, G. (2002). Structure-function relationships of shared-stem and conventional molecular beacons. *Nucleic Acids Res.* **30**(19), 4208–4215.

Ugwoke, M. I., Agu, R. U., Vanbilloen, H., Baetens, J., Augustijns, P., Verbeke, N., Mortelmans, L., Verbruggen, A., Kinget, R., and Bormans, G. (2000). Scintigraphic evaluation in rabbits of nasal drug delivery systems based on carbopol 971p((R)) and carboxymethylcellulose. *J. Control Release* **68**(2), 207–214.

van de Vijver, M. J., He, Y. D., Van't Veer, L., Dai, H., Hart, A. A. M., Voskuil, D. W., Schreiber, G. J., Peterse, J. L., Roberts, C., Marton, M. J., Parrish, M., Atsma, D., *et al.* (2002). A gene-expression signature as a predictor of survival in breast cancer. *N. Engl. J. Med.* **347**(25), 1999–2009.

Voura, E. B., Jaiswal, J. K., Mattoussi, H., and Simon, S. M. (2004). Tracking metastatic tumor cell extravasation with quantum dot nanocrystals and fluorescence emission-scanning microscopy. *Nat. Med.* **10**(9), 993–998.

Warren, S. J., and Kellaway, I. W. (1998). The synthesis and *in vitro* characterization of the mucoadhesion and swelling of poly(acrylic acid) hydrogels. *Pharm. Dev. Technol.* **3**(2), 199–208.

Weiner, L. M. (1999). An overview of monoclonal antibody therapy of cancer. *Semin. Oncol.* **26**(4, Suppl. 12), 41–50.

White, W. M., Rajadhyaksha, M., Gonzalez-Serva, A., White, W. M., and Anderson, R. R. (1999). Noninvasive imaging of human oral mucosa *in vivo* by confocal reflectance microscopy. *Laryngoscope* **109**(10), 1709–1717.

Wilenzick, R. M., Russell, D. C., and Morriss, R. H. (1967). Uniform microcrystals of platinum and gold. *J. Chem. Phys.* **47**, 533–536.

Yasuda, R., Noji, H., Yoshida, M., Kinosita, K., Jr., and Itoh, H. (2001). Resolution of distinct rotational substeps by submillisecond kinetic analysis of F1-ATPase. *Nature* (London, United Kingdom) **410**(6831), 898–904.

Yguerabide, J., and Yguerabide, E. E. (2001). Resonance light scattering particles as ultrasensitive labels for detection of analytes in a wide range of applications. *J. Cell. Biol.* **37**(Suppl.), 71–81.

Yun, P. L., Kenneth, A. A., and Anderson, R. R. (2002). "Q-Switched Laser-Induced Chrysiasis Treated With Long-Pulsed Laser," pp. 1012–1014. Wellman Laboratories of Photomedicine, Massachusetts General Hospital, Boston, MA.

Zeman, E. J., and Schatz, G. C. (1987). An accurate electromagnetic theory study of surface enhancement factors for Ag, Au, Cu, Li, Na, Al, Ga, In, Zn, and Cd. *J. Phys. Chem.* **91**(3), 634–643.

Zhang, W., Ran, S., Sambade, M., Huang, X., and Thorpe, P. (2002). A monoclonal antibody that blocks VEGF binding to VEGFR2 inhibits vascular expression of Flk-1 and tumor growth in an orthotopic human breast cancer model. *Angiogenesis* **5**, 35–44.

Personalized Medicine for Cancer: From Molecular Signature to Therapeutic Choice

Karol Sikora

Faculty of Medicine, Hammersmith Hospital Imperial College, London, United Kingdom; and Cancer Partners UK, London, United Kingdom

I. Summary
II. Introduction
  A. The Past
  B. The Future
III. Prevention and Screening
IV. Detecting Cancer
V. New Treatment Approaches
VI. The Development of Personalized Medicine
VII. Barriers to Innovation
VIII. Patient's Experience
IX. Conclusions
References

I. SUMMARY

In the field of cancer medicine, great strides have been made in understanding the fundamental biology of cancers and impressive treatments have emerged resulting in markedly prolonged survival for many patients. These advances mean that cancer could well become a chronic disease within the next 20 years, but that promise depends on sustained investment in innovation in both diagnostics and therapies as well as society's willingness to pay for both.

The two great challenges facing cancer medicine in the future will be understanding the biology of the very wide range of cancers affecting different organs and the increased prevalence of the disease that can be expected in an aging population. How will biomedical science and healthcare systems rise to these challenges? An understanding of the way in which advances have been applied in personalizing treatments in the past points a way ahead to address future challenges.

Our cancer future will emerge from the interaction of four factors: the success of new technology, society's willingness to pay, future healthcare delivery systems, and the financial mechanisms that underpin them. The only

Advances in CANCER RESEARCH

0065-230X/07 $35.00
DOI: 10.1016/S0065-230X(06)96013-8

way to reduce the costs of cancer care is to ensure that the right patient gets the right treatment. Investing in sophisticated diagnostics is a clear imperative in making personalized medicine for cancer a reality.

II. INTRODUCTION

The age of the world's population is rising dramatically. This will increase the total burden of cancer with many patients living with considerable comorbidity. At the same time, new technology in many areas of medicine is bringing improvements to the quality and length of life. Major innovations in the following six areas are likely to have the greatest impact on cancer.

Molecularly targeted drugs with associated sophisticated diagnostic systems to personalize care
Biosensors to detect, monitor, and correct abnormal physiology and to provide surrogate measurements of cancer risk
Our ability to modify the human genome through systemically administered novel targeted vectors
The continued miniaturization of surgical intervention through robotics, nanotechnology, and precise imaging
Computer driven interactive devices to help with everyday living
The use of virtual reality systems which together with novel mood control drugs will create an illusion of wellness

Over the last 20 years, a huge amount of fine detail of the basic biological processes that become disturbed in cancer has been amassed. We now know the key elements of growth factor binding, signal transduction, gene transcription control, cell cycle checkpoints, apoptosis, and angiogenesis (Sikora, 2002). These have become fertile areas to hunt for rationally based anticancer drugs. This approach has already led to a record number of novel compounds currently in trials. Indeed, targeted drugs such as rituximab, trastuzumab, imatinib, erlotinib, lapatinib, bevacizumab, and cetuximab are now all in widespread clinical use. Over the next decade, there will clearly be a marked shift in the types of agents used in the systemic treatment of cancer.

Because we know the precise targets of these new agents, there will be a revolution in how we prescribe cancer therapy. Instead of defining drugs for use empirically and relatively ineffectively for different types of cancer, we will identify a series of molecular lesions in tumor biopsies. Future patients will receive drugs that target these lesions directly. The human genome project provides a vast repository of comparative information about normal and malignant cells. The new therapies will be more selective, less toxic, and be given for prolonged periods of time, in some cases for the rest of the patients' life. This will lead to a radical overhaul of how we provide cancer care (2020 Vision, 2003).

A considerably increased investment in more sophisticated diagnostics is now urgently required. Holistic systems such as genomics, proteomics, metabolomics, and methylomics provide fascinating clues as to where needles can be found in the haystack of disturbed growth. By developing simple, reproducible, and cheap assays for specific biomarkers a battery of companion diagnostics will emerge (Nicolette and Miller, 2003). It is likely that for the next decade, these will be firmly rooted in tissue pathology making today's histopathologist essential to move this exciting field forward. Ultimately, the fusion of tissue analysis with imaging technologies may make virtual biopsies of any part of the body—normal and diseased a real possibility (Adam *et al.*, 2002).

Individual cancer risk assessment will lead to tailored prevention messages and a specific screening program to pick up early cancer and have far reaching public health consequences. Cancer preventive drugs will be developed to reduce the risk of further genetic deterioration. The use of gene arrays to monitor serum for fragments of DNA containing defined mutations could ultimately develop into an implanted gene chip. When a significant mutation is detected, the chip would signal the holder's home computer and set in train a series of investigations based on the most likely type and site of the primary tumor.

There will be an increase in the total prevalence of cancer as a result of improved survival as well as change in cancer types to those, such as prostate cancer, with longer survival. This will create new challenges in terms of assessing risks of recurrence, designing care pathways, use of IT, and improving access to services. There will be new opportunities for further targeting and development of existing therapies as experience grows with risk factors over the longer term. Careful monitoring of patient experiences could help in improving results. Cancer could soon be a long-term management issue for many patients where they enjoy a high quality of life even with a degree of chronic illness (Tritter and Calnan, 2002).

The funding of cancer care will become a significant problem (Bosanquet and Sikora, 2006). Already we are seeing inequity in access to the taxanes for breast and ovarian cancer and gemcitabine for lung and pancreatic cancer. These drugs are only palliative, adding just a few months to life. The emerging compounds are likely to be far more successful and their long-term administration considerably more expensive. Increased consumerism in medicine will lead to increasingly informed and assertive patients seeking out novel therapies and bypassing traditional referral pathways through global information networks. It is likely that integrated molecular solutions for cancer will develop, but unless issues related to access are addressed, this will lead to far greater inequity than at present. Cost effectiveness analyses will be used to scrutinize novel diagnostic technology as well as therapies.

A. The Past

The personalization of cancer therapy is not new. The first recorded reference to cancer was in the Edwin Smith Papyrus of 3000 BC where eight women with breast cancer are described. The writings of Hippocrates in 400 BC contain several descriptions of cancer in different sites. But our understanding of the disease really began in the nineteenth century with the advent of cellular pathology and the beginnings of modern surgery.

Successful treatment by radical surgery became possible in the later part of that century, thanks to advances in anesthetics and antiseptics. Radical surgery involved the removal of the tumor-containing organ and its draining lymph nodes in one block. Halstead in Johns Hopkins was the main protagonist of the radical mastectomy, Wertheim the hysterectomy, Trotter the pharyngectomy, and Miles the abdomino-perineal resection of the rectum. These diverse surgical procedures all followed the same principles. The twentieth century ended with the conservation of organs by minimizing the destruction caused by surgery and replacing it with radiotherapy and for some sites effective adjuvant therapy with drugs. The surgical staging of cancer was one of the first personalized approaches. It led to tailoring the aggression of surgery to the likely sites of spread of the disease. The development of conservative breast surgery was based on logical stepwise clinical trials and has led to a revolution in the individualization of surgery and adjuvant treatment based on tumor size, stage, grade, and lymph node involvement. The advent of sentinel node biopsy as a surrogate for auxiliary involvement and the use of polymerase chain reaction (PCR) technology to detect micrometastases in nodal biopsies represents a modern extension of this work. Gene expression studies are now being used to select patients for more aggressive adjuvant postsurgical chemotherapy regimens based on the likely predicted natural history of their tumor for both breast and lung cancer.

Radiotherapy has come a long way since the first patient with a nasal tumor was treated in 1899, only a year after the discovery of radium by Marie Curie. Although radiobiology developed as a research discipline, it has really contributed little to clinical practice. The rationale behind modern fractionated radiotherapy comes as much from empirical trial and error as from experimental results. Radiotherapy is remarkably successful for certain areas of the body. Increasing sophistication in equipment coupled with dramatic strides in imaging have led to great precision in planning and execution of treatment so sparing critical normal tissues and increasing the dose to the tumor. Again, the high dose volumes treated with radiotherapy are highly individualized based on structural anatomy of tumor and critically sensitive normal tissues. Less success has been achieved in tailoring the total radiation dose and fractionation. Molecular radiobiology has really had minimal impact on clinical practice so far but this could change dramatically over the next decade. It is unlikely that molecular signatures will have significant impact on the practice of radiotherapy—which will eventually be used for fewer and fewer patients as systemic therapies become more successful.

The sinking of the US battleship John B. Harvey in Bari Harbor by the Germans in 1942 led to the development of effective chemotherapy. The warship was carrying canisters of mustard gas for use in chemical warfare. Survivors developed leucopenia and this led Goodman and others back in the United States to experiment with halogenated alkylamines in patients with high white cell counts—lymphomas, leukemias, and Hodgkin's disease. From the first publication in 1946, the field has blossomed with over 200 drugs now available in our global pharmacopoeia. But as with radiotherapy our clinical practice is based mainly on empiricism (Symonds, 2001). Most currently used drugs were found serendipitously from plants or fungi—taxol, vincristine, doxorubicin—and not by rational drug design. Although very successfully used in combination for lymphoma, leukemia, choriocarcinoma, testicular cancer, and several childhood cancers, results in metastatic common solid tumors have been disappointing with little more than palliative benefit (Fig. 1). The advent of molecularly targeted drugs promises to change this dramatically.

B. The Future

Within 20 years cancer will be considered a chronic disease, joining conditions such as diabetes, heart disease, and asthma. These conditions impact on the way people live but will not inexorably lead to death. The model of prostate cancer, where many men die with it rather than from it, will be more usual. Progress will be made in preventing cancers. Even greater progress will be made in understanding the myriad causes of cancer. Our concepts will be different to today's and the new ways in which cancer will be detected, diagnosed, and treated will be crucial to understanding the future.

Chemotherapy for advanced cancer

High CR	*High CR*	*Low CR*
High cure	*Low cure*	*Low cure*
5%	*40%*	*55%*
HD	AML	NSCLC
ALL	Breast	Colon
Testis	Ovary	Stomach
Chorio	SCLC	Prostate
Childhood	Sarcoma	Pancreas
BL	Myeloma	Glioma

Fig. 1 Chemotherapy for advanced cancer. There are three groups of cancer. The first is frequently cured by drugs with a high complete response (CR), the second where although there is a high CR but most patients relapse with resistant disease, and a third group where CR is rare. Five percent of cancer patients are in the first group, 40% in the second, and 55% in the third.

When a cancer does develop, refinements of current technologies and techniques—in imaging, radiotherapy, and surgery—together with the availability of targeted drugs will make it controllable. Cure will still be sought, but will not be the only satisfactory outcome. Patients will be closely monitored after treatment, but fear that cancer will definitely kill, still prevalent in the early years of the twenty-first century, will be replaced by an acceptance that many forms of cancer are a consequence of old age.

Looking into the future is fraught with difficulties. Who could have imagined in the 1980s the impact of mobile phones, the internet, and low-cost airlines on global communication? Medicine will be overtaken by similarly unexpected step changes in innovation. For this reason, economic analysis of the impact of developments in cancer care is difficult. The greatest benefit will be achieved simply by assuring that the best care possible is on offer to the most patients. This would be irrespective of their socioeconomic circumstances and of any scientific developments. But this is unrealistic. Technologies are developing fast, particularly in imaging and the exploitation of the human genome. Well-informed patients, with adequate funds, will ensure that they have rapid access to the newest and the best—wherever it is in the world. More patients will benefit from better diagnosis and newer treatments, with greater emphasis on quality of life (Laing, 2002). Innovation will bring more inequality to health if the parties do not work together to ensure they address the challenges of access. The outcome of the same quality of care differs today between socioeconomic groups and will continue to do so.

Table I The Challenges of Cancer Care

Increasing the focus on prevention
Improving screening and diagnosis and the impact of this on treatment
New targeted treatments—how effective and affordable will they be?
How patients and their carers' expectations will translate into care delivery?
Reconfiguration of health services to deliver optimal care
The impact of reconfiguration on professional territories
Will society accept the financial burden of these opportunities?

Clinicians in Europe will continue to be dependent on technologies primarily designed for the major health market in the world—the United States which currently consumes nearly 55% of cancer medication but contains less than 5% of the population. European legislation covering clinical trials could bring research in the United Kingdom to a grinding halt, while ethicists—zealously interpreting privacy legislation—could impose restrictions on the use of tissue. Targeted niche drugs will be less appealing to industry as the costs of bringing each new generation of drugs to market will not be matched by the returns from current blockbusters. The delivery of innovation will be underpinned by patient expectation. The well-informed will be equal partners in deciding the health care they will receive. Much of it will take place close to their homes using mechanisms devised by innovative service providers (World Cancer Report, 2003).

This has huge implications for the training of health professionals and the demarcations between specialties. Emerging technologies will drive the change. Intraprofessional boundaries will blur—doctors from traditionally quite distinct specialties may find themselves doing the same job. And clinical responsibilities will be taken up by health professionals who will not be medically qualified. All professionals are likely to find challenges to their territory hard to accept. Table I shows the challenges that need to be addressed in order to deliver most health benefits.

III. PREVENTION AND SCREENING

At the beginning of the twenty-first century, 10 million people in the world develop cancer each year (Blackledge, 2003). The cause of these cancers is known in roughly 75% of cases: 3 million are tobacco related, 3 million are a result of diet, and 1.5 million are caused by infection. In the United Kingdom, 120,000 people die from cancer each year, even though many are preventable—with a third related to smoking. But cancer prevention absorbs only 2% of the total funding of cancer care and research.

Antismoking initiatives are considered to be successful—although it has taken 50 years from the time the association between smoking and cancer was first identified. In the 1960s, 80% of the population smoked; by 2005, the average was under 30%. This masks real health inequality—the percentage of smokers in the higher socioeconomic classes are in low single figures, while the percentage in the deprived is still about 50% in parts of the country. Despite the known risks, if friends and family smoked and there was no social pressure to stop, there was no incentive. Banning smoking in public places will lead to a further drop of about 4%. Increases in tax had been a powerful disincentive to smoke but the price of a packet of cigarettes is so high that smokers turn to the black market: as many as one in five cigarettes smoked is smuggled into the country. Lung cancer, for example, is a rare disease in higher socioeconomic groups—it is a disease of poverty.

Lessons from antismoking initiatives will be instructive for prevention in the future. Although the link between poor diet, obesity, and lack of exercise and cancer has not been confirmed, there is sufficient circumstantial evidence to suggest that strong associations will be found. There will be bans on advertising for crisps, sweets, and soft drinks on television, the introduction of a health tax on these products, and a ban on sponsorship of any public event by manufacturers of these products. By 2010, obesity among the middle classes will be socially unacceptable, but it will remain common among the economically disadvantaged. Creating meaningful, imaginative incentives for people to adopt healthy lifestyles will be a major challenge.

The future prevention picture will be colored by post-genomic research. It is now accepted that about 100 genes are associated with the development of a whole range of cancers. The detection of polymorphisms in low-penetrance cancer-related genes—or a combination of changed genes—will identify people of increased risk. Within 20 years most people will be genetically mapped. The information—gained from a simple blood test—will be easily stored on a smart card. Legislation will be required to prevent this information being used to determine an individual's future health status for mortgage, insurance, and employment purposes. However, the process of mapping will reveal that every person who has been screened will carry a predisposition to certain diseases. People will learn to live with risk.

Today the average age of diagnosis of cancer in the United Kingdom is 68. Improvements in screening, detection, and diagnosis will reduce this. A predisposition for some cancers, which manifests itself in a patient's 70s or 80s, will be found in young adult life and detected and corrected successfully in the patient's 30s. Increasing age will remain the strongest risk predictor. Little of what has been described is not happening already in some form but the computing power of the future will bring accurate calculation of risk and predictions will take place on an unimaginable scale. Screening programs will be developed on a national basis if they are simple,

Table II Balancing Cancer Risk

Great health inequity exists in smoking related diseases
Novel prevention strategies are likely to lead to similar inequity
Creating meaningful incentives to reduce risk will be essential
Individually tailored messages will have greater power to change lifestyles
Biomarkers of risk will enhance the validation of cancer preventive drugs
Novel providers of risk assessment and correction will emerge

robust, and cheap. Patients will expect the screening to take place at a convenient venue for them—in shopping malls and not be painful or overly time consuming. Health professionals will demand that any program is accurate and does not give misleading results, and governments will demand that its costs will lead to more effective use of other resources. Novel providers of risk assessment services are likely to emerge (Table II).

IV. DETECTING CANCER

Cancers are fundamentally somatic genetic diseases that result from several causes: physical, viral, radiation, and chemical damage. There are other processes implicated, for example chronic inflammatory change, immunosurveillance, and failure of apoptosis. In the future, cancer will no longer be understood as a single entity—it will be considered to be a cellular process that changes over time. Many diseases labeled as cancer today will be renamed, as their development will not reflect the new paradigm. Patients will accept that cancer is not a single disease and increasingly understand it as a cellular process. Many more old people will have increased risk or a precancer. This has huge implications for cancer services. Today, most diagnoses of cancer depended on human interpretation of changes in cell structures seen down a microscope. Microscopes will be superseded by a new generation of scanners to detect molecular changes. These scanners will build up a picture of change over time, imaging cellular activity rather than just a single snapshot. We will have the ability to probe molecular events that are markers for early malignant change. This dynamic imaging will lead to more sensitive screening and treatments; imaging agents which accumulate in cells exhibiting telltale signs of precancer activity and will be used to introduce treatment agents directly (Brumley, 2002).

Imaging and diagnosis will be minimally invasive and enable the selection of the best and most effective targeted treatment (Table IV). Even better imaging will be able to pick up predisease phases and deal with them at a stage long before they are currently detectable. These techniques will also be crucial in successful follow-up. A patient who has a predisposition to a

Table III Delivering New Diagnostics for Personalized Therapy

Radiology and pathology will merge into cancer imaging
Dynamic imaging will create a changing image of biochemical abnormalities
Cancer will be detected prior to disease spread from primary site
Greater precision in surgery and radiotherapy will be used for precancer
Molecular signatures will determine treatment choice
Cost control will be essential for healthcare payers to avoid inefficient diagnostics

certain cancer process will be monitored regularly and treatment offered when necessary. Not all cancers will be diagnosed in these earliest of stages—some patients will inevitably fall through the screening net. Nevertheless, there will be opportunities to offer less invasive treatment than at present. Surgery and radiotherapy will continue but in greatly modified form as a result of developments in imaging. Most significantly, surgery will become part of integrated care. Removal of tumors or even whole organs will remain necessary on occasion. However, the surgeon will be supported by 3D imaging, by radio-labeling techniques to guide incisions, and by robotic instruments. And although many of the new treatments made possible by improved imaging will be biologically driven, there will still be a role for radiotherapy—the most potent DNA-damaging agent—to treat cancer with great geographical accuracy. The targeting of radiotherapy will be greatly enhanced enabling treatment to be more precise.

In addition to the reconfiguration and merging of the skills of clinicians, the delivery of care will also change. Minimally invasive treatments will reduce the need for long stays in hospital. As more patients are diagnosed with cancer, the need to provide the care close to where patients live will be both desirable and possible—and, as this report will show later—expected. The prospect of highly sophisticated scanning equipment and mobile surgical units being transported to where they are required is not unrealistic. Technicians, surgical assistants, and nurses would provide the hands-on care, while technical support will be provided by the new breed of clinician—a disease-specific imaging specialist working from a remote site. Cost control will be an essential component of the diagnostic phase. Healthcare payers will create sophisticated systems to evaluate the economic benefits of innovative imaging and tissue analysis technology (Table III).

V. NEW TREATMENT APPROACHES

Future cancer care will be driven by the least invasive therapy consistent with long-term survival. Eradication, although still desirable, will no longer be the primary aim of treatment. Cancers will be identified earlier and the disease process regulated in a similar way to chronic diseases such as diabetes. Surgery and radiotherapy will still have a role but how much will depend on the type of cancer a patient has and the stage at which disease is identified. It will also depend on how well the drugs being developed today perform in the future.

Cancer treatment will be shaped by a new generation of drugs. What this new generation will look like will critically depend on the relative success of agents currently in development and the willingness to pay for innovation. Over the next 3–5 years, we will understand more fully what benefits compounds such as kinase inhibitors are likely to provide. It is estimated that there are about 500 drugs currently being tested in clinical trials. Of these, around 300 inhibit specific molecular targets (Melzer, 2003). But this number is set to rise dramatically. Two thousand compounds will be available to enter clinical trials by 2007 and 5000 by 2010. Many of these drug candidates will be directed at the same molecular targets and industry is racing to screen those most likely to make it through in the development process. Tremendous pressures are coming from the loss of patent protection from the majority of high-cost chemotherapy drugs by 2008. Unless new premium-priced innovative drugs are available, cancer drug provision will come from global generic manufacturers currently gearing up for this change.

So what will these drug candidates look like? Small molecules are the main focus of current research—most of which are designed to target specific gene products that control the biological processes associated with cancer such as signal transduction, angiogenesis, cell cycle control, apoptosis, inflammation, invasion, and differentiation. Treatment strategies involving monoclonal antibodies, cancer vaccines, and gene therapy are also being explored. Although we do not know exactly what these targeted agents will look like, there is growing confidence that they will work. More uncertain is their overall efficacy at prolonging survival. Many could just be expensive palliatives. In future, advances will be driven by a better biological understanding of the disease process (Fig. 2).

Already we are seeing the emergence of drugs targeted at a molecular level—trastuzumab, directed at the HER-2 protein, imatinib which targets the BCR/ABL tyrosine kinase, and gefitanib and erlotinib directed at epidermal growth factor receptor (EGFR) tyrosine kinase. These therapies will be used across a range of cancers. What will be important in future is whether a person's cancer has particular biological or genetic characteristics. Traditional categories will continue to be broken down and genetic profiling will enable treatment to be targeted at the right patients. Patients will understand that treatment options are dependent on their genetic profile. The risks and benefits of treatment will be much more predictable than today (Table IV).

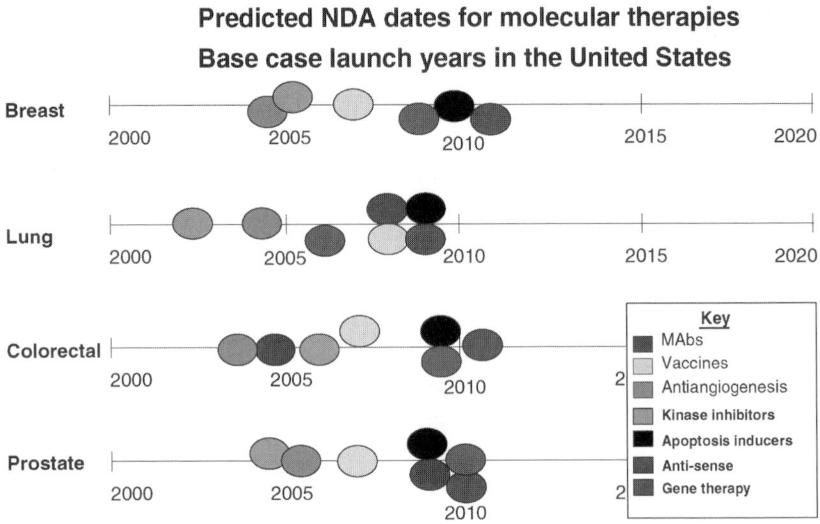

Fig. 2 Predicted new drug application dates for molecular therapies in the United States. The years 2005–2010 will see an explosion of novel therapies coming into clinical use outside the research setting. The costs to healthcare payers will be huge unless better methods can be developed to select the correct drugs for the correct patients. (See Color Insert.)

Table IV Drivers of Molecular Therapeutics

HGP and bioinformatics
Expression vectors for target production
In silico drug design
Robotic high-throughput screening
Combinatorial chemistry
Platform approach to drug discovery
Huge increase in number of molecular targets

Therapies will emerge through our knowledge of the human genome and the use of sophisticated bioinformatics. Targeted imaging agents will be used to deliver therapy at screening or diagnosis. Monitoring cancer patients will also change as technology allows the disease process to be tracked much more closely. Treatment strategies will reflect this and drug resistance will become much more predictable. Biomarkers will allow those treating people with cancer to measure if a drug is working on its target. If it is not, an alternative treatment strategy will be sought. Tumor regression will become less important as clinicians look for molecular patterns of disease and its short-term response to novel agents. Eventually, only those patients showing a validated surrogate response will continue with treatment so speeding up and increasing the statistical power of pivotal studies (Fig. 3).

There will be more of a focus on therapies designed to prevent cancer. A tangible risk indicator and risk-reducing therapy, along the lines of cholesterol and statins, would allow people to monitor their risk and intervene. Delivering treatment early in the disease process will also be possible because subtle changes in cellular activity will be detectable. This will lead to less aggressive treatment. The role of industry in the development of new therapies will continue to change. Smaller more specialized companies linked to universities will increasingly deliver drug candidates and innovative diagnostics to "Big Pharma" to develop and market.

People will be used to living with risk and will have much more knowledge about their propensity for disease. Programs will enable people to determine their own predisposition to cancer. This in turn will encourage health-changing behavior and will lead people to seek out information about the treatment options available to them. Patients will also be more involved in decision making as medicine becomes more personalized. Indeed, doctors may find themselves directed by well-informed patients. This, and an environment in which patients are able to demonstrate choice, will help drive innovation toward those who will benefit. However, inequity based on education, wealth, and access will continue (Table V).

VI. THE DEVELOPMENT OF PERSONALIZED MEDICINE

The era of molecularly personalized medicine for cancer has already begun. Herceptin can only work in erbB2 positive breast cancer. Similarly, the humanized monoclonal antibody Rituximab can only bind to CD20-expressing lymphoma cells. Molecular phenotyping prior to drug use is now accepted clinical practice. But this is just the beginning. It is likely that increasing use of sophisticated diagnostics will revolutionize, we use all our therapies (Watters and McLeod, 2003). Figure 4 examines the six diagnostics needed for effective cancer care. Each is important for different parts of controlling cancer. To those involved in drug development the three most important are identifying pharmacodynamic biomarkers, validating effective early surrogates of tumor response, and the predictive reclassification of disease. This last diagnostic has two strands. First, it can be used to predict the relative aggression of the disease so selecting patients for more intensive therapy and second, it can be used to identify those patients who are likely to respond to a specific molecularly targeting agent. Figure 5 considers the likely

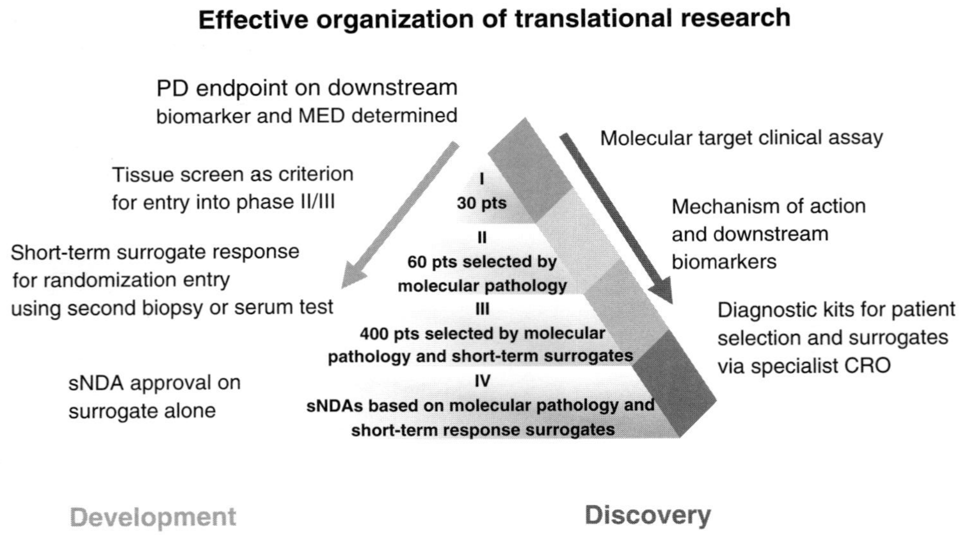

Fig. 3 The future of cancer drug development. Drugs will enter patients for the first time accompanied by effective biomarkers. These will be used to choose the maximum effective dose (MED). They will also be used to identify surrogate markers of response so selecting patients early in pivotal studies to either continue or stop a specific trial. This will enhance the speed and statistical power of pivotal studies. In addition, continued laboratory research will be used to create diagnostic kits to identify signatures of response.

Table V The Uncertainty of Novel Drugs for Cancer

Will the new generation of small molecule kinase inhibitors really make a difference or just be expensive palliation?
How will big pharma cope with most high value cytotoxics becoming generic by 2008?
Can expensive late stage attrition really be avoided in cancer drug development?
How will sophisticated molecular diagnostic services be provided?
Will effective surrogates for cancer preventive agents emerge?
Will patient choice involve cost considerations in guiding therapy?

Cancer diagnostics for personalised medicine

Diagnostic	Value
Predisposition screen	Identify patients for chemo—prevention
Screen for presence of cancer	Increase in patients—earlier disease
Pharmacodynamic biomarker	Establish pharmacological dose
Surrogate marker of clinical efficacy	Early indication of proof of concept
Predictive reclassification of disease	Target therapy to those likely to respond
Patient-specific toxicity prediction	Avoid adverse events, adjust dose

Fig. 4 Cancer diagnostics for personalized medicine. There are six areas where diagnostics will be helpful in personalizing cancer medicine.

impact versus the technological uncertainty behind them. So toxicity prediction is of low uncertainty as it is already available for some drugs, but really of very little impact. Effective surrogates look less certain at the moment but would have a huge clinical impact. Figure 6 looks at two future scenarios over a 15-year time frame—one conservative and one optimistic. Inevitably the real future will be somewhere in between—with some unpredicted step changes leading to greater successes than expected and some failures.

It is likely that the next decade will be focused on getting more information from smaller and smaller pieces of tumor tissue. Molecular histopathology will be the core discipline. Eventually, developments in functional imaging and perhaps serum proteomics will drive nontissue-based methods of obtaining the same information. The potential technologies are listed.

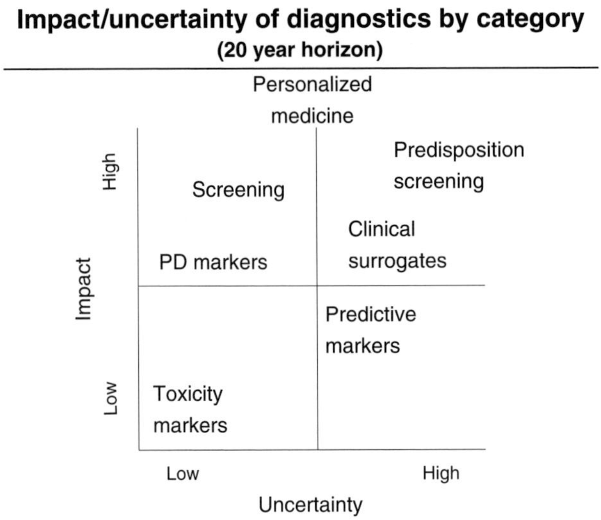

Fig. 5 The impact of cancer diagnostics versus their relative uncertainty over a 20 year horizon. Personalized medicine has the highest impact and will almost certainly be in routine practice by 2020.

Cancer diagnostics in 2020

Baseline	Upside
Risk prediction in small subsets	Population risk banding for cancer Identify people for chemo-prevention
Examples of early cancer detection	Massive expansion in patients with early cancer
Used for dose determination for some mechanistically based drugs	Universal use
Accepted by regulators in some diseases	Short-term surrogates used to register and obtain sNDA's
Microbrands targeting specific pathways reclassify disease	Routine molecular phenotyping prior to treatment decision
Low impact as drugs less toxic	Used to avoid adverse events and to adjust dose

Fig. 6 A pessimistic and optimistic prediction for cancer diagnostics in 2020. The baseline predictions are for little real change. Given the current efforts in this area seems unlikely.

Genetics
Genomics
Proteomics
Peptidomics
Metabolomics
Methylomics
Acetylomics
Integromics
Histopathology
Immunohistochemistry
Serum markers

Each has its own start-up costs, running costs, throughput capacity, accuracy, potential for automation, data handling problems, drawbacks, and of course utility. The holistic technologies such as gene expression analysis or proteomics generate huge amounts of raw data that need to be sifted for patterns. But the intersample variability can be massive leading to false conclusions. Furthermore, promising observations using relatively small sample cohorts have a tendency to disappear as the sample size increases. Ultimately, it is likely that the holistic approach using complex number-crunching techniques will be superseded by precision assays for defined biochemical constituents just as in current clinical chemistry. Those biomarkers that can be used as early response surrogates will have huge value in reducing the costs of targeted therapy as drugs can be stopped in nonresponding patients. It is likely that eventually biomarkers will be used at every stage on cancer from diagnosis to palliative care (Table VI).

VII. BARRIERS TO INNOVATION

Innovation in cancer treatment is inevitable (Dixon *et al.*, 2003): its nature and intensity critically influenced by the way innovation is rewarded.

Table VI The Clinical Use of Biomarkers

Diagnosis of early disease including molecular precancer
Providing prognostic information to choose appropriate therapy
Identifying drug sensitivity so the right drug goes to the right patient
Early surrogate markers of tumor response
Monitoring quality of response
Monitoring effectiveness of adjuvant therapy
Monitoring chronic drug dosage
Monitoring length of drug administration

However, there are certain prerequisites for the introduction of new therapies. First, innovation has to be translated into usable therapies. These therapies must be deliverable to the right biological target and to the right patient in a way that is acceptable by patient, healthcare professional, and society. Innovation must also be marketed successfully so that professionals, patients, and those picking up the cost understand the potential benefits. Those making the investment in research will inevitably create a market for innovation even if the benefits achieved are minimal. The explosion of new therapies in cancer care is going to continue and pricing of these drugs will remain high. The cost of cancer drugs in 2005 is estimated to be US$24 billion globally, of which US$15 billion is spent in the United States. If effective drugs emerge from the research and development pipeline, the cancer drug market could reach US$300 billion globally by 2025, with this cost spreading more widely around the world (Fig. 5).

But parallel to this explosion in therapies and increase in costs, a number of confounding factors will make markets smaller (Locock, 2003). The technology will be there to reveal which patients will not respond to therapy so making blockbuster drugs history. Doctors will know the precise stage of the disease process at which treatment is necessary. And as cancer transforms into a chronic disease, people will have more comorbidities, which will bring associated drug–drug interactions and an increase in care requirements.

How do we balance this equation? The pharmaceutical companies will not necessarily want to do the studies to fragment their market. Research leading to rational rationing will need to be driven by the payers of health care. There is a risk that pharmaceutical companies will stop developing drugs for cancer and focus instead on therapeutic areas where there is less individual variation and therefore more scope for profit. Furthermore, development costs are rising. Ten years ago, the average cost of developing a new cancer drug was around US$400 million. Now it is US$1 billion. At this rate of growth, the cost of developing a new drug could soon reach US$2 billion, an amount unsustainable in a shrinking market. With this in mind, the process of developing drugs needs to be made faster.

However, instead of research being made simpler, changes in legislation concerned with privacy and prior consent are making it more difficult. The EU Clinical Trials Directive will make quick hypothesis testing trials impossible. Other challenges exist, as well, such as obtaining consent for new uses of existing human tissue—following political anxiety when consent for removing and storing tissues had not been obtained in the early years of the twenty-first century. However, surveys have shown that patients who gave consent for tissue to be used for one purpose were happy for it to be used for another. They do not wish to be reminded of their cancer years later. To overcome these constraints, regulators will have to start accepting surrogate markers rather than clinical outcomes when approving therapies.

Table VII Barriers to Innovation

The drug industry will continue to compete for investment in a competitive, capitalist environment
Blockbuster drugs drive profit—niche products are unattractive in today's market
Personalized therapies are difficult for today's industry machine
Surrogate endpoints will be essential to register new drugs
Novel providers will emerge providing both diagnostic and therapy services
Payers will seek robust justification for the use of high-cost agents

Outcome studies may well move to postregistration surveillance of a drug's efficacy similar to cholesterol lowering agents today.

The rise of personalized medicine will mean the temptation to overtreat will disappear. Doctors and patients will know whether a particular treatment is justified. The evidence will be there to support their decisions. As a consequence of this, treatment failure—with all its associated costs—will be less common (Table VII).

VIII. PATIENT'S EXPERIENCE

Two separate developments will determine the patient's experience of cancer care in future. Increasing expectations of patients as consumers will lead health services to become much more responsive to the individual, in the way that other service industries have already become. Targeted approaches to diagnosis and treatment will individualize care. People will have higher personal expectations, be less deferential to professionals, and more willing to seek alternative care providers if dissatisfied. As a result, patients will be more involved in their care (Wanless, 2002). They will take more responsibility for decisions rather than accepting a paternalistic "doctor knows best" approach. This will partly be fuelled by the internet and competitive provider systems. By 2025, the overwhelming majority of people in their 70s and 80s will be familiar with using the internet to access information through the massive computing power that they will carry personally (Institute of Medicine, 2001).

With patients having access to so much health information, they will need someone to interpret the huge volumes available, helping them assess the risks and benefits as well as determining what is relevant to them. These patient brokers will be compassionate but independent advocates who will act as patients' champions, guiding them through the system. They will be helped by intelligent algorithms to ensure patients understand screening and the implications of early diagnosis. They will spell out what genetic susceptibility means and guide patients through the treatment options. Patients and health professionals will have confidence in computer-aided decision making because they will have evidence that the programs work.

How the service will be designed around patients' needs and expectations will be determined by the improved treatments available and their individualization. Care in the early stages will be provided near to where patients live. Even the most sophisticated diagnostic machinery or robotic surgeon will be mobile so much of this intervention will be carried out by technicians and nurses, with the most highly trained professionals in audiovisual contact from a distant base. When cancer centers developed in mid twentieth century, the diseases were relatively rare and survival was low. Although distressing for patients when they were referred to a center, their existence concentrated expertise. Cancer will be commonly accepted chronic conditions that even when inpatient care is required, patients will be able to choose many places in the world where they will receive care at a "cancer hotel." But for many patients even that option will not be necessary. Most new drugs will be given orally, so patients will be treated in their communities. However, this approach to cancer and other concomitant chronic conditions will place a huge burden on social services and families. Systems will be put in place to manage the ongoing control of these diseases and conditions—psychologically as well as physically. Pain relief and the control of other symptoms associated with cancer treatment will be much improved.

Today, 70% of the cancer budget is spent on care associated with the last 6 months of people's lives. Although many recognized that such treatment was more to do with the management of fear rather than the management of cancer, medical professionals have relatively few treatment options available and there was limited awareness of which patients would benefit. There is also an institutional reluctance to destroy patients' hopes that led to confusion between the limits of conventional medicines and reluctance to face the inevitable—by both patients and their families and doctors. There is a widespread perception that if patients were continuing to be offered anticancer treatment there was the possibility that their health might be restored.

With better treatments, consumers of services will be able to focus on quality of life. Much of the fear now associated with cancer will be mitigated. Demand for treatments with few side effects or lower toxicity will be high, even if there are only quite modest survival gains. The transition between active and palliative care is often sudden, but in future, because patients will be in much greater control of their situation, the change in gear will not be as apparent (Table VIII).

Table VIII Experiencing Cancer in Future

Patient brokers will guide people with cancer through the system
Choice will be real and will involve cost decisions
Patients will make a contribution to their care costs
Complementary therapies will be widely available and well regulated
Themed death chosen by patients will be possible

IX. CONCLUSIONS

Cancer will become incidental to day-to-day living. Cancers will not necessarily be eradicated but that will not cause patients the anxiety that it does today. People will have far greater control over their medical destinies. Patients in all socioeconomic groups will be better informed. In addition, surgery and chemotherapy will not be rationed on grounds of age since all interventions will be less damaging—psychologically as well as physically. Patients will want to know more about the likely progression of their cancer and how different treatments will affect it. We can already see the beginnings of patient-empowered risk analysis using relatively crude, mainly clinical data driven programs for the choice of adjuvant therapy after breast cancer surgery (http://www.adjuvantonline.com). Eventually, this concept will apply to most clinical situations and be driven by far more sophisticated measurements of biomarkers in clinical samples and their changes following treatment.

How true this picture will be will depend on whether the technological innovations will emerge. Will people, for example, really live in smart houses where their televisions play a critical role in monitoring their health and well-being. It is also dependent on health care professionals working alongside each other, valuing the input of carers who, even more than today, will provide voluntary support because of the number of people in older age groups compared with those of working age. The reality for cancer care may be rather different. The ideal will exist for a minority of patients, but the majority may not have access to the full range of services. Old people, having been relatively poor all their lives, may suffer from cancer and a huge range of comorbidities that will limit their quality of life. Looking after them all—rich and poor—will place great strains on younger people: will there be enough of them to provide the care? As with all health issues the question of access will be determined by cost and political will. In 2005, a cancer patient consumes about £25,000 worth of direct medical care costs with 70% spent in the last 6 months of life. Conservatively, with patients living with cancer, rather than dying from it, and with access to new

Marketed targeted therapies

Drug	Generic	Manufacturer	Per annum cost
Herceptin	Traztuzumab	Roche	$100,000
Mabthera	Rituximab	Roche	$80,000
Glivec	Imatinib	Novartis	$80,000
Erbitux	Cetuximab	BMS	$80,000
Avastin	Bevacizumab	Genentech	$100,000
Tarceva	Erlotinib	Roche	$60,000
Iressa	Gefitinib	AZ	$60,000

Fig. 7 The high annual costs of molecularly targeted drugs. Included here are the costs of administration and its supervision. As cancer therapy becomes more successful the prevalence of the disease will increase further increasing its overall cost.

technologies this could reach £100,000 per patient per year by 2025. Figure 7 shows the current annual cost of currently marketed targeted therapies. In theory, cancer care could absorb an ever-increasing proportion of the health care budget. Would this be a reflection of what patients want? Probably "yes." Surveys reveal that three quarters of the population believed cancer care should be the National Health Service (NHS) priority with no other disease area even a close second.

But to achieve that expenditure—and assuming that part of the health service will be funded from taxation—the tax rate might have to rise to 60%. Inevitably, there will be conflicting demands on resources: the choice may be drugs or care costs. And how are the costs computed? Although the technology will be expensive, it will be used more judiciously since it will be better targeted. Another argument suggests that when patients are empowered they use less and fewer expensive medicines, in effect lowering the overall costs. An extension of that argument is that although costs will increase for treating each individual patient, the overall costs will decrease because more care will be delivered at home. But because people will live longer, the lifetime costs of cancer care will rise along with comorbidity costs. Politicians will be faced with a real dilemma: if the prevalence of cancer increases, the cost of delivering innovative care could be massive. Will cancer care need to be rationed in a draconian way?

One dilemma for the future will be the political power of old people. More will be living longer and their chronic problems will not necessarily incapacitate them physically or mentally. This educated gerontocracy will have high expectations that will have been sharpened through the first two decades of the twenty-first century and they will not tolerate the standards of care now offered to many old people. They will wield considerable influence. Will a tax-based health system be able to fund their expectations? Politicians will have to consider the alignment between patients' requirements, and taxpayers' and voters' wishes. Fewer than 50% of voters now pay tax, and the percentage of tax-paying voters is set to fall as the population ages. Will the younger taxpayers of the future tolerate the expensive wishes of nontaxpayers? The interests of voters may be very different to the interests of taxpayers. It seems likely, therefore, that the days of an exclusively tax-funded health service are numbered. Copayments and deductibles will be an inevitable part of the new financial vocabulary. Figure 8 shows the four components of cancer's future—innovation, delivery, finances, and society.

Whatever system is put in place there is the prospect of a major socioeconomic division in cancer care. A small percentage of the elderly population will have made suitable provision for their retirement—both in terms of health and welfare, but the vast majority will not be properly prepared.

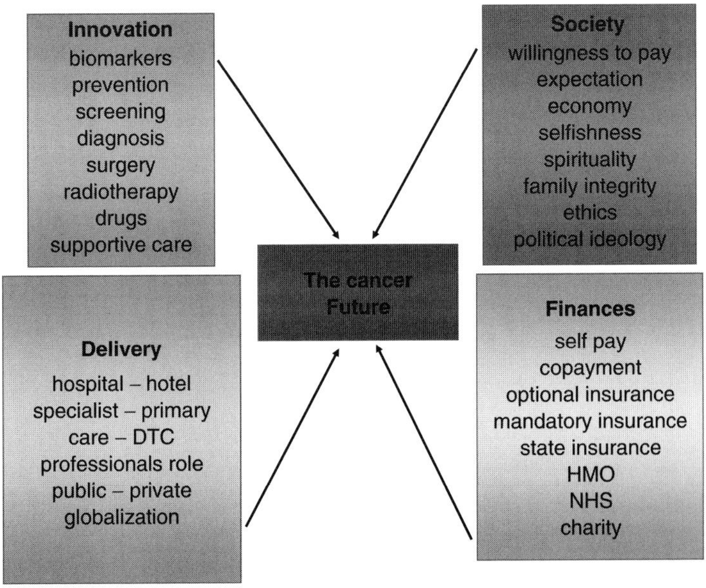

Fig. 8 The four building blocks of cancer's future—innovation, society, delivery, and finances. Cancer is predominantly a disease of retired relatively low taxpayers so putting the financial burden increasingly on younger people in society. If costs escalate then at some point resistance will come leading to rationing and inequity.

Policymakers need to start planning now as they are doing for the looming pensions crisis. The most productive way forward is to start involving cancer patient and health advocacy groups in the debate to ensure that difficult decisions are reached by consensus. Societal change will create new challenges in the provision of care. A decline in hierarchical religious structures, a reduction in family integrity through increasing divorce, greater international mobility, and the increased selfishness of a consumer-driven culture will leave many lonely and with no psychological crutch to lean on at the onset of serious illness. There will be a global shortage of carers—the unskilled, low paid but essential component of any health delivery system. The richer parts of the world are now harnessing this from the poorer but eventually the supply of this precious human capital will evaporate.

New financial structures will emerge with novel consortia from the pharmaceutical, financial, and healthcare sectors enabling people to buy into the level of care they wish to pay for. Cancer, cardiovascular disease, and dementia will be controlled and join today's list of chronic diseases such as diabetes, asthma, and hypertension. Hospitals will become attractive health hotels run by competing private sector providers. Global franchises will provide speciality therapies through these structures similar to the internationally branded shops in today's malls. Governments will have long ceased to deliver care. Britain's NHS, one of the last centralized systems to disappear, will convert to UK Health—a regulator and safety net insurer by the end of this decade.

This vision presents huge financial challenges for all societies—rich and poor. The only way to reduce cancer care costs will be to ensure that expensive medicines are only given to patients who are predicted to really benefit from them and to confirm their response as soon as possible. Molecular signatures to guide therapy choice will be sought after by those paying for care. Over 55% of cancer drugs are sold in the United States which houses less than 5% of the world's population. It is significant that the Food and Drug Administration and the National Cancer Institute have this year teamed up with the Center for Medicare Services to form the Oncology Biomarker Quantification Initiative (OBQI). Regulator, researcher, and payer are working together for the first time to reduce the overall costs by the development of novel strategies for patient selection. This suggests the death knell of the blockbuster approach to cancer drugs with its multimillion dollar advertising and marketing strategy.

The ability of technology to improve cancer care is assured. But this will come at a price—the direct costs of providing it and the costs of looking after the increasingly elderly population it will produce. We will eventually simply run out of things to die from. New ethical and moral dilemmas will arise as we seek the holy grail of compressed morbidity. Living long and dying fast will become the mantra of twenty-first century medicine.

REFERENCES

2020 Vision. (2003). "Our Future Healthcare Environments." The Stationery Office, Norwich.
Adam, B. L., Qu, Y., Davis, J. W., Ward, M. D., Clements, M. A., Cazares, L. H., Semmes, O. J., Schellhammer, P. F., Yasui, Y., Feng, Z., and Wright, G. L., Jr. (2002). Serum protein fingerprinting coupled with a pattern-matching algorithm distinguishes prostate cancer from benign prostate hyperplasia and healthy men. *Cancer Res.* **62,** 3609–3614.
Blackledge, G. (2003). Cancer drugs: The next ten years. *Eur. J. Cancer* **39,** 273.
Bosanquet, N., and Sikora, K. (2006). "The Economics of Cancer Care." Cambridge University Press, Cambridge.
Brumley, R. D. (2002). Future of end of life care: The managed care organisation perspective. *J. Palliat. Med.* **5,** 263–270.
Dixon, J., Le Grand, J., and Smith, P. (2003). "Shaping the New NHS: Can Market Forces be Used for Good?" King's Fund, London.
Institute of Medicine (2001). "Crossing the Quality Chasm: A New Health System For the 21st Century." National Academy Press, Washington, USA.
Laing, A. (2002). Meeting patient expectations: Healthcare professionals and service re-engineering. *Health Serv. Manage. Res.* **15,** 165–172.
Locock, L. (2003). Redesigning health care: New wine from old bottles? *J. Health Serv. Res. Policy* **8,** 120–122.
Melzer, D. (2003). "My Very Own Medicine: What Must I know?" Cambridge Luxemburg Press, Cambridge, UK.
Nicolette, C. A., and Miller, G. A. (2003). The identification of clinically relevant markers and therapeutic targets. *Drug Discov. Today* **8,** 31–38.
Sikora, K. (2002). The impact of future technology on cancer care. *Clin. Med.* **2,** 560–568.
Symonds, R. P. (2001). Radiotherapy. *Br. Med. J.* **323,** 1107–1110.
Tritter, J. Q., and Calnan, N. (2002). Cancer as a chronic illness? Reconsidering categorisation and exploring experience. *Eur. J. Cancer* **11,** 161–165.
Wanless, D. (2002). "Securing Good Health for the Whole Population." Department of Health, London.
Watters, J. W., and McLeod, H. L. (2003). Cancer pharmacogenomics: Current and future applications. *Biochim. Biophys. Acta* **1603,** 99.
World Cancer Report IARC Press, Lyons.

Cancer Drug Approval in the United States, Europe, and Japan

R. A. V. Milsted*

Oncology Regulatory Affairs, AstraZeneca Pharmaceuticals, Alderley Park, Cheshire SK 4TF, United Kingdom

I. USA
A. The Nixon Years
B. The Reagan Years
C. The Bush Years
D. The Clinton Years
E. The Other Bush Years
II. Europe
A. The Decentralized Procedure
B. The Centralized Procedure
C. Changes in the European Environment
D. Scientific Advice
III. Japan
A. The Role and Function of the DO
B. The Move Toward Evidence-Based Medicine
IV. Discussion

I. USA

A. The Nixon Years

The declaration by Richard Nixon of a war on cancer in 1971 gave birth in due course to the National Cancer Act of 1971. This Act established a number of National Cancer Centers and was the start of a spend of $27,000,000,000 of US taxpayers' money over the ensuing 25 years. It effectively set up a quadrangle of forces, namely academia as represented by National Cancer Institute (NCI) and other cooperative groups, the Food and Drug Administration (FDA), industry, and politicians. This set the scene for 30 years of interplay between these four factions as they attempted to reconcile the conundrum of making new therapies for cancer available to patients in the United States quickly, while maintaining an adequate regulatory surveillance to protect patients.

*MBBS, M.Sc., MRCP, M.D., FFPM, and Vice-President of Oncology Regulatory Affairs.

Advances in CANCER RESEARCH

0065-230X/07 $35.00
DOI: 10.1016/S0065-230X(06)96012-6

B. The Reagan Years

The main domestic focus of the Reagan administration was to attempt to reduce government expenditure and thereby reduce taxation levels. Against this broader agenda, the problems of bringing new cancer medicines to patients quickly might seem rather small. However, there were two other emerging issues which would ensure that some action was required.

First, it was becoming apparent that there was both a general and an oncology-specific lag time between new drug approvals in Europe and the United States, which had been estimated as being nearly 3 years for oncology drugs during the period from 1977 to 1987. Second, the emergence of AIDS and the high profile being accorded to that disease by the media clearly required some response.

The administration followed a simple well-tried maneuver, which was to be repeated by successive administrations: first, set up a committee and second, have the Vice-President chair it. Thus was born the Task Force on Regulatory Relief, chaired by Vice-President George Bush. This Task Force had a remit far wider than health care and was charged with addressing the scope of Federal Regulation, generally. It did, however, incorporate advice from a body called the President's Cancer Panel.

From this exercise came three things:

1. 1987 TREATMENT IND REGULATIONS

FDA was called on to bring forth regulations to allow patients earlier access to experimental medicines. Its response was the Treatment IND regulations of 1987 allowing unapproved drugs to be made available to patients under patient-specific INDs.

2. 1988 SUB-PART E REGULATIONS

The agency was required to bring forth regulations outlining an accelerated approval procedure for serious and life-threatening conditions. In 1988 it published the sub-part E regulations. These were designed to expedite the clinical development and review of drugs for life-threatening disease for which treatment did not exist. This was to be achieved by early and close FDA-sponsor collaboration (pre-IND and End of Phase 1 meetings) and by *permitting approval based on expanded Phase II trials.*

It is worth noting that both of these initiatives were amendments to the Code of Federal Regulations and did not require changes to the law.

Finally, the Task Force gave rise to the Lasagna Committee. Set up in 1989, the official title was the National Committee to Review Current Procedures for the Approval of Drugs for Cancer and Aids. Chaired by

Professor Lasagne, the committee's report arrived during the office of the former Vice-President, now President Bush.

C. The Bush Years

To place the recommendations of the Lasagna Committee in context, it is necessary first to understand the potential causes of the US drug lag. Many observers felt that apart from long review times there were two obvious causes, one of which was a general feature of the regulations and one specific to oncology products.

The general cause was held to be the insistence on two efficacy studies. The 1938 Food, Drugs, and Cosmetic Goods Act had required FDA to assess the safety of new pharmaceuticals, but not efficacy. The 1962 amendment to the act added a requirement for *substantial* evidence of efficacy. It defined this as "*evidence consisting of adequate and well-controlled investigations, including clinical investigations.*"

FDA's interpretation of this wording in the regulations, particularly the use of the plural in "*investigations*," was that "*at least two adequate and well-controlled studies*" were required for efficacy. While this is clearly one reasonable interpretation of the Act many people argued that it was an unnecessarily restrictive interpretation. Furthermore, given that by the 1980s drug development programs routinely involved multicenter efficacy trials, which allowed an analysis of by-center effects, the need for replicate clinical experiments was not scientifically valid. Nevertheless, the agency stuck to its interpretation stubbornly in the face of criticism which commenced with the Lasagna Committee and continued until the politicians determined that a change in behavior could only be brought about by changing the law governing the regulations. This change formed an important part of the FDA Modernization Act of 1997 (FDAMA).

The oncology-specific cause for the drug lag was perceived to be the insistence on mature survival data even where the medical community did not believe that existing treatment affected survival. This seemed unnecessarily restrictive to many in academia.

1. 1989 THE LASAGNA COMMITTEE

The Lasagna Committee made three recommendations:

FDA should use non-FDA reviewers in order to speedup review times.
FDA should exhibit flexibility in the use of surrogate endpoints.
There should be no FDA review of noncommercial trials of marketed products.

Thus, the focus of successive administrations was to be a war on the drug lag comprising seeking mechanisms to decrease review times, increase the use of surrogate endpoints, and ultimately to challenge the insistence on two adequate and well-controlled studies.

For any administration seeking to change the performance and behavior of such a large institution, there is a hierarchy of levels at which change can occur. The highest and most direct route is to change the law, since no individual or organization is above the law and all must act within it. Beneath the law come the regulations, essentially the FDA's interpretation of the law in a codified form. As we have seen from the Reagan era, it is sometimes possible to make important changes, such as the sub-part E provisions, by amending the regulations without requiring any change in the law. Given that regulations are more detailed than the law, they can frequently appear more restrictive and questions can arise, as with the need for two studies, over interpretations of the law. Beneath the regulations come the guidances put out by FDA, ostensibly to assist sponsors, but which can, on occasion, appear even more restrictive than the regulations on which they are supposedly based. Finally, beneath these three layers, comes the actual behavior of divisions and individuals within FDA. Perhaps not surprisingly, opinions offered can appear on occasion to be at variance with one or more of the previous levels of the hierarchy of control.

The Bush administration used the first two levels to bring about change. However, first it was necessary to go through the well-tried political process, to which President Bush was no stranger. He set up a committee and had the Vice-President chair it. Thus was born in 1991 the Council on Competitiveness, chaired by Dan Quayle and charged with improving the nation's drug approval process.

2. 1991 THE COUNCIL ON COMPETITIVENESS

a. Overarching Goals

This set 1994 as a time by which FDA would be required to reduce average development times for all new drugs, particularly those eligible for accelerated approval. It also envisaged reduced review times of 12 and 6 months respectively, timelines which later appear as part of Prescription Drug User Fee Act I (PDUFA I).

The Council made 11 further recommendations which are summarized below. Some of them have a direct line of descent from the Lasagna Committee and several run as threads right through to FDAMA.

External review

This proposed that FDA increases the use of external reviewers, although retaining final approval authority. The agency had used external reviewers for parts of submissions, but not to the extent of some other countries. This proposal, perhaps of all others, remains the most significant reform which FDA has successfully managed to avoid for well over a decade.

Expanded use of advisory committees

This envisaged earlier use of advice from advisory committees in design of trial programs leading to earlier approvals.

Expanded role for Institutional Review Boards

This suggested that sponsors should be able to submit INDs for Phase I studies of commercial drugs and all INDs for noncommercial drugs to Institutional Review Boards (IRBs) without the need for FDA review. This concept in a more limited form reappears many years later in the Reinventing Government (REGO) initiatives for noncommercial trials of new indications of approved drugs.

Flexible interpretation of the efficacy standard

FDA was to make a deliberate effort to interpret the statutory requirement of efficacy in a manner that maximized rather than limited a drug's potential for approval and took into account the risks to human life and health that might result from delay of new treatments. This proposed reform also called on FDA to develop and adopt surrogate endpoints to measure efficacy in life-threatening diseases.

Accelerated approval

This provision called on FDA to reduce the number of studies required for approval in life-threatening diseases and those for which no satisfactory treatment existed, shifting the approval point to before Phase III. It also mentioned use of surrogate endpoints and agreement between FDA and the sponsor to conduct postmarketing studies. It foreshadowed the revised withdrawal procedures in sub-part H which allow FDA to readily withdraw a drug as a result of further data from such studies.

Mutual recognition

This provision called on FDA to initiate discussions to establish reciprocity for approvals on a country-by-country basis. It indicated that enabling legislation would be required. It also required FDA to agree common standards for clinical trials with other countries, develop a common format for submissions and a common set of requirements for animal testing, thus foreshadowing several outcomes of the International Conference on Harmonization (ICH) process.

Enhanced computerization

This provision called on FDA to establish a plan for fully computerizing New Drug Application (NDA) reviews, with a uniform format and also a system for tracking the status of reviews. Although the dates set out were not met, FDA certainly upgraded its project management capability over the next few years and has moved progressively toward electronic submissions. The provision noted that industry had shown a willingness to pay for computerization, perhaps opening the door to Prescription Drug User Fee Act (PDUFA).

Priority review

This proposed a new system for classifying all applications into "routine" or "expedited."

Internal systems of accountability

This also referred to the need for a tracking system for applications and called on FDA to reduce the use of clinical holds.

Product liability

This supported efforts to exempt drug manufacturers from punitive damages for products with FDA approval.

Staff and financial resources

This accepted that FDA would need staff expansion to meet its goals and directed that these should be directed first toward new drug approvals. It again envisaged industry paying for computerization at FDA.

In the wake of these recommendations the Bush administration sought two changes, one legislative and one an amendment to the regulations. The 1992 PDUFA was aimed squarely at reducing review times by tying FDA funding to performance. The sub-part H regulations were aimed at accelerated approval for serious or life-threatening diseases.

b. 1992 PDUFA I

This act introduced a "fee for service" contract between industry and FDA. There were three categories of fees: drug and biological applications and supplements, annual establishment fees, and annual marketed product fees. The service was provided by using fees to hire additional staff and to improve performance by setting a date (12 months) by which an action letter was to be issued. In addition, if a project was designated for priority review, an action letter was to be issued within 6 months of the application being submitted. The act enabled FDA to recruit 600 additional staff between 1992 and 1997. Certainly, the actions taken appeared to have the desired effect as over the same period NDA review times decreased from a median of 30 to 14 months and there was an increase in NDA approvals from 25 to 47 per year.

c. 1992 Sub-Part H Regulations

These regulations outlined a mechanism and the conditions for granting accelerated approval for serious and life-threatening diseases. They required a meaningful therapeutic benefit over existing therapy, for example patients unresponsive to or intolerant of available therapy, or improved patient response. The basis of approval was adequate and well-controlled trials (note the plural), but could utilize a surrogate endpoint that was "*reasonably likely based on epidemiologic, therapeutic, pathophysiologic or other evidence to predict clinical benefit.*" Thus, the clinical endpoint could be something other than survival or irreversible morbidity. Thus, these regulations had a direct lineage from the sub-part E regulations through the recommendations of the Lasagna committee and the Council on Competitiveness.

D. The Clinton Years

The Clinton administration had set off with a zeal to reform the US healthcare system. However, the first lady's initiative to do so soon ran into the sand. There was only one thing to do; set up a committee and have the Vice-President chair it. Enter Al Gore and the initiative on Reinventing the Regulation of Cancer Drugs.

1. 1996 REINVENTING THE REGULATION OF CANCER DRUGS

Despite the grand title, the initiative was remarkable only for the fact that it introduced no major new features into the regulation of cancer drugs in the United States. The REGO document described four areas.

Accelerated approval for cancer drugs

This bundled together the existing provisions under sub-part E and sub-part H and those for priority review.

Additional uses of approved cancer drugs

This removed the need for investigators performing noncommercial trials of new indications of approved drugs to open an IND. Essentially, it placed the review of suitability of the trial with the IRB and the decision as to whether an IND should be sought with the investigator. The lineage of this modification can be traced back through the Council on Competitiveness to the Lasagna Committee.

Expanded access for drugs approved abroad

This described how FDA would seek out cancer drugs approved abroad, but for which there was no open IND and seek to discuss with the sponsor making them available, for example, through a treatment IND.

Patient representative at FDA advisory committees

In addition to the consumer representative on Oncology Drug Advisory Boards, this added a patient representative as a voting member.

2. 1997 THE FDA MODERNIZATION ACT

The passage of new law governing how FDA conducts itself implies that there is a body of political opinion which has been convinced by academia, industry, patient groups, or others that desired changes could not be achieved other than by legislation. As will be seen, some of the provisions within FDAMA are echoes of recommendations made some years earlier, but which for various reasons had not been effected.

In brief, probably the most important component of FDAMA was the technical one of providing for reauthorization of PDUFA. In addition, it enshrined the REGO initiatives on fast track approval, added a provision for FDA to promote and protect public health by reviewing clinical research promptly and required FDA to bring forth a final rule within 1 year of the act being signed into law. The following focuses only on FDAMA provisions of particular relevance to the approval of cancer indications.

a. Performance Standards

Required FDA to establish objectives and review its performance opposite its mission, meeting deadlines, staffing and resources, and ICH progress.

b. New Drug Approval

Established that data from one well-controlled clinical trial together with confirmatory evidence obtained either before or after that trial are sufficient to establish effectiveness. Thus, after a number of years of trying to get FDA to move away from a rigid adherence to the "two trial" maxim, the politicians have finally altered the efficacy requirement by legislation.

c. Expediting Approval of Fast Track Drugs

Mandated accelerated approval for serious or life-threatening conditions with potential to address unmet medical needs. It also requires FDA to establish a program that will encourage the development of surrogate endpoints.

d. Information Program on Clinical Trials for Serious or Life-Threatening Disease

Required FDA to establish a clinical trial data bank for IND (or government sponsored) trials, although inclusion of results is only with the permission of the sponsor.

e. Approval of New Uses of Approved Drugs and Biological Products

Required FDA to issue new guidance on requirements for Supplementary New Drug Applications (SNDAs).

Directed FDA to adopt a consistent efficacy standard for NDAs and SNDAs.

Required FDA to identify individuals who will facilitate SNDAs.

Required FDA to collaborate with external organizations, for example NIH.

f. Dissemination of Information on New Uses

Authorized dissemination of information on unapproved uses, subject to certain provisions.

3. 1997 PDUFA II

PDUFA I had only been authorized for 5 years. The provision in FDAMA allowed for its renewal. It also increased fees by 21% and set goals to reduce NDA review time from 12 to 10 months as well as reducing overall drug development time.

4. 1998 FINAL RULES AND GUIDANCES

FDAMA required FDA to bring forth a number of guidances within a year of the act being signed into law (November 1997). This section will describe one final rule, that is, a change in the regulations and three guidances which are of relevance to the rapid approval of cancer indications.

Providing clinical evidence of effectiveness for human drugs and biological products

This guidance addresses both the quantity and quality of evidence required for effectiveness. It discusses extrapolation from existing studies, use of a single study supported by related studies and use of a single stand alone study. It also describes circumstances in which data may be acceptable even though there is less than usual access to data, for example in publications and where monitoring of studies has not been to good clinical practice (GCP) standards.

FDA approval of new cancer treatment uses for marketed drug and biological products

This guidance describes types and quantity of data required, again citing examples where a single trial may be sufficient and also looks at alternative sources of data, such as independent cancer clinical trial organizations or reports of well-controlled studies in peer reviewed journals.

It also describes FDA initiatives to encourage sponsors to submit SNDAs for unapproved indications where there is substantial use in the medical community.

Fast track drug development programs, designation, development, and application review

This guidance essentially describes the REGO "initiatives," detailing in a single document provisions of sub-part E, sub-part H, and priority review for anyone that could not locate them before in the Code of Federal Regulations. It does, however, introduce one potentially interesting concept, that of rolling review of submissions. It has sometimes been possible in the past to negotiate with a division the submission of a part of an NDA prior to the main package. In line with FDAMA, the guidance sets out how this may now be done on a more routine basis. Sadly, however, the agency's interpretation is that although it may be compelled to accept parts of an NDA ahead of time, it is not obliged to review them. At a stroke, therefore, the most important potential means of reducing the review time for products still further is thrown away.

Dissemination of information on unapproved uses for marketed drugs, biologics, and devices

This final rule details as regulations the FDAMA provision allowing dissemination of information on unapproved uses. According to the rule, articles must be from peer reviewed journals which appear in Index Medicus and must be submitted to FDA 60 days before dissemination. The sponsor must maintain a list of all recipients, furnish them with a full bibliography, and attach an unapproved disclaimer to the front of the article. There should also be an intention to submit an SNDA for the unapproved indication within 36 months.

E. The Other Bush Years

Apart from the renewal of PDUFA there have no attempts at legislative reform. During this time however, the previously separate biological organization, CBER has been largely subsumed within the Oncology Division. This division has then been elevated into a distinct Oncology Office to oversee all cancer applications, biological or otherwise. There have been two key initiatives, one undertaken by the FDA as a whole and one by the Oncology Division.

Workshops on oncology endpoints

The Oncology Division has held several workshops to discuss the use of endpoints other than survival in the approval of cancer drugs. So far, workshops have been held on prostatic, colorectal and lung cancer, and on acute myeloid leukemia. The quality of the discussions has been somewhat variable and the chairmanship has sometimes left something to be desired, but there have been some positive outcomes from the discussions. Since the only body that can legally give expert advice to FDA is the Advisory Committee, the discussions are rerun at an Oncology Drug Advisory Committee and the stated intention is that the outcomes would form the basis of tumor-specific guidance, although none of these has so far seen the light of day.

The critical path initiatives

The FDA has commenced an initiative designed to improve the speed and quality of drug development in all therapeutic areas by fostering discussion between academia, industry, and FDA. The published list of proposed topics is long, but will certainly impact on oncology drug development and approval since it includes the use of surrogate endpoints, the use of biomarkers for patient selection, and issues of trial design and analysis.

1. THE ADVISORY COMMITTEE SYSTEM

In the United States, the reviewing divisions of FDA have been able to call on advisory committees of experts to answer any specific questions on which they wished to receive advice. The advisory meetings are held in public and various interest groups or individuals may also make submissions to the meeting. The committees have clinicians and statisticians but also an industry and a patient representative. The Oncology Division has tended to make considerable use of its Oncology Drug Advisory Committee both to answer questions on individual applications and more general aspects of clinical practice and drug approval. The committee decisions are not binding on the division, but do leave a clear public statement of how doctors who are actually treating patients view the matter under discussion.

II. EUROPE

Prior to 1995, all new medicines were approved in the European Union (EU) by the various National Agencies. While there were no specific regulations for cancer medicines, several agencies, notably those of France, Italy, and the United Kingdom made a practice of approving many new cancer drugs on the basis of limited packages, often with only Phase II data. These countries frequently approved cancer drugs in around 6 months of submission of the application.

In 1995 the member states of the EU stopped approving new drugs on a national basis and adopted two mechanisms for approval: a mutual recognition procedure and a centralized procedure. A European Medicines Evaluation Agency (EMEA) was set up to oversee the centralized procedure and a committee (CPMP—since renamed CHMP) comprising two members from each member state was formed to perform the technical evaluation and advise the EMEA on whether to approve the drug or not.

A. The Decentralized Procedure

In this procedure, the applicant submits the dossier to a selected national agency (the Reference Member State). This Agency then reviews and approves the application and agrees the Prescribing Information. The company then provides any data needed to update the dossier and the dossier, review, and proposed prescribing information are sent to the other "Concerned Member States" who should be able to avoid the need for undertaking a full review themselves, but can make comments and objections as well as proposing amendments to the prescribing information. This latter part of the process should complete in 90 days, following which each Concerned Member State should issue a national license with the agreed prescribing information.

B. The Centralized Procedure

The centralized procedure differs from the decentralized procedure in that the company submits the dossier not to national agencies, but to the EMEA. Two member states are then selected to act as the rapporteur and corapporteur, undertaking a full review of the dossier and making a recommendation to the CHMP on approval. The intended time from start of the review process for a normal application to questions being received by the company is 210 days. There is then a "clock stop," while the company responds and a further period of review before the recommendation to approve or not is made. The CHMP then makes a recommendation to the EMEA, which in turn makes a recommendation to the European Commission. This august body then in turn makes a recommendation to the Council of Ministers who rubber stamp the decision and send it back. This unbelievably bureaucratic process of stepwise recommendations to a higher authority has normally added 3 months to the approval process without appearing to add any value at all to the quality of the decision. At the end of the process, the EMEA issues a European license which is valid in all member states and puts in place a single European Prescribing Information. The whole centralized process normally takes 15–18 months, which obviously contrasts rather poorly with the 6-month approval times which some member states used to manage for cancer drugs when their national agencies were doing the review.

Since 1995 all biological agents have had to use the centralized procedure and since 2005 all new cancer drugs have to use this procedure rather than the decentralized one. Following the expansion of the EU with the accession of a further 10 states from Eastern Europe and the Mediterranean, the CHMP now has a single representative from each of the 25 member states, plus 5 other members. Norway and Iceland also participate as observers and implement the CHMP decisions.

Since 1995 there has been a major flaw in the European system in that there was no codified mechanism for granting accelerated or conditional approval for drugs intended for serious or life-threatening diseases. The only possible mechanism which the European Regulators could use was called the "Exceptional Circumstances" provision. This was intended to be used only in circumstances where it was not technically possible to provide the amount of information normally contained in a full dossier. This mechanism was not, therefore, intended for circumstances where one wanted to expedite public access of a drug which showed promise in an area where there was either no existing treatment or current treatments were not very effective.

Between January 1995 and July 1999 EMEA approved 13 drugs under exceptional circumstances, only one of which (docetaxel) was an oncology drug. Since 1999 only a further three oncology drugs were approved under exceptional circumstances.

Thus, the European Regulators have had no equivalent procedure to the sub-part H provisions of the FDA regulations which allow accelerated approval on early data for drugs which are reasonably likely to be effective where there is either no existing treatment or current treatment is unsatisfactory. For cancer patients in Europe, this proved to be a disastrous gap since over the last 10 years, the drug lag in the approval times for new cancer drugs between Europe and the United States has reversed in favor of the United States, as the FDA has been able to use its provisions for accelerated review. This has become more apparent as the new wave of novel agents has emerged. The table below shows in stark detail how European patients have seen European approvals for exciting novel agents lag behind those in the United States. It is worth pointing out that the nonvalue adding cascade of recommendations to the Council of Ministers in Europe for Gleevec took longer than it took the FDA Oncology Division to complete its entire review of the submission.

Drug	FDA approval time in days	EMEA approval time in days	Date of first approval
Glivec/Gleevec	72	224	10/05/01 (USA)
Herceptin	144	551	25/09/88 (USA)
Avastin	150	404	26/02/04 (USA)

C. Changes in the European Environment

Recent directives have sought to improve the time taken for the centralized procedure as well as introducing a priority review system which would shorten the initial part of the process to 180 days. The bureaucratic steps from EMEA upward will also be shortened, although these still remain unacceptably long.

However, at last in 2006, a regulation which would allow the issue of a Conditional Marketing Authorization has been published in Europe. The system in Europe allows the applicant to make a request for a conditional approval either at the time of assessment or during the review of the application. As with the accelerated approval system in the United States there would be commitments to conduct further studies post-approval, but in the European system the approval would only be valid for 1 year and would need to be renewed annually.

The circumstances for using the Exceptional Circumstances route have also been separately clarified. Thus, the conditional approval route would be used where it was expected that data could be provided at some point which would allow conversion to a full approval, whereas the exceptional circumstances route would apply where it was technically impossible to provide a full dossier.

The correction of this gap in the procedures available to European regulators is very welcome and long overdue. It is of concern therefore that some European Regulators have already chosen to distance themselves from the procedure by making public statements that they do not intend to use it very often. Such an attitude would be in contrast to FDA who have used their provisions for accelerated approval for new cancer drugs widely over the last 10 years and would lead to a continuation of the disparity in approval times and therefore access to new medicines for European cancer patients.

A long overdue revision of the European guideline on the evaluation of anticancer medicinal products in man has also been issued and this recognizes that the novel agents now emerging from development are significantly different from traditional cytotoxic drugs.

D. Scientific Advice

Another notable deficiency of the European system has been that it has been much more difficult to get timely and sensible interactions with the CHMP on technical issues in a particular drug development. However, this situation has been much improved over the last few years and a new process for obtaining scientific advice through the Scientific Advice Working Group (SAWG) has been issued. Also, the CHMP now has in place a number of Therapy Area Groups (TAGs), including one for oncology which it can use for advice during the process of reviewing submissions, although it is not clear to what extent if at all, they will be used during the Scientific Advice process. Unlike the US system, there is no forum for public discussion of issues arising from an application or in which the advice given to CHMP by the TAG is debated in public.

III. JAPAN

Approval for cancer drugs in Japan has traditionally been given on the basis of Phase IIb studies with a post-approval commitment to supply survival data within 6 years. In effect, therefore the whole of cancer has been regarded as an area of unmet medical need with all cancer products potentially having an accelerated approval route open to them. However, the review process in Japan was typically long, taking at least 2 years and was made more complicated by the fact that it was difficult to discuss the development with the Japanese health authority and that non-Japanese data was not thought useful in the evaluation.

It is recognized that there are some ethnic differences between different populations and racial groups. This has caused, "*concern that ethnic differences may affect [a] medication's safety efficacy, dosages and dose regimen.*" This in turn has meant that, traditionally, different regions have been unwilling to rely on foreign clinical data generated in other regions. In Japan, the perceived ethnic differences between Western and Japanese patients, combined with the fact that the Japanese population is on average smaller (by body weight) than the Western population, have led to a perceived risk that Japanese patients will respond to drugs differently in terms of safety and efficacy to Western patients. While these differences have been shown to be rarely of as much significance as interpatient variation within any population, the Japanese regulatory authorities have traditionally required a stand-alone development program of clinical trials in Japan, believing clinical data from foreign patients to be of limited applicability to Japanese patients.

However, in 1997, the Japanese Government undertook a radical revision of the structure, function, and philosophy of the Japanese drug approval process. This resulted in new legislation together with a new framework for obtaining regulatory approval. Of particular importance in this revision was:

1. The creation of the Kiko or Drug Office (DO); and
2. The move toward employing international standards of "evidence-based" medicine.

A. The Role and Function of the DO

From 1997, the government body responsible for health and welfare in Japan, the Ministry of Health and Welfare (MHW), used a body called the DO to administer a new consultation process for pharmaceutical companies seeking the approval of products in Japan. The DO offered consultations during which it would review development strategies and data put before it and provide feedback and advice. While the DO was independent of the MHW [which, in conjunction with the Pharmaceuticals and Medical Device Evaluation Center (PMDEC), retained control of the approval process itself], a pharmaceutical company could be confident that, if it successfully developed a product in line with the DOs feedback, it would ultimately obtain approval from the MHW.

The DO offered both official and unofficial consultations. An unofficial consultation could be used to check on the process to follow in an informal way, whereas an official consultation was minute. The minutes of formal consultations could subsequently be attached to a Japanese New Drug Application (J-NDA) to demonstrate that the DO was satisfied that the application was compliant with the applicable regulatory requirements. This consultation process became the first real channel for a dialogue between the regulatory authorities and pharmaceutical companies and meant that a pharmaceutical company could remove some of the risks of development by obtaining feedback from the DO.

B. The Move Toward Evidence-Based Medicine

In the past, pharmaceutical products only had to demonstrate that they were safe (e.g., not harmful) in order to obtain registration in Japan and the West. However, since the early 1960s in the West, pharmaceutical products have also had to provide evidence of efficacy, based on data obtained from clinical trials in patients. Approving pharmaceutical products for use on the basis of clinical data is referred to as "evidence-based medicine." During the 1990s, the Japanese regulatory authorities wanted to bring the Japanese system more into line with the rest of the world, by attaching increased importance to clinical data, rather than the opinions of clinicians. There was a recognition in Japan and beyond that, "*requirements for extensive duplication of clinical evaluation for every compound [could] delay the availability of new therapies and unnecessarily waste drug development resources.*" This led to a move toward evidence-based medicine in Japan in order both to reduce the duplication of studies and to ensure that Japanese pharmaceutical products were acceptable and, therefore, competitive in the West (which demanded evidence-based medicine).

The move toward evidence-based medicine culminated in the Japanese regulatory authorities issue of the "*ICH Harmonised Tripartite Guideline: Ethnic Factors in the Acceptability of Foreign Clinical Data*" (E5 Guideline). The International Conference on Harmonization of Technical Requirements for registration of Pharmaceuticals for Human Use (ICH) comprised the regulatory authorities of the EU, Japan, and the United States along with the Pharmaceutical Trade Organizations of the three regions, and had been set up to harmonize a number of key requirements for the approval of treatments across the three regions. The Japanese regulatory authorities issued the E5 Guideline on August 11, 1998. However, its principles and draft language had been adopted by the DO a number of years previously.

The E5 Guideline sought to provide a framework which would "permit adequate evaluation of the influence of ethnic factors while minimizing duplication of clinical studies and supplying medicines expeditiously to patients for their benefit." It sought to address concerns relating to potential ethnic differences through the use of "bridging studies." A bridging study is an additional clinical study designed to compare the responses of different racial populations. During a bridging study, comparable dose(s) of a compound are given to both Western and Japanese subjects and pharmacodynamic, pharmacokinetic, and/or safety and efficacy data, demonstrating how the subject handles or responds to a compound, are generated. The results of this study can then be compared either to demonstrate that there are no ethnic differences between the two populations or to correlate the differences so that foreign data can be extrapolated and relied on. This means that there is no need to carry out completely separate clinical development programs in Japan and in the West, resulting in savings of both time and resources.

The acceptance of the concept of "bridging" between Western and Japanese data along with the restructuring of the regulatory body changed the dynamics of drug development in Japan. Now it became possible to have dialogue with the Japanese agency and to agree the clinical plan, including the use of any Western data. At the same time, the review process was shortened with a target of completing reviews in a year. There is also provision for priority review within 6 months.

Since 2000 the Japanese authorities have tended to become more conservative in terms of their acceptance of bridging plans for Western data. Additionally, in 2005, a revised Oncology Guideline was issued which will undoubtedly lengthen development and approval times for new cancer medicines in Japan. Importantly, the guideline now stipulates the need for Phase III data at the time of submission and emphasizes the need for survival data, particularly in the common tumors. Although intended to bring the Japanese regulatory practice in line with Western agencies, this change happens at just the time that the FDA is seeking to lessen its dependence on survival data for approval.

Sadly, the combination of a more conservative attitude to bridging studies and the greater insistence on survival from Phase III studies will inevitably mean that approval of cancer drugs in Japan will once again be several years behind approval in Western countries.

IV. DISCUSSION

The observation of the evolution of cancer drug regulation in the United States over the last two decades provides a fascinating window into the resolution of a number of forces with different goals and opinions. One can think of each group, politicians in and out of power, regulators, academics, industry, and patient groups as a large tectonic plate, slowly, but inexorably putting pressure on the others until some minor or major cataclysmic event has to take place to relieve the situation.

The political approach to health care of any administration is likely to be determined by the guiding rule of medico-politics, namely that while it is not possible to win votes on healthcare, it is most certainly possible to lose them. Any reforming zeal is likely to be tempered by this recognition. Nevertheless, it is important to appear to have a coherent policy on health. Health care provides a ready means of embarrassing politicians and this is particularly true for the care of patients with serious illnesses such as cancer and AIDS. The emergence of evidence of a drug lag in the 1980s was sufficient to convince many politicians that the FDA, far from fulfilling a role as the protector of the public health, was in fact now a barrier to it. The Reagan administration recognized that it needed to do something in this area and the familiar vehicle of a vice-presidential committee was used to bring forth some new initiatives. The view of the regulatory agency is inevitably different, viewing the protection of the public as a responsibility too precious to be left to the vagaries of medico-politics. It is not a big step from here to developing entrenched positions and resisting change. This is most obvious in the rigid adherence to the need for two studies as the minimum efficacy standard and the insistence on survival data. In an era of multicenter trials, the former made no scientific sense and the latter was clearly not supported by the practicing medical community. Thus, begins over a decade of rearguard defending of the status quo by the agency pursued by a succession of administrations determined to effect change in the areas of surrogate endpoints and the efficacy standard, culminating in the enactment of FDAMA to effect change.

The emergence of AIDS brought patient pressure into a new era of effectiveness and the voice of academia was a constant presence both through NCI and other cooperative groups and after their creation, Advisory Boards. Meanwhile, lobbying by industry to loosen regulation inevitably arouses deep suspicion among regulators and some politicians. Other parts of the political spectrum, appalled to find new medicines unavailable in the United States, have been prepared to accept the need for reform.

What is remarkable looking retrospectively at the various attempts to effect change is less how long it has taken as an iterative process, than how remarkably resilient many of the proposals have been, surfacing and resurfacing whichever administrative vice-presidential committee has taken up the gauntlet of reform. The appearance of the various guidances post-FDAMA clarifies and cements the changes. Only the provisions for dissemination of information on unapproved uses appear unnecessarily bureaucratic. Indeed, an observer could be forgiven for wondering if the intent was not to subvert the law by making the whole process so unwieldy it was effectively useless. Across the various FDA divisions there is no doubt that the greatest single factor in abolishing the drug lag has been PDUFA. However, it should be pointed out that the Oncology Division never had prolonged review times even before the enactment of PDUFA. Nevertheless, the division has reduced its mean new chemical entities (NCEs) review time albeit on the basis of small numbers of NCEs. From a mean review time of 14 months in 1995 for four NCEs, the mean of two NCEs in 1998 was 7 months, second only to the antivirals division with a mean of 4.9 months.

It is difficult to assess what impact changing attitudes to the need for survival data and the provisions for conditional approval have had. However, from the outside it would appear that gains made have been largely as a result of improved project management of reviews. This has allowed the division to set itself an internal target of completing all reviews within 6 months at a time when other divisions were struggling to meet the provisions of PDUFA for an action letter within 12 months. With some clear views on the use of conditional approval mechanisms and a staff of clinical reviewers with relevant therapeutic area experience, the Oncology Office of FDA has become an efficient, expert group capable of giving consistent and well-reasoned opinions on projects and data, although also capable of being inconsistent and doctrinaire.

This contrasts interestingly with the two other ICH regions. In Japan, approval for cancer drugs has traditionally been on the basis of Phase IIb studies with a post-approval commitment to supply survival data within 6 years. In effect, therefore all cancer products had an accelerated approval route open to them. This sensible and satisfactory approach was revolutionized even further by the initiation of a formal consulting mechanism, the acceptance of Western data with suitable bridging studies and targets to reduce review time below 1 year. Indeed, the commitment to change in Japan over this period did the MHW enormous credit. It is particularly disappointing then to observe the retreat into conservatism over the last few years and the emergence of a guideline for the approval of cancer products which ceases to place cancer as an area of special need and will lead to lengthier development times for new agents.

In Europe, the extraordinary upheaval of introducing two European systems for drug approval was successfully negotiated, but taking stock of cancer drug approval the picture is less rosy. A number of applications which FDA felt able to approve had to be withdrawn in Europe during the 1990s and approval times in Europe have remained slower than the United States. The European system, with its multiple nationalities and need for translations is rather bureaucratic and cumbersome, but proper use of the new conditional approval mechanism could improve approval times and benefit European cancer patients. On the plus side, the procedure for scientific advice has been improved and it is to be hoped that greater use will be made of the TAGs and that these interactions can be more open.

In the United States, the clinical reviewers will normally have been involved in clinical practice within the therapeutic area and will have some relevant expertise. In the European system, the reviewers or CHMP may well not have the relevant specialist expertise and will therefore be reliant on the advice of external experts. In the United States, when FDA seeks specialist advice from the Advisory Committee, the discussion, advice, and decisions are held in public and become a matter of public record. However, in the European system the CHMP and EMEA tend to prefer levels of secrecy normally only adopted by the Vatican when choosing a new Pope. Thus, not even the company making the application knows what external experts are giving advice or what that opinion is. Indeed, EMEA initially attempted to keep even the membership of the Therapy Advisory Groups secret.

The FDA has a cancer liaison program to provide cancer patients and others with information on the drug approval process and therapies for cancer. They have a process for recruiting and training patient representatives who serve on the Oncology Drug Advisory Committee (ODAC). So, all ODAC meetings include at least one ad hoc member who has personal experience of the disease for which a new drug is being assessed. In addition, there is one other consumer representative who sits as a full member. In Europe, there is no involvement of patients as stakeholders in the decision-making process. EMEA has recently appointed two patient representatives to their board, although what their role is at this level is not clear.

The FDA has over the last decade or more used the provisions for accelerated approval widely to ensure that new medicines which offer hope where current treatment is unsatisfactory are made commercially available to patients. Without such a mechanism, the approval of novel cancer agents in Europe has generally lagged behind the United States. It remains to be seen whether the CHMP/EMEA will now make full use of the conditional approval mechanism which has been introduced.

There is no question, therefore, that with its codified mechanisms for rapid approval, post-PDUFA and -FDAMA and with good project management in place, the Oncology Division of FDA represents the standard for agreeing development plans, reviewing, and approving new cancer agents rapidly. However, there are areas for improvement. Over the last 25 years, the Oncology Division has doggedly adhered to a need for survival data, even though in practice they have often had to approve drugs on other criteria. Indeed, only 12% of their full approvals have used survival as the primary endpoint and none of their accelerated approvals. The Europeans, by contrast, have always had a pragmatic approach to what endpoints they were prepared to accept as a basis for approval.

Uniquely among health authorities, the FDA clinical and statistical review effectively involves completely reanalyzing and reconstructing the submission.

There appears scant evidence that this approach is either necessary or beneficial. Indeed, more often than not, it merely leads to either the statistician or the clinical reviewer excluding from the analysis certain patients, sometimes on quite inappropriate grounds, only to present a revised analysis which while quantitatively different does not alter what the data are telling us qualitatively. This practice does, however, reinforce the tyranny of the *p* value and make FDA the agency that is most likely to allow the statistical treatment of the data to take primacy over clinical common sense. The Europeans, including the statistical reviewers are more interested in asking the question "what are the data telling us?" and on a number of issues (such as active comparator studies) have adopted much more pragmatic and sensible positions than their US counterparts.

Finally, accelerated or conditional approval brings an increased risk of future problems in either the efficacy or safety areas post-approval. Sponsors need to find the balance between getting to market quickly and providing enough data to enable those charged with guarding the public health to make a reasoned assessment. Otherwise the risk of a public backlash against early approvals may threaten any environment which facilitates the rational, but rapid flow of new cancer drugs through development to the clinic.

Index

A

ACP. *See* Automated compound profiling
Acute lymphoblastic leukemia (ALL), 68
Acute myeloid leukemia (AML), 66–68
Acute promyelocytic leukemia (APL), 68
Adaptive immunity, 176
Adjuvant therapy, 348
  patients identification and, 61–62
Advisory Committee System, 380–381
Affymetrix GeneChip® analysis, 119–120
AGO2, argonaute protein, 77
ALL. *See* Acute lymphoblastic leukemia
American Association of Cancer Research (AACR), 229
American Society of Clinical Oncology (ASCO), 25, 229
AML. *See* Acute myeloid leukemia
Anaplastic thyroid carcinoma (ATC), 63
Anergy induction, by tumor antigens in absence of costimulatory molecules, 178
Angiogenesis, 195, 220
  assays of, 225
Animal antitumor efficacy, 193
Animal models, purpose in drug development, 203
Animal modes, post-genomics age, 203
Animal welfare issues, 224
Annexin V, 89
Anthracycline antibiotic doxorubicin, 61
Antiangiogenic agents, 233
Anticancer agents, 216
Antineoplastic NCEs, 193
Antineoplastic therapies, 191
Anti-prostatespecific membrane antigen (PSMA), 329
Antisense technology, ribozymes, 76
Antitumor activity, 195
  National Cancer Institute (NCI) screening program, 192
Antitumor effects, evaluation of, 224
Antitumor efficacy, 200, 206
  molecular analysis, biomarker, 202
  predictive value of, 200
Antitumor therapeutics
  animal antitumor efficacy, 193
  clinical efficacy, 193
Antivascular agents, 233
  combretastatin-A4 phosphate, 234
  5,6-dimethylxanthenone-4-acetic acid, 233
  ZD6126, 234
APL. *See* Acute promyelocytic leukemia
Apoptosis, 220
  assays of, 225
  assessment of, [^{124}I]-annexin V, 231
Apoptosis induced by external stimuli, activation of IAP and, 127
  proapoptotic genes identification and survival selection
    FAS-induced apoptosis in cervical cancer HeLa cells, genes involved in, 133–134
    TNF-α-induced apoptosis, genes involved in, 134–135
  ribozymes identification and enhancement of extrinsic apoptotic pathway
    FAS-Induced apoptosis activation in colon carcinoma DLDI cells, 129–132
    TRAIL activation of apoptosis in AsPC-1 and A2058 cells, 132–133
Aptamer beacon technology, 332–334
ARNT. *See* Aryl hydrocarbon receptor nuclear translocator
Aryl hydrocarbon receptor nuclear translocator (ARNT), 154
ASCO. *See* American Society of Clinical Oncology

Assays of circulating tumor markers
CA-125 for ovarian cancer, 217
PSA for prostate cancer, 217
ATC. *See* Anaplastic thyroid carcinoma
Athymic nude mouse, 192
Automated compound profiling (ACP), 157–159

B

BAC. *See* Bacterial artificial chromosome
Bacterial artificial chromosome (BAC), 8
BCL-XL, 85
BCR-ABL, 84–85
BCR/ABL, 84–85
imatinib and, 355
Biochemical assay, 222
Biomarker(s), 204, 207
of activity, phosphorylation as a marker of activation state of kinases, 205
biochemical and microscopic techniques
immunohistochemistry, 202
Western blot, 202
and cancer drug development, 272–275
discovery tool
ELISA and, 34
mass spectrometry and, 34–35
measurements, 221
qualitative assays of, 284
quality assurance (QA) and, 290–291
quality control (QC), and, 290
types of, 216
Biomarker methodology
gene expression microarrays and proteomics, 236–237
immunohistochemistry (IHC), 235–236
invasive molecular endpoints, 235
minimally invasive imaging
dynamic contrast-enhanced magnetic resonance imaging (DCE-MRI), 234
magnetic resonance spectroscopy (MRS), 232–233
positron emission tomography (PET), 231–232
minimally Invasive Imaging, 230–231
Biomarker research challenges, 275–277, 281–285
clinical qualification
biomarker assay commercialization, 293–294
clinical trials, 291–293
methods of choice, 279–281
method validation
assay dynamic range, 288–289
parallelism and dilution linearity to evaluate matrix effect, 289–290
reference standard and calibrators, 288
regulatory issues, 290–291
validation samples and quality controls, 290
technology translation, 279
translation, considerations prior to, 277–278
translation from research to bioanalytical environment
methods selection and feasibility test, 286–287
research work plan, 285–286
sample integrity, 287–288
Blood–tumor barrier (BTB), 195
Blood volume, [^{15}O]-CO, 231
Brain tumors, models, 195
Breast cancer, 57, 61–62
HER1, 68–69
HER2, 68–69
HER3, 68–69
PSAT1, 69
TMA analysis and, 69
VEGF-A, 69
VEGFR2, 69
Breast tumors, 200. *See also* Breast cancer
BRICHOS domain (137–231AA), 117
BRI/ITM2B, 117
Bush Years
Act of 1992 PDUFA I, 376
Council on Competitiveness, 374–376
Lasagna Committee, 373
Sub-Part H Regulations of 1992, 376

C

Cadmium oxide (CdO), 319
Caenorhabditis elegans, 77, 153
dsRNA, injection of, 76
Calibrators, 284
Caliper measurement, 204
loss of, 197
Cancer
animal models
genetically defined tumor models, 191

hollow fiber models, 191
orthotopic models, 191
subcutaneous xenograft models, 191
biomarkers, current, 24–25
ASCO and, 25
EGTM and, 25
NACB and, 24
biomarkers and therapeutic targets, 303–304
epidermal growth factor receptor, 304–305
matrix metalloproteinases, 305
VEGF and its receptors, 305
diagnostic tool
mass spectrometry and, 35–37
SELDI and, 35–36
early detections
of ovarian cancer, 25–26
palliative care and, 25
screening tests and, 25–27
SEER and, 26
WHO and, 26
genes
exploitation of , 215
expression analysis, 28
genome project, 214
liaison program, 390
patient's experience and, 363–365
targets, Inverse Genomics® screenings for, 137
therapy
cytotoxic, 199, 203
targeted therapies, 201, 203
types of brain, 193
breast, 193
colon, 193
gastric, 193
head and neck, 193
lung cancer, 193
melanoma, 193
non-small cell, 193
ovary, 193
pancreas, 193
renal, 193
Cancer, personalized medicine for barriers to innovation, 361–363
detection, 353–354
apoptosis, failure of, 353
chronic inflammatory change, 353
3D imaging, 354
immunosurveillance, 353
development of. *See* Personalized medicines, development of future of, 349–351
socioeconomic circumstances and, 350
past of
adjuvant therapy and, 348
Hodgkin's disease and, 349
leukemias and, 349
lymphomas and, 349
radical surgery and, 348
radiotherapy and, 348
patient's experience and, 363–365
prevention and screening, 351–353
antismoking initiatives and, 352
mapping, 352
treatment approaches and, 354–357
risk-reducing therapy, 357
tangible risk indicator, 357
Cancer, therapeutic strategies identification using microarrays
disease subtypes, therapeutic targets identification in
breast cancer, 68–70
leukemias, 65–68
lymphomas, 65–68
drug development and
pharmacogenomics, 59
toxicity and pharmacogenomic evaluation, 59–60
technologies
cDNA microarray, 52–54
comparative genomic hybridization, 56–57
Nonsense-Mediated mRNA Decay (NMD), 57–58
small molecule microarray, 58–59
tissue microarray, 54–56
use of, 60
adjuvant therapy, patients identification, 62
DNA methylases, epigenetic silencing, and tumor suppressors, 64–65
farnesyltransferase inhibitor, R115777, 63–64
histone deacetylase inhibitors and, 64
imatinib mesylate (Gleevec) and, 62–63
Cancer antigens, 177, 179–180
Cancer-based genomics, 146
discovery of novel targets, 148–151
drugs, druggability, and target validation, 147–148

Cancer-based genomics (*continued*)
oncology, small molecule screens in
data analysis of multidimensional cellular datasets, 162–165
mechanisms for new drugs, 166–167
MOA determination, 161–162
parallel cellular screens, 155–160
post-genomic discovery of novel targets, 148–151
functional cDNA screening and oncology, 151–152
RNA interference screening and oncology, 152–154
Cancer biology, RNAi as discovery tool in
CML and, 84–85
hairpin RNAs and, 84
p53-hypomorphs and, 84
Cancer gene target discovery and validation, ribozyme-based genomic technology in, 113
apoptosis induced by external stimuli, activation of proapoptotic genes identification and survival selection, 133–135
ribozymes identification and enhancement of extrinsic apoptotic pathway, 129–133
HeLa/HeLaHF cervical cancer cell system and anchorage-independent growth oncogenes identified through gene expression profiling, 118–127
tumor suppressors identification through Inverse Genomics®, 117–118
Inverse Genomics® screenings for cancer targets, 137
in vitro invasion assay, genes identification involved in cell invasion using, 135–137
Cancer Genome Anatomy Project (CGAP), 27
Cancer therapeutics, limitations of siRNAs as, 95–96
Cancer therapeutics, microarrays to direct the use of
adjuvant therapy, patients identification, 61–62
anthracycline antibiotic doxorubicin, 61
cDNA microarrays and, 61–62
farnesyltransferase inhibitor, R115777, 63–64
5-fluorouracil (5-FU), 61
histone deacetylase inhibitors and, 64
imatinib mesylate (Gleevec) and, 62–63
oxaliplatin and, 61
Cancer vaccines, 184
allogeneic cell line-based vaccines, 185
DC vaccines, 185–186
Carbopol–PVP, 323
Carcinogenesis, role of molecular-specific optical imaging of, 301–302
monitoring effectiveness of therapy, 302
rational selection of targeted therapeutics, 302–303
CBER's Oncology Division, workshops on oncology endpoints, conduction of, 380
CCAAT/enhancer-binding protein-α (C/ebpα), 65
CCD camera, 307
CCR7, 178
CDK4, 223
CDK6, 88
CDK inhibition, 223
cDNA microarrays, 56, 61–62, 65–66, 68
global cellular transcriptional states and, 54
LMPC and, 52
Cell-based assay, 222
Cell cycle
assays of, 225
control, 220
Cell transformation, mechanisms of
hematologic malignancies, 198
ovarian cancer, 198
Cellular proliferation
^{11}C-thymidine, 231
[^{18}F]-fluorothymidine, 231
CGAP. *See* Cancer Genome Anatomy Project
CGH. *See* Comparative genome hybridization
Chemotherapy, for advanced cancer, 348–350
Chitosan–Carboxymethylcellulose, 323
Chronic lymphocytic leukemia (CLL), 68
Chronic myeloid leukemia (CML), 62–63, 84–86, 146, 303
CID. *See* Collisional-induced dissociation
Ciphergen Biosystems TOF, mass spectrometry and, 35
Cisplatin, 146
Cisplatin, *in vivo* sensitivity, 196
c-KIT inhibitor, 231
Clinical biomarkers, use of. *See also* Biomarkers, 192

Clinical trials, lack of efficacy in, 191–192
Clinton Years
FDA Modernization Act of 1997, 377–378
Final Rules and Guidances of 1998, 379
PDUFA II of 1997, 379
Reinventing the Regulation of Cancer Drugs of 1996, 377
CLL. *See* Chronic lymphocytic leukemia
CML. *See* Chronic myeloid leukemia
CMV-pol II promoters, 107
Coisogenic wild-type mice, 197
Collaborative projects
International Genomics Consortium genomic array analyses for cancer, 230
NCI's Cancer Bioinformatics Information Grid (caBIG) project, 230
NCI's Specialized Programs of Research Excellence (SPORE)
biomarker identification and validation, 230
UK National Translational Cancer Research Network, 230
Collisional-induced dissociation (CID), 31
Combinatorial ribozyme gene library, drug target discovery using
hairpin ribozyme gene library and, 111–112
hammerhead ribozyme gene library and, 112
Inverse Genomics® for gene identification and, 112–113
rationale and procedures, 110–111
Combretastatin-A_4 phosphate, 233
Comparative genome hybridization (CGH), 7–11
COMPARE, 163–164, 166
Computerized tomography (CT), 217
Conditional Marketing Authorization, 383
Costimutatory signaling, 177
Council on Competitiveness
goals of, 374
origin of, 374
recommendations of, 374–376
C-RAF, 223
CRE-lox system, 197
Crystallography, 223
c-SRC, 88–89
CTL. *See* Cytotoxic T-lymphocyte
CTL activation, mechanism of, 176
CXCR4, chemokine receptor, 126
Cyclin-dependent kinase (CDK) inhibitors, 223
Cyclooxygenase-2 (COX-2), 302–303
inhibitor drugs, 270
CYLD, 93
CYP450, amplichip, 59
CYP2C19, 59
CYP2D6, 59
Cytotoxic therapies, 207
Cytotoxic T-lymphocyte (CTL), 176
Cytotoxic traditional clinical trial design, 225

D

DC. *See* Dendritic cells
Dendritic cells (DC), 177
cancer vaccines and, 185
Developmental Therapeutics Program at NCI, 193
Diagnostic strategies, need of, 27–29
EDRN and, 27
ELISA and, 27
gene expression analysis and, 28
hybrid strategies and, 27
imaging and, 27
Dicer, 76–78, 83
Diffuse large B-cell lymphoma (DLBCL), 65–66
DISCOVERY, 164
Distinct disease subtypes, therapeutic targets identification in
breast cancer, 68–70
leukemias, 65–68
lymphomas, 65–68
DKK-1, 126
DLBCL. *See* Diffuse large B-cell lymphoma
DLD1 cells, 129–131
DNA barcode, 91
DNA expression vectors
polymerase III promoters, advantage of, 82–83
DNA methylases, epigenetic silencing, and tumor suppressor
CCAAT/enhancer-binding protein-α (C/ebpα), 65
cDNA microarray and, 65
ishikawa cells, 65
RUNX3, 65
DNMT1, 81
DNMT3B, 81
DOBI, 87

Dose
confirmation, 227
estimation, two-stage approach of , 227
Doselimiting toxicities (DLTs), 225
Double-stranded RNA (dsRNA), 76, 93
Drosophila, embryo protein of, 81
Drosophila ppan gene, 117
Drug(s)
activity, biological readouts of, 223
bioavailability, 215
biological activity of, 225
clinical trials, 215
concentrations
in blood, 219
in plasma, 219
in tumor tissue, 219
development
Food and Drug Administration (FDA), 191
new chemical entities (NCEs), 191
development, cost of, 214–216
development, microarrays in pharmacogenomics, 59
toxicity and pharmacogenomic evaluation, 59–60
discovery, cancer models, 191
effect, dynic assessment of, 206
functional or molecular imaging, 206
pharmacokinetic (PK), 215
therapeutic activity, 215
toxicity, 215
Drug discovery flow, target
epidermal growth factor receptor (EGFR), mutations of, 203
role of telomerase in transformation, 203
dsRNA. *See* Double-stranded RNA

E

Early Detection Research Network (EDRN), 27
ECD. *See* Extracellular domain
Ectopic secreted protein, β-hCG, 200
EDRN. *See* Early Detection Research Network
EGFR. *See* Epidermal growth factor receptor
EGFR gene, patients with mutations in, 229
EGTM. *See* European Group on Tumor Markers
E5 Guideline, 386–387
Electrospray ionization (ESI), 30, 35
ELISA. *See* Enzyme-linked immunosorbant assay
EMEA. *See* European Medicines Evaluation Agency
Endpoints
measurement, 204
strategies and guidance on, 229
Enzymatic protein degradation
peptides and, 30
by trypsin, 30
Enzyme-linked immunosorbent assay (ELISA), 27, 34, 38, 205, 223, 235, 277, 280
Enzyme targets
histone acetylation substrates, 205
phosphorylation state of kinase substrate, 205
Ep-CAM. *See* Epithelial cell antigen
EphB4, erythropoietin-producing hepatocellular, 120–123
Epidermal growth factor receptor (EGFR), 4, 330
erlotinib and, 355
gefitanib and, 355
inhibitors, 229
Epithelial cell antigen (Ep-CAM), 12
Epithelial ovarian tumors
clinical features of, 2–5
mucinous tumors, 3
ovarian cancer. *See* Ovarian cancer, epithelial ovarian tumor
pathological features of, 2–5
ERBB2, 217–218
ErbB2 promoter, 81
ERK 1/2 phosphorylation, 223
Erlotinib, EGFR and, 355
ESI. *See* Electrospray ionization
EST. *See* Expression sequencing tag
European Group on Tumor Markers (EGTM), 25
European Medicines Evaluation Agency (EMEA), 381
European Union's approval mechanism of new drugs
centralized procedure, 382
decentralized procedure, 381
Evidence-based medicine, 386–387
Exceptional Circumstances, provision of, 382
Expression sequencing tag (EST), 11–12
Extracellular domain (ECD), 123

F

FACS. *See* Fluorescence-activated cell sorting
Farnesyltransferase inhibitor, R115777, 63–64
Farnesyltransferase inhibitors, 229
FDAMA. *See* FDA Modernization Act
FDAMA provisions, for approval of cancer indications, 378
FDA Modernization Act of 1997 (FDAMA), 373
FDA's clinical and statistical review, 390–391
FDA's critical path initiatives, 380
Fee for service, contract of, 376
2-[^{18}F]-fluoro-2-deoxyglucose (FDG), 231
Fiber optic confocal microscopes, 311
Flow cytometry, 205
FLT3-induced myeloproliferative disease, 198
Fluorescence-activated cell sorting (FACS), 127
Functional cDNA screening and oncology, 151–152

G

Gal4-inducible system, hairpin RNAs and, 83
Gastrointestinal stromal tumor (GIST), 231
Gefitinib, 229
  EGFR and, 355
Gene expression
  microarrays, 220
  RNAi inhibition of, 152–153
Gene inactivation, ribozymes as tools for
  delivery of ribozymes into cells, 106–108
  design of ribozymes for targeting specific mRNA, 105–106
  effectiveness of ribozyme-mediated gene inactivation, factors influence, 108–109
Gene microarrays, 296
Gene ontology (GO), 12
Gene target discovery and validation, ribozymes as tools in
  combinatorial ribozyme gene library, drug target discovery using
    advantages of Inverse Genomics® for gene identification, 112–113
    hairpin ribozyme gene library, 111–112
    hammerhead ribozyme gene library, 112
    rationale and procedures, 110–111
  principles of ribozyme-based gene target validation, 109–110
Genetically engineered mouse (GEM) models, 197
Genetic dissection of signaling pathways in human disease, 151
Genome functionalization through arrayed cDNA transduction (GFAcT), 148–149
GFAcT. *See* Genome functionalization through arrayed cDNA transduction
GFP-based reporter constructs, 203
GLP-compliant, 291
Glucose consumption
  ^{18}F-fluorodeoxyglucose (FDG), 202
Glucose utilization, [^{18}F]-FDG, 231
Glycerophosphocholine (GPC), 234
Glycerophosphoethanolamine (GPE), 234
GO. *See* Gene ontology
GOG. *See* Gynecologic oncology group
Gold nanoparticles, 314
Gold standard treatment, 226
Gynecologic oncology group (GOG), 4

H

Hairpin ribozyme gene library, 111–112, 117
Hairpin RNAs, 84, 93
  Gal4-inducible system and, 83
Hammerhead ribozyme gene library, 112
HCA. *See* Hierarchical clustering analysis
HeLa/HeLaHF cervical cancer cell system and anchorage-independent growth adenocarcinoma and, 115
  inverse genomics, tumor suppressors identification through hairpin ribozyme library and, 117–118
  oncogenes identification through gene expression profiling, 118–127
    Affymetrix GeneChip® analysis, 119–120
  soft-agar, 115, 117
HER1, 68–69
HER2, 68–69
HER3, 68–69
Herceptin, 294
  erbB2 positive breast cancer and, 357
HER2 gene, 217
HER-2 protein, trastuzumab and, 355
Hierarchical clustering analysis (HCA), 163
Histone
  deacetylase (HDAC), 64
  deacetylase inhibitors, 235
HIV-1 envelope glycoprotein, 82

HIV-LTR, 107
Hodgkin's disease, 349
Hollow fiber assay, 194, 202
Hollow fiber models, 194
  predictive value, 194
HSP70, 223–225
HSP90
  client proteins
    CDK4, 225
    C-RAF, 225
  inhibition, 223
  inhibitor, 17-allylamino-17-demethoxygeldanamycin (17-AAG), 225, 234
  molecular chaperone, 223
  molecular chaperone inhibitor 17-AAG, 233
hTERT, 94. *See* Human telomerase reverse transcriptase
HT29 human colon tumor xenografts, ^{31}P MR spectra of, 234
HTK-317, 120
Human breast cancer, treatment of, 233
Human colon
  cancer xenograft model, 224
  cell, 233
  xenografts, 233
Human disease, genetic dissection of signaling pathways in, 151
Human ovarian tumor xenograft model, 224
Human telomerase reverse transcriptase (hTERT), evaluation of, 181–182
Human tumor(s)
  cells, 193
  in situ, 199
  xenografts, 192, 224
Human xenografts, 199
Hyphenated MS methods, 296

I

IAP. *See* Inhibitors of apoptosis
IEOC. *See* Invasive epithelial ovarian cancers
IGFBP2, 133
IGFBP3, 126
IHC. *See* Immunohistochemistry
Imatinib, BCR/ABL tyrosine kinase and, 355
Imatinib mesylate (Gleevec)
  CML and, 62–63
  2-phenylaminopyrimidine derivative, 62
Immune response
  adaptive (or acquired) immunity, 176
    cell-mediated, 176
    humoral, 176
  advancements in research of, 176
  innate immunity, 176
Immunodeficient mice, 193
Immunohistochemistry (IHC), 63, 205, 296
  TMA and, 56
Immunostaining, 205
INF-α. *See* Interferon-α
Inhibition of kinase, 219
Inhibition of protease, 219
Inhibitors of apoptosis (IAP), 127
Inorganic phosphate (Pi), 232
Institutional Review Boards (IRB), role of, 375
Interferon-α (INF-α), 80
Intracranial implantation, 196
Invasion, 219
  assays of, 224
In silico screening, 222
In vitro invasion assay, genes identification involved in cell invasion using, 135–137
Invasive epithelial ovarian cancers (IEOC), 5
Inverse Genomics®, 126
  advantages of, for gene identification, 112–113
  screenings for cancer targets, 137
  target discovery process of, 111
  tumor suppressors identification through, 117–118
IRB. *See* Institutional Review Boards
Ishikawa cells, 65
ITM2B, 266-amino acid type II membrane protein, 118
  BH3 domain and, 117

J

Japanese drug approval process, revision of, Drug Office (DO), role and function of, 385–386
  evidence-based medicine, 386–387

K

K562
  cells, 85
  human erythroleukemic cell line, 91
Ki67 expression, 218
Kiko or Drug Office (DO), role and function of, 385–386

Kinase inhibitor, sorafenib, 228
KLF4, Krüppel-like transcription factor, 94
Knockout mouse models, 224

L

Lasagna Committee, recommendations of, 373
Laser capture microdissection (LCM), 32
Laser-microdissected cryostat sectioning, 296
Laser microdissection and laser pressure catapulting (LMPC), 52
LCM. *See* Laser capture microdissection
LCSM in multiple reaction mode (MRM-LCMS), 280–281
Leukemias
  ALL, 68
  AML, 66–68
  APL, 68
  cDNA microarrays and, 66, 68
  CLL, 68
  FLT3 mutations and, 66–68
Limit of detection (LOD), 289
L1210 leukemia model , 192
LMO2, 84
LMPC. *See* Laser microdissection and laser pressure catapulting
LNCaP, human prostate cancer cells, 126
LOD. *See* Limit of detection
LOH. *See* Loss of heterozygosity
Loss of heterozygosity (LOH), 5–6
Low-molecular-weight contrasts, gadolinium diethyltriaminepentaacetic acid (Gd-DTPA), 233
LPR-TL3, 131
Luciferase-based reporter constructs, 203
Luciferase-mediated bioluminescence, 201
Luminescent organic dyes, 311
Lymphoid organs, secondary, 177–178
Lymphomas, DLBCL and, 65

M

MAB. *See* Monoclonal antibodies
Magnetic resonance imaging (MRI), 285
MALDI. *See* Matrix-assisted laser desorption/ionization
Malignant progression, molecular basis of, 214
Marker of apoptosis, 1,6-bisphosphate, 233
Markers of specific pathway inhibition, phosphodiesters (PDEs), 233
Mass spectrometry
  Ciphergen Biosystems and, 35
  diagnostics, based on biomarker discovery tool, 34–35
    cancer diagnostic tool, 35–37
    tissue imaging tool, 32–33
  future direction, 43
  ionization source
    ESI, 30
    MALDI, 30
  limitations of. *See* Mass spectrometry, limitations of
  mass analyzers, 31
  mass-to-charge ratio (*m/z*) and, 29, 31, 33, 41
  peptide and, 30–31
  protein identification, 31
    peptide mass fingerprinting, 31
    peptide sequencing, 31
  quantitation, 31–32
  subproteome and, 29
  suggestions, 42
  surface-enhanced laser desorption/ionizationtime-of-flight (SELDI-TOF), 29, 35–42
  tandem, 31
Mass spectrometry, limitations of
  analytical
    bias toward high-abundance molecules, 39–40
    cancer biomarkers, identification of, 39
    discriminatory peaks, identity and origin of, 40
    dynamic range, 38–39
    ionization efficiency, 40
    reproducibility, 40–41
    robustness, 41
    sensitivity, 39
  postanalytical
    bioinformatic artifacts, 41
    external validation, 41–42
  preanalytical, 38
Matrix-assisted laser desorption/ionization (MALDI), 30, 35
  imaging, 277
Matrix metalloproteinases (MMP), 135, 304
  inhibitors, 229
Maximum effective dose (MED), 358
Maximum tolerated dose (MTD), 225
MDA-MB-435 human breast cancer cells, 308

Mdr1 expression, tissue-specific modulation of, 196
MDS. *See* Multidimensional scaling
MED. *See* Maximum effective dose
Medico-politics, rule of, 388
MEK 1/2, inhibition of, 233
MEK inhibitor, 223
Mercaptopropionic acid (MPA), 319
Mesoscale discovery (MSD) platform, 235
Metastasis, 196, 220
  assays of, 225
M-fold, 108
Microarray comparative genomic hybridization (mCGH), 58
  in breast cancer, 57
  CCNE1, 57
  cDNA microarrays and, 56
  EMS1, 57
  ERBB2, 57
  MOSE and, 56
  TGCTs and, 57
  TOP2A, 57
Microarray technology, use of, 223
Mie theory, 316
MIRSA, 87
MISSION™ TPX–Hs 1.0, 158
MK-STYX, 88
MLL/AF4, 85
MMP. *See* Matrix metalloproteinase
MMTV. *See* Mouse mammary tumor virus
Modern genomic technologies
  combinatorial chemistry, 192
  high information content screening, 192
  robotics, 192
Modification-specific antibodies
  acetylation, 205
  phosphorylation, 205
Molecular beacons, 321
Molecular biomarkers, 218
Molecular cancer therapeutics, 213. *See also* Molecular therapeutic
  pharmacological audit trail, 230
  stages involved in, 222
  trial design, 225–229
  tumor response, 225
Molecular classification of tumor, 214
Molecular effects of drugs, molecular imaging, 202
Molecular hallmarks of self-sufficiency in growth signals, reporters of, 331–332
Molecular imaging
  Bioluminescence, 202
  fluorescence, 202
  micro-positron emission tomography (micro-PET), 202
Molecularly targeted agents
  combining chemotherapy with, 252–253
  PD biomarkers, 230
  pharmacological audit trail, 218–220
Molecularly targeted therapeutics, 224
Molecular mechanism of action (MOA), 160
Molecular-specific optical imaging
  metal nanoparticles
    gold and fluorescent labeling, comparison of, 325–327
    imaging of metal nanoparticles, 327
    metal nanoparticles and EGFR, 324–325
    safety of metal nanoparticles for *in vivo* use, 328
    signal intensity, 327
  quantum dots, 328–329
    aptamer-targeted, 329
    biocompatibility and cytotoxicity of, 329–330
  role of, of carcinogenesis, 301–302
    monitoring effectiveness of therapy, 302
    rational selection of targeted therapeutics, 302–303
Molecular-targeted cancer therapeutics, 330
Molecular targets
  achievement of desired biological effect, 220, 223
  activity of the pathway, 218–220
  desired activity, 219
  expressed, 218
  modulation, 205
  mutated, 218
  relevant clinical response, 219
Molecular therapeutics, 215, 216, 224, 226 *See also* Molecualr cancer therapeutics
Moloney murine stem cell virus (MoMuLV), 83, 108
MoMuLV. *See* Moloney murine stem cell virus
Monoclonal antibodies (MAB), 182
  for cancer therapy, 183
MOSE. *See* Mouse ovarian surface epithelial
Mouse cancer models in drug discovery, practices and pitfalls of, 191–212
Mouse mammary tumor virus (MMTV), 107
Mouse ovarian surface epithelial (MOSE), 56

Mouse syngeneic transplant models, 199
Mouse tumor cell lines
B16 melanoma, 197
Lewis lung carcinoma, 197
MRI. *See* Magnetic resonance imaging
MSCV. *See* Murine stem cell virus
MTD, 225–227
Mucinous tumors, 3
Multidimensional scaling (MDS), 163
Multiplex binding assays, 296
Multiplex immunoassays, 219
Murine stem cell virus (MSCV), 83
M-VAC neoadjuvant chemotherapy, 61–62
Myc-induced lymphomagenesis, 84
Mytomycin C, *in vivo* sensitivity, 196

N

NACB. *See* National Academy of Clinical Biochemistry
National Academy of Clinical Biochemistry (NACB), 24
National Cancer Institute (NCI), 25–27, 229
NCE. *See* New chemical entities
NCI. *See* National Cancer Institute
NCI 60 cell lines, 193
NDA. *See* New Drug Application
New chemical entities (NCE), 191, 389
New Drug Application (NDA) reviews, computerization of, 375
NF-AB signaling pathway, 93
N-hydroxysuccinimide (NHS) ester, 324
NIH3T3 cell, 137
Nixon Years, National Cancer Act of 1971, 371
NLK, 88
NMD. *See* Nonsense-mediated mRNA decay
Noninvasive assessment of enzyme activities, 203
Noninvasive fluorescence imaging, 204
Nonmalignant stromal cells, 195
Nonsense-mediated mRNA decay (NMD)
CGH and, 58
Non-small cell lung carcinoma (NSCLC), 193, 304–305
Novel technologies
functional imaging, 229
genomic, 229
proteomic, 229
Novel therapeutics, mechanism of action
degradation of ERBB2 receptors, 232
VEGF/VEGF receptor expression, 232
Novel therapeutic strategies, small molecule kinase inhibitors, 198
NR4A1, 123–127
NR4A2, 123–126
NR4A3, 123–126
NSCLC. *See* Non-small cell lung carcinoma (NSCLC)
N-tectradecylphosphonic acid (TDPA), 319
Nuclear magnetic resonance, 223
α-, β-, γ-Nucleoside triphosphates (α-, β-, γ-NTP), 234
Nur77/NGFIB family, 126

O

OCT. *See* Optical coherence tomography
ODAC. *See* Oncology Advisory Committee
Oncogene, 75
LMO2, 84
RAS, 84, 86, 94
RNAi and, 76
Oncology, 191, 214
bevacizumab (Avastin), 214
cetuximab (Erbitux), 214
erlotinib (Tarceva), 214
gefitinib (Iressa), 214
imatinib (Gleevec), 214
sorafenib (Nexavar), 214
sunitinib (Sutent), 214
targeted agents in. *See also* Anticancer agents
trastuzumab (Herceptin), 214
Oncology Advisory Committee (ODAC), 390
Optical coherence tomography (OCT), 306
Optically active contrast agents and caner, 311–314
metal nanoparticles
conjugation, 317–318
optical properties, 314–316
synthesis, 316–317
quantum dots
optical properties, 318
synthesis and conjugation strategies, 318–320
smart contrast agents, 320–322
Optical technologies and caner, 305–306
fiber optic, 311
high-resolution microscopy, 309–311
widefield microscopy, 306–309

Optimum biological dose, 226, 227
determination of, 226
Orthologous target, 199
Orthotopic implantation, 196–197
Orthotopic models, 195, 204
in lung, 196
Orthotopic SCLC model, 196
Ovarian cancer
cDNA array and, 8, 11, 14
early detection and, 25–26
endometrioid and, 3, 5, 9, 13–15
epithelial ovarian tumor features of, 2–4
gene ontology (GO), 12
genomic approaches and
biomarker discovery, 5–18
comparative genome hybridization (CGH), 7–11
loss of heterozygosity (LOH), 5–6
transcription profiling, 11–18
GOG and, 4
ovachip, 12
p53 overexpression, 4
prognosis and, 3
prognostic markers for, 4–5
RAB25 mRNA, 8
TP53 mutation, 4
Ovarian Cancer Prognostic Profile (OCPP), 16

P

Parallel cellular screens, 155–160
PCR. *See* Polymerase chain reaction
PDCD6, proapoptotic calcium-binding protein, 94
PD–PK relationships, 227
PDUFA I Act of 1992, 376
PEG. *See* Polyethylene glycol
PEGylation, 330
Peptide
enzymatic protein degradation, 30
mass fingerprinting, 31
mass spectrometry and, 30
sequencing
collisional-induced dissociation (CID), 31
by tendem mass spectrometry, 31
Personalized medicine, development of
Herceptin, erbB2 positive breast cancer and, 357
nontissue-based methods and, 359
potential technologies and, 360
gene expression analysis, 361
rituximab, CD20 and, 357
PET. *See* Positron emission tomography
PET, methodological issues, 232
PET probes
HER2 expression, 202
integrin expression , 202
Pharmacodynamic markers, 194, 205–206
Pharmacodynamic (PD) biomarkers, 213, 215, 218, 228, 272–273
choline compounds, [^{1}H]-MRS, 233
inorganic phosphate, using [^{31}P]-MRS, 232
inositol compounds, [^{1}H]-MRS, 232
phosphocholine, 232
phosphomonoesters (PMEs), 232
potential applications of, 218
for specific new drug classes, 237
EGFR inhibitors, 240–244
HSP90 molecular chaperone, inhibitors of, 244–247
imatinib mesylate as paradigm for development of molecular
inhibitors of the PI3K-AKT-mTOR pathway
mTOR inhibitors, 249–252
PI3K inhibitors, 247–249
therapeutics, 238–240
Pharmacodynamic (PD) endpoints, 218, 225, 227, 230
assessment of blood flow, 233
assessment of permeability, 233
Pharmacodynamic (PD) markers, 225. *See also* Pharmacodynamic (PD) biomarkers
clinical trials, 225–229
methodological issues
assay-related factors, 220
marker- and sample-related factors, 220
statistics and clinical trial issues, 221
preclinical drug development, 222–225
Pharmacodynamic/Pharmacokinetic Technologies Advisory Committee (PTAC) of Cancer Research UK, 230
Pharmacogenomics, cytochrome P450 (CYP) pathway
CYP2C19, 59
CYP2D6, 59
Pharmacological audit trail, 218–220
2-Phenylaminopyrimidine, 62
Phosphocholine (PC), 233–234
Phosphocreatine (PCr), 234

Phosphodiester (PDE), 233
Phospho-ERK 1/2, inhibition of, 233
Phosphoethanolamine (PE), 234
Phosphomonoester (PME), 234
Phosphorylation-state-specific
atibodies, 206
Phosphoserine aminotransferase (PSAT), 69
P53-hypomorphs, 84
PI3 kinase (PI3K) inhibitors LY294002, 233
PITX1, homeodomain pituitary transcription
factor, 94
PK endpoints, 215, 218
PK–PD
endpoints, 225
relationships, 218, 219, 224–227
technologies, 230
Plasmacytoid dendritic cells, INF-α and, 80
PLK1-shRNA, plasmid, 85
PME, 233
Poly (ADP-ribose) polymerase, 233
Polycarbophil–PVP, 323
Polyethylene glycol (PEG), 317
Polymerase chain reaction (PCR), 111, 348
Positron emission tomography (PET), 285
Preclinical models, 207
Prediction of activity spectra for substances
(PASS), 167
Primary *in vitro* endpoint, 199
Proliferation assays, 224
Prostate-specific antigen (PSA), 186
Protease-sensitive nanoparticle contrast
agents, 331
Protein assay, 280
Protein biomarkers, 284–285
ProteinChip, 296
Proteomics, 220
PSA. *See* Prostate-specific antigen
PSAT. *See* Phosphoserine aminotransferase
PSMA. *See* Anti-prostatespecific membrane
antigen

Q

Qualitative assays of biomarkers, 284
Quality assurance (QA) and biomarkers,
290–291
Quality control (QC), and biomarkers, 290
Quantum dots, 318, 328–329
aptamer-targeted, 329
biocompatibility and cytotoxicity of,
329–330

R

Radical surgery, 348
Radiological disease response, 220
Radiotherapy
3D imaging, 354
nasal tumor and, 348
RAp594, 132
Rapid screening, 223
Ras, 303
Ratio of treated to control tumor volume
(% T/C), 193
RB phosphorylation, 223
Reagan Years
Reagan administration, focus of, 372
Task Force on Regulatory Relief
Lasagna Committee, 372–373
Sub-part E regulations, 372
Treatment IND regulations of
1987, 372
Real-time polymerase chain reaction
(RT-PCR)genomics, 280
Receptor tyrosine kinase, 206
Recommended phase II dose (RPIID), 226
REGO. *See* Reinventing Government
Reinventing Government (REGO), 375
document
accelerated approval for cancer
drugs, 377
additional uses of approved cancer
drugs, 377
expanded access for drugs approved
abroad, 377
patient representative at FDA advisory
committees, 377
Reinventing the Regulation of Cancer Drugs
of 1996, 377
Renal cell carcinoma, 228
Response evaluation criteria in solid tumors
(RECIST) guidelines, 226
Reverse immunology, 184
Ribozyme
based genomic technology in cancer
gene target discovery and
validation, 113
activation of apoptosis induced by
external stimuli, 127–135
HeLa/HeLaHF cervical cancer cell
system and anchorage-independent
growth, 115–127
Inverse Genomics® screenings for cancer
targets, 137

Ribozyme (*continued*)
in vitro invasion assay, gene identification involved in cell invasion using, 135–137
biology of, 103–105
hairpin ribozymes and, 103–104
hammerhead ribozymes and, 103–104
tools for gene inactivation
delivery of ribozymes into cells, 106–108
design, for specific mRNA target, 105–106
influencing factors and, 108–109
tools in gene target discovery and validation
combinatorial ribozyme gene library, drug target discovery using, 110–113
principles of ribozyme based gene target validation, 109–110
RISC. *See* RNA-induced silencing complex
Risk biomarkers
neoplastic progression
ERBB2/HER2 gene amplification, 217
Rituximab, 357
RNA-induced silencing complex (RISC), 76–79, 83
RNA interference (RNAi)
consortium, 154
as discovery tool in cancer biology, 84–87
inhibition of gene expression, 152–153
mechanism of
dicer, 76–78
dsRNA and, 76–79
RISC and, 76–80
virus infection and, 77
oncogenes and, 76
screening and oncology, 152–154
screens
siRNA expression vectors and, 91–95
synthetic siRNAs and, 87–89
transfected cell assays and, 89–91
RNase H-sensitive sites, 108
ROCK1, 136
ROCK2, 136
ROR1, 88
RPIID, 226
RPS6KL1, 88
RSV -pol II promoters, 107
RTK, receptor tyrosine kinase, 122
RT-PCR, 235
RUNX3, 65
RUNX1/CBFA2 T1, 85
Rz568, 117
Rz619, 117, 119, 123
Rz-HFSC1, 117

S

SEER. *See* Surveillance Epidemiology and End Results
SELDI-TOF. *See* Surface-enhanced laser desorption/ionizationtime-of-flight
Self-organizing maps (SOM), 163
Semiquantitative immunohistochemistry (IHC), 217
SEREX. *See* Serological analysis of tumor antigens by recombinant cDNA expression cloning
Serological analysis of tumor antigens by recombinant cDNA expression cloning (SEREX), 180
SID-1, transmembrane protein, 77
SiRNA
delivery
DNA expression vectors and, 82–83
synthetic siRNAs and, 81–82
viral vectors and, 83–84
limitations of, as cancer therapeutics, 95–96
transcriptional gene silencing by, 80–81
SMAD7, 94
Small molecule microarray (SMM)
cyclin-dependent kinases and
Cdk1, 58
Pho85, 58
Smart contrast agents, and cancer
systematic delivery, 323–324
topical delivery, 322–323
"Smart" contrast agents for cancer, 331–335
SMM. *See* Small molecule microarray
Solid tumor(s)
isolation
fimmunomagnetic separation, 205
fluorescence-activated cell sorting, 205
laser-capture microdissection, 205
tumor-specific (or species-specific) surface antigens, 205
measurements of, 204
SOM. *See* Self-organizing maps
Stable disease, duration of, 220

Standard endpoint, attenuation of tumor growth, 205
Study endpoints
efficacy endpoints, 191, 199
functional and molecular endpoints, 191, 201
Subcutaneous (SQ)
(ectopic) models, 195–196
models, 204
or intraperitoneal (IP) space, 195
tumors, 200
xenograft models, 193
clinical antitumor efficacy, 194
of NSCLC, 194
Sub-Part H Regulations of 1992, 376
Surface-enhanced laser desorption ionization (SELDI)-MS, 277
Surface-enhanced laser desorption/ ionizationtime-of-flight (SELDI-TOF), 29, 35–42
Surrogate endpoints, bone marrow suppression, 226
Surrogate marker, 186
TAA, 187
Surrogate normal tissues, 221
Surrogate tissues
buccal mucosa, 225
PBMCs, 225
skin, 225
Surveillance Epidemiology and End Results (SEER), 26
SV40 LT, 94
Syngeneic naive mice, 198
Syngeneic transplant, 197
Syngeneic tumor–host compartments, 197
Synthetic siRNAs
cell culture models and, 81
cholesterol, attachment of, 81–82
Drosophila, embryo protein of, 81
electroporation and, 82
RNAi screens and TRAIL-induced apoptosis, 88

T

TAA. *See* Tissue-associated antigen
Targeted immunotherapy for cancer
by immune-mediated specific anti-TAA response, 179–180
Targeted therapies, 207, 214, 218
imatinib, 194
rituximab, 194
testing of, 198
toxicity of, hypersensitivity pneumonitis, gefitinib, 227
trastizumab, 194
Target modulation, 206
Taxol, 146
T-cell receptor (TCR), 126
TCR. *See* T-cell receptor
T47D, 120
Tellurium-trioctylphosphine (TOP), 319
TEL/PDGFRβ 85
Testicular germ-cell tumor (TGCT), 57
Tetrazoliumbased colorimetric assay, 195
TGCT. *See* Testicular germ-cell tumor
TGFBR2, 94
TGS. *See* Transcriptional gene silencing
Th-1 cells, 178
Therapeutic targets, cancer biomarkers and, 303–304
epidermal growth factor receptor, 304–305
matrix metalloproteinases, 305
VEGF and its receptors, 305
Therapeutic targets, identification of
breast cancer and, 68–70
in distinct disease subtypes, 65–70
DNA methylases, epigenetic silencing, and tumor suppressors, 64–65
leukemias and, 65–68
lymphomas and, 65–68
Tissue-associated antigen (TAA)
clinical effective trageting of, 182–184
identification technique of, SEREX, 180–181
as therapeutic targets, 184
Tissue imaging tool
LCM and, 32
mass spectrometry and, 32–33
Tissue microarray (TMA), 57, 63–64, 69
cDNA microarrays and, 56
IHC and, 56
Tissue perfusion, [^{15}O]-H_2O, 231
TLR. *See* Toll-like receptor
TMA. *See* Tissue microarray
TNF-related apoptosis inducing-ligand (TRAIL), 154
TNF-α-related apoptosis-inducing ligand (TRAIL), 132, 134
Tobacco ringspot virus, hairpin ribozymes and, 103–104
Toll-like receptor (TLR), 80, 95

TRAIL. *See* TNF-related apoptosis inducing-ligand; TNF-α-related apoptosis-inducing ligand
TRAIL-induced apoptosis, 88
Transcriptional gene silencing (TGS)
  chromatin, epigenetic changes in, 80
  erbB2 promoter and, 81
  SIRNAS and, 80–81
Transfected cell array (TCA)
  Annexin V, 89
  cell lethality and, 91
  nonadherent cells and, 90
  RNAi screens and, 89–91
  tissue culture dishes and, 89
Transgenic mouse models, 224–225
Trastuzumab, 217
  HER-2 protein and, 355
$tRNA^{Lys3}$-ribozyme, 107
$tRNA^{Met}$-ribozyme, 107
$tRNA^{Val}$-ribozyme, 107
Tumor(s)
  angiogenesis, 196
  asynchronous development of, 197
  biopsies, 230
  burden, 217
  cells, response to doxorubicin, 196
  formation, mechanisms for, 199
  growth, 200
    in vivo, effect assessment, 203–204
  implant, 194
  implantation or transplant, location of, 204
  microvasculature
    initial area under the gadolinium curve (IAUC), 233
    leakage space (ve), 233
    transfer constant (K^{trans}), 233
  progression, 216
  regression, 216
  regression agents
    Ara-C, 200
    cyclophosphamide, 200
    5-fluorouracil, 200
    melphalan, 200
    N-phosphon-acetylaspartat, 200
    6-thioguanin, 200
  size, radiological assessment of, 225
Tumor antigens, 185–186
  classification of, 177
  hTERT, 181–182
  immune system of, 175–176
  NY-ESO-1, as a target, 182
  recognition of, 176–178
  as surrogate markers and targets, 186–187
  TAA, 176
  tumor-rejection antigen, 180
Tumor burden
  assessment of caliper measurement of superficial tumor, 193
  quantification, 201, 204
  quantification, optical imaging
    bioluminescence, 201
    fluorescence, 201
Tumorigenesis, 195, 197
Tumor markers
  CA-125, 226
    ovarian cancer, 217
  prostate-specific antigen (PSA)
    prostate cancer, 217
  PSA, 226
Tumor models
  cell types, 191, 198
  genetically defined, 197
Tumor quantification, noninvasive methods
  magnetic resonance imaging (MRI), 200
  plain radiographs, 200
  ultrasound (US), 200
  X-ray computed tomography (CT), 200
Tumor rejection antigens, 180
TUNNEL assay, 218

U

US National Institutes of Health, definitions
  biological marker or biomarker, 217
  clinical correlates
    disease-free survival, 217
    time to progression, 217
    tumor response rate, 217
  clinical endpoint, 217
  PD biomarkers, 218
  prognostic biomarkers
    biological progression markers, 217
    risk biomarkers, 217
  surrogate endpoint, 217

V

Valid biomarker, 273
Vascular endothelial growth factor (VEGF), 69, 82, 95, 183, 303
  and receptors, 305

Vasodilator hydralazine, 196
VEGF. *See* Vascular endothelial growth factor
VEGF-targeted therapies, 234
Viral vectors
  LMO2 and, 84
  MSCV and, 83
  siRNAs delivery and, 83–84

W

Western blotting, 205, 223–225
WHO. *See* World Health Organization
Whole cell vaccines, 179
World Health Organization (WHO), 26
Wortmannin, 233

X

Xenograft
  models, 199
    lack of predictability, 198
  tumor model, 117, 131
  tumors, 199
    sensitivity to therapeutics, 196
Xenotransplant model, 85

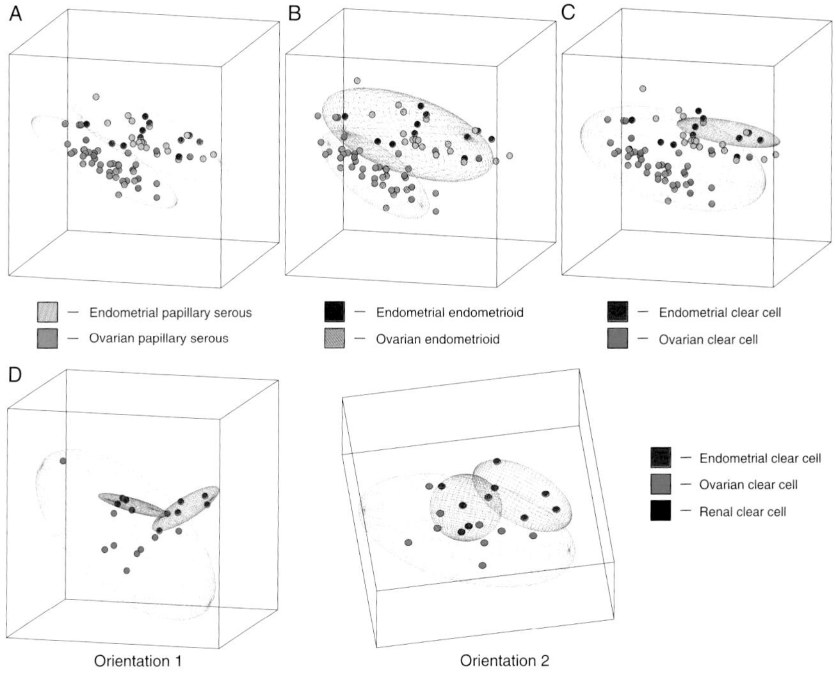

Please refer to Fig. 1.4 in text for figure legend.

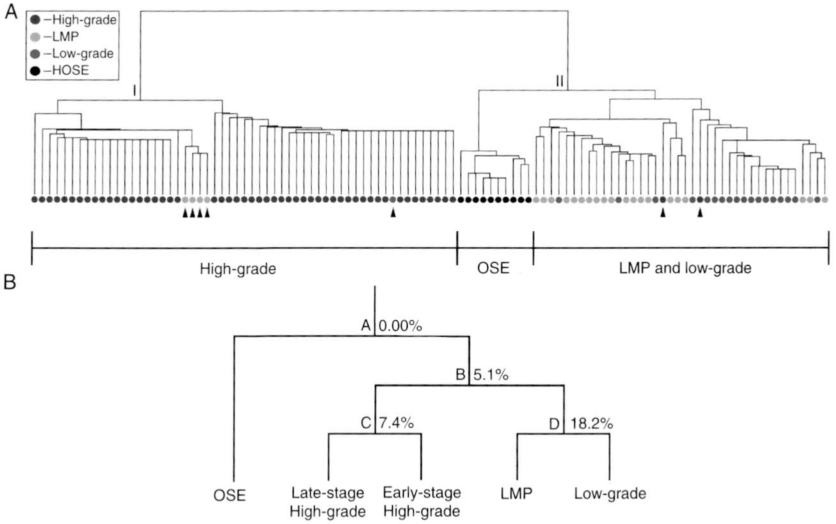

Please refer to Fig. 1.5 in text for figure legend.

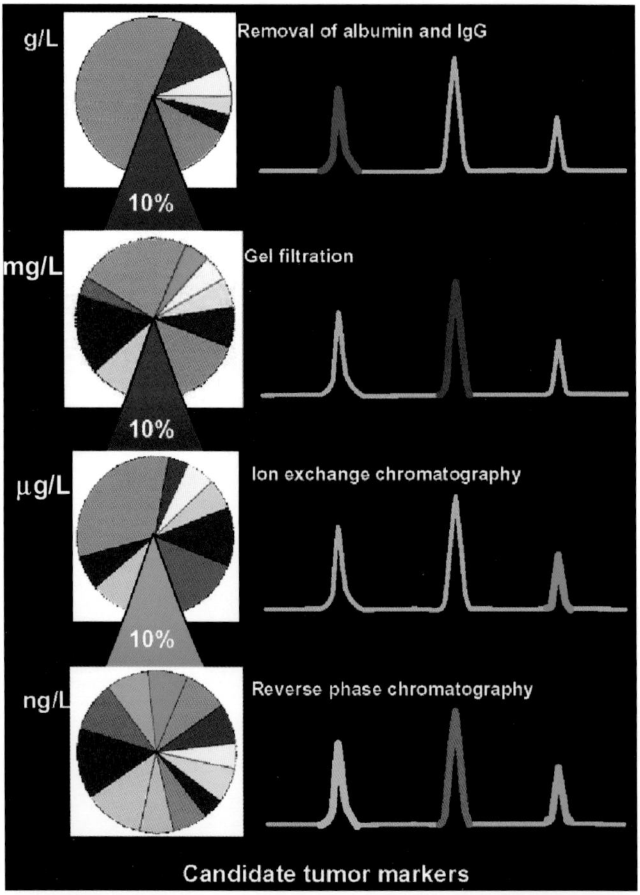

Please refer to Fig. 2.3 in text for figure legend.

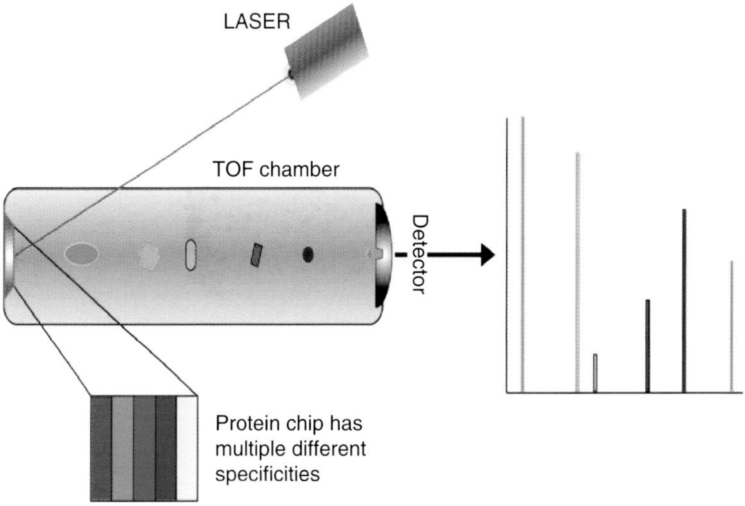

Please refer to Fig. 2.4 in text for figure legend.

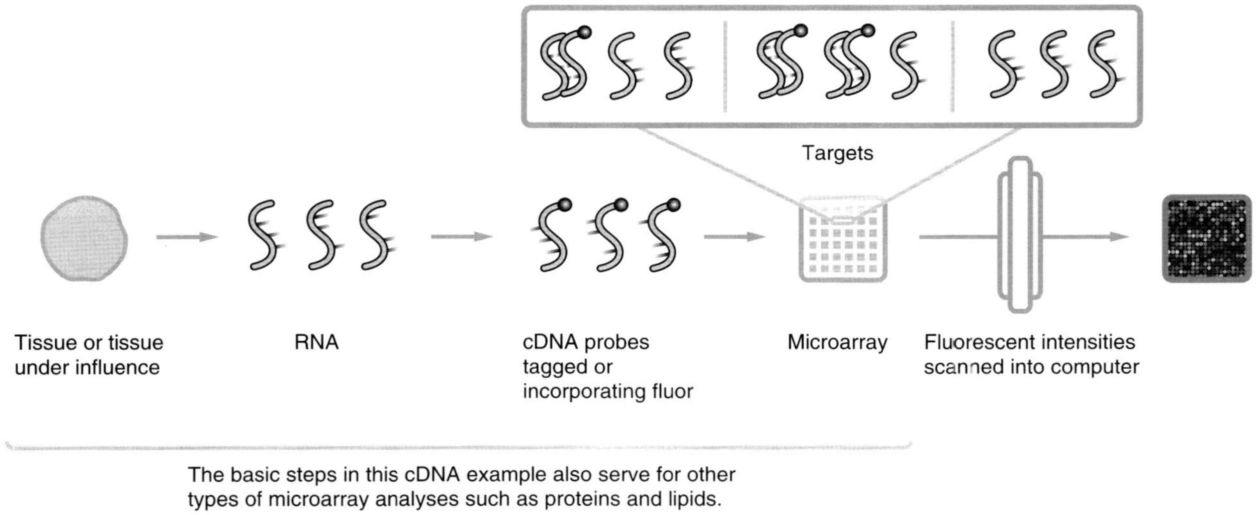

Please refer to Fig. 3.1 in text for figure legend.

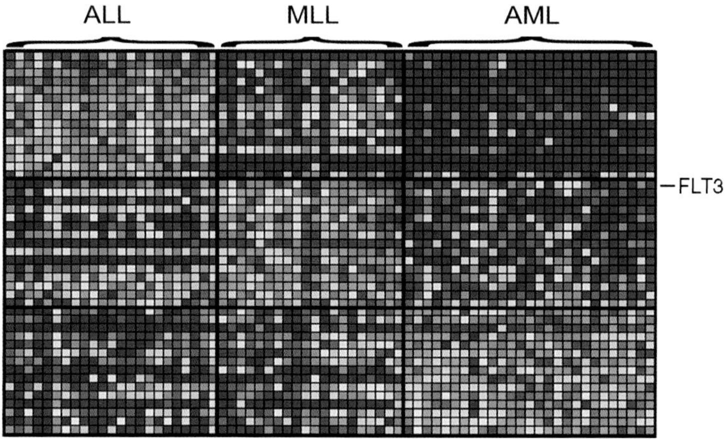

Please refer to Fig. 3.3 in text for figure legend.

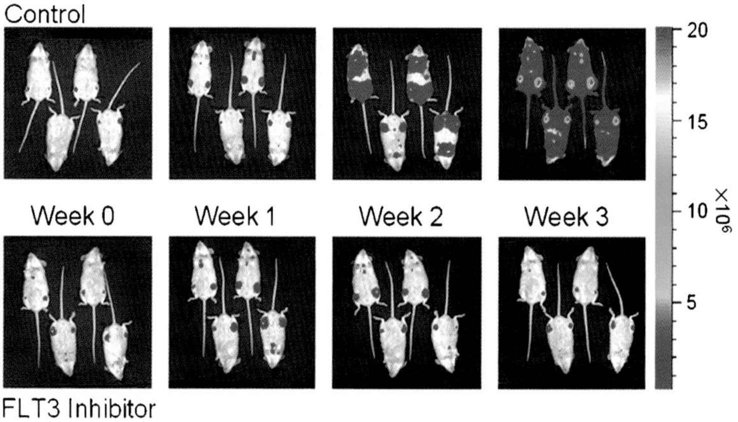

Please refer to Fig. 3.4 in text for figure legend.

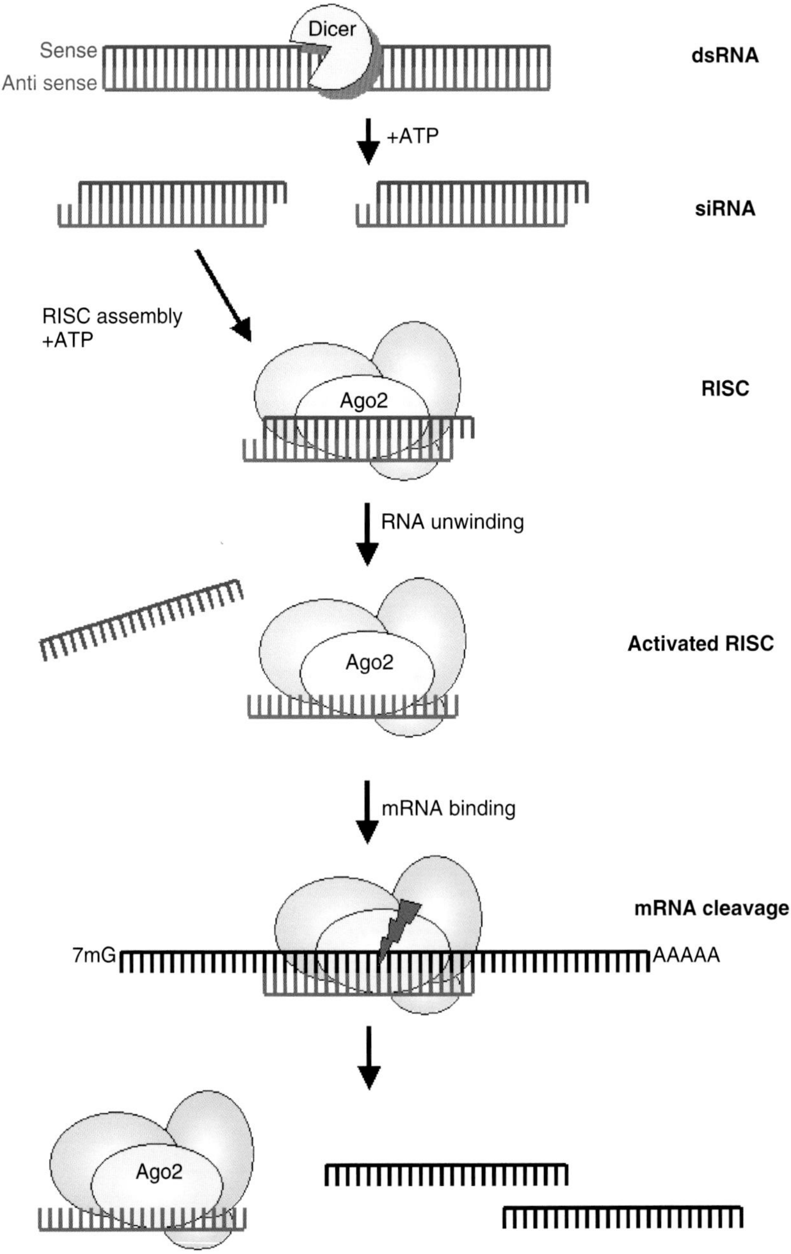

Please refer to Fig. 4.1 in text for figure legend.

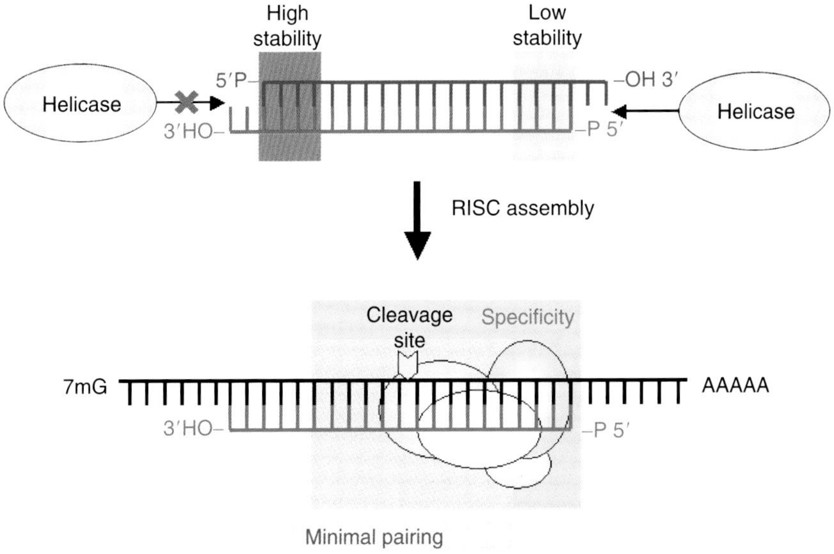

Please refer to Fig. 4.2 in text for figure legend.

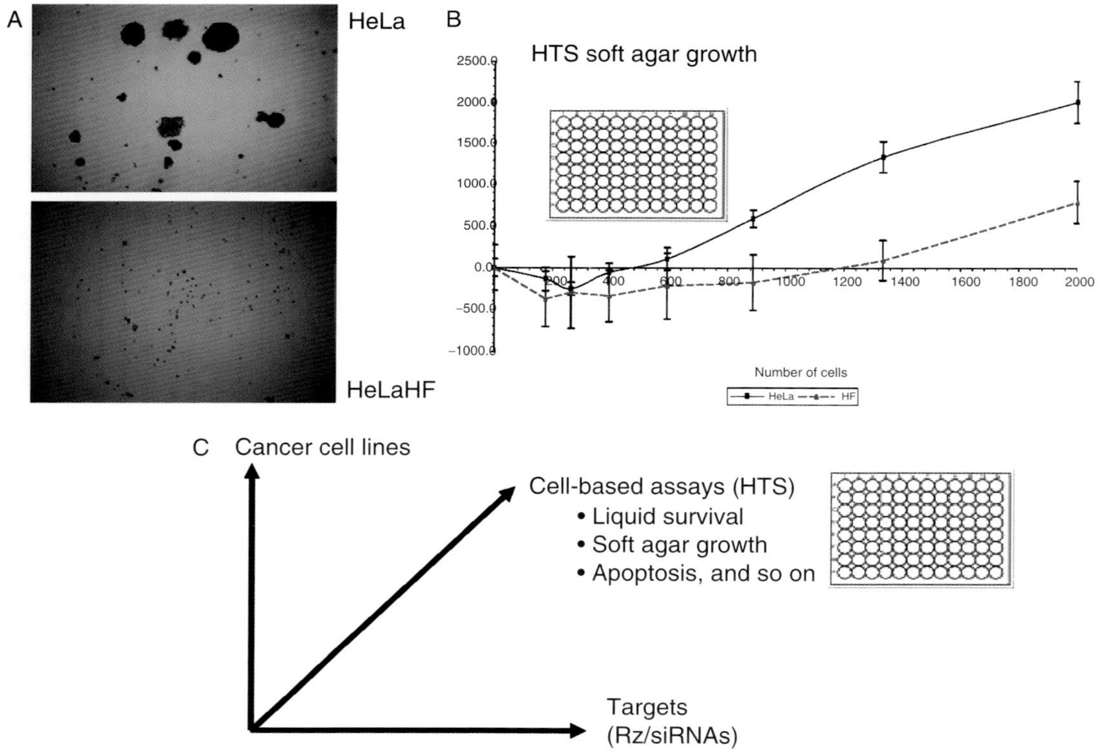

Please refer to Fig. 5.4 in text for figure legend.

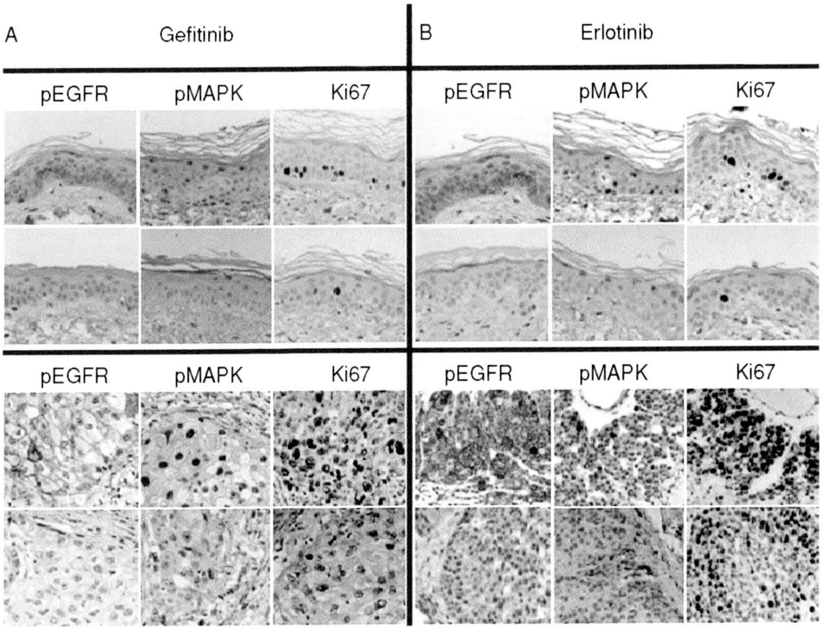

Please refer to Fig. 9.5 in text for figure legend.

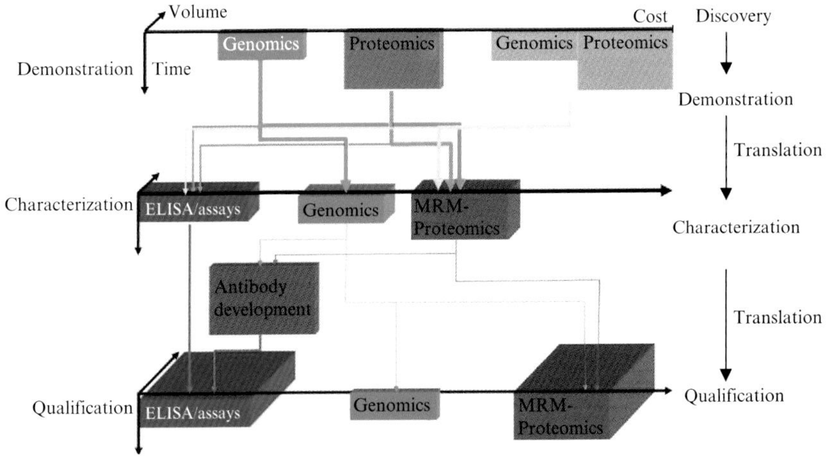

Please refer to Fig. 10.3 in text for figure legend.

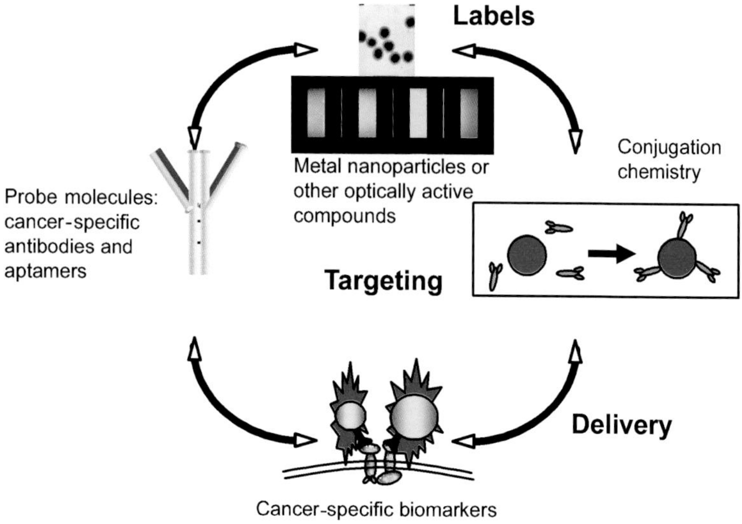

Please refer to Fig. 11.1 in text for figure legend.

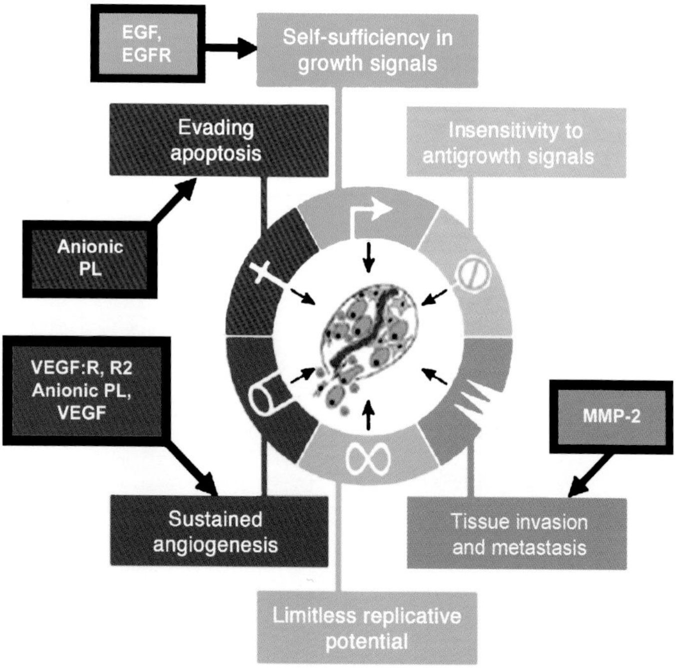

Please refer to Fig. 11.2 in text for figure legend.

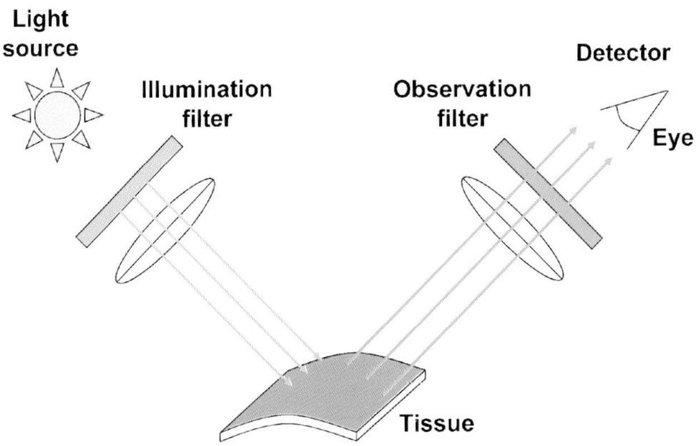

Please refer to Fig. 11.3 in text for figure legend.

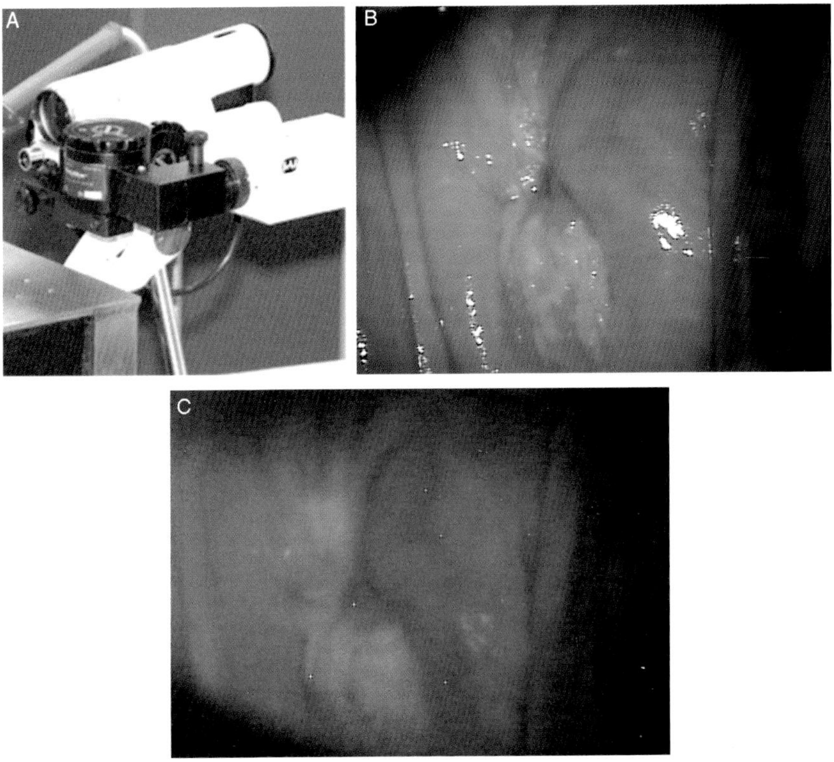

Please refer to Fig. 11.4 in text for figure legend.

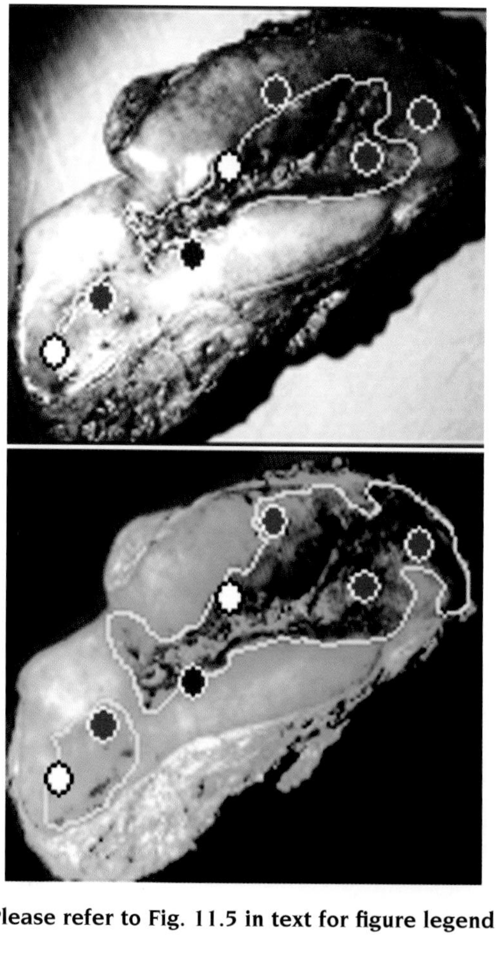

Please refer to Fig. 11.5 in text for figure legend.

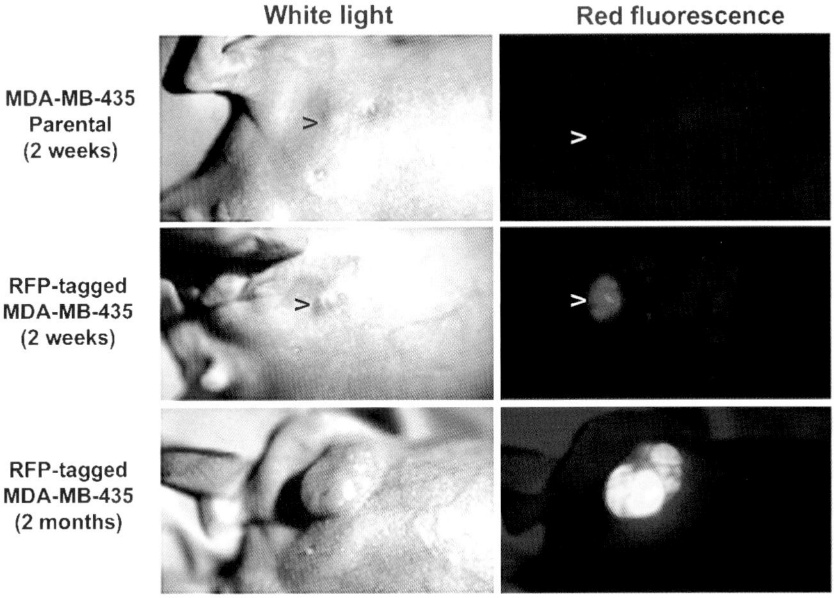

Please refer to Fig. 11.6 in text for figure legend.

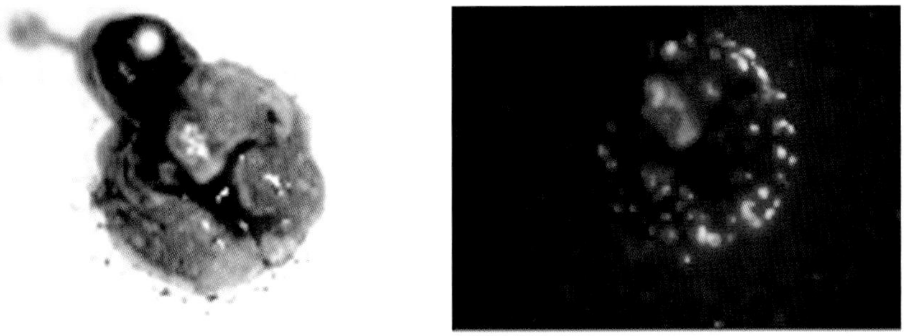

Please refer to Fig. 11.7 in text for figure legend.

Please refer to Fig. 11.9 in text for figure legend.

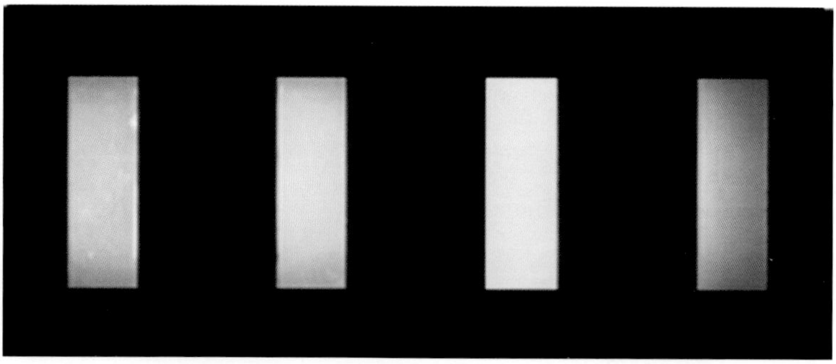

Please refer to Fig. 11.12 in text for figure legend.

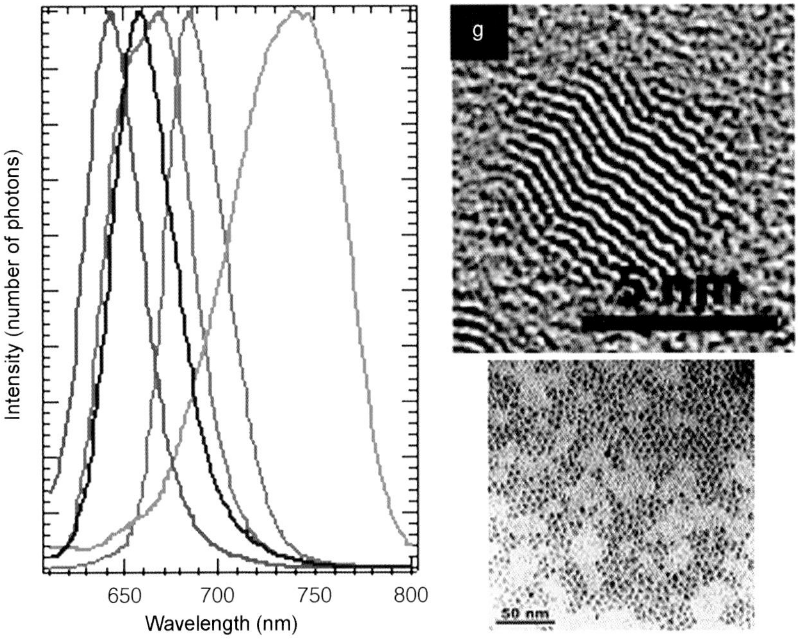

Please refer to Fig. 11.13 in text for figure legend.

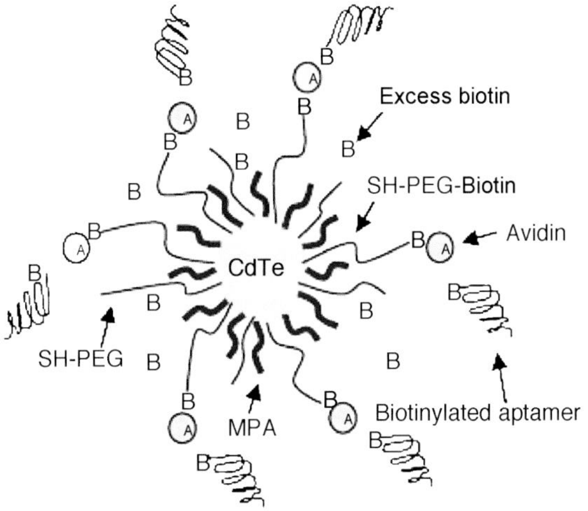

Please refer to Fig. 11.14 in text for figure legend.

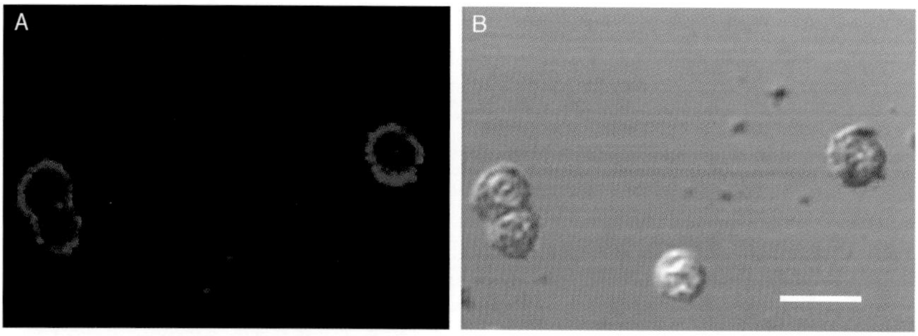

Please refer to Fig. 11.15 in text for figure legend.

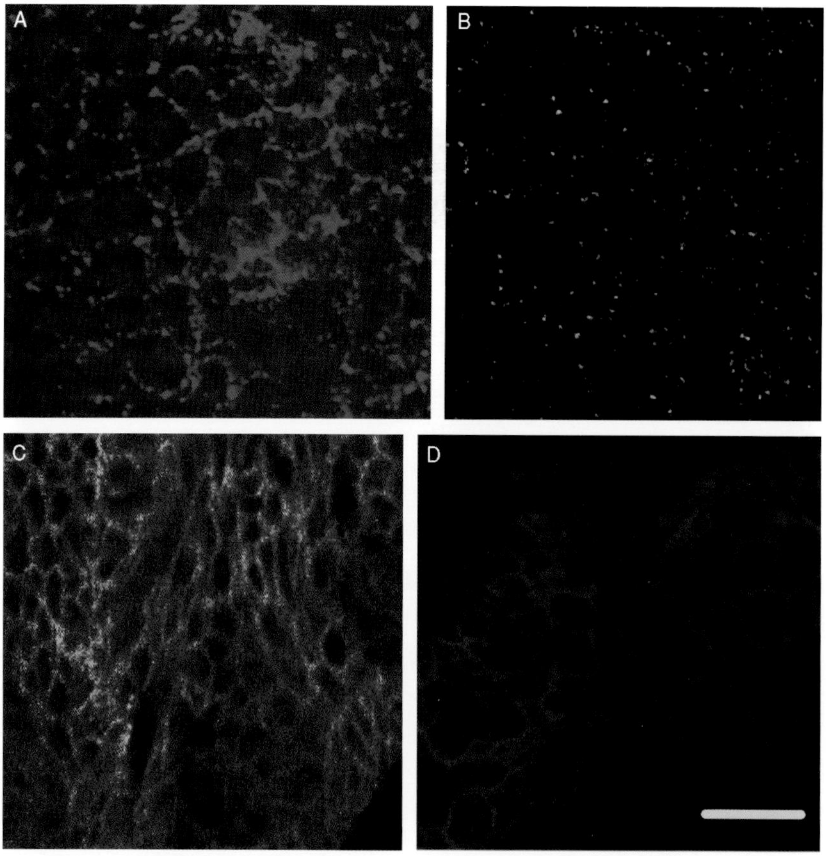

Please refer to Fig. 11.16 in text for figure legend.

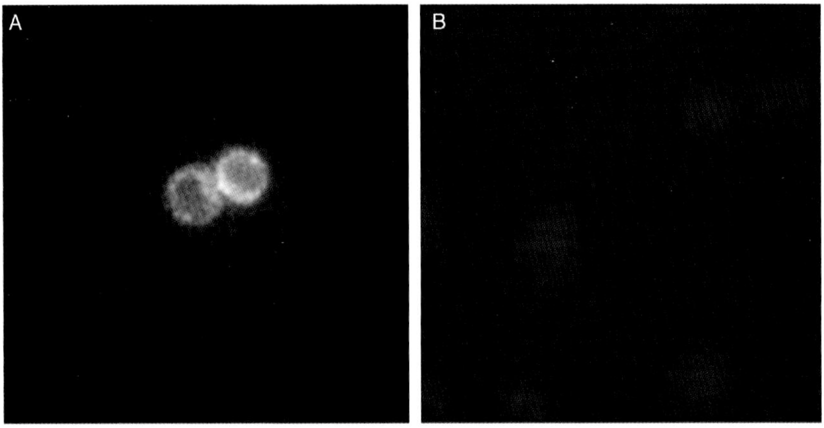

Please refer to Fig. 11.17 in text for figure legend.

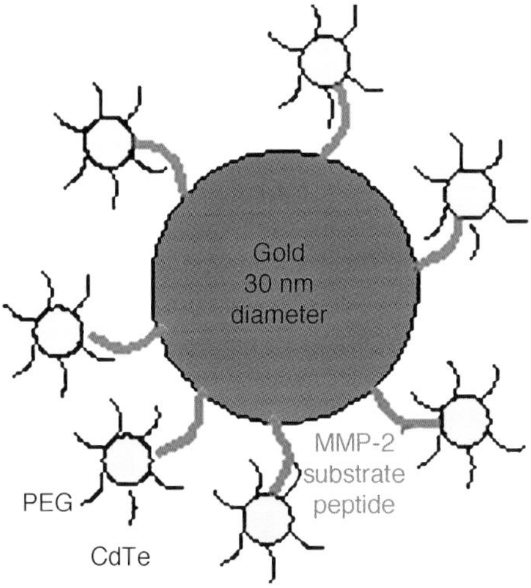

Please refer to Fig. 11.18 in text for figure legend.

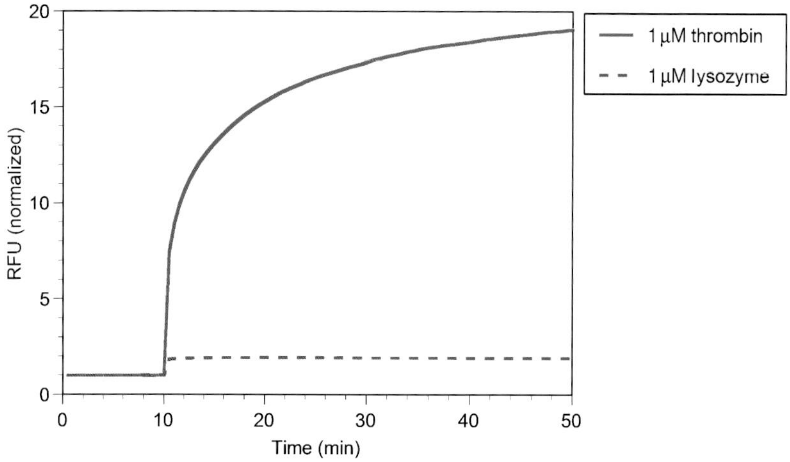

Please refer to Fig. 11.20 in text for figure legend.

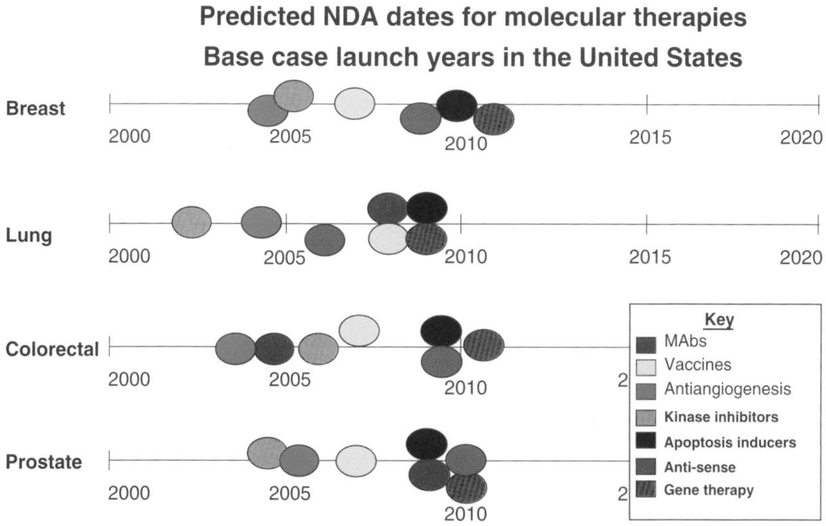

Please refer to Fig. 12.2 in text for figure legend.